CORINNE 90
Nantes

**INTERNATIONAL WORKSHOP ON
PARTICLE CORRELATIONS AND INTERFEROMETRY
IN NUCLEAR COLLISIONS**

CORINNE 90
Nantes

INTERNATIONAL WORKSHOP ON PARTICLE CORRELATIONS AND INTERFEROMETRY IN NUCLEAR COLLISIONS

Nantes, France   June 28–30, 1990

Editor
D Ardouin
Université de Nantes
France

World Scientific
Singapore • New Jersey • London • Hong Kong

Published by

World Scientific Publishing Co Pte Ltd
5 Toh Tuck Link, Singapore 596224
USA office: 27 Warren Street, Suite 401-402, Hackensack, NJ 07601
UK office: 57 Shelton Street, Covent Garden, London WC2H 9HE

British Library Cataloguing-in-Publication Data
A catalogue record for this book is available from the British Library.

CORINNE 90
International Workshop on Particle Correlations &
Interferometry in Nuclear Collisions

ISBN-13 978-981-02-0317-7
ISBN-10 981-02-0317-9

CORINNE 90
Nantes

INTERNATIONAL WORKSHOP ON PARTICLE CORRELATIONS AND INTERFEROMETRY IN NUCLEAR COLLISIONS

June 28–30, 1990 Nantes, France

ORGANIZING COMMITTEE

J.P. ALARD (Clermont)
D. ARDOUIN (Nantes)
R. DAYRAS (Saclay)
H. DOUBRE (Ganil)
F. MERCHEZ (Grenoble)
J. QUEBERT (Bordeaux)
B. REMAUD (Nantes)
P. SCHUCK (Grenoble)
Y. SCHUTZ (Ganil)
B. TAMAIN (Caen)

Foreword

The CORINNE 90 International Workshop on "Particle Correlations and Interferometry in Nuclear Collisions" was held at Chateau de Clermont (Le Cellier) in June 1990. It appears now as one of the rare meetings dedicated to that subject besides the recent series of the LESIP workshops which started in 1984 at Bad-Honnef.

Soon after the famous work of the astronomers Hanbury-Brown and Twiss, the interferometry method was discovered by Goldhaber, Goldhaber, Lee and Pais and has been largely used for the study of boson correlations. Since the end of the 70's, we have also witnessed a growing interest in this method in the case of baryon correlations for probing the space-time properties of hadron-hadron collisions. Simultaneously and following the renowned works of Kopylov, Podgorestkii and Koonin, a considerable progress has been made in the theoretical treatment of final state interactions and their link with non-static source origins dealing with dynamical models of hadron collisions.

The main purpose of this workshop was to gather practitioners of this field from different energy domains. The exchange of results, ideas and aims was thought to bring mutual benefits in the domain, particularly for unsolved questions like the importance of coherent effects or life-time influences.

In the field of hadron collisions and especially heavy-ion physics, the correlation technique has recently started to be used for the determination of emission temperatures or multi-particle correlations in the nuclei itself. Another goal of this workshop was the comparison of these results and the theoretical frameworks of their interpretation.

It is our hope that this workshop will contribute to the development and the progress of such important problems and can be renewed in a near future.

The participants and the speakers who have made alive this meeting are warmly thanked. We are grateful to our colleagues M. Podgorestkii, A. Giovannini, J. Alexander, R. Weiner, J. Bondorf, Y. Abe, and J. Peter for having accepted to chair our sessions. We express our thanks to the discussion leaders B. Remaud and K. Gelbke, and members of the panel who made the round table a success; to A. Labeyrie and A. Giovannini for presenting interesting informative talks. We regret that the lecture of J. Pochodzalla was not available for inclusion in these proceedings.

It is our personal pleasure to acknowledge all the people who have contributed in organizing this workshop:

Mrs. Evelyne Gerbaud for her great and appreciated efficiency. She deserves our gratitude and special thanks for taking care of many administrative and organization problems.

M. M. Rio, Rigalleau and Ganachaud for their help in solving practical problems.

Michele De Claverie, Marie France Baudet, and Catherine Ardouin for their assistance at the secretariat.

Mrs. Barbara Erazmus for her help at the conference site and for collecting the manuscripts.

All our colleagues and staff members from Laboratoire de Physique Nucleaire, University of Nantes for their assistance to the participants.

M. Stephan for welcoming us at Chateau de Clermont.

This Workshop was supported by Institut National de Physique Nucléaire et Physique des Particules (IN2P3/CNRS), Commissariat à l'Energie Atomique (IRF/CEA), University of Nantes and Société Française de Physique. The CORINNE 90 was also sponsored by Ganelec, Le Croy, Société Générale, Creative Electronics Systems, Conseil General de Loire-Atlantique, Ville de Nantes and Commune du Cellier.

For the organizing committee
D. ARDOUIN
Laboratoire de Physique Nucleaire
University of Nantes

Contents

† Manuscript not received.

* These talks were considered as informative seminars and the written version does not exist.

I. Boson Interferometry — Baryon Correlations

2

Speckle and Intensity Interferometry at Optical Wavelengths

A. Labeyrie

Boson Interferometry at High Energy

B. Lörstad
Department of Particle Physics, Lund University
S-22362 Lund, Sweden

Abstract

The pion interferometry has developed into an important tool for the understanding of the interaction mechanism in different reactions and in this review I will discuss some results from high energy physics. In hadronic jets, but also in general minimum-bias triggered hadron-hadron interactions, the correlation function of identical particles is found to depend only on the absolute value of the four-momentum transfer, a Lorentz invariant quantity. The shape of the correlation is more peaked than a Gaussian in accordance with a string model. The implication is that the interferometry effect in these data measures a local size only, a restricted region where particles of approximately the same momentum are produced.

In larger systems formed by nucleus-nucleus collisions global sizes are measured. The radii measured are found to depend on the particle density in the final state, providing evidence for a transversal expansion of hydrodynamical nature within these systems before pion "freeze-out". Experiments dedicated for interferometry studies are now being prepared promising great progress in the field during the next years.

1 Introduction

The study of interferometry at high energy physics started already 3 decades ago in 1959 when Goldhaber *et al.* discovered that pairs of equally charged particles show a higher probability for emission under small opening angles than pairs of oppositely charged particles[1]. This totally unexpected effect was soon interpreted as a reflection of pions obeying Bose-Einstein statistics[2].

Bose-Einstein correlations manifest themselves as an enhanced probability for identical bosons to be emitted with small relative momenta, as compared to non-identical particles under otherwise similar conditions. The effect originates in the quantum mechanical interference of the wave functions of the particles and is a consequence of a fundamental physics principle, namely that the wave function of a system of bosons is symmetric under particle exchange. This requirement influences the wave function to yield an effective clustering of the particles in phase space, which clearly illustrates their preference of occupying the same quantum state. From the characteristics of the arisen pattern it is possible, at least principally, to determine directly the space-time dimensions of the boson source.

These principles of universal validity find applications on the macroscopic as well as microscopic level. The effect was originally recognized in astronomy as the so-called Hanbury-Brown-Twiss effect[3] and offers an extensively used method to measure the sizes of various stellar objects, both in the visible and radio frequency ranges. This is done by detecting, simultaneously and in different telescopes, photons that are emitted from different parts of the distant object.

While in astronomy interference effects occur in the space near the detectors, hadronic interference effects will have to be looked for near the source, when the bosons leave the interaction region. The characteristic interaction times of hadronic processes beeing very short, one has to reconstruct the space-time description of the events by measurements in energy-momentum space. Apart from extraction of the spatial extension of the boson emitting mechanism, it seems to be possible to get information about its time evolution and the degree of chaoticity. Many results have been obtained from studies of different interactions, from events with different signatures however mostly with weak statistical significance and large systematic errors. Up to now there has been no experiment dedicated to the study of this effect. In general experimental set-ups the need for high resolution and high statistics in a limited part of the phase space has hindered the progress of the interferometry studies until recently.

In this talk I will present some significant progress from studies of hadronic jets produced in pp interactions and e^+e^- annihilations at high energy. The shape measurements of the boson emitting source in these jets and in general proton-proton interactions show that the basic system seems to be the $\pi\pi$ center of mass system and that, subsequently, shape effects are merely due to Lorentz boosts from this $\pi\pi$ -system to other systems. A string model can accomodate these results.

At very high energy there are experimental evidence for a dependence of the Bose-Einstein effect on the particle density, en effect which might be coupled to the occurence of multiple strings, or could be interpreted as due to a transversal expansion in a hydrodynamical mode. This developement has stimulated the interest in pion interferometry during the last years. The results reported by the NA35-collaboration on ^{16}O and ^{32}S reactions with different nuclei have been of great importance. Their first results show effects in these large systems which could indicate a sought-for signal of quark-gluon plasma.

The topics I discuss here on interferometry is a subjective selection. There are many interesting results which I will not touch upon here and I present my excuses to their authors. The topics selected is part of the development during the last years which has resulted in that interferometry is now considered an important tool for understanding the interaction mechanism in many different interactions.

2 Evidence of simple Q-dependence

The phenomenon of Bose-Einstein correlations is usually studied in terms of the ratio R of the joint probability for a pair of identical particles to be emitted, $P(p_1, p_2)$ to the product $P(p_1) \cdot P(p_2)$ of the single-particle probabilities, where p_1, p_2 are the four-momenta of the two particles. For fully chaotic sources this ratio is given by

$$R = \frac{P(p_1, p_2)}{P(p_1) \cdot P(p_2)} = 1 + \rho^2(Q), \tag{1}$$

where $Q = \vec{p_1} - \vec{p_2}$ is the four-momentum difference between the identical bosons, and $\rho(Q)$ is the Fourier transform of the space-time source distribution, normalized to unity as $Q \to 0$. This ratio R is identified with the two-particle correlation function, C(Q).

A source density distribution fixed in space-time and described by Gaussians separable in space and time is mostly assumed in hadronic interactions

$$\rho(x) = const.\exp(-\frac{x_T^2 r_T^2}{2} - \frac{z^2 r_z^2}{2} - \frac{t^2 \tau^2}{2}), \tag{2}$$

where $r_T(r_z)$ is the transverse (longitudinal) radius and τ is the spread in production times. The corresponding correlation function is

$$C(Q_T, Q_z, Q_0) = 1 + \lambda \exp(-Q_T^2 r_T^2 - Q_z^2 r_z^2 - Q_0^2 \tau^2) \tag{3}$$

You may remark that this formalism is not Lorentz covariant.One often omits the "time" factor regarding C as a function of the transverse and longitudinal components of the pair momentum difference only. The factor λ, which can take values between 0 and 1, is the so-called incoherence factor which takes into account that the interference, for various reasons, may not be complete.

At high energy hadronic jets are produced. Their mechanism of production is described phenemenologically by string models[4]. Hadronic jets have a very different structure longitudinal to the jet axis as compared to transversal to the jet- axis. It is tempting to try to visulize the size of the jet by interferometrical methods. Strings are described by a Lorentz covariant formulation, why jet-data are fitted to a pair of variables invariant to boosts along the string direction, Q_T^2 and Q_L^2. Q_T^2 is the square of the four-momentum difference transverse to the jet-axis, whereas Q_L^2 is given by

$$Q_L^2 = Q_z^2 - Q_0^2 \tag{4}$$

where Q_z is the longitudinal momentum difference and Q_0 the energy difference between the two particles. Q_L^2 is the variable conjugated to $t^2 - z^2$, which describes the source density in a one dimensional string model and is thus related to the longitudinal development of the jet.

The data from the study of hadronic jets by the AFS-collaboration[5] are divided into a two dimensional array in Q_T^2 and Q_L^2 and the correlation function is fitted to the form

$$C = N(1 + \lambda \exp(-r_T^2 Q_T^2 - r_L^2 Q_L^2)), \tag{5}$$

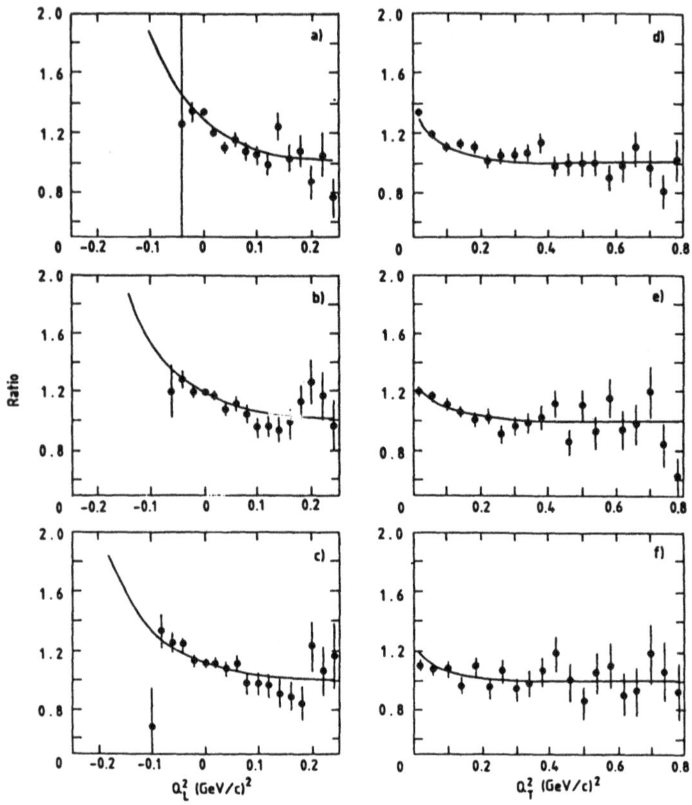

Figure 1: **a-f** Representation of a part of the fit in (Q_T^2, Q_L^2). **a-c** show tha data versus Q_L^2 in slices of Q_T^2. **d-f** show tha data versus Q_T^2 in slices of Q_L^2.

where N is a normalization factor. A good fit is obtained with the parameters $r_T \approx r_L$. In Fig.1 some slices in Q_T^2 and Q_L^2 are shown with the fit superimposed upon the data. The data given in the figures have been normalized so that N=1. We remark that the correlation function follows the Gaussian form even for negative Q_L^2-values, providing evidence that this combination, $Q_z^2 - Q_0^2$, is the relevant combination to describe the data.

Thus the correlation function shows the same dependence on Q_T^2 as on Q_L^2, consequently, it is well parametrized as a Gaussian in the single variable $Q^2 = Q_T^2 + Q_L^2$. Similar results have been obtained from jet studies in e^+e^- annihilations[6].

Also general pp events, collected with a so-called miminum bias trigger, have been analysed in a similar way, i.e. the interesting axis is here the beam axis instead of the jet axis. Fitting Eq.(5) using the boost-invariant variables Q_T^2 and Q_L^2 it is found also in these events that $r_T \approx r_L$ [5,7].

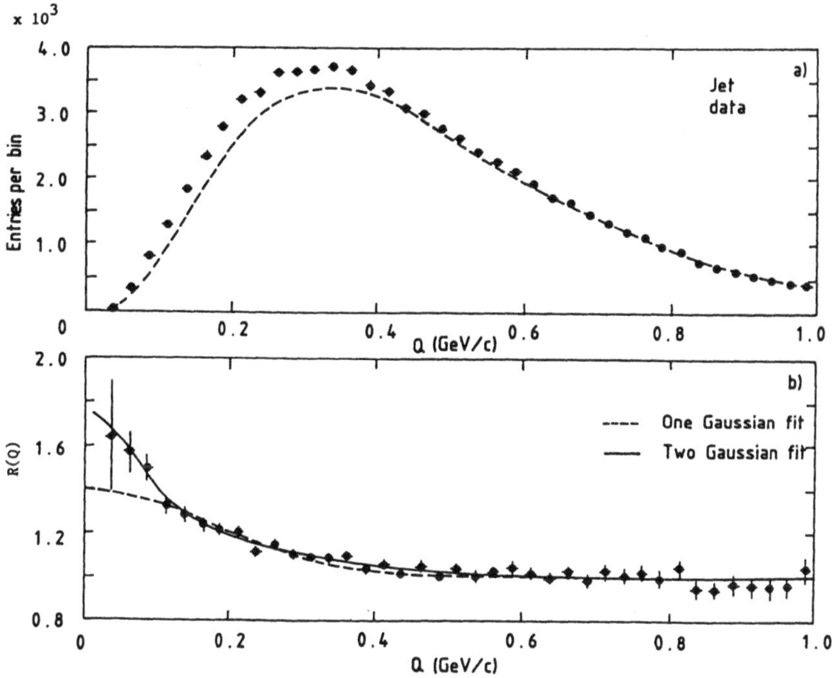

Figure 2: **a** The two-particle and the reference distributions for particle combinations within a hadronic jet as a function of Q, the four-momentum transfer. **b** The correlation function R(Q), i.e. the ratio of the two-particle distribution to the reference distribution. The full line shows the fit with the sum of two Gaussians, whereas the dashed line is the fit to one Gaussian.

In order to expand the interesting region of small momentum transfers we will display the correlation function as a function of Q rather than the most often used Q^2. Figure 2 a shows the two-particle and reference distributions as a function of Q, normalized to the tail of the two-particle distribution. Figure 2 b displays the ratio R of these distribution and it clearly shows that the correlation is more peaked than a simple Gaussian (dashed line). It can be parametrized as a sum of two Gaussians (full line) or with an exponential function.

Figures 3 a and 3 b show that the correlation function in minimum bias data is very similar to the one in hadronic jets even though the two-particle and reference distributions are quite different.

Earliar interferometry studies have not had enough statistics to show this peaked structure[8], why most results are given using a single Gaussian to extract the parameters λ and r_G:

$$R = 1 - \lambda \exp[-(r_G \cdot Q)^2] \tag{6}$$

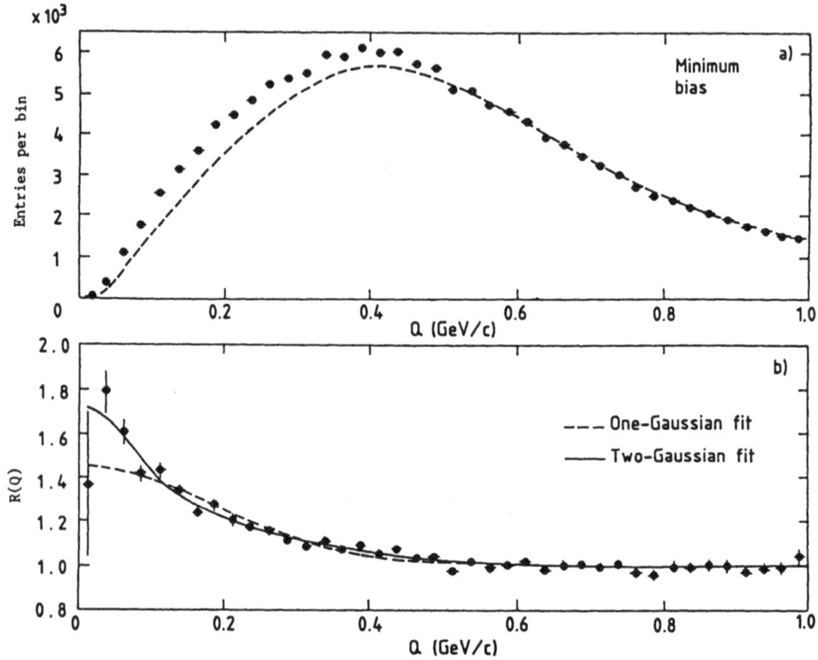

Figure 3: **a** The two-particle and the reference distributions for minimum-bias-triggered data as a function of Q, the four-momentum transfer. **b** The correlation function R(Q), i.e. the ratio of the two-particle distribution to the reference distribution. The full line shows the fit with the sum of two Gaussians, whereas the dashed line is the fit to one Gaussian.

Another used formulation, in fact corresponding to a similar shape of the measured experimental correlation function, is

$$R = 1 - \lambda \frac{\left(\frac{2J_1(r_B \cdot q_T)}{(r_B \cdot q_T)}\right)^2}{(1 - (Q_0 \cdot \tau)^2)} \tag{7}$$

Eq.(7) corresponds to pion sources uniformly distributed over the surface of a sphere with radius r_B; in some models τ represents the lifetime of the sources, in others it is ascribed to the depth of the emission layer[9]. Here q_T is the component of $\vec{p}_1 - \vec{p}_2$ transverse to $\vec{p}_1 + \vec{p}_2$, Q_0 is the energy difference between the two identical bosons, and J_1 is a Bessel function of first order.

Eq.(7) has mostly been used to investigate the structure of the interaction region in hadron-hadron collisions, whilst Eq.(6) is preferred in the analysis of the jet events in e^+e^- collisions.

Both formulae turn out to be nearly indistinguishable in shape, and both fit the measured correlations quite well, formula(7) giving a radius parameter r_B approximately twice as large as r_G of Eq.(6), as can be seen from the exponential approximation of Eq.(7),see also [22].

$$R \approx 1 - \lambda \exp(-r_B^2/4 \cdot q_T^2 - \tau^2 Q_0^2) \tag{8}$$

The determination of the parameter λ, the incoherence or chaoticity parameter, is subject to large correction factors and systematic errors, whereas the determination of the radius parameter is not that much influenced. The mutual Coulomb interaction of the two pions gives sizable corrections for low Q values. On the other hand, the Coulomb interactions between the π's and the residual hadronic system results in rather small corrections. For detectors with restricted acceptance the iterative generation of the reference distribution can introduce systematic errors. Other effects could affect the value of λ, such as the creation of pions through the decay of resonances, final state interactions and the prescence of several different pion sources[8]. Most sources of experimental difficulties, like particle misidentification and problems with close tracks, tend to dilute the enhancement why measured λ-values are to be considered as lower limits.

Furthermore we have here evidence for another important point: the shape of the

correlation function must be measured with high accuracy at low Q-values as the supposed Gaussian shape do not fit the data. The factor λ, parametrizing the strength of the correlation, has been extracted by single Gaussian fits with typical values in the region $0.4 - 0.6$. The more peaked exponential form gives larger values of λ. Generally we can conclude that even if many experiments report values of $\lambda < 1$, no clear evidence for coherence has been observed.

R. Weiner[10] has shown that in the picture of partial coherence data should follow a somewhat more peaked structure than the single Gaussian, the correlation function being a sum of two coupled Gaussians. However this predicted peaking is much weaker that what is seen in these data[11].

The peaked structure parametrized with a second Gaussian could be due to the resonance effects. It is well known that many of the pions in the jets are produced via resonances. Short-lived resonances, such as the ρ-resonance, will have decay lengths of only a few fermi, and their decay pions will show interference with directly produced pions as well as with those produced via other short-lived resonances. This might explain the observed structure of the two-particle correlation.

Another interpretation is possible using a string model. Andersson and Hofmann[12] have made an estimate of interference effects for particles produced in a jet within the Lund string model. Their approximate form ressembles the measured correlation function. We will discuss this model in further detail below.

3 Shape measurements of the pion emission region

The simple dependence on the absolute value of Q of the Bose-Einstein effect in these data can be interpreted as evidence that the pion emission is spherical in the $\pi\pi$ centre-of-mass system. The shapes in other systems as the total centre-of-mass system should then be given by Lorentz boosts from the $\pi\pi$ system. We will see that this is the case, at least qualitatively.

The shape of the emission region in space can be determined by a two-dimensional fit to the function

$$R = 1 + \lambda \exp(-r_T^2 Q_T^2 - r_z^2 Q_z^2) \tag{9}$$

where z-axis is a significant axis and T stands for transversal to this axis. This formula is not in contradiction to Eq.5, as you naively might think. The radius r_z is in fact connected to the Lorentz invariant radius r_L via a γ-factor, between the $\pi\pi$-system and the new system: $r_z = r_L/\gamma$.

In the new system the line of flight of the $\pi\pi$-system is given by

$$\vec{n} = \frac{\vec{p}_1 + \vec{p}_2}{\mid \vec{p}_1 + \vec{p}_2 \mid} \tag{10}$$

The component of Q parallel to $\vec{n}$, $Q_{||}$, and Q^0 are related:

$$Q_{||} = (\vec{p}_1 - \vec{p}_2) \cdot \vec{n} = \frac{Q_0}{\beta} \tag{11}$$

where β is the velocity of the $\pi\pi$-systems:

$$\beta = \frac{\mid \vec{p}_1 + \vec{p}_2 \mid}{E_1 + E_2} \tag{12}$$

This gives immediately that the Q_L-component, invariant to boosts along $\vec{n}$, will be given as

$$Q_L^2 = Q_{||}^2 - Q_0^2 = \frac{Q_{||}^2}{\gamma^2} \tag{13}$$

From this formula the relation $r_z = \frac{r_L}{\gamma}$ follows directly when Eqs.(9) and (5) are used for the fitting.

Qualitatively this effect is found: in the hadronic jet system the $\pi\pi$-system moves longitudinally with the jet-axis and it is found that $r_{jet-axis} = 0.43 \pm 0.06$ fm$< r_T =$ 0.70 ± 0.04 fm[5]. In the minimum bias sample, restricted to the central region $y \leq 0.8$, the $\pi\pi$-system moves mainly transversal to the beam-axis and $r_{beam-axis} = 1.3 \pm 0.1$ fm is found to be larger than $r_T = 0.72 \pm 0.04$ fm, ie an prolate shape[13]. In another pp experiment[14] at somewhat lower energy an oblate shape is found. This can in fact be consilient considering that most $\pi\pi$-systems in these events go forward thus contracting the longitudinal radius.

Quantitatively we can calculate mean γ-factors of the samples to check the relation $r_z = \frac{r_L}{\gamma}$. Using Gaussian fits the consistency is not very great, however with exponential functions instead of Gaussians for the extraction of the radii the values are consistent[15]. This result emphasizes that exponential functions are preferable: they fit the data better with fewer parameter which are consistent with each others.

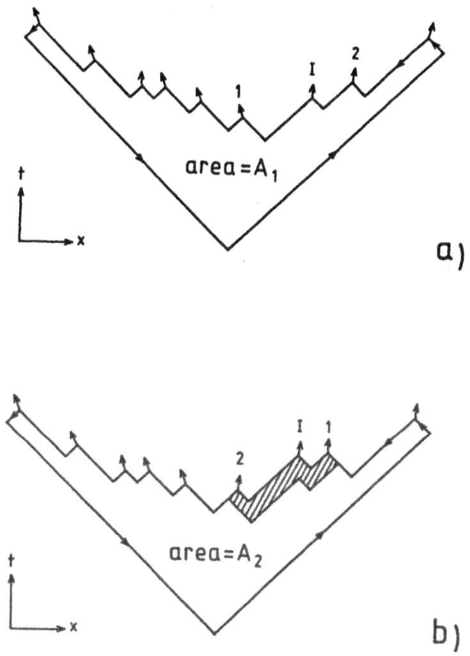

Figure 4: Space-time diagram for particle production in e^+e^- annihilations, based on a string model. $A_{1,2}$ denotes the space-time area of a color field enclosed by the quark loop. The particles 1 and 2 are separated by the system I, the intermediate system. When the particles 1 and 2 are identical, the configuration in b) is indistinguishable from that of fig a and their amplitudes for production must be added. The probability of production will depend on the difference in area $\Delta A = A_1 - A_2$, shown as the hatched area in b.

4 String model interpretation of Bose-Einstein correlations

String models have been very successful in describing hadronic jets. The total source size is rather large: of the order of 10-30 fermi which is much larger than the 1-fermi effect seen by interferometry experiments. The reason is the strong correlation between particle production points and particle momentum in a string model. Hadrons with approximately the same momentum are in fact produced only in a region of invariant size of about 1 fermi, which is what can be measured by interferometry.

The space-time structure of an e^+e^- annihilation event is shown in Figure 4 for a string model. Final states with two identical particles are indistinguisable and can be obtained in different ways. Suppose that the two particles indicated as 1 and 2 in Figure 4a are identical, then in Figure 4b (where 1 and 2 are interchanged) the hadron state can also be considered as being the same. The amplitude should be the sum of these two final states for bosons. The analysis leads to a Bose-Einstein correlation which, as a function of the four-momentum transfer, reproduces the data

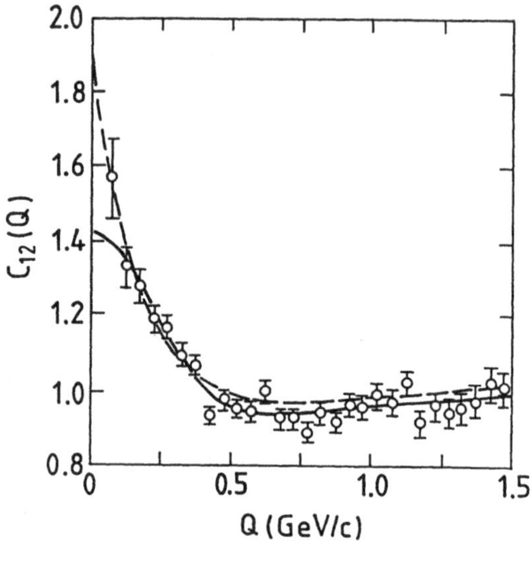

Figure 5: The correlation function of two identical pions displayed as a function of the absolute value of the four-momentum transfer Q, for two-jet data sampled from e^+e^- annihilations. The full line is a fit to a gaussian function whereas the dashed line is the result of a string model calculation.

rather well but shows a steeper dependence at small Q than a Gaussian function. The effect is furthermore approximately proportional to Q. A comparison between data[16] and this model is shown in Figure 5.

In this calculation the influence of resonances and final state interactions was not taken into account. M. Bowler[17] has analysed these effects in great detail as they can have a large impact on the measured correlation. It turns out that the resonance-effect is rather small and that final state interactions tend to cancel, even though single terms can be quite important.

The difference in space-time ΔA will depend on the identity of the particles creating a different interference pattern for different boson pairs. For $\pi^0\pi^0$ combinations, no intermediate system is necessary, which ought to result in a wider correlation peak in Q. On the other hand, for $K^\pm K^\pm$ combinations, the intermediate system is generally heavier than for $\pi^\pm\pi^\pm$ combinations, implying a narrower correlation peak. Such predictions are specific to the string model and future measurements could be a good test of its validity.

Such measurements can be performed by e^+e^- annihilation experiments at LEP, accumulating large eventsamples with identified hadronic jets. $K^\pm K^\pm$ correlations,

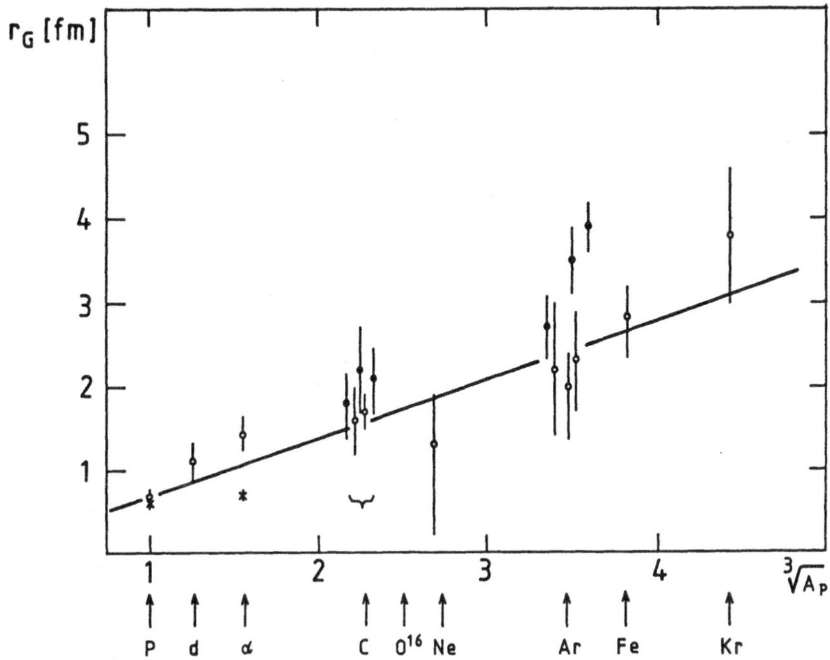

Figure 6: The radius parameter R_{Gauss} as a function of the cubic root of the number of nucleons in the projectile $A_p^{1/3}$. Open circles show results from general event samples, filled circles from events of higher multiplicity selected by a trigger on central collisions, and stars show measurements restricted to the central region. The line indicates the radius of the projectile as given by $R = 0.7A_p^{1/3}$ fm.

as reported from hadronic reactions[18], have similar characteristics to $\pi^{\pm}\pi^{\pm}$ correlations, but within rather large statistical uncertainties.

5 Larger systems

In nucleus-nucleus interactions the width of the two-particle correlation function has been shown to depend on two parameters: i) the projectile size, and ii)the particle multiplicity density.

5.1 The Projectile size

The projectile size has been measured by interferometry methods in nucleus-nucleus collisions in the energy domain of a few GeV/nucleon[19,8]. Figure 6 shows a graphical representation of r_{Gauss} as a function of A_p, the atomic number of the beam.

Different radii measurements have been renormalized to r_{Gauss} which is the parametrization of the correlation function corresponding to a Gaussian source distribution, Eq.(6). Only measurements with negatively charged particle pairs , ie a poor mans pion identification, are introduced on the Figure 6. You could remark that the projectiles are always smaller than the target nucleus in these data. One can see that r_{Gauss} is close to the nuclear radius of the projectile itself, $R = 0.7A_p^{1/3}$ fm, shown in Fig. 6 with a straight line. This is true for the measurements of general event samples, shown with open circles. The filled circles show the results from experiments selecting more central collisions than the ordinary event. The larger value of the radius parameter in these events could be connected with a multiplicity density effect. In fact, the rise of radius with multiplicity is compatible with the slow increase with multiplicity density at high energy as discussed below.

5.2 Multiplicity density

The AFS collaboration used different data samples produced in $pp, p\bar{p}$ and

$\alpha\alpha$ at $\sqrt{s} = 52$ GeV to show that the parameter r_{Gauss} (or equivalently r_B) depends only on the multiplicity of particles produced in the central region $-0.8 < y < 0.8$, not on the initial particles[20]. The ABCDHW-collaboration [21] confirmed this behaviour with a much larger data sample from pp collisions at the same energy. Both show a slow rise of r_{Gauss} with multiplicity density. The ABDCHW-collaboration finds no rise at lowest ISR energy, $\sqrt{s} = 30$ GeV, as confirmed by other experiments at the CERN SPS [7,23]. On the other hand, the UA1-collaboration[22] reports a clear dependance of r_{Gauss} with multiplicity density in data from a large energy range, $200 < \sqrt{s} < 900$ GeV. No energy dependence is found over this energy range. The UA1 and ABCDHW results are however inconsistent at higher particle densities, probably due to Lorentz contraction effects. The UA1 experiment covers a long range in rapidity, $-3 < y < 3$. In this interval the mean Lorentz factor, $\gamma \approx 4$, if we assume a flat rapidity distribution of the pions. This will have a significant contraction effect on the longitudinal radius and, consequently, also on the average r_{Gauss} value, roughly as $< \gamma >^{1/3}$. Similar contraction effects for the ABCDHW-experiment are small as they cover only $-1.5 < y < 1.5$. In the Figure 7 I have plotted the AFS and SFM measurements of r_{Gauss} as a function of the charged particle density. The higher density UA1 radius estimates are multiplied by a the factor $< \gamma >^{1/3}$. The lowest UA1 data point corresponds to the simple dependence on the absolute value of Q of minimum bias events and ,hence, should not be corrected.

These r_{Gauss} values can be compared to the CERN NA35 study of events with extremely high particle density produced in O on Au reactions at a beam energy of 200GeV/nucleon[24]. These measurements result from fitting the correlation function to the parametrization of Eq.(9), where $r_T(r_L)$ is the transverse (longitudinal) radius parameter of the pion source with respects to the beam axis. The r_T value can be compared to the r_{Gauss} values from the hadron-hadron experiments as plotted on Fig.7. The density of positive particles is assumed to be the same as for negative particles. We can see that these transversal radii measurements is close to a square

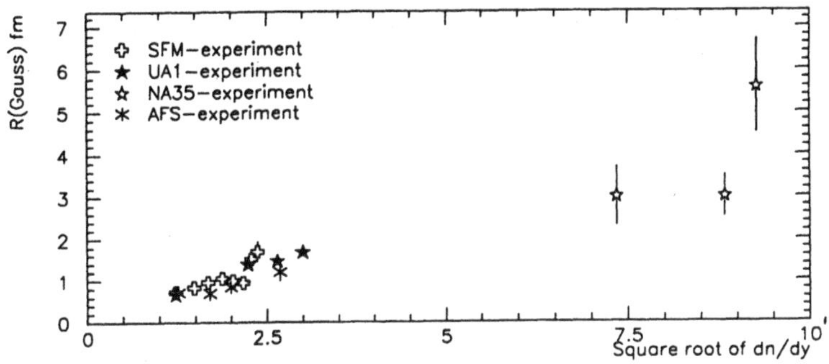

Figure 7: Transverse source radii from hadron-hadron and nucleus-nucleus collisions plotted against charged particle rapidity density.

root dependence on the particle density irrespective of the initial state. The radius values are much larger than the sizes of the projectiles providing evidence for a transversal expansion before the pions are liberated. A square root dependence follows naturally in hydrodynamical models including transversal expansion[25].

It will be interesting to test this dependence with upcoming new data from both heavy ions beams and hadron-hadron collisions from FNAL.

6 Summary and Outlook

I have presented some of the results which have made interferometry an important tool for understanding the reaction mechanism in some different reactions. Simple hadron-hadron collisions as well as hadronic jets seem to follow a string model. Future large data samples from LEP and FNAL will be able to better test this model.

Most interest goes into the study of nucleus-nucleus systems where interferometry analysis can be a key instrument for understanding of the reaction mechanism. I have discussed the evidence for a transversal expansion before pion "freeze-out". Two dedicated experiments, NA44 ar CERN SPS and E859 at BNL, are now being set up, promising new data of high accuracy with systematic problems under better control than earliar. With such data it will be possible to test dynamical models requiring quark-gluon plasma formation[26,27]. Future $Pb+Pb$ collisions in upcoming CERN SPS experiments constitute a fascinating experimental issue.

References

[1] G.Goldhaber*et al.* ,*Phys. Rev. Lett.* **3**, 181 (1959)

[2] G.Goldhaber,S. Goldhaber, W. Lee and A. Pais, *Phys. Rev.* **120**, 300 (1960)

[3] R. Hanbury-Brown and R.Q. Twiss, *Phil. Mag.* **45**, 663 (1954); *Nature* **178**, 1046 (1956)

[4] B. Andersson et al., *Phys. Rep.* **97**,31 (1983)

[5] T.Åkesson et al., *Z.Phys. C - Particles and Fields* **36**, 517 (1987)

[6] M. Althoff et al., *Z.Phys.C - Particles and Fields* **30**,355 1986)

[7] J.L. Bailly et al. *Z. Phys. C - Particles and Fields* **40**, 215 (1988)

[8] Recent reviews are

B. Lörstad, *Int. Journal Mod. Phys.* **A4**, 2861 (1989)

W. Zajc, *in Hadronic Multiparticle Production, P. Carruthers ed., World Scientific Publ. 1988*

G. Goldhaber in *Proc. First Int. Workshop on Local Equilibrium in Strong Interaction Physics*, eds. D.K. Scott and R.M. Weiner, (World Publishing Co. 1985)

[9] G. Kopylov and M.I. Podgoretsky,*Sov. J. Nucl. Phys.* **18**, 336 (1974) and **19**,215 (1974); G. Kopylov, *Phys. Lett.* **50b**,472 (1974); G. Cocconi, *Phys. Lett* **49B**, 459 (1974)

[10] R.M. Weiner *Phys. Lett.* **232B**, 278 (1989)

[11] K. Kulka and B. Lörstad, *Z. Phys. C - Particles and Fields* **45**, 581 (1990)

[12] B. Andersson, W. Hofmann, *Phys. Lett.***169B** , 364.1986)

[13] T.Åkesson et al., *Phys. Lett.* **187B**, 420 (1987)

[14] C. De Marzo et al. *Z. Phys. C - Particles and Fields* **33**, 363 (1984)

[15] B. Lörstad in *Proc. XVIII Symp. on Multiparticle Dynamics*, Tashkent, USSR, 1987, eds. I. Dremin and K. Gulamov (World Scientific Publishing Co., 1988) p. 111.

[16] H. Aihara et al., *Phys. Rev.* **D31**, 996 (1985)

[17] M.G. Bowler, *Phys. Lett.* **180B**,299 (1986); *Z. Phys. C - Particles and Fields* **39** ,81 (1988) and *Z. Phys. C - Particles and Fields* **46** , 305 (1990)

[18] T.Åkesson et al., *Phys. Lett.* **155B**, 128 (1985)

[19] J. Bartke and M. Kowalski, *Phys. Rev.* **C30**, 1341 (1984)

[20] T.Åkesson et al., *Phys. Lett.* **129B**, 269 (1983)

[21] A. Breakstone et al., *Z. Phys. C - Particles and Fields* **33**, 333 (1987)

[22] C. Albajar et al., *Phys. Lett.* **226B**, 410 (1989)

[23] M. Adamus et al. *Z. Phys. C - Particles and Fields* **37**, 347 (1988)

[24] A. Bamberger et al. *Phys. Lett.* **203B**, 320 (1988)

[25] R. Stock, preprint *IKF* **90-3**, to be published in *Annalen der Physik*

[26] G.F. Bertsch, *Nucl. Phys.* **A498**, 173c (1989)

[27] Yu.M. Sinyukov, V.A. Averchenkov and B. Lörstad, preprint *Lund University* LUNFD6(NFFL-7055)1990

PION INTERFEROMETRY IN INTERACTIONS OF ^{16}O AND ^{32}S NUCLEI AT THE CERN SPS

Jerzy BARTKE

High Energy Physics Laboratory
Institute of Nuclear Physics
ul. Kawiory 26A
30-055 Cracow, Poland

NA35 COLLABORATION,CERN

J.Bartke[5], H.Białkowska[12], R.Bock[6], R.Brockmann[6], S.I. Chase[3], I.Derado[10], V.Eckardt[10], J.Eschke[7], C.Favuzzi[2], D.Ferenc[13], B.Fleischmann[6], M.Fuchs[6], M.Gaździcki[12], H.J.Gebauer[10], E.Gładysz[5], C.Guerra[6], O.Hansen[7,14], J.W. Harris[3], W.Heck[7], T.Humanic[6], K.Kadija[13], S.Kabana[7], A.Karabarbounis[1], R.Keidel[9], J.Kosiec[7], M.Kowalski[5,10], A.Kühmichel[7], M.Lahanas[7], J.Y.Lee[7], M.LeVine[7,14], A.Ljubicic[13], S.Margetis[7], E.Nappi[2], G.Odyniec[3], G.Paic[6,13], A.D.Panagiotou[1,4], A.Petridis[1], J.Pfennig[4], A.Piper[9], F.Posa[2], H.G.Pugh[3], F.Pühlhofer[9], G.Rai[3], W.Rauch[3,10], R.Renford[7], D.Röhrich[7], H.Rothard[7], K.Runge[8], A.Sandoval[6], E.Schmidt[7], N.Schmitz[10], E.Schmoetten[8], I.Schneider[7], L.S.Schroeder[3], P.Seyboth[10], J.Seyerlein[10], E.Skrzypczak[11], P.Spinelli[2], R.Stock[7], H.Ströbele[7], L.Teitelbaum[3], A.Thomas[7], S.Tonse[3], G.Vassileiadis[1], G.Vesztergombi[10], D.Vranic[13], S.Wenig[7]

[1]Physics Dept.,Univ.of Athens, [2]Dipartimento di Fisica, Univ.di Bari, [3]Lawrence Berkeley Lab., [4]CERN, [5]Institute of Nuclear Physics, Cracow, [6]GSI, Darmstadt, [7]Fachbereich Physik, Univ.Frankfurt, [8]Fakultät für Physik, Univ.Freiburg, [9]Fachbereich Physik, Univ.Marburg, [10]Max-Planck Institut für Physik, München, [11]Inst.of Experimental Physics, Univ.of Warsaw, [12]Inst.for Nuclear Studies, Warsaw, [13]Rudjer Bošković Inst., Zagreb, [14]Physics Dept., Brookhaven National Lab., Upton, N.Y.

ABSTRACT

We report results from pion interferometry analysis of O+Au, S+Ag and S+S central collisions at 200 GeV/nucleon in a wide rapidity range. Several models are used to fit the experimental correlation functions. Dependence of longitudinal and transverse source size and of chaoticity parameter on rapidity is obtained. For O and S collisions with a heavy target, but not for S+S collisions, the transverse size parameter and the chaoticity parameter are found to be large around and beyond midrapidity. Discussion of these results within the frame of existing models is given.

1. INTRODUCTION

Intensity, or second-order interferometry of identical particles can give information on the size of the emitting source, as the correlation function in momentum space is linked via the Fourier transform to the radial density distribution of the source. For identical bosons, such as like-charge pions, and for the simplest assumption of the totally chaotic source, the two-pion correlation function should be enhanced by a factor of two at zero relative momentum of the pair, due to Bose-Einstein symmetrization of the two-pion wave function. The width of the peak is inversely proportional to the size of the source at freeze-out, i.e. at the time when interaction between particles ceases. This approach can be refined so as to give information also on the shape of the source (transverse and longitudinal dimensions) and on its lifetime by using suitably chosen variables. Various parametrizations of the correlation function have been proposed - for a review on relevant theory see ref. [1].

Pion interferometry has been extensively used in studies of various high-energy elementary reactions involving leptons as well as hadrons, and also for relativistic nuclei. The results are reviewed in refs. [2-4] . For nucleus-nucleus collisions at ultrarelativistic energies interferometric studies seem especially interesting as they can shed light on the space-time development of the collision process, possible quark-gluon plasma formation, etc. In this connection central (small impact parameter) collisions in which all, or almost all, nucleons of both colliding nuclei participate in the interaction, are likely to lead to the highest energy density and thus appear especially attractive. It should, however, be pointed out that in a collision of two nuclei larger source sizes are expected because of larger sizes of colliding objects (typical sizes for e^+e^-, πp and pp collisions are about 1 fm), what implies the necessity for a corresponding improvement in experimental momentum resolution and in the ability to resolve close tracks. The wide-acceptance streamer chamber spectrometer of the NA35 Collaboration meets these requirements and the feasibility of Bose-Einstein correlation studies was demonstrated already at the early stage of data analysis [5].

2. SAMPLES OF CENTRAL COLLISIONS FROM THE NA35 EXPERIMENT

The NA35 detector consists of a big ($2.0 \times 1.2 \times 0.7$ m^3) streamer chamber placed in a 1.5 T field of a superconducting magnet and a set of trigger devices. Thin solid targets are mounted directly in front of the chamber. The chamber volume is viewed by three cameras with Zeiss optics, equipped with two-stage high gain magnetic focussing ITT image intensifiers. Events are recorded on high sensitivity 70 mm film. A typical event is shown in Fig.1. The spatial resolution is about 2 mm, what provides the accuracy of momentum measurements of the order of 1%. The film is then analyzed off-line in the collaborating laboratories by means of digitized measuring devices working under computer control with interactive programmes facilitating track matching in all three views. The trigger devices consist of a set of beam defining counters upstream from

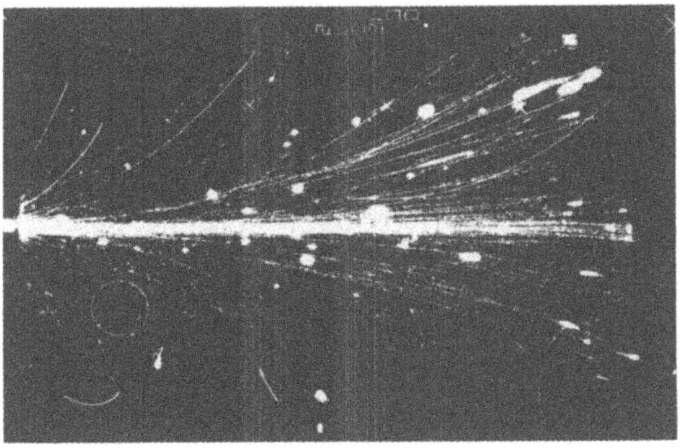

CERN – SPS E_{BEAM} 3200 GeV

Fig.1 A streamer chamber picture of an interaction of 200 GeV/nucleon oxygen nucleus with a heavy target (NA35 Collaboration)

the target and three types of detectors downstream of the streamer chamber: a scintillation counter, and segmented electromagnetic and hadronic calorimeters. For selecting the small-impact parameter ("central") collisions two kinds of trigger have been used: the veto trigger requiring less than two (for ^{16}O beam) or less than six (for ^{32}S beam) non-interacting fragments of the projectile, and the E_T trigger requiring the transverse energy exceeding a certain threshold value. Both triggers work at a few percent level. The NA35 Collaboration took data in the ^{16}O beam at 60 and 200 A GeV in 1986 and in the ^{32}S beam at 200 A GeV in 1987. Table 1 gives characteristics of the samples of central collision events used in pion interferometry studies (all at 200 A GeV). The results for the ^{32}S beam should be considered preliminary.

Table 1. Event samples used for interferometric analysis

Beam	Target	Trigger	No.events	$(dn_-/dy)_c$	Measured at
^{16}O	Au	hard veto	105 40	~40	CERN MPI
^{32}S	S	hard veto	408	~30	Frankfurt
^{32}S	Ag	semi-hard E_T	275	~60	Zagreb
^{32}S	Au	very hard E_T	85	~85	Zagreb

3. RESULTS FOR ^{16}O + Au COLLISIONS

Results of pion interferometry analysis of 105 ^{16}O+Au events measured with the CERN ERASME facility were published in ref. [5]. We recall the main points of this study. Only tracks of negatively charged particles were measured and these were all assumed to be pions, the particle misidentification (e^-, K^-) should not exceed 10%. All pair combinations of pions from different events were taken as the "background" distribution. The analysis was done in the rapidity interval $1<y<4$, the c.m. rapidity being $y_c \cong 2.5$ (this value corresponds to the "16+50" collision, as expected from the simple geometrical model: a cylinder containing about 50 nucleons in the Au target is struck by the ^{16}O projectile). The standard Gamov correction [6] was applied to the experimental correlation function in order to correct for the final state Coulomb repulsion between pions. Two theoretical models were fitted to the experimental correlation function using the multidimensional maximum likelihood procedure. The first was the simple static Gaussian model [7] in which the pion source density is assumed to be described by a gaussian, with time dependence omitted. The correlation function in this model can be expressed in terms of the invariant momentum difference Q_{inv}

$$C(Q_{inv}) = A[1 + \lambda \exp(-Q_{inv}^2 R^2/2)] ,$$

or separated into transverse and longitudinal components

$$C(Q_T, Q_L) = A[1 + \lambda \exp(-Q_T^2 R_T^2/2) \exp(-Q_L^2 R_L^2/2)] .$$

The fitted parameters are: transverse and longitudinal radii R_T and R_L, and the so-called "chaoticity parameter" λ. The second model applied was that due to Kolehmainen and Gyulassy (K-G) [8] which incorporates the relativistic dynamics of the reaction. Here the correlation function depends in a rather complicated way on three parameters: transverse radius R_T, proper time for pion freeze-out τ_0, and chaoticity λ

$$C(Q_T, \Delta y, \ldots) = C(R_T, \tau_o, \lambda) \ .$$

The temperature T of the source is also a model parameter, and it was fixed at 130 MeV as determined from our experimental p_T distribution for pions.

The fit in Q_{inv} in the interval $1 < y < 4$ yielded $R = 4.0 \pm 0.2$ fm, a value slightly greater than the radius of the ^{16}O projectile, and $\lambda = 0.29 \pm 0.03$.

The fit in Q_T, Q_L yielded different values of R_T, R_L and λ in various rapidity intervals, with substantial increase of the values of all the parameters in the interval $2 < y < 3$ located at the c.m. rapidity y_c. Correlation functions in Q_T variable are shown in Fig.2. The K-G fit gives

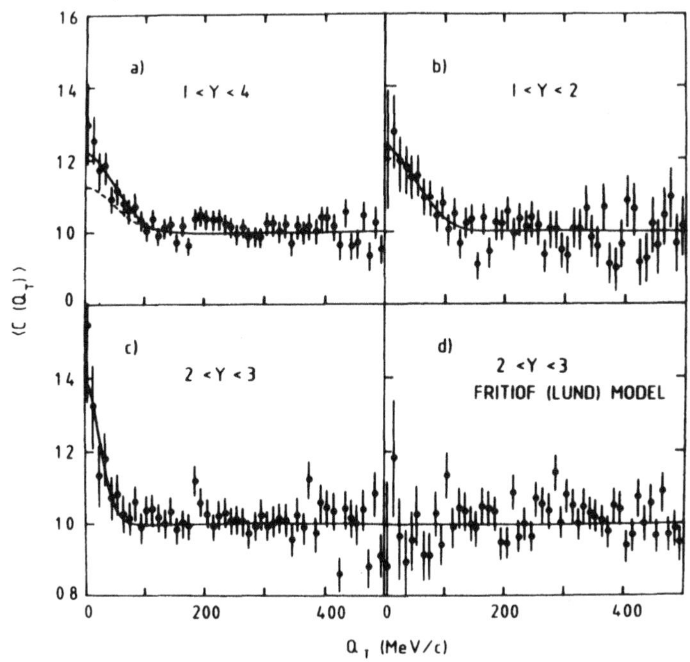

Fig.2 Correlation function in the Q_T variable for O+Au collisions in three different rapidity intervals a-c), d)- FRITIOF calculation. Curves are Gaussian. Ref.[5].

similar results. The obtained numerical values of the parameters are collected in Tables 2 and 3, and visualized in Figs. 3 and 4 for an easier judgement on their statistical significance. One can see that the data indicate the exis-

Table 2. Pion source parameters for ^{16}O + Au collisions in various rapidity intervals for the Gaussian model. From ref. [5], values of R_T^{side} and R_T^{out} from ref.[11].

y-interval	R_T, fm	R_L, fm	λ
1< y< 4	4.1 ± 0.4	3.1 +0.7/-0.4	0.31 +.07/-.03
1< y< 2	4.3 ± 0.6 R_T^{side} = 4.0±1.0 R_T^{out} = 4.4±1.0	2.6 ± 0.6	0.34 +.09/-.06
2< y< 3	8.1 ± 1.6 R_T^{side} = 6.6±1.8 R_T^{out} =11.2±2.3	5.6 +1.2/-0.8	0.77 ± .19

Table 3. Pion source parameters for ^{16}O + Au collisions in various rapidity intervals for the K-G model. From ref.[5].

y-interval	R_T, fm	τ_o, fm/c	λ
1< y< 4	3.6 ± 0.3	2.9 ± 0.7	0.29 ± .05
1< y< 2	4.0 ± 0.7	2.5 ± 1.0	0.30 ± .12
2< y< 3	7.3 ± 1.6	6.4 ± 1.0	0.84 ± .15

tence of a large, chaotic, and relatively long-lived (i.e. supposedly thermalized) fireball at $y=y_c$. Its parameters are roughly:

$$R_T \cong 7\text{-}8 \text{ fm} , \quad \tau_o \cong 6 \text{ fm/c} , \quad \lambda \cong 0.8 .$$

Outside this interval, i.e. for $y<2$ or $y>3$ we obtain

$$R_T \cong 4 \text{ fm} , \quad \tau_o \cong 3 \text{ fm/c} , \quad \lambda \cong 0.3 .$$

Another parametrization of the correlation function has been proposed by Pratt [9] and by Bertsch et al. [10]. In this approach the transverse momentum difference Q_T is decomposed into two orthogonal components, labelled "outward"

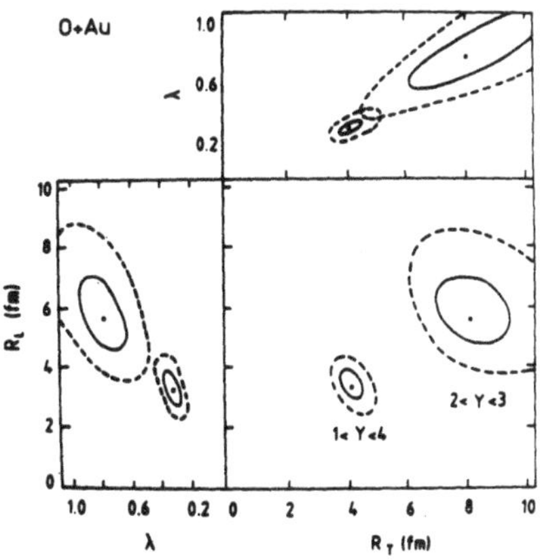

Fig.3 Comparison of maximum likelihood contours for Gaussian fits to the intervals $1<y<4$ and $2<y<3$ for projections on to three planes in (R_T, R_L, λ) parameter space. From ref.[5].

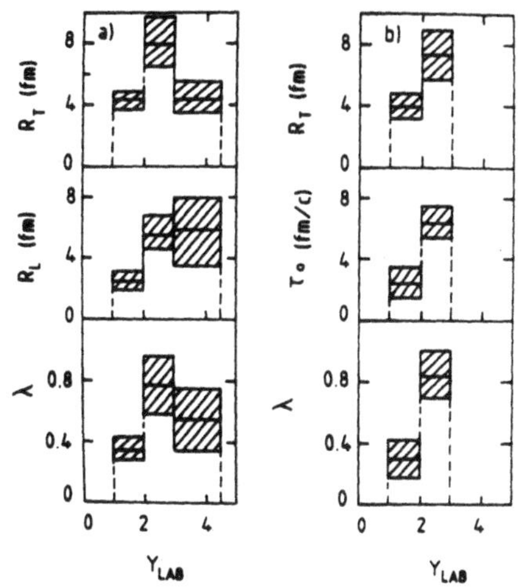

Fig.4 Comparison of a) Gaussian and b) K-G source parameters fitted to the data for different rapidity intervals. From ref. [5].

Q_T^{out} (along the pair momentum vector component transverse to the beam) and "sideward" Q_T^{side} (perpendicular to both the beam and transverse direction). The sideward component measures the transverse radius of the source, while the outward component is related to its time evolution. The sample of ^{16}O+Au collisions has been analyzed in these parameters near midrapidity $(2<y<3)$ and away from midrapidity $(1<y<2)$ [11]. The fit has been performed with R_L and λ parameters kept fixed at the best values from the previously described fits. The results are shown in Table 2. Away from midrapidity we find $R_T^{out} \cong R_T^{side}$, whereas at midrapidity $R_T^{out} = 11.2 \pm 2.3$ fm and $R_T^{side} = 6.6 \pm 1.8$ fm. This indicates the creation of a long-lived state which emits pions in the outward direction.

Finally, the dependence of the size parameters on the transverse momentum of the pair has been investigated, yielding an increase in all parameters for $p_T > 150$ MeV/c as compared with $p_T < 150$ MeV/c - see Table 4.

Table 4. Pion source parameters for ^{16}O+Au collisions for two ranges of transverse momentum of the pion pair. From ref.[12].

p_T interval	R_T , fm	R_L , fm	λ
$p_T < 150$ MeV/c	4.5 ± 0.4	4.1 ± 1.0	0.32 ± .05
$p_T > 150$ MeV/c	7.9 ± 0.9	6.9 ± 1.2	0.85 ± .30

4. RESULTS FOR ^{32}S + Ag AND ^{32}S + Au COLLISIONS

Pion interferometry studies have also been performed for ^{32}S+Ag and ^{32}S+Au central collisions at 200 A GeV, yielding the following results [13].

A fit to the exponential (better than Gaussian!) function in the invariant momentum difference Q_{inv}

$$C(Q_{inv}) = A[1 + \lambda \exp(-Q_{inv}R)]$$

in the wide rapidity interval $1 < y < 4$ yielded

for S+Ag $R = 6.8 \pm 0.4$ fm , $\lambda = 0.82 \pm 0.09$
for S+Au $R = 6.9 \pm 1.1$ fm , $\lambda = 0.7 \pm 0.2$

what is larger than the radius of the projectile sulphur nucleus and comes closer to the radius of the target nuclei.

Fits to the Gaussian function in Q_T for various rapidity intervals and $Q_L < 300$ MeV are shown in Fig.5. A rise of transverse source size (decrease of the width of the peak in $C(Q_T)$) is observed beyond the midrapidity, i.e. at much higher rapidities than for ^{16}O+Au collisions. In Fig.6 the obtained values of R_T and R_L are displayed versus rapidity using a sliding rapidity interval: an interval one unit of rapidity wide being displaced along the y-axis in steps of 0.5 units. The plot clearly shows the increase of R_T beyond midrapidity. The interpretation of R_L is more complicated as this quantity, in contrast to

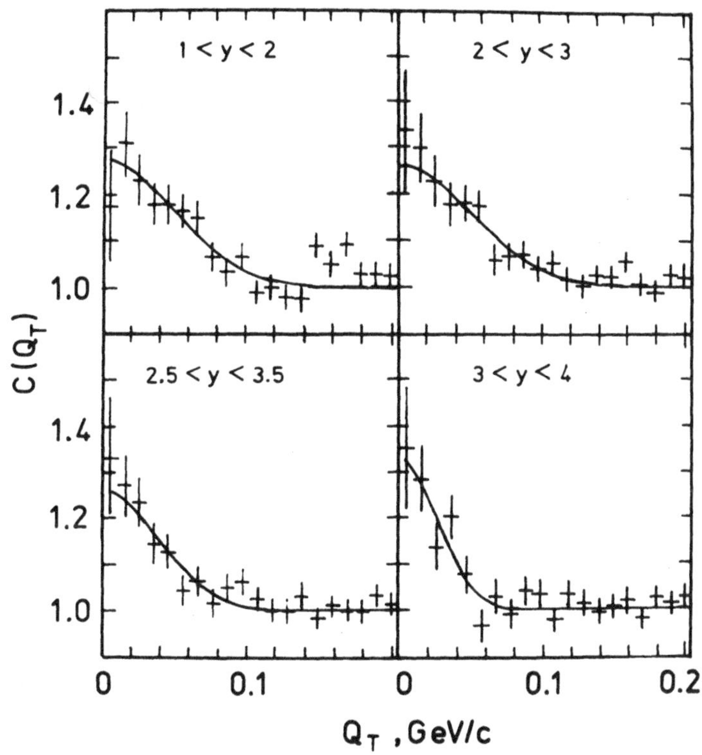

Fig.5 Correlation function in the Q_T variable for ^{32}S+Ag collisions and $Q_L < 300$ MeV in four different rapidity intervals. Curves are Gaussian fits. From ref.[13]

R_T, is not Lorentz-invariant. For a stationary source one would expect a flat rapidity dependence, the real size of the source being measured at any value of y. This does not seem to be the case and the data point towards another scenario, which might well be a cylindrical emitting volume expanding along the longitudinal direction [14]. In this model there is a strong correlation between the position of local sources and their longitudinal velocity, pairs emitted from two distant points of such a cylinder will not be able to contribute to the interferometric peak because of their large momentum difference, and we measure

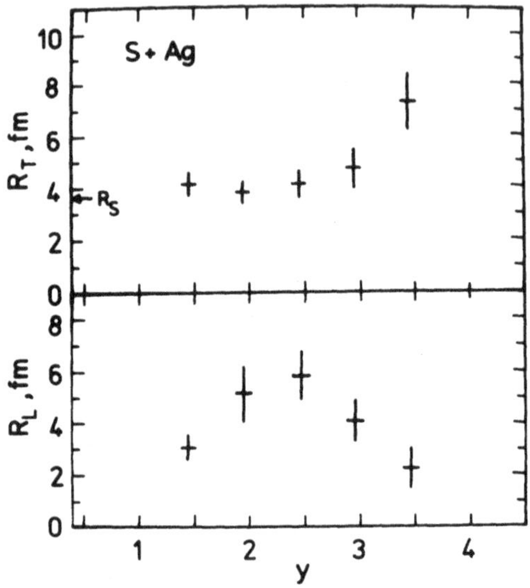

Fig.6 Dependence of pion source parameters for S+Ag collisions on rapidity [13].

only a section of the total cylinder length which is short enough as to radiate two pions of close momenta (the real pion correlation length). Away from midrapidity the values of R_L should be recalculated in the local reference frame in order to have the same meaning (see below).

An additional selection of ^{32}S+Ag events with highest multiplicities leads to slightly higher values of the size parameters, the trend of the data remains the same. Results for events with $n_- > 110$ are shown in Fig.7 [12]. Here again the sliding rapidity interval of unit length has been used,being displaced in steps of 0.2 . For the longitudinal parameter R_L two sets of points are shown: full points are calculated for a stationary observer located at midrapidity, points shown with crosses are for a "sliding observer" (local frames). They go up symmetrically to the left and to the right of midrapidity as $R_L' = \gamma R_L$ where γ is the Lorentz factor of the reference frame considered relative to midrapidity.

The comparison of the transverse size pattern for S+Ag with that for O+Au is not clear. One might have expected a certain shift in the position of maximal R_T due to a different value of the effective midrapidity which for S+Ag collisions is about 2.7 , but the observed shift is large and thus difficult to explain. The longitudinal size pat-

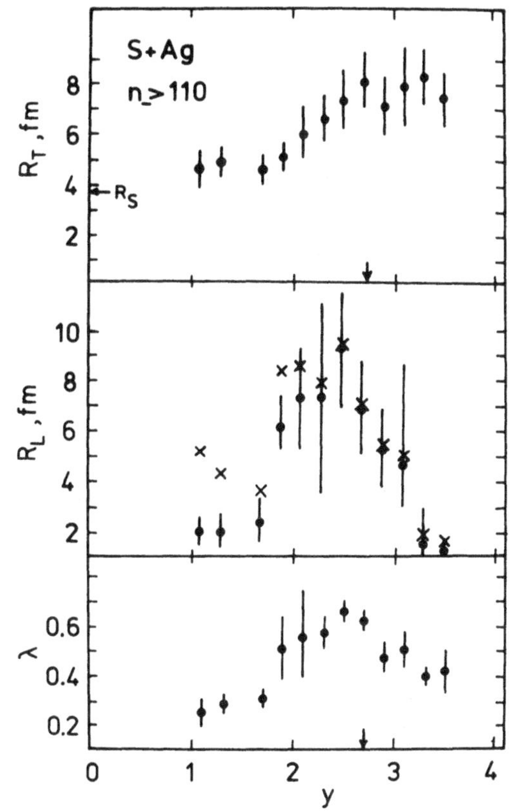

Fig.7 Dependence of pion source parameters for high-multiplicity S+Ag collisions on rapidity. For details see text. From [12].

tern is similar in both cases.

The analysis of S+Ag data in terms of Bertsch parameters yields a large value of $R_T^{out} = 11 \pm 2$ fm for $p_T > 200$ MeV/c versus 5.1 ± 0.5 fm for $p_T < 200$ MeV/c [15].

5. RESULTS FOR ^{32}S+S COLLISIONS

Finally, we present results of pion interferometry for the sample of central S+S collisions [12]. The values of transverse and longitudinal size parameters obtained from the Gaussian fits to these data are displayed in Fig.8. For S+S collisions the midrapidity coincides with that for N+N collisions at 200 GeV, i.e. $y_c = 3.0$. Taking advantage of the symmetry of the collision, track measurements have been done only in the backward hemisphere, and the points shown to the right of $y = y_c$ are just reflected from below y_c. They are marked with open circles, while other symbols have the same meaning as in Fig.7. The numerical values are collected in Table 5. One can see that, contrary to our preliminary results [11] which

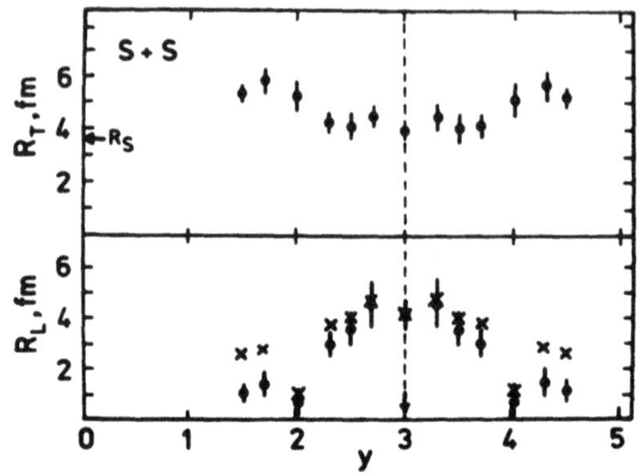

were based on a small sample of events, for S+S collisions no large source emerges at midrapidity. There is also no marked dependence of size parameters on transverse momentum – see Table 6.

Fig.8 Pion source parameters for S+S collisions as function of rapidity. For details see text. From ref.[12].

Table 5. Pion source parameters for S+S collisions for two different rapidity intervals [12]

y-interval	R_T , fm	R_L , fm	λ
$0.5 < y < 2.0$	5.2 ± 0.3	1.2 ± 0.3	0.50 ± 0.02
$2.0 < y < 3.0$	4.3 ± 0.4	3.0 ± 0.4	0.38 ± 0.06

Table 6. Pion source parameters for S+S collisions for two ranges of transverse momentum [12] (for rapidity interval $0.5 < y < 3.0$)

p_T-interval	R_T , fm	R_L , fm	λ
$p_T < 150$ MeV/c	5.1 ± 0.4	1.6 ± 0.2	0.31 ± 0.01
$p_T > 150$ MeV/c	4.4 ± 0.4	1.2 ± 0.2	0.46 ± 0.02

We close the presentation of the experimental data with Fig.9 which shows the lifetime parameter for O+Au, S+Ag and S+S collisions [12], plotted as function of rapidity. The lifetime increases when passing from light to heavier nuclei, reaching the maximum value of about 20 fm/c for O+Au collisions.

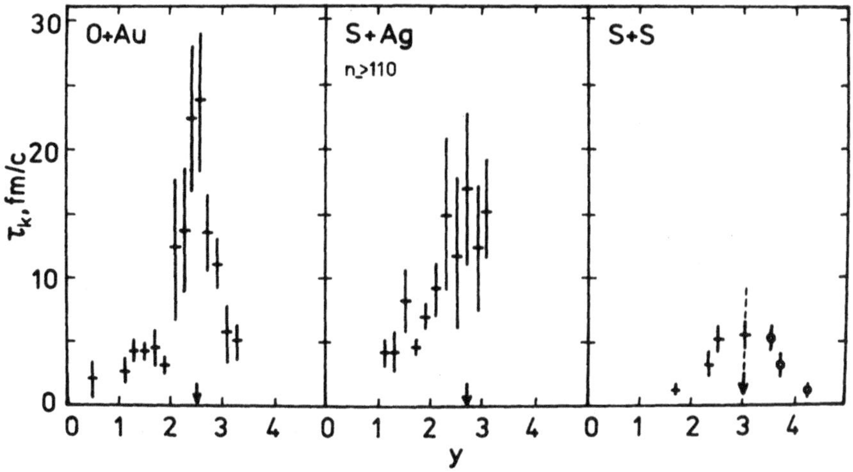

Fig.9 Dependence of the time parameter on rapidity of the pion pair for various colliding nuclei [12].

6. SUMMARY AND DISCUSSION

Pion interferometry applied to nucleus-nucleus interactions at 200 A GeV indicates that for a light projectile (^{16}O or ^{32}S) colliding with a heavy target (Ag or Au) a large pion source is formed at midrapidity or beyond it. This source exhibits a high degree of chaoticity and a relatively long lifetime, and thus could be considered a thermalized "fireball".

The large radius of the pion emission source in the central region is no longer compatible with the $A^{1/3}$ scaling [16], nor with the conjecture that the size of the pion emission region is about the size of the smaller of the two colliding nuclei [17]. The observed large source could be regarded as a volume which should be big enough as to contain all produced pions (the freeze-out condition) and thus its radius should be proportional to $(dN/dy)^{1/3}$. This seems to be verified by the data [18].

The change of the radius and chaoticity of the source towards midrapidity has been interpreted within the frame of the model in which two kinds of pion sources: chaotic and coherent contribute to the emission of pions [19]. The analysis of the O+Au data along these lines leads to redefined chaoticities $p_1 = 0.22$ and $p_2 = 0.88$ for the regions $1<y<2$ and $2<y<3$ respectively. The radii in these two rapidity intervals should then be in the ratio $R_2/R_1 = \sqrt{2}$, what might be compatible with the data within errors. Unfortunately, a direct fit of the relevant "constrained-two-Gaussian" correlation function to the experimental data is not successful which also seems to be the case for some other experiments and might be due to insensitivity of this formula to the shape of the distribution [18]. There are also some basic objections raised against models with coherent states [1]. Besides, one should keep in mind that many various factors tend to lower the value of the λ-parameter. These are: pion production via intermediate states (resonances), presence of misidentified kaons or other particles among the "pions", finite momentum resolution, and other methodical aspects of the particle detection and of the data analysis. They should all be well understood and accounted for, otherwise the λ-parameter obtained from the fit to the experimental distribution represents only a lower limit for the true "chaoticity".

Analysis of the longitudinal size parameters suggests that the pion source is expanding in longitudinal direction, as considered in relativistic hydrodynamics or cascade models [14, 8, 20]. However, ref.[20] predicts smaller effective radii for pion pairs with higher transverse momenta, whereas our data show the opposite trend.

New data with increased statistics should help to establish the space-time picture of the collision process (here relativistic hydrodynamic models look most relevant), and possibly relate the experimental results to underlying physics at a deeper level, such as the QCD string tension [21].

The author is grateful to his collegues from the NA35 Collaboration D.Ferenc and M.Lahanas for having made available to him their preliminary results prior to publication. He would also like to express his gratitude to Prof. R.Weiner for the invitation to the CAMP Workshop in Marburg where these problems were extensively discussed.

This report has been prepared under a contract from the Polish Government Research Programme CPBP 01.09.

REFERENCES

1. M.I.Podgoretsky, Sov.J.Part.Nucl.(EChaYa) 20,628 (1989).
2. W.Zajc, in "Hadronic Multiparticle Production", ed. P.Carruthers, World Scientific, 1988, p.235.
3. J.Bartke, Int.Journ.Mod.Phys. A4, 1319(1989).
4. B.Lörstad, Int.Journ.Mod.Phys. A4, 2861(1989).
5. A.Bamberger et al., Phys.Lett. B203, 320(1988).
6. M.Gyulassy et al., Phys.Rev. C20, 2267(1979).
7. F.B.Yano and S.E.Koonin, Phys.Lett. B78, 556(1978).
8. K.Kolehmainen and M.Gyulassy, Phys.Lett.B180,203(1986).
9. S.Pratt, Phys.Rev.Lett. 53, 1219(1984); Phys.Rev. D33 1314 (1986).
10. G.Bertsch, M.Gong and M.Tohyama, Phys.Rev. C37, 1896 (1988).
11. J.Harris, NA35 Collab., Nucl.Phys. A498, 133c(1989).
12. M.Lahanas, NA35 Collab., QUARK MATTER 90, Menton, to be published and CAMP Marburg 1990, to be published.
13. D.Ferenc, NA35 Collab., Rencontres de Moriond 1990, to be published.
14. J.D.Bjorken, Phys.Rev. D27, 140(1983).
15. D.Ferenc, NA35 Collab., CAMP Marburg 1990, to be publ.
16. W.A.Zajc et al., Phys.Rev. C29, 2173(1984).
17. J.Bartke, Phys.Lett. B174, 32(1986).
18. W.Zajc, QUARK MATTER 90, Menton, to be published in Nucl.Phys.A, and CAMP Marburg 1990, to be published
19. R.M.Weiner, Proc.Workshop on Multiparticle Dynamics, Perugia 1988, World Scientific, 1989.
20. Y.Hama and S.S.Padula, Phys.Rev. D37, 3237(1988).
21. B.Andersson and W.Hofmann, Phys.Lett. B169, 364(1986).

BOSE-EINSTEIN CORRELATIONS IN THE TARGET FRAGMENTATION REGION IN 200 A GEV ^{16}O + NUCLEUS COLLISIONS

T. Peitzmann[4], R. Albrecht[1], T. C. Awes[5], M. Bahr[4], P. Beckmann[4,a], F. Berger[4], R. Bock[1], G. Claesson[3], G. Clewing[4], L. Dragon[4], A. Eklund[3], R. L. Ferguson[5], A. Franz[5,b], S. Garpman[3], R. Glasow[4], H. Å. Gustafsson[3], H. H. Gutbrod[1], G. Hölker[4], J. Idh[3], P. Jacobs[2], K. H. Kampert[4], B. W. Kolb[1], P. Kristiansson[3], H. Löhner[4,a], I. Lund[1], F. E. Obenshain[5,b], A. Oskarsson[3], I. Otterlund[3], S. Persson[3], F. Plasil[5], A. M. Poskanzer[2], M. Purschke[4], H. G. Ritter[2], H. Rothgänger[4], S. Saini[5], R. Santo[4], H. R. Schmidt[1], R. Schmidt[4], S. P. Sørensen[5,b], K. Steffens[4], E. Stenlund[3], D. Stüken[4], M. L. Tincknell[5] and G. R. Young[5]

WA80 Collaboration

1. *Gesellschaft für Schwerionenforschung, D-6100 Darmstadt, Fed. Rep. of Germany*
2. *Lawrence Berkeley Laboratory, Berkeley, California 94720, USA*
3. *University of Lund, S-22362 Lund, Sweden*
4. *University of Münster, D-4400 Münster, Fed. Rep. of Germany*
5. *Oak Ridge National Laboratory, Oak Ridge, Tennessee 37831, USA*

a. *now at: KVI, University of Groningen, NL-9747 AA Groningen, Netherlands*
b. *University of Tennessee, Knoxville, Tennessee 37996, USA*

Abstract

Correlations of positive pions are investigated in the target fragmentation region in 200 A GeV ^{16}O + nucleus collisions. The pions are measured with the Plastic Ball detector in the WA80 experiment at the CERN SPS. The detailed behaviour of the fitted parameters on rapidity, centrality and the number of spectator nucleons is studied. The data indicate a large source in central O + Au collisions for $y_{lab} \leq 0$. The observations are compatible with a substantial amount of rescattering in the target.

1 Introduction

Pion interferometry allows to study the space-time properties of the particle emitting source in high energy collisions [1]. An enhancement in the production of identical pions with similar momentum - known as Bose-Einstein Correlations, the Hanbury-Brown - Twiss effect (HBT) [2] or the Goldhaber - Goldhaber - Lee - Pais effect (GGLP) [3] - has been observed in various experimental studies. A simple theoretical picture of multi-particle production yields the following prediction for the two-particle correlation function:

$$C_2 \equiv \frac{< n^2 >}{< n(n-1) >} \cdot \frac{d^6N/dp_1^3dp_2^3}{d^3N/dp_1^3 \cdot d^3N/dp_2^3} = 1 \; + \; \lambda \cdot |\tilde{\rho}(p_1^\mu - p_2^\mu)|^2.$$

$\tilde{\rho}$ is the Fourier transform of the distribution of emitters; the most commonly chosen analytic expressions for $\tilde{\rho}$ contain a suitable correlation length R. While this parameter seems to measure dynamical properties of the produced strings in e^+e^- or pp collisions, it can provide geometrical information in medium energy heavy ion collisions.

The importance of rescattering of secondaries in the target spectator matter in high energy nuclear reactions has been emphasized in earlier publications [4]. It is therefore of interest to study pion correlations in the target fragmentation region to clarify both the general role of the target spectators and their influence on pion interferometry. In this paper we will present a study of π^+ correlations in reactions of 200 A GeV ^{16}O ions with various targets in the target rapidity region ($y_{lab} \leq 1$).

2 Data analysis

Data were taken in the WA80 experiment [5] at the CERN SPS during the ^{16}O run in fall 1986. The pions were identified with the Plastic Ball detector in $30° \leq \theta_{lab} \leq 160°$ using both ΔE-E-information and the time delayed signal of the π^+-decay. Details on the detector design and performance and the identification are given in [6]. Only pions, which are stopped in the detector modules ($E_{kin} \leq 120\,\text{MeV}$), are used. The contamination of misidentified particles was estimated to be of the order of 5 % for all samples. The resolution in relative momentum is $\delta q \approx 10\,\text{MeV/c}$.

Reactions with C, Cu, Ag and Au targets were investigated. Background distributions were obtained by the event-mixing procedure: A pion from one event was combined with a pion from another event with the same apparent π^+ multiplicity. The correlation functions were corrected for the Coulomb-interaction with the standard Gamow-factor [7]. The data were analyzed both in terms of the absolute value of the four momentum difference: $Q \equiv \sqrt{-(p_1^\mu - p_2^\mu)^2}$ and in terms of the energy- and momentum-difference in the lab-system: $(|\vec{q}|, q_0) \equiv (|\vec{p}_1 - \vec{p}_2|, |E_1 - E_2|)$. For simplicity, gaussian functions have been used both for one- and two-dimensional analysis:

$$C_2(Q) = 1 + \lambda_{inv} \cdot \exp(-\frac{1}{2}R_{inv}^2 Q^2)$$

$$C_2(|\vec{q}|, q_0) = 1 + \lambda_{lab} \cdot \exp(-\frac{1}{2}R_{lab}^2|\vec{q}|^2 - \frac{1}{2}\tau^2 q_0^2).$$

For the analysis in Q an exponential function:

$$C_2(Q) = 1 + \lambda_{exp} \cdot \exp(-R_{exp}Q)$$

has also been used for comparison. Fits were performed in the region of $Q \leq 400\,\mathrm{MeV}$ by maximizing the logarithm of a maximum likelihood.

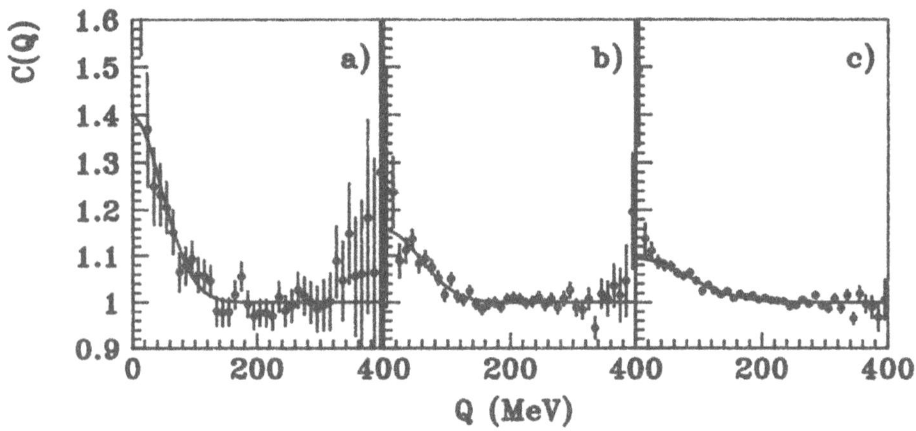

Fig. 1: π^+ correlation functions in dependence of Q for $-1 \leq y_{lab} \leq 1$. a) O + C, b) O + Cu, c) O + Au.

3 Results and discussion

Fig. 1 shows correlation functions in dependence of Q for central reactions of ^{16}O with different targets in the rapidity region $-1 \leq y_{lab} \leq 1$. A clear enhancement for small values of Q is visible in all cases. A fit with a gaussian parametrization is shown as a solid line. Counterintuitively the radii R_{inv} decrease with increasing target mass, also the intercept λ_{inv} goes down for the heavier systems. The correlation functions have also been fitted with an exponential distribution to investigate systematic effects connected to the choice of the parametrization. In addition, the same data sets have been analyzed in $(|\vec{q}|, q_0)$ and fitted with a two-dimensional gaussian. The resulting fit parameters are summarized in table 1.

R_{inv} and λ_{inv} (R_{exp} and λ_{exp}) show a significant decrease with increasing target size. The correlation strength λ_{lab} shows a similar decrease for the heavier systems, but the space-component of the correlation length R_{lab} seems to be only weakly dependent on the target mass, the values are close to the charge radius of the projectile: $R_{proj}^{rms} \approx 2.9\,\mathrm{fm}$, while the time parameter τ exhibits a drastic decrease.

Table 1: Pion Correlation Fit Parameters ($-1 \leq y_{lab} \leq 1$)

target	Au	Ag	Cu	C
R_{inv} (fm)	$2.5^{+0.1}_{-0.1}$	$3.1^{+0.2}_{-0.2}$	$3.5^{+0.1}_{-0.1}$	$4.1^{+0.3}_{-0.3}$
λ_{inv}	$0.094^{+0.003}_{-0.003}$	$0.14^{+0.01}_{-0.01}$	$0.16^{+0.01}_{-0.01}$	$0.40^{+0.05}_{-0.02}$
R_{exp} (fm)	$2.5^{+0.1}_{-0.1}$	$3.3^{+0.4}_{-0.4}$	$4.2^{+0.3}_{-0.3}$	$5.7^{+0.6}_{-0.6}$
λ_{exp}	$0.162^{+0.006}_{-0.006}$	$0.22^{+0.02}_{-0.03}$	$0.29^{+0.03}_{-0.01}$	$0.92^{+0.14}_{-0.03}$
R_{lab} (fm)	$2.8^{+0.1}_{-0.1}$	$3.2^{+2.2}_{-0.2}$	$3.5^{+0.3}_{-0.3}$	$3.3^{+0.1}_{-0.1}$
τ (fm/c)	$1.8^{+0.5}_{-0.7}$	$0.0^{+1.2}$	$3.4^{+0.8}_{-0.8}$	$12.1^{+1.4}_{-1.4}$
λ_{lab}	$0.116^{+0.003}_{-0.003}$	$0.16^{+0.01}_{-0.01}$	$0.20^{+0.10}_{-0.10}$	$0.60^{+0.04}_{-0.04}$

How can the variations in the parameters be understood in our current picture of ultrarelativistic nuclear collisions? A reduction of the value of λ may be caused by a variety of mechanisms. Most of these, as e.g. final state interaction, resonance production or coherence, should not show such a drastic variation for different targets.

It is however easy to realize that pions observed in the target rapidity region might undergo very different evolutions. They may be

- directly produced by the participant nucleons or
- rescattered in the target spectator matter.

It has been noted that data from the target fragmentation region indicate rescattering effects. Such rescattering need not affect the pion multiplicities, but it would alter the volume measured by interferometry to the "freeze-out" volume of the excited spectator matter. The rescattered pions might come from very large volumes or late times, and their correlation could thus be hidden at very small values of Q hardly accessible to experiment. In fact, there is an indication of a steep rise in the high statistics O + Au correlation function for $Q \leq 30\,\text{MeV}$ (see fig. 1). These data points do however depend crucially on the Coulomb correction, whereas the overall fit is not very sensitive to this correction.

Rapidity dependence: Fig. 2 shows the parameters R_{inv} and λ_{inv} for the rapidity intervals $-1 \leq y_{lab} \leq 0$ and $0 \leq y_{lab} \leq 1$ for all targets. The variation in the parameters is much stronger for the backward interval, indicating that we observe a physical effect connected to the target and not created by e.g. accidental double hits or other detector effects. The λ_{inv}-values are systematically smaller for the more forward interval.

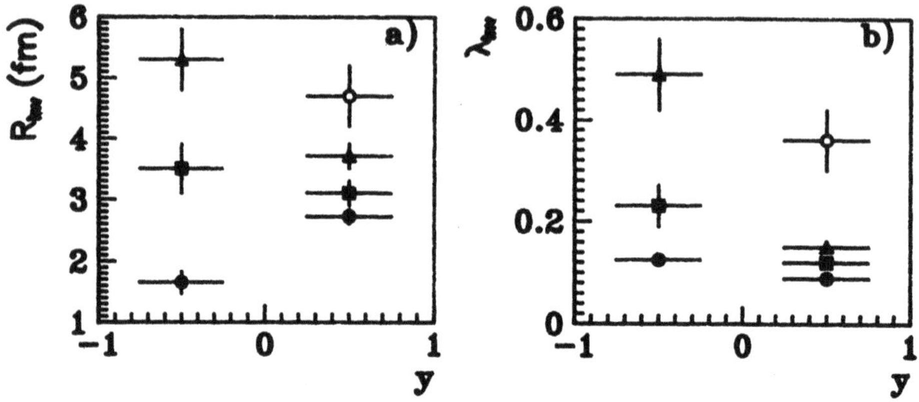

Fig. 2: Fit parameters – a) R_{inv}, b) λ_{inv} – of the π^+ correlation as a function of the pion rapidity. The filled circles show O + Au, the squares O + Ag, the triangles O + Cu and the open circles O + C data.

Centrality dependence: For the system O + Au at 200 *A* GeV we have studied the centrality dependence. The data were divided into three subsamples selected by the forward energy flow ($\theta_{lab} \leq 0.3°$) measured with the Zero-Degree Calorimeter [8]. We have then used the FRITIOF Monte-Carlo to extract the number of participant nucleons from the forward energy E_{ZDC} event by event. From this we can calculate an effective participant radius $R_{part} \equiv < N_{part} >^{1/3}$ which characterizes the size of the participant volume. We obtained R_{part} = 2.77, 3.34 and 3.82 for the peripheral, medium central and central reactions respectively.

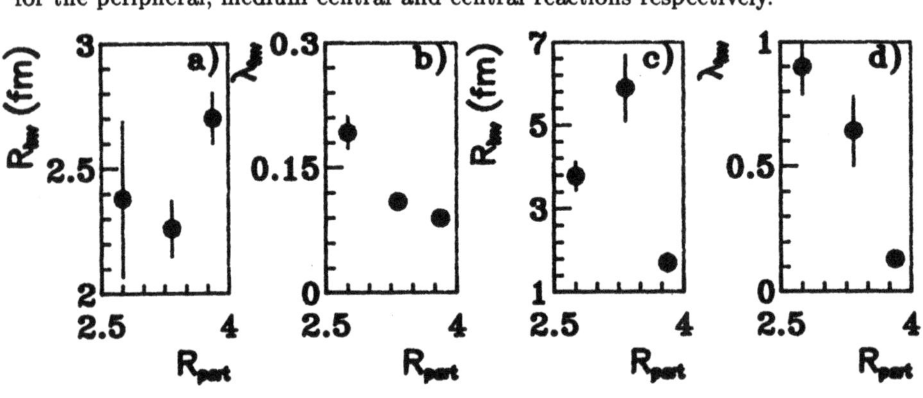

Fig. 3: Fit parameters – a) R_{inv} $(0 \leq y_{lab} \leq 1)$, b) λ_{inv} $(0 \leq y_{lab} \leq 1)$, c) R_{inv} $(y_{lab} \leq 0)$, d) λ_{inv} $(y_{lab} \leq 0)$ – of the π^+ correlation as a function of the participant radius (see text).

Fig. 3 shows the dependence of R_{inv} and λ_{inv} on the participant radius, analyzed seperately in the rapidity intervals $0 \leq y_{lab} \leq 1$ and $y_{lab} \leq 0$. In the more forward region the R_{inv}-parameter increases from $R_{inv} \approx 2.3-2.4$ fm for the peripheral and medium central reactions to 2.7 fm for the central collisions. The λ_{inv}-value decreases with centrality - it is small ($\lambda_{inv} < 0.2$) for all cases. In the backward rapidity region the R_{inv}-parameter increases from 3.8 fm to 5.9 fm when going from peripheral to medium central collisions and the λ_{inv}-parameter decreases slightly, but is rather large (0.90 and 0.64 resp.). Both parameters drop to small values for the central reactions – if there is a two-component source, this might be due to the fact that the fit is more sensitive to the smaller component. Indeed, if we divide the data sample again into pions of low and high transverse momentum (fig. 4), we observe a larger component ($R_{inv} = 7.0 \pm 0.8$ fm and $\lambda_{inv} = 0.52 \pm 0.08$) for the lower p_T-pions, while only a small source ($R_{inv} = 1.5 \pm 0.2$ fm and $\lambda_{inv} = 0.14 \pm 0.01$) is visible from the high p_T-particles.

The parameters are consistent with the assumptions that we observe in the more forward interval only the fraction of pions coming directly from the participant volume and that in the backward region the full source including spectator matter is seen.

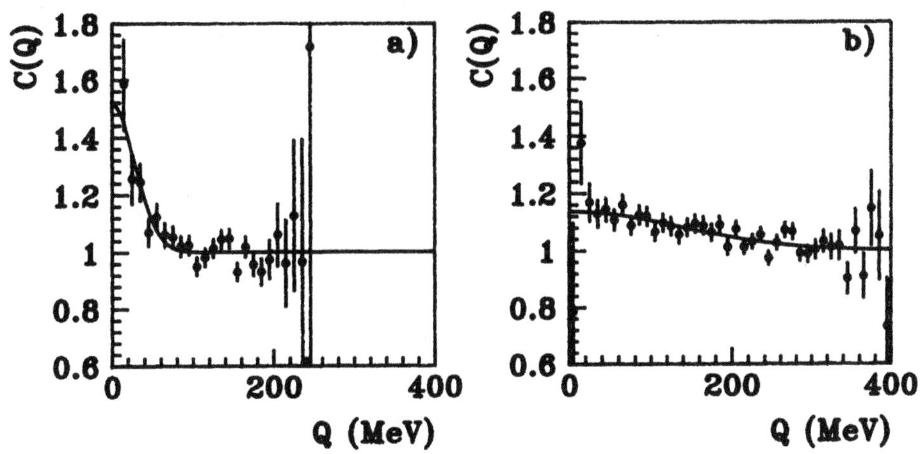

Fig. 4: π^+ correlation functions in dependence of Q for central O + Au in $-1 \leq y_{lab} \leq 0$.
a) $p_T \leq 100$ MeV/c, b) $p_T \geq 100$ MeV/c

Spectator dependence: Finally we can compare systems, which are similar according to the number of participants and the effective center-of-mass rapidity, but which differ by the number of target spectators. We know from analysis with the FRITIOF Monte-Carlo that the data sets given in table 2 are pairwise comparable in this sense. The fit parameters for these systems vary significantly. Both parameters are substantially smaller for the systems with larger number of spectators, when all other relevant parameters are fixed. The presence of spectator matter seems to diminish the "visible" fraction of the pions, and their apparent radius is reduced.

Table 2: Spectator Dependence of Pion Correlation ($-1 \leq y_{lab} \leq 1$)

system	Cu central	Au medium	C central	Au peripheral
$< N_{part} >$	39.5	37.4	19.2	21.2
$< y_{cm} >$	2.8	2.7	3.1	2.9
$< N_{spect} >$	38.4	172.9	3.2	184.3
R_{inv} (fm)	$3.5^{+0.1}_{-0.1}$	$2.2^{+0.2}_{-0.2}$	$4.1^{+0.3}_{-0.3}$	$2.8^{+0.2}_{-0.2}$
λ_{inv}	$0.16^{+0.01}_{-0.01}$	$0.12^{+0.01}_{-0.01}$	$0.40^{+0.05}_{-0.05}$	$0.25^{+0.02}_{-0.02}$

4 Summary and Conclusions

We have observed correlations of positive pions in the target fragmentation region for reactions of 200 A GeV ^{16}O on C, Cu, Ag and Au targets. The interpretation of the enhancement in terms of Bose-Einstein correlations parametrized as a gaussian yields the following results:

1. Applying a simple one component fit gives a decrease in R_{inv} and λ_{inv} with increasing target mass. A two-dimensional (R_{lab}, τ) analysis indicates a more or less constant R_{lab} and a decrease in the τ-parameter.

2. This decrease in the parameters is obviously connected to the number of spectators. It is consistent with a substantial amount of rescattered pions emerging from a large freeze-out space-time volume of the excited spectator matter in the case of the heavy targets.

3. At backward rapidities in O + Au reactions we observe a large source whose size increases from 3.8 fm for peripheral to 7.0 fm for central collisions. The large component in central collisions is most easily seen with an additional cut in transverse momentum: $p_T \leq 100\,\mathrm{MeV/c}$. The λ_{inv}-parameters are above 0.5 in all these cases.

4. In the rapidity interval $0 \leq y_{lab} \leq 1$, which gives the dominant contribution to our overall correlation functions, we see in O + Au a smaller source increasing slightly for central collisions. The measured λ_{inv}-values are smaller than 0.2.

This analysis emphasizes the need to study the dynamics of the so-called "spectator nucleons", which in fact seem to participate strongly in a later stage of the collision. Care should be taken, if one compares pion interferometry in different rapidity regions, when the presence of target spectators may obscur the interpretation.

[1] G.I. Kopylov and M.I. Podgoretskii, Sov. J. Nucl. Phys. 15 219 (1972); 18 336 (1974)

[2] R. Hanbury-Brown and R.Q. Twiss, Nature 178 1046 (1956)

[3] G. Goldhaber et el., Phys. Rev. 120 300 (1960)

[4] R. Albrecht et al., Z. Phys. C - Particles and Fields 45 529-537 (1990)

[5] R. Albrecht et al., Report No. GSI-85-32, Gesellschaft für Schwerionenforschung, D-6100 Darmstadt (1985)

[6] A. Baden et al., Nucl. Instr. Meth. 203 189 (1982)

[7] M. Gyulassy et al., Phys. Rev. C 20 2267 (1979)

[8] G. R. Young et al., Nucl. Instr. Meth. A 279 503-517 (1989)

FINAL STATE INTERACTION EFFECT ON CORRELATIONS IN NARROW PARTICLES PAIRS

R. LEDNICKY, V.L. LYUBOSHITZ
Joint Institute for Nuclear Research, Dubna, USSR

Abstract

The dependence of the two-particle correlation function on the space-time dimensions of the particle production region is discussed. The basic formulae, taking into account the effects of quantum statistics and final state interaction, and the conditions of their applicability are given.

1. As a narrow particle pair we understand two particles with a small relative momentum in their c.m.s.; this means that in the laboratory these particles have near-by velocities (or near-by momenta, in the case of equal masses). It is well-known that due to the effects of quantum statistics (QS) and final state interaction (FSI) the correlations in the narrow particle pairs contain an important information on the space-time characteristics of the multiparticle production process [1-6]. This circumstance has been extensively and successfully used for the determination of the space-time dimensions of the pion, kaon and nucleon production regions in various multiparticle processes (see, e.g., reviews [7,8] and refs. [9-12]). In the present paper we briefly review the main ingredients of the method.

2. It is convenient to define the two-particle correlation function (CF) $R(p_1,p_2)$ as the ratio of the double inclusive distribution of the particle 4-momenta p_1,p_2 to the one which would be observed in the

case of no QS and FSI effects. In heavy ion collisions, or at sufficiently high energies, we may neglect the kinematic constraints and construct the CF, e.g., with the help of the particles from different events.

According to the general principles of the relativistic quantum theory, the CF of two nonidentical particles may be represented in the form:

$$R(p_1,p_2) = \sum_S \int \tilde{\rho}_{S,P}(x_1,x_2;x_1',x_2')\,\psi^{(S)}_{p_1p_2}(x_1,x_2)\,\psi^{(S)*}_{p_1p_2}(x_1,x_2)\cdot d^4x_1 d^4x_2 d^4x_1' d^4x_2', \tag{1}$$

where the space-time density matrix $\tilde{\rho}_{S,P}$ determines the production probability of two noninteracting particles with the same 4-velocities p_j/m_j, the total 4-momentum $P \equiv 2p = p_1 + p_2$ and the total spin S; $\psi^{(S)}$ is the Bethe-Salpeter amplitude corresponding to a given interaction potential and describing the relative motion of the particles (see, e.g., [6,13]):

$$\psi^{(S)}_{p_1p_2}(x_1,x_2) = e^{-iPX}[e^{-i\tilde{q}x/2} + \varphi^{(S)}_{p_1p_2}(x)]. \tag{2}$$

Here X is the two-particle c.m.s. 4-coordinate, $x = x_1 - x_2$ and $\tilde{q} = q - P(q\cdot P)/P^2$, $q = p_1 - p_2$. In the case of particles with equal masses: $X = (x_1 + x_2)/2$, $\tilde{q} = q$. It is implied in Eqs. (1),(2) that the main contribution to the double inclusive distribution comes from the configurations $1,2,\ldots,j,\ldots$ with large relative c.m.s. momenta in pairs $(1,j)$ and $(2,j)$ as compared with a small one $|\vec{p}_1^{\,*} - \vec{p}_2^{\,*}| = |\vec{q}^{\,*}| \equiv 2\vec{k}^*$. In such a case the FSI may be neglected in all pairs except $(1,2)$; also, the density matrix $\tilde{\rho}$ is insensitive to a small change of the momentum $\vec{k}^*$ in the vicinity of zero. The $\tilde{q}$-dependence of the CF is then determined by the FSI of particles $1,2$, as well as by symmetrization of the Bethe-Salpeter amplitudes if these particles are identical:

$$\psi^{(S)}_{p_1p_2}(x_1,x_2) \to \frac{1}{\sqrt{2}}[\psi^{(S)}_{p_1p_2}(x_1,x_2) + (-1)^S\psi^{(S)}_{p_1p_2}(x_2,x_1)]. \tag{3}$$

The function $\varphi^{(S)}$, responsible for the interaction, may be represented in an integral form [6]:

$$\varphi^{(S)}_{p_1p_2}(x) = \frac{\sqrt{P^2}}{4\pi^2 i} e^{iPx(1+Pq/P^2)/2} \int \frac{e^{-i\varkappa x} f^{(S)}(p_1,p_2,\varkappa,P-\varkappa)}{(\varkappa^2-m_1{}^2+i0)[(P-\varkappa)^2-m_2{}^2+i0]} d^4\varkappa, \quad (4)$$

where $f^{(S)}$ is the scattering amplitude analytically continued to the unphysical region; on mass-shell it coincides with the usual scattering amplitude $f^{(S)}(\vec{k}^*)$. We will consider such small values of k^* that the contribution of a short-range s-wave interaction dominates in the absence of Coulomb forces. For the scattering amplitude at sufficiently small values of k^* the effective-range approximation of Landau-Smorodinsky is valid:

$$f^{(S)}(\vec{k}^*) = [1/f_0^{(S)} + \tfrac{1}{2}d_0^{(S)}k^{*2} - ik^*]^{-1}, \quad (5)$$

where $f_0^{(S)}$ and $d_0^{(S)}$ are the scattering length and the effective radius, respectively, the latter measuring the range of the strong FSI (e.g., $d_0^{(0)} = 2.7$ fm for a two-neutron system).

For k^* much smaller than the reduced mass $m = 2m_1m_2/(m_1+m_2)$ the Bethe-Salpeter amplitude may be expressed in a form of 3-dimensional integral [6]:

$$\varphi^{(S)}_{p_1p_2}(x) = \int \varphi^{(S)}_{\vec{k}^*}(\vec{r}\,')g(\vec{r}^{\,*}-\vec{r}\,',t^*)\, d^3\vec{r}\,', \quad (6)$$

where $\varphi^{(S)}_{\vec{k}^*}(\vec{r}\,') = \psi^{(S)}_{\vec{k}^*}(\vec{r}\,') - exp(-i\vec{k}^*\vec{r}\,')$ measures a deviation of the two-particle wave function from the plane wave, and, in the absence of Coulomb interaction, has the asymptotics of a diverging spherical wave:

$$\varphi^{(S)}_{\vec{k}^*}(\vec{r}\,') \to f^{(S)}(\vec{k}^*)\cdot exp(ik^*r')/r'. \quad (7)$$

The g-function is given by:

$$g(\vec{\Delta},t^*) = \tfrac{1}{8}(m/\pi\tilde{t}^*)^{3/2}(1-i)^3 exp[i(\Delta^2 m/2\tilde{t}^* + k^{*2}\tilde{t}^*/2m)], \quad (8)$$

$\tilde{t}^* = |t^*|+t^*(m_1-m_2)/(m_1+m_2)$. The integration in (6) of the spherical wave leads to the following asymptotics for the function $\varphi^{(S)}_{p_1p_2}(x)$ [6]:

$$\varphi^{(S)}_{p_1p_2}(x) \to f^{(S)}(\vec{k}^*)\cdot\{i\sin(k^*r^*) + \tfrac{1-i}{2}[E_1(z_-)e^{ik^*r^*} + E_1(z_+)e^{-ik^*r^*}]\}/r^*, \quad z_\pm = \sqrt{m/2\tilde{t}^*}(r^* \pm k^*\tilde{t}^*/m), \quad (9)$$

where $E_1(z) = \sqrt{2/\pi} \int_0^z e^{iy^2} dy$ is the Fresnel integral. It can be shown that, on the condition

$$m \langle r^{*2}/2\tilde{t}^{*} \rangle \gg 1, \tag{10}$$

the g-function may be replaced by $g(\vec{\Delta},0) = \delta^{(3)}(\vec{\Delta})$, so that we arrive at the usual approximation (see, e.g. [5]) of equal emission times in c.m.s. of the particles $1,2$, i.e.:

$$\varphi^{(S)}_{p_1 p_2}(x) \cong \varphi^{(S)}_{\vec{k}^*}(\vec{r}^*). \tag{11}$$

In general case: $|\varphi^{(S)}_{p_1 p_2}(x)| \leqslant |\varphi^{(S)}_{\vec{k}^*}(\vec{r}^*)|$; $\varphi^{(S)}_{p_1 p_2} \to 0$ in the limit $\tilde{t}^* \to \infty$.

3. Since it is unlikely that the particle pairs produced in different space-time points are acompanied by the same multiparticle system, we may neglect the corresponding nondiagonal interference terms in the density matrix and put:

$$\tilde{\rho}_{S,P}(x_1,x_2;x_1',x_2') \cong \rho'_{S,P}(x_1,x_2)\cdot\delta^{(4)}(x_1-x_1')\delta^{(4)}(x_2-x_2'). \tag{12}$$

In the case of unpolarized particle production we can factorize the dependence of the density matrix on the total spin S:

$$\rho'_{S,P}(x_1,x_2) = \rho_S \cdot W_P'(x_1,x_2), \tag{13}$$

where $\rho_S = (2S+1)/[(2j_1+1)(2j_2+1)]$. For noninteracting identical particles with spin $j = j_1 = j_2$ we then get the well-known result of Kopylov-Podgoretsky [3], originally obtained in the model of independent point-like one-particle sources, i.e.:

$$R(p_1,p_2) = 1 + c_0\langle\cos(qx)\rangle, \quad c_0 = \sum_{S\text{-even}} \rho_S - \sum_{S\text{-odd}} \rho_S, \tag{14}$$

$\langle\cos(qx)\rangle = \int W_P'(x_1,x_2) \cos(qx)\, d^4x_1 d^4x_2$. If the particles are characterized by one-particle spin-density matrix $\hat{\rho}$, and the factorization in Eq. (13) still holds, then [6]: $c_0 = (-1)^{2j} Sp\hat{\rho}^2$; in particular, $c_0 = -(1+\vec{P}^2)/2$ for nucleons produced with polarization $\vec{P}$. For unpolarized particles, $c_0 = (-1)^{2j}/(2j+1)$.

The emission points x_1, x_2 are usually supposed to be independent. For simplicity we further assume that they are distributed according to the product of Gaussian laws corresponding to the mean-square radius and emission time equal to $\sqrt{3}r_0$ and τ_0, respectively. This yields:

$$W_p(x) = \int W'_p(x_1,x_2)\, \delta^{(4)}(x_1-x_2-x)\, d^4x_1 d^4x_2 =$$
$$= exp(-r^2/4r_0{}^2-t^2/4\tau_0{}^2)/[(4\pi)^2 r_0{}^3\tau_0] \tag{15}$$

and

$$R(p_1,p_2) = 1 + c_0\ exp(-r_0{}^2\vec{q}^2-\tau_0{}^2q_0{}^2). \tag{16}$$

Eq. (16) demonstrates the appearance of the interference peak (dip) due to the effect of QS, with the width inversly proportional to the space-time dimensions of the particle production region. Of course, the parametrization in Eq. (16) may happen to be oversimplified. Thus the production process may be characterized by two or several essentially different space-time parameters [14] (a special case corresponds to the admixture of the multiparticle sources [15], or the coherent states [13,16]), the form of the production region may not be spherically symmetrical, particle momenta, space and time coordinates of the emission points may be correlated (resonances [2,17-19], correlated tube [20], expanding sources [21], inside-outside cascade [22], colour string [23]). In the former case, the description of the interference effect by a single set of the space-time parameters, as in Eq. (16), effectively leads to its suppression. This is usually taken into account by an additional parameter λ multiplying the interference term.

4. According to Eqs. (1), (12), (13), the CF of two nonidentical interacting particles takes the form:

$$R(p_1,p_2) = \sum_S \rho_S <|\psi^{(S)}_{p_1p_2}(x_1,x_2)|^2>. \tag{17}$$

It is convenient to consider the symmetrized CF $R_{sym}(p_1,p_2) = [R(p_1,p_2) + R(p_2,p_1)]/2$ which, in terms of the φ-amplitudes, takes the form:

$$R_{sym}(p_1,p_2) = 1 + \sum_S \rho_S\, b^{(S)}(p_1,p_2), \tag{18}$$

$$b^{(S)}(p_1,p_2) = <|\varphi^{(S)}_{p_1p_2}(x)|^2 + 2Re\varphi^{(S)}_{p_1p_2}(x)cos(\tilde{q}x/2)>. \tag{19}$$

Eq. (19) assumes that the φ-amplitude is symmetric under the interchange $p_1 \leftrightarrow p_2$, i.e. dominated by the s-wave.

The CF of interacting identical particles is also given by Eq. (17), but with the Bethe-Salpeter amplitudes symmetrized according to Eq. (3). In the case of the s-wave FSI we have:

$$R(p_1,p_2) = 1 + c_0\langle cos(qx)\rangle + 2\sum_{S\text{-even}} \rho_S\, b^{(S)}(p_1,p_2), \qquad (20)$$

where the $b^{(S)}$-function is given in Eq. (19). Note that only even values of the total spin of identical particles contribute at zero orbital momentum due to the equality $(-1)^{L+S} = 1$. In particular, $S = 0$ and $2\rho_0 = 1 + c_0 = (1-\vec{P}^2)/2$ for identical spin-1/2 particles with polarization $\vec{P}$.

In the equal emission time approximation (11), valid on the condition (10) which can be written in the form [6]:

$$m\gamma\rho r_0/2\tau_0 \gg 1, \qquad \rho = \sqrt{r_0^{\,2}+v^2\tau_0^{\,2}}, \qquad (10')$$

and for sufficiently large radius of the particle production region as compared with the range of the strong interaction, we may replace the φ-amplitude by the spherical wave (7) and obtain [6]:

$$b^{(S)}(p_1,p_2) = |f^{(S)}(\vec{k}^*)|^2\langle 1/r^{*2}\rangle + 2Re\langle f^{(S)}(\vec{k}^*)cos(\vec{k}^*\vec{r}^*)e^{ik^*r^*}/r^*\rangle -$$
$$- \frac{2\pi}{k^*}\frac{d}{dk^*}[1/f^{(S)}(\vec{k}^*)]\overline{W}_P(0), \qquad (21)$$

where the averaging over $\vec{r}^*$ is determined by the distribution $\overline{W}_P(\vec{r}^*) = \int W_P(\vec{r},t)dt^*$. For the Gaussian distribution in Eq. (15), we have:

$$\overline{W}_P(\vec{r}^*) = exp(-r_T^{*2}/4r_0^{\,2}-r_L^{*2}/4\gamma^2\rho^2)/(8\pi^{3/2}\gamma r_0^{\,2}\rho), \qquad (22)$$

where r_T^* and r_L^* are the transverse and longitudinal components of $\vec{r}^*$ with respect to the direction of the particle pair velocity $\vec{v}$; $\gamma = (1-v^2)^{-1/2}$ is the Lorentz factor. The last term in Eq. (21) represents a correction taking into account the deformation of the spherical wave (7) within the range of strong forces. The integration in Eq. (21) can be carried out analytically at $k^* = 0$, leading to [6]:

$$b^{(S)}(p,p) = \{|f_0^{(S)}|^2[A_1-d_0^{(S)}/2\sqrt{\pi}r_0]/2r_0 + A_2Ref_0^{(S)}/\sqrt{\pi}\}/\gamma\rho, \quad (23)$$

where $A_1 \approx A_2 \approx 1$ for nonrelativistic particles. The correction term $\propto d_0^{(S)}$ allows one to use the approximation of a large radius of the production region down to $r_0 \sim d_0^{(S)}/2$. E.g., it provides the accuracy of the CF for a two-neutron system better than 10% if $r_0 > 1.5$ fm [24]. For slow particles ($v \ll 1$, r_0/τ_0) and on the conditions: $|f_0^{(S)}| \gg r_0 \gg d_0^{(S)}$, $mr_0^2 \gg \tau_0$, the function $b^{(S)}$ may be approximated at small values of k^* by [6]:

$$b^{(S)}(p_1,p_2) \cong |f^{(S)}(k^*)/r_0|^2 exp(-4r_0^2k^{*2}). \qquad (24)$$

It may be seen from Eqs. (21)-(24) that the correlation due to the FSI effect is sufficiently sensitive to the space-time dimensions of the particle production region. The effect becomes stronger with decreasing the dimensions. E.g., its contribution to the two-neutron CF ($f_0^{(S)} = 17$ fm) at small k^* and r_0, $c\tau_0$ of 2-3 fm is several tens as large as the contribution of the QS effect (fig. 1). The FSI is unessential in the

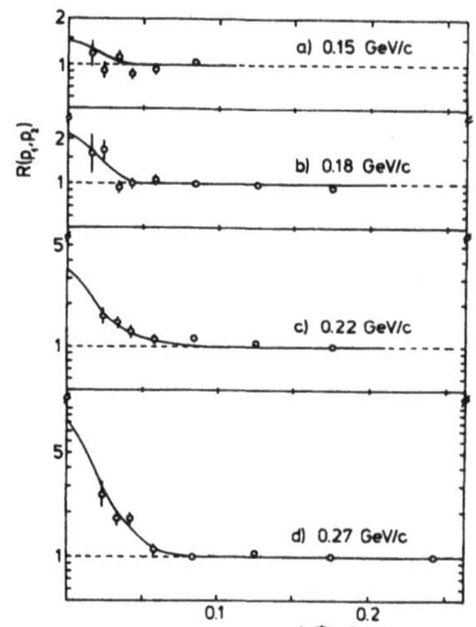

Fig. 1. The two-neutron CF measured in p-Pb interactions at 7.5 GeV/c for various mean neutron momenta [10]. The curves are calculated according to Eq. (20) using the Gaussian distribution of the space-time coordinates with (a) r_0= 3.25 fm, $c\tau_0$= 350 fm; (b) r_0= 3.25 fm, $c\tau_0$= 120 fm; (c) r_0= 3.25 fm, $c\tau_0$= 46 fm; (d) r_0= 3.55 fm, $c\tau_0$= 1 fm.

case of large distances between production points or large emission times, e.g. in the case of neutron evaporation by excited heavy nuclei [3] (if $r_0 \approx 3$ fm, $v\tau_0 \approx 10^4$ fm, $b^{(0)} < \sim 0.02$ at $k^* = 0$), or, for ultralativistic particles: $b^{(S)} \propto 1/\gamma$.

Note that the function $b^{(S)}$ in Eq. (23) depends on the emission time only through the variable ρ, which appears due to the Lorentz transformation into the particle pair rest frame. However, at a large time-scale τ_0, when the condition (10') is not fulfilled, an additional time dependence arises due to the amplitude $\varphi^{(S)}_{p1p2}(x)$, which vanishes at $\tilde{t}^* \to \infty$ (see Eq. (9)); in particular, $b^{(S)} \to 0$ for large values of τ_0 even if $v\tau_0$ is small. The eventual use of the equal time approximation in the case when the condition (10') is not fulfilled thus would lead to an overestimation of the space-time dimensions of the production region.

It follows from Eqs. (21), (23) that, on condition $k^* r_0 \ll 1$, the contribution of the s-wave FSI into the CF vanishes with the volume $V \propto r_0^3$ of the particle production region as $V^{-2/3}$ if $|f_0^{(S)}| \gg r_0$, or as $V^{-1/3}$ if $|f_0^{(S)}| \ll r_0$. In the opposite limit of large $k^* r_0$, this contribution vanishes as V^{-1}; e.g., the CF of spin-0 particles takes the form [25,26]:

$$R(p_1,p_2) = 1 + \frac{2\pi}{k^{*2}} \sum_L (2L+1) \frac{d\delta_L}{dk^*} \bar{W}_P(0). \tag{25}$$

The first measurement of the space-time dimensions of the neutron production region with the help of two-neutron correlations has been recently performed in p-Pb interactions at 7.5 GeV/c [10] (fig. 1). The correlations in proton-neutron pairs are somewhat weaker than in two-neutron ones, but still sufficiently sensitive to the dimensions of the production region [6], while the netron-deuteron correlations are rather weak: $R^{nD} < 1$ for $r_0 > 1$ fm.

5. Proceeding to the case of interacting charged particles one observes that the (long-range) Coulomb FSI leads to the distortion, even in the asymptotic region $r^* > d_0^{(S)}$, of both the sperical and plane waves:

$$\psi^{(S)}_{\vec{k}^*}(\vec{r}^*) = \sqrt{A_c^{\pm}}\, e^{i\delta_0^c} \{ e^{-i\vec{k}^*\vec{r}^*} F[-i/k^* a_c, 1, i(\vec{k}^*\vec{r}^* + ik^* r^*)] + $$
$$+ f_c^{(S)}(\vec{k}^*) \tilde{G}(k^* r^*, 1/k^* a_c)/r^* \}. \tag{26}$$

Here A_c is the Gamow factor.

$$A_c^{\pm}(k^*) = \pm(2\pi/k^*a_c)\{exp[\pm(2\pi/k^*a_c)]-1\}^{-1}, \quad (27)$$

the sign + (-) corresponds to repulsion (attraction), $a_c = 2\hbar^2/me^2$ is the Bohr radius for particles with unit charges (for protons a_c = 57.5 fm), $\delta_0^c = arg\Gamma(1+i/k^*a_c)$ is the Coulomb s-wave phase shift, $F(\alpha,1,z)$ is the confluent hypergeometric function [27] and $\tilde{G}(\rho,\eta)$ is a specific combination of the confluent hypergeometric functions [24]; in the limit $\eta\rho = r^*/a_c \ll 1$,

$$F(\alpha,1,z) = 1, \quad \tilde{G}(\rho,\eta) = cos\rho + iA_c sin\rho. \quad (28)$$

The s-wave scattering amplitude $f_c^{(s)} = (e^{2i\delta_0} - 1)/2ik^*A_c$ is due to the short-range interaction renormalized by the Coulomb force. In the effective-range approximation it can be represented in the form of Eq. (5) with the substitution

$$ik^* \rightarrow ik^*A_c(k^*) + 2h(k^*a_c)/a_c, \quad (29)$$

where the h-function has been introduced in [27].

Similarly to the case of neutral particles, the asymptotic form in Eq. (26) is a reasonable approximation to the actual wave function provided the range of the strong interaction potential is smaller than the radius r_0 of the particle production region and the correction proportional to $d_0^{(s)}$ is included. If further r_0 is small as compared with the Bohr radius, i.e. $d_0^{(s)} < r_0 \ll a_c$, practically all the long-range effects are factorized in the Gamow factor and we get [6]:

$$R(p_1,p_2) = A_c(k^*)\,\tilde{R}(p_1,p_2). \quad (30)$$

Here the CF $\tilde{R}(p_1,p_2)$, corresponding to the "neutral particles", should be calculated using the effective amplitude $f_c(k^*)$ and modifying the spherical wave in Eq. (7) by substitution $e^{i\rho} \rightarrow cos\rho + iA_c sin\rho$. Since the Gamow factor essentially differes from 1 only for $k^* < 2\pi/a_c$ (= 21.6 MeV/c for protons), the Coulomb effects are not important outside the low-k^* region. At $k^* \rightarrow 0$ the Coulomb repulsion (attraction) leads to $R(p_1,p_2) \rightarrow 0$ (∞). For same-signed charged particles the combined effect of Coulomb repulsion and short-range forces leads to a peak in the CF at $k^* \cong 2\pi/a_c$ (see fig. 2). The accuracy of Eq. (30) is better than

10-15% for r_0 > 1.5 fm. This formula is not valid at large space-time dimensions of the production region when the CF tends to 1 (the correlation vanishes) and not to the Gamow factor.

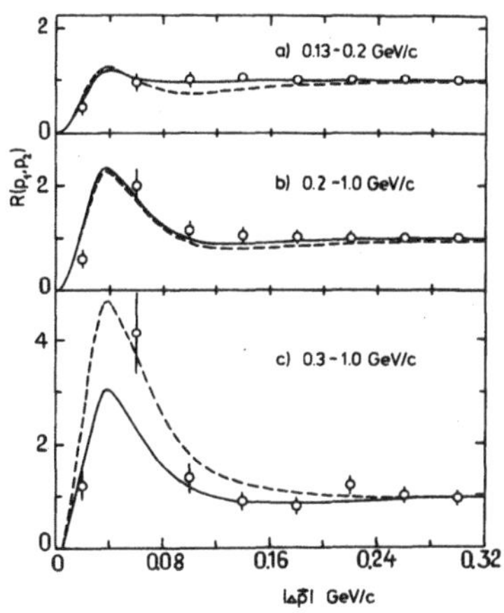

Fig. 2. The two-proton CF measured in p-Ne interactions at 300 GeV/c for various momenta of produced protons [9]. The curves are calculated according to Eqs. (17), (3) using the square-well approximation for the short-range potential and the Gaussian distribution of the space-time coordinates with (a) r_0= 4 fm, $c\tau_0$= 1 fm, (b) r_0= 2 fm, $c\tau_0$= 1 fm, (c) r_0= 1.5 fm, $c\tau_0$= 1 fm for the full curves and (a) r_0= 1.5 fm, $c\tau_0$= 60 fm, (b) r_0= 1.5 fm, $c\tau_0$= 7 fm, (c) $r_{0\parallel}$= 1.5 fm, $r_{0\perp}$= 0.5 fm, $c\tau_0$= 1 fm for the dashed ones.

At a small relative distance between the production points $r^* < d_0^{(s)}$, the two-particle CF is determined by the behaviour of the wave function $\psi^{(s)}$ inside the range of the strong interaction potential and becomes sensitive to the form of this potential. The deviation of the wave function from the asymptotics in Eq. (26) can be calculated analytically in the case of a square-well potential. The numerical comparison with more realistic nucleon-nucleon potentials, with and without repulsive core, has shown [24] that the square-well approximation is reasonably accurate ($\Delta R/R$ < 10%) at r_0 > 1 fm, i.e. at an r.m.s. radius larger than 1.5-2 fm.

The effect of FSI on correlations in narrow proton pairs has been studied in a number of experiments. It was found that at fixed time-scale parameter $\tau_0 < \sim r_0$ the proton production region has at least two characteristic dimensions. One, larger than the size of a fragmenting nucleus, corresponding to the generation of slow protons (p < 0.2 GeV/c), the other, much smaller, characterizing the emission of fast protons (see, e.g., fig. 2). The large effective size is probably due to a substantial emission (evaporation) time, while the small one may be understood in the tube model [9]. The question however arises to what extent are these results affected by the assumption of the Gaussian distribution of the emission points. It appears that more realistic distributions yield the same r.m.s. radius of the production region within 5-12% [9].

For charged pions the Coulomb effects are important at $k^* < 3$ MeV/c. The effect of FSI outside this region is practically negligible for $\pi^+\pi^+$ or $\pi^-\pi^-$ pairs ($f_0 \approx -0.03$ fm) and rather small for $\pi^+\pi^-$ ones ($f_0 \approx 0.2$ fm) [6].

6. Due to the $\cos(\vec{k}^*\vec{r}^*)$-terms in Eqs. (14) and (19), the CF is sensitive to the form of particle production region. E.g., if the Gaussian distribution of the emission points is characterized by different transverse and longitudinal dispersions $r_{0\perp}^2$ and $r_{0\parallel}^2$ with respect to the reaction axis, the CF of two noninteracting particles takes the form:

$$R(p_1,p_2) = 1 + c_0\,exp(-r_{0\perp}^2 q_\perp^2 - r_{0\parallel}^2 q_\parallel^2 - \tau_0^2 q_0^2) \equiv$$
$$\equiv 1 + c_0\,exp(-r_T^2 q_T^2 + r_{LT} q_L q_T - r_L^2 q_L^2), \tag{31}$$

$$r_T^2 = r_{0\perp}^2 + (r_{0\parallel}^2 - r_{0\perp}^2)\sin^2\theta\cos^2\varphi, \quad r_{LT} = (r_{0\parallel}^2 - r_{0\perp}^2)\sin 2\theta\cos\varphi,$$
$$r_L^2 = v^2\tau_0^2 + r_{0\parallel}^2 - (r_{0\parallel}^2 - r_{0\perp}^2)\sin^2\theta. \tag{32}$$

Here φ, $\vec{q}_T = 2\vec{k}_T^*$ and $q_L = q_0/v = 2\gamma k_L^*$ are the azimuthal angle, transverse and longitudinal components of the vector $\vec{q}$ with respect to $\vec{v}$, θ is the pair production angle. Note that the CF depends on the direction of the vector $\vec{k}^*$ also in the case of spherically symmetric form of the production region:

$$R(p_1,p_2) = 1 + c_0\,exp(-4r_0^2 k_T^{*2} - 4\rho^2\gamma^2 k_L^{*2}). \tag{33}$$

In principle, the dependence of the CF on the $\vec{k}^*$-orientation may be used to measure the form of the production region and the emission time. For a two-nucleon system this seems to be possible in the case of identicle nucleons produced at sufficiently large space-time distances so that the short-range FSI, dominated by the "isotropic" terms $\propto$ $\langle |f^{(S)}/r^*|^2 \rangle$, is suppressed. On the other hand, the FSI may allow one to determine the form of the production region of sufficiently relativistic particles with the help of the CF measured at different production angles θ [9]. This may be illustrated in the limiting case of the tube model when $r_\perp = t = 0$ and $r^{*2} = r_\parallel^{\,2}(\sin^2\theta + \gamma^2\cos^2\theta)$. It is seen that the corresponding CF is minimal for the pairs emitted in the forwad or backward direction and achieves its maximal value at 90°. Even larger anisotropy is obtained in the limiting case of the correlated tube model corresponding to the correlated particle emission from the points on the trajectory of the beam particle moving in the nuclear matter with a velocity u: $r_\parallel = ut$, $r_\perp = 0$ and $r^{*2} = r_\parallel^{\,2}[\sin^2\theta + \gamma^2(v/u - \cos\theta)^2]$; the corresponding CF decreases with the increasing production angle θ.

7. In conclusion we summarize the results. Simple analytical formulae for the CF of narrow particle pairs are given taking into account the effects of QS and FSI. They are valid for arbitrary particles (like or unlike, charged or neutral, relativistic or nonrelativistic, polarized or unpolarized) at sufficiently large dimensions of the particle production region as compared with the range of nuclear forces. For charged particles, the effect of the long-range Coulomb interaction is practically factorized in the Gamow factor at distances smaller than the Bohr radius. The calculations are substantially simplified on condition (10) or (10') guaranting the usual equal time approximation. The computer code is available allowing one to calculate the CF for various distributions of the emission points.

References

1. G.Goldhaber et al., Phys. Rev. 120 (1960) 300.
2. V.G.Grishin et al., YaF 13 (1971) 1116.
3. G.I.Kopylov, M.I.Podgoretsky, YaF 15 (1972) 392; 18 (1973) 656; G.I.Kopylov, Phys. Lett. 50B (1974) 472.
4. G.Cocconi, Phys. Lett. 49B (1974) 459.

5. S.E.Koonin, Phys. Lett. 70B (1977) 43.
6. R.Lednicky, V.L.Lyuboshitz, YaF 35 (1982) 1316.
7. B.Lörstad, Int. J. Mod. Phys. A4 (1989) 2861.
8. D.Ardouin et al., Nantes Univ. Report LPN 90-03, Nantes, 1990.
9. M.L.Allaberdin et al., YaF 46 (1987) 1785.
10. Yu.D.Bayukov et al., Phys. Lett. 189B (1987) 291.
11. S.A.Azimov et al., Phys. Rev. D29 (1984) 1304.
12. S.N.Agakishev et al., Z. Phys. A327 (1987) 443.
13. M.Gyulassy et al., Phys. Rev. C20 (1979) 2267.
14. R.Lednicky, M.I.Podgoretsky, YaF 30 (1979) 837.
15. R.Lednicky et al., YaF 38 (1983) 261.
16. G.N.Fowler, R.M.Weiner, Phys. Rev. D17 (1978) 3118.
17. G.H.Thomas, Phys. Rev. D15 (1977) 2636.
18. P.Grassberger, Nucl. Phys. B120 (1977) 231.
19. R.Lednicky, JINR B2-3-11460, Dubna, 1978.
20. R.Lednicky, M.I.Podgoretsky, JINR P2-12302, Dubna, 1979.
21. S.Pratt, Phys. Rev. D33 (1986) 1314.
22. K.Kolehmainen, M.Gyulassy, Phys. Lett. 180B (1986) 203.
23. B.Andersson, W.Hofmann, Phys. Lett. 169B (1986) 364.
24. M.Gmitro et al., Czech. J. Phys. B36 (1986) 1281.
25. V.L.Lyuboshitz, YaF 41 (1985) 820.
26. B.K.Jenning et al., Phys. Rev. C33 (1986) 1303.
27. L.D.Landau, E.M.Lifshitz, "Quantum Mechanics", Nauka, Moscow, 1974.

PARTICLE SOURCE PARAMETERS FROM $\pi\pi$ AND pp CORRELATIONS IN RELATIVISTIC COLLISIONS AT DUBNA

Jan Pluta

Institute of Physics, Warsaw Technical University,
Koszykowa 75, 00-662 WARSAW, Poland

ABSTRACT

A short review of Dubna experimental results on $\pi\pi$ and pp correlations at small relative momenta is given. Some recent results are discussed in detail. Interference $\pi\pi$ correlations in π-p and π-C collisions at 40 GeV/c show a dependence of the source sizes on the region of pion production. The dependence of the size of proton emission volume on the momenta of emitted protons has been analyzed assuming limited emission time intervals and non-spherical shape of the proton emission volume. A compilation of different experimental results is included.

1. INSTEAD OF INTRODUCTION - SHORT CHRONOLOGICAL REVIEW OF DUBNA EXPERIMENTAL RESULTS

"CORINNE'90" is the first meeting where the word "interferometry" is used to indicate the main subject of consideration in the field of nuclear physics. It can stimulate certain reflection about the history of interferometrical investigations. Exactly thirty years passed after the pioneer work of Goldhabers and co-workers[1], where interference effect between identical pions was observed and interpreted as a consequence of Bose-Einstein statistics.

Two-particle correlations and the space-time characteristics of particle emission process have been investigated during many years at JINR, Dubna. Experimental studies have been performed using bubble chamber technics. Recently, some experimental groups have extended their investigations with interferometrical measurements. A few results will be presented at this workshop.

The first experimental evidence of two-proton correlations at small relative momenta was found by Siemiarczuk and Zielinski more than twenty years ago (1967)[2]. Investigations were performed using experimental data from the xenon bubble chamber irradiated at Dubna in the beam of negative pions at 9 GeV/c. Two-pion interference was observed two years later. Strong angular correlations between neutral pions have been found in 1969 by K.Eskreys[3] using the same experimental material.

Theoretical base for applying the interference effects as a tool to study the space-time mechanism of multiparticle production was formulated by Kopylov and Podgoretsky at the beginning of seventies.[4,5] They introduced a set of variables ($q_\perp, q_0$), related to the radius of production region r and the duration of generation process τ, widely used to describe experimental distributions. Indeed, in 1975 Grishin reported first Dubna results[6] and in a following series of experimental works[7-11] the space-time dimensions of pion emission process were evaluated using Kopylov-Podgoretsky parameterization. Some results will be presented here.

Fig. 1. shows two examples of the correlation distributions for pions produced in $\pi^- C$ interactions at 40 GeV/c[10]. It is seen in Fig.1a that the size of pion emission volume increases in the so called "multinucleon collisions". The emission region decreases with the increase of perpendicular pion momenta, Fig.1b.

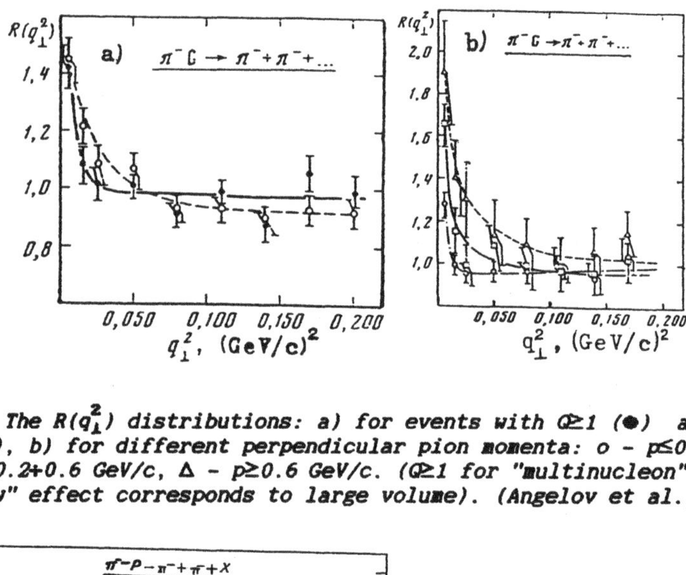

Fig.1. The $R(q_\perp^2)$ distributions: a) for events with Q≥1 (●) and with Q<1 (o), b) for different perpendicular pion momenta: o - p≤0.2 GeV/c, □ - p=0.2÷0.6 GeV/c, Δ - p≥0.6 GeV/c. (Q≥1 for "multinucleon" events, "narrow" effect corresponds to large volume). (Angelov et al., 1981)[10]

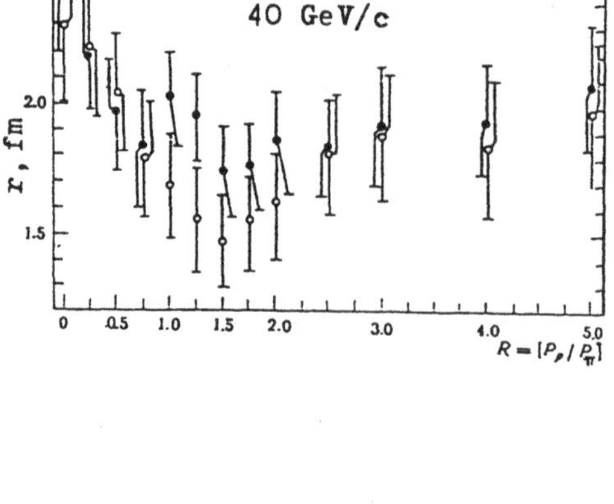

Fig.2. The size of pion emission volume in different reference frames. $R=|P_p/P_{\pi^-}|$ - where P_p, (P_{π^-}) - momentum of proton (π^-) in a given reference frame; R=0 for lab. system, r=1 for cms, r=1.5 for "quark system". (Angelov et al., 1982)[11]

Fig. 2. visualizes the interferometric method allowing one to find the rest frame of particle sources[11]. It was shown by Podgoretsky that the size of pion emission volume reaches his extreme value when correlation distribution is considered in the rest frame of the sources[12]. In the presented example the minimal value of r was found for the "quark system"; for details see Ref. 11.

Interesting effect has been observed while analyzing $R(q_\perp^2)$- distribution for different q_0 cut-off[13], Fig. 3. The radius of negative pion emission volume rapidly decreases for the q_0 cut-off equal to 0.2 GeV/c. It has been shown that this effect appears as a result of resonance production.

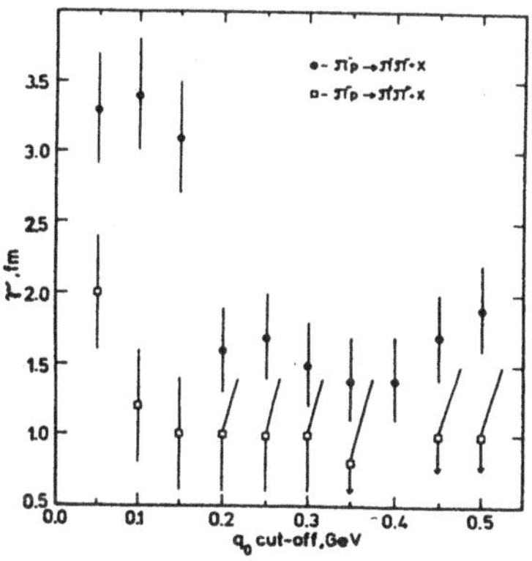

Fig. 3. The radii of pion emission volume in $\pi^- p$ collisions at 40 GeV/c for positive (□) and negative (●) pions and for different q_0 cut-off (Akhababian et al. 1983)[13]

The study of nucleus-nucleus collisions at 4.2 GeV/c per nucleon has shown that the RMS radii of the pion emission region increase with the increasing atomic number of the projectile nucleus, as well as with the collision "centrality"[14]; Fig. 4.

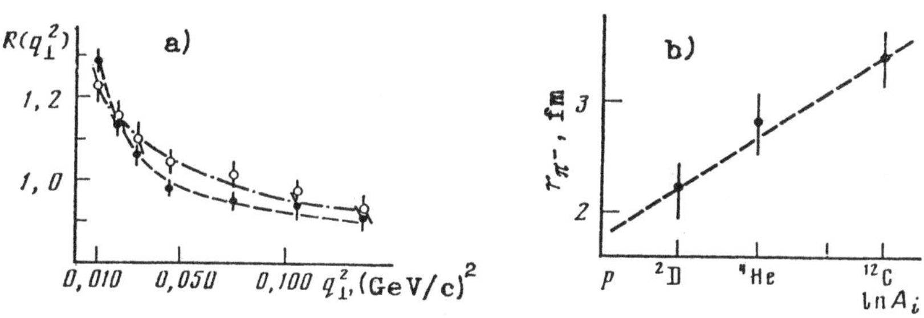

Fig. 4. a) - The $R(q_\perp^2)$ distribution for negative pions produced in d-Ta (o) and C-Ta (●) collisions at 4.2 Gev/c per nucleon. b) - Dependence of the radius of pion emission volume on the projectile mass A_i. (Agakishiev et al. 1984)[14]

The first attempt to determine the size of neutral pion emission volume was made using experimental data from 180-liter xenon bubble chamber[15]. Fig. 5. presents an example of the correlation function plotted as a function of Q, where $Q^2 = M_{\pi\pi} - 4m_\pi^2$. Relatively small value of the radius of neutral pion emission volume, $\langle r^2 \rangle^{1/2} = 1.2 \pm 0.3$ fm[16], confirm the general rule[17] that the pion source size in relativistic nuclear collisions is determined by the smaller colliding object.

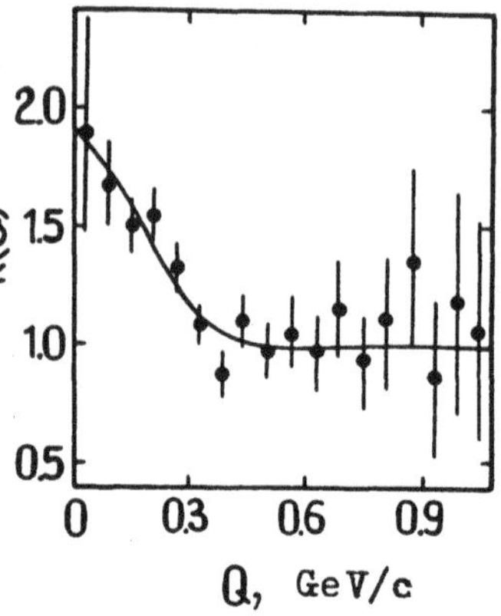

Fig.5. The R(Q) distribution for neutral pions emitted in π^--Xe collisions at 3.5 GeV/c. (J.Pluta 1986)[15]

At the same time the two-proton correlations were investigated. Theoretical approach was developed by Lednicky and Lyuboshitz[18]. First experimental analysis, performed in 1980, showed that the correlation effect strongly depends on the momentum of correlated protons[19]. Fig.6. demonstrates this observation for two classes of protons emitted in π^-p collisions at 40 GeV/c. For small proton momenta destructive correlation is seen which corresponds to large space-time intervals; for higher momenta the form of correlation changes, which indicate the decreasing of related space-time characteristics.

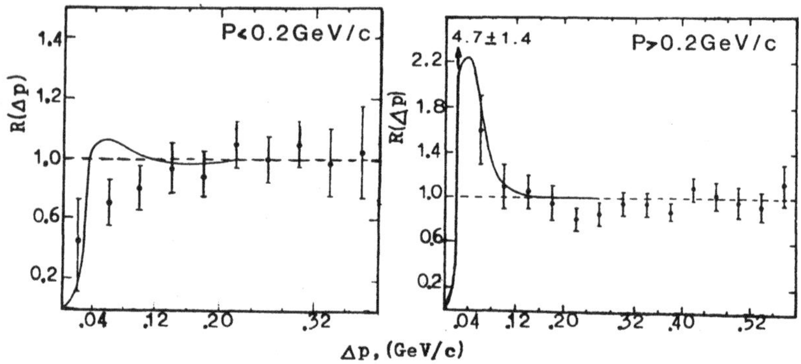

Fig.6. The R(Δp) distributions for protons emitted in π^-C collisions at 40 GeV/c for different momentum intervals. (Angelov et al. 1980)[19]

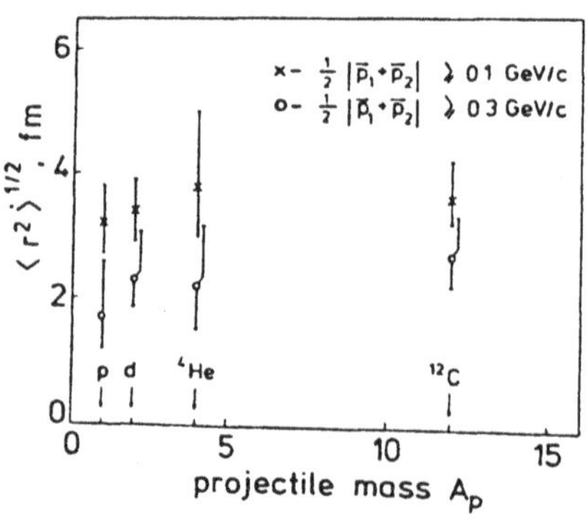

This dependence was confirmed later.[20-25] Some results are collected in Fig.7.[23]

Fig.7. RMS radii of the proton emission region for interactions of light nuclei with the carbon target plotted versus the mass number of the projectile nucleus. (Agakishiev et al. 1987)[23]

This figure shows also that the size of proton emission volume does not depend on the mass of the projectile nucleus. Detail analysis of these effects will be presented later on.

2. SOME GENERAL DEFINITIONS AND QUANTITIES

It will be convenient to define some useful quantities before performing detailed analysis of the experimental data.

Two-particle correlations in the region of small relative momenta can be expressed as

$$R(\vec{q},q_0) = \rho(\vec{q},q_0)/\rho_0(\vec{q},q_0), \tag{1}$$

where: $\rho(\vec{q},q_0)$ - density of identical particle pairs in a phase space, $\rho_0(\vec{q},q_0)$ - pair density in the absence of interference. Here, $\vec{q} = (\vec{p}_1 - \vec{p}_2)$, $q_0 = E_1 - E_2$; $\vec{p}_1,\vec{p}_2$ and E_1,E_2 - momenta and energies of considered particles. The emitting sources are usually assumed to be distributed according to the Gaussian law:

$$W(\vec{r},t) \sim \exp(-\vec{r}^2/2r_0^2 - t^2/2\tau^2). \tag{2}$$

where r_0 and τ are parameters which define the space-time dimensions of particle emission process. The RMS radius is then: $\langle r^2 \rangle^{1/2} = \sqrt{3}r_0$.

Two-pion correlations arise mainly due to quantum-symmetrization. If particle sources are distributed according to eq. (2) the correlation function can be written as

$$R(\vec{q},q_0;\, r_0,\tau_0) = A\,(\,1 + \lambda\, e^{-\vec{q}^2 r_0^2 - q_0^2 \tau_0^2}\,), \tag{3}$$

where λ - phenomenological parameter, A - normalization coefficient.

In the case of protons the correlations appear as a result of Coulomb repulsion, final state interaction and quantum statistical effects.[18] The correlation function takes the form:

$$R(q,p) = A_c(k^*)[\,1 + B_0(q,p) + B_i(q,p)] \tag{4}$$

where p and q are sum and difference of proton four-momenta; $A_c(k^*)$, $B_0(q,p)$, $B_i(q,p)$ describe three effects mentioned above; k^* is a half-difference of the proton momenta in the rest frame of the pair. A stronger effect corresponds to smaller space-time parameters of the emitting sources.

The average value of the correlation function calculated as

$$\langle R(r_0)\rangle = \int_{k_1^*}^{k_2^*} R(k^*;r_0)\, \rho(k^*)\, dk^*, \tag{5}$$

can be used to find the r_0 value from the data.[26] The values of k_1^* and k_2^* define the "region of effect"; $\rho(k^*)$ is the pair density in the background distribution normalized to unity in the interval $(k_1^* \le k^* \le k_2^*)$; for details see Refs. 23,26.

3. INTERFERENCE CORRELATIONS IN FOUR-DIMENSIONAL HADRON JETS PRODUCED IN $\pi^- p$ AND $\pi^- C$ COLLISIONS AT 40 GEV/C*)

The characteristic dimensions resulting from the interference correlation method in hadron-hadron collisions are about (1÷2)fm; in lepton reactions smaller sizes are usually obtained; in nuclear collisions the typical radii are comparable with those of the smaller colliding nucleus.

Here, we report the results which join the mentioned features of different kinds of interactions. Interference correlations are studied in the hadron jets, selected using invariant method in soft collisions of π^--mesons with protons and carbon nuclei at 40 GeV/c.[27] Some papers on interference correlations in jets registered in e^+e^- annihilation have been published so far[28-30]; interference effects have been investigated also in "central jets" separated in pp-collisions at $\sqrt{s}$ = 63 GeV.[31] There are no similar data for "soft jets" in hadron-hadron collisions and nuclear reactions.

The process of jet production has been described using relativistic invariant variables:[32]

$$b_{ik} = -(P_i/m_i - P_k/m_k)^2 = -(u_i - u_k)^2, \qquad (6)$$

where P_i, P_k - four-momenta, m_i, m_k - masses, u_i, u_k four- dimensional velocities of considered particles. Jets are defined as groups of particles with small values of b_{ik}. The center of jet is defined as

$$V_\alpha = \sum u_i / \sqrt{(\sum u_i)^2} . \qquad (7)$$

It has been shown[32,33] that in the reactions considered, two jets can be selected in general. In order to separate them, the following quantity has been minimized:

$$A_2 = \min [-\sum_k (V_\alpha - u_k^\alpha)^2 - \sum_i (V_\beta - u_i^\beta)^2]. \qquad (8)$$

The values of V_α and V_β are simultaneously determined. The distance between jets in four-velocity space is defined by

$$b_{\alpha\beta} = -(V_\alpha - V_\beta)^2. \qquad (9)$$

To characterize a region of jet production, the invariant variables have been used, which express the part of colliding particles four-momenta taken by a given jet:

$$x_{sn} = \frac{M_{\alpha(\beta)}}{m_I} \frac{(V_{\alpha(\beta)} u_{II})}{(u_I u_{II})}, \quad x_{sm} = \frac{M_{\alpha(\beta)}}{m_{II}} \frac{(V_{\alpha(\beta)} u_I)}{(u_I u_{II})}, \qquad (10)$$

---

*) Definition of jets, used here, is based on a relativistic invariant approach to the description of multiparticle process. Jet is defined as a cluster of particles with small four-dimensional relative velocities[32].

where $M_{\alpha(\beta)}$ - effective mass of the cluster α or β, m_I - mass of the projectile, m_{II} - target mass.

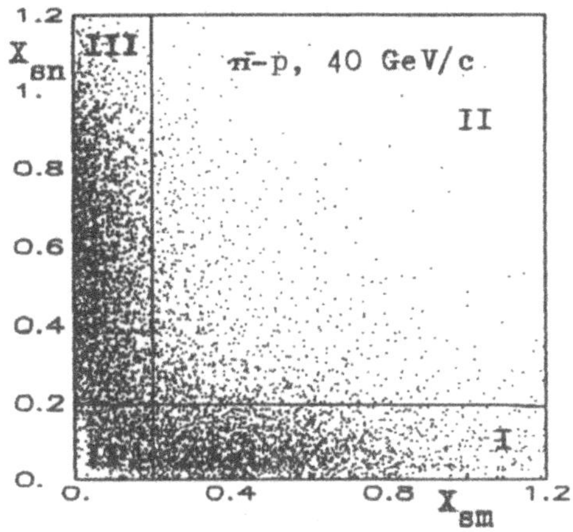

Fig. 8. Distribution of pion jets over the x_{sn} and x_{sm} variables. in π^-p interactions at 40 GeV/c.

Jets were selected in events with secondary pion multiplicity $n_{\pi^\pm} \geq 4$. The value of $b_{\alpha\beta}$ was calculated for all considered events and the quantities x_{sn} and x_{sm} were determined for each of selected jets. It has been accepted that two jets are separated in four-velocity space if $b_{\alpha\beta} \geq 10$. Fig. 8. presents two-dimensional distribution over x_{sn} and x_{sm} for jets selected in π^-p interactions. Large x_{sn} values correspond to beam fragmentation region, large x_{sm} denote the target fragmentation.

Selection of different regions has been made using following criteria:

I) target fragmentation region: ($x_{sm} \geq 0.2$ and $x_{sn} < 0.2$)

II) central region: ($x_{sn} \geq 0.2$ and $x_{sm} \geq 0.2$)
or ($x_{sn} < 0.2$ and $x_{sm} < 0.2$).

III) beam fragmentation region: ($x_{sn} \geq 0.2$ and $x_{sm} < 0.2$)

These regions are indicated in Fig. 8. The statistics of events and pion numbers in each of selected regions are listed in Table 1.

Table 1. Statistics of experimental data

region	π^-p		π^-C	
	N_{events}	$N_{\pi^\pm}$	N_{events}	$N_{\pi^\pm}$
all $\pi^\pm$	17352	88434	8307	58900
beam fragmentation	5187	17199	1483	5049
target fragmentation	4209	13865	2605	10388
central region	1896	11528	799	6453

In the frame of relativistic invariant approach[32] the origin of jets in the fragmentation regions (I, III) can be considered as a result of hadronization of quark and diquark not participating in the πp interaction. Pions from the central region appear as a result of hadronization of interacting quarks.

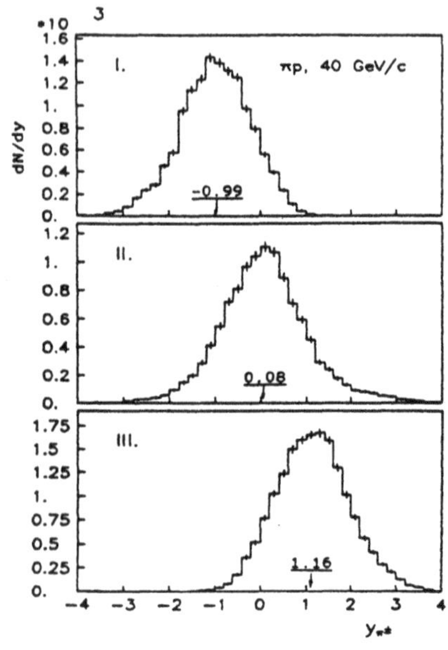

Fig. 9. Rapidity distributions for different regions of pion production.

Fig. 9. presents rapidity distributions in $\pi^- p$ cms for pions from three selected regions - (I,II,III).

The analysis of interference correlations was performed for one-dimensional distributions $R(\vec{q};\ r_o)$, $\langle q_o \langle 0.2$ GeV$\rangle$ due to limited statistics of our data. Experimental distributions were fitted with the formula

$$R(\vec{q};\ r_o) = A(\ 1 + \lambda\ e^{-\vec{q}^2 r_o^2}). \quad (11)$$

To construct background distributions pairs of different charge pions were used. This method allows to remove interference effects but preserves other kinds of correlations due to jet structure of events.

The examples of correlation distributions for two groups of pions selected in $\pi^- p$ - interactions at 40 GeV/c are presented in Fig.10. The fitted parameters for all selected groups are collected in Table 2.

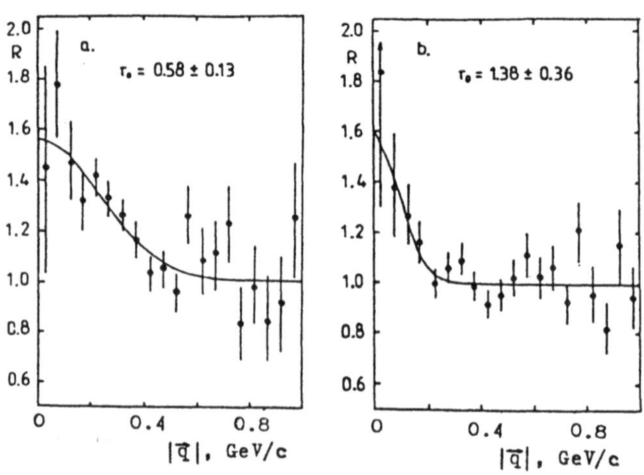

Fig.10. Correlation distributions R(q) for pions from two selected regions: a - beam fragmentation, b - *central region.*

Table 2. $\pi^- p$ interactions at 40 GeV/c

region	r_0 , fm	λ	χ^2/NDF
all $\pi^\pm$	0.93 ± 0.14	0.42 ± 0.07	1.05
beam fragmentation	0.58 ± 0.13	0.56 ± 0.12	0.82
target fragmentation	0.56 ± 0.12	0.37 ± 0.16	1.10
central region	1.38 ± 0.36	0.58 ± 0.27	0.95

The sizes of pion emission volume, obtained for the regions of beam and target fragmentation, are considerably smaller than those obtained for all pions in a given reaction, and are smaller than a half of the size obtained for events, where clear two-jet structure is not seen (central region). The observed effect is similar to that in NA35 experiment in nucleus-nucleus collisions at 200 GeV,[34] though absolute values differ considerably. For the projectile fragmentation region these values are consistent with the size, characteristic for e^+e^--annihilation.[28-30] Interference correlations in this process are usually described using Q-variable; $Q^2 = -(P_1 - P_2)^2$ where P_1, P_2 are four momenta of correlated pions.
Fig.11.a. presents the correlation distribution obtained using Q-variable for pions from the beam fragmentation region. Analogous distribution for pions registered in e^+e^- annihilation at $\sqrt{s}$ = 29 GeV[28], is shown on the right hand side on this figure. Both distributions were approximated with the formula, which takes into account a slow increase of correlation function beyond the interference region.

$$R(Q,r) = A\ (\ 1 + \lambda\ e^{-Q^2 r^2}\ )\ (\ 1 + \gamma\ Q\ ). \tag{12}$$

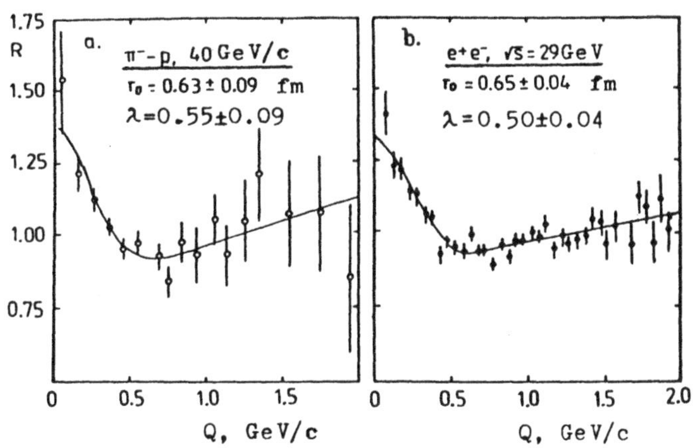

Fig.11. Correlation distributions R(Q). a - pions from the beam fragmentation region in $\pi^- p$ interactions at 40 GeV/c , b - *pions from the reaction of e^+e^- annihilation at $\sqrt{s}$ =29 GeV.*[28]

Table 3. presents the results, obtained for π^-C collisions at 40 GeV/c. Here, the size of pion emission volume for beam fragmentation region is also close to the sizes, characteristic for lepton processes. For other regions the results show greater values than in the case of π^-p collisions, which may reflect grater mass of the target nucleus.

Table 3. π^-C interactions at 40 GeV/c

region	r_0, fm	λ	χ^2/NDF
all $\pi^{\pm}$	1.32 ± 0.20	0.34 ± 0.08	1.62
beam fragmentation	0.65 ± 0.28	0.68 ± 0.28	0.85
target fragmentation	1.20 ± 0.51	0.33 ± 0.20	1.15
central region	2.5 ± 1.2	1.3 ± 0.7	1.35

Presented results show a new property of multiparticle production processes in soft hadron-hadron and hadron-nucleus collisions: space-time characteristics of hadron jets, selected by the invariant method in the incident hadron fragmentation region, are the universal property of jet production process not depending on the type and nature of colliding objects.

4. MOMENTUM DEPENDENCE OF THE TWO-PROTON CORRELATIONS

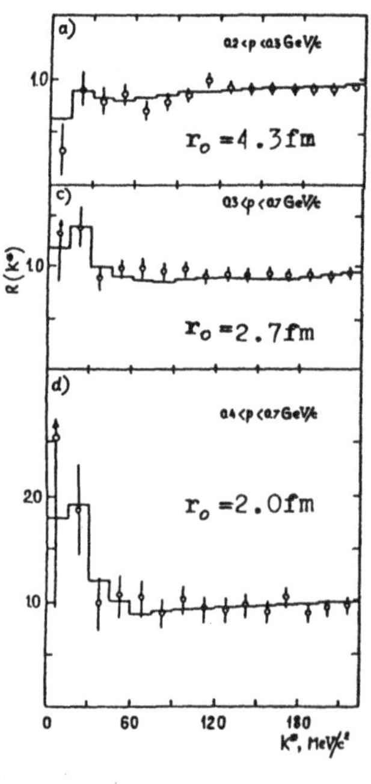

Let us look now at some regularities seen in proton-proton correlations. One of the widely discussed effects is the dependence of the size of proton emission volume on the momenta of emitted protons. Faster protons seem to be emitted from smaller sources. This effect has been investigated at JINR for various reactions and in the secondary proton momentum region from 150 MeV/c up to 1.0 Gev/c.[19-25] Fig.12. presents the observed dependence for protons emitted in π^-Xe collisions at 3.5 GeV/c[22].

Fig.12. Two-proton correlation distributions for different intervals of the proton momenta. Histogram - theoretical distributions for the values of r_o indicated in the figure.[22]

To compare different results we will present them in the unified form as a dependence of the RMS radius of the proton emission volume on the mean proton momenta; Fig.13:[35]

$$\langle r^2 \rangle^{1/2} = f(\langle p_p \rangle), \tag{13}$$

where: $p_p = (\vec{p}_1 + \vec{p}_2)/2$; $\vec{p}_1$, $\vec{p}_2$ are the momenta of correlated protons.

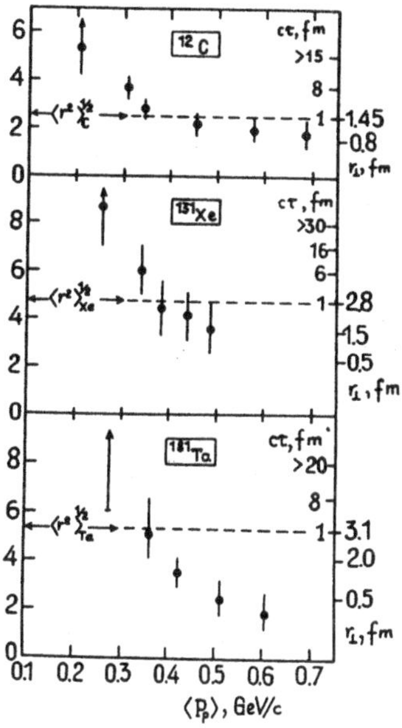

Fig.13. Space-time parameters of the proton emission volume for different mean values of the proton momenta[35]. *The reactions for target indicated are: [C]-(p,d,α,C)-C, 4.2 GeV/c per nucleon*[23], *[Xe]-π⁻-Xe, 3.5 GeV/c*[22], *[Ta]-p-Ta, 10 GeV/c*[25].

A similar behavior is seen in each set of the data. In the momentum range (0.3 - 0.4) GeV/c the size of the proton emission volume turns out to be close to that of the target nucleus in all considered reactions. For lower proton momenta the RMS radii are systematically larger; for higher momenta smaller radii have been obtained (see left hand side scale). The theoretical correlation functions, used by us to find the RMS values, were computed assuming τ = 1fm. The velocity of proton pairs was taken $v/c = 0.4$. The emission points were assumed to be independent and distributed in the space and time according to the four-dimensional Gaussian law (see formula (2)).

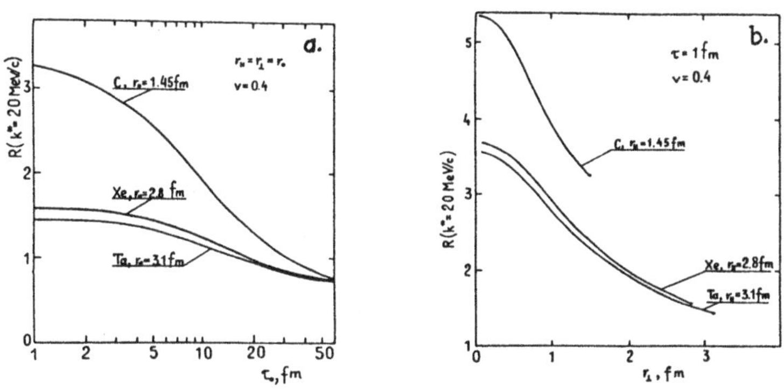

Fig.14. Time dependence (a) and perpendicular size dependence (b) of the height of two-proton correlation function (k^*=20MeV/c) for different source sizes of the emission volume[35].

Let us assume now that the changes of the two-proton correlations, giving growth to the proton emission volume above the RMS radius of the target nucleus, are due to the increase of the emission time intervals, without changing the space sizes. The opposite changes are supposed to be due to decreasing of the perpendicular source size with respect to the incoming particle direction, without changing the longitudinal one. These dependencies are shown in Fig.14. The chosen values of r and v correspond to our experimental conditions: $r_o, r_{\parallel}$ values are $\sqrt{3}$ times smaller than the RMS radii of given target nuclei and v corresponds roughly to the mean velocity of registered protons.

The right hand side scales in Fig.13. (non linear) indicate the values of time intervals and perpendicular dimensions which should be used to obtain the correlation effect corresponding to the RMS radius on the left hand side scale.

The dependencies presented separately for different targets in Fig.13. can be simply compared by changing the RMS radius of the proton emission volume with the ratio:

$$r_{rel} = \langle r^2 \rangle^{1/2} / \langle r^2 \rangle_T^{1/2}, \qquad (14)$$

where $\langle r^2 \rangle_T^{1/2}$ is the RMS radius of the target nucleus; Fig. 15.

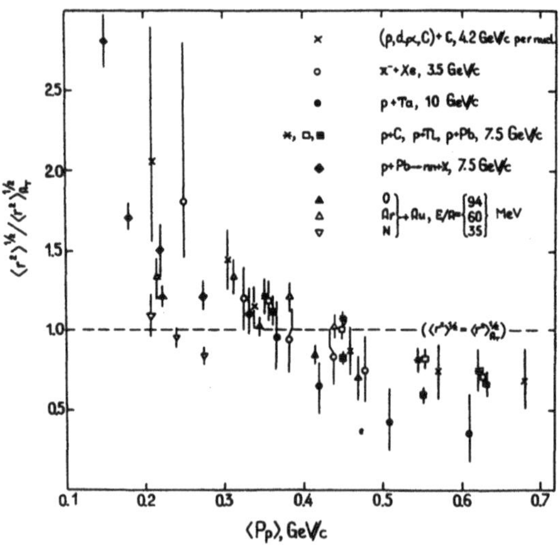

It is interesting to compare these results with the other ones obtained elsewhere. All the data in Fig.15.[22,23,25,36-40] were converted to the RMS radii, where it was necessary, before presenting in the figure. The RMS radii of the nuclei were obtained from the experimental data[41]. One can see that all the results follow the common rule.

Fig.15. Dependence of the source sizes on the mean proton momenta expressed in units of the RMS radius of the target nucleus[35].

For the mean proton momenta about 0.4 GeV/c, the source dimensions are comparable to those of the target nucleus independently of the target mass. The only exception are data for the very low incident energy (E/A=35MeV[36]). Similar dependencies have also been observed qualitatively in several other works.[21,43,44]

Fig.16. presents a direct comparison of the RMS radii obtained for "fast" protons ($\langle p_p \rangle$ > 0.3 GeV/c) in different experiments with the corresponding RMS radii of the target nuclei. The reactions where the mass of the target nucleus is much greater than that of the projectile were selected. The straight line presents the approximation of experimental data obtained for the RMS radii of different nuclei[41].

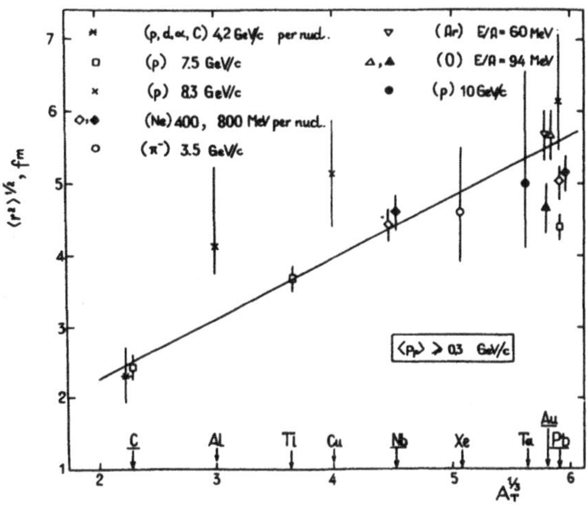

Fig.16. Dependence of the RMS radius of the fast proton emission volume on the mass number of the target nucleus[35]. The line represents the RMS radii of the target nuclei[41].

One can see that for fast protons the RMS radii of the emission volume are compatible with the RMS radii of the target itself. Certain dispersion, seen for heaviest nuclei, reflects the momentum and angular dependencies due to different experimental conditions. Three points are located relatively high, though their positions follow the overall dependence. These data correspond to backward emitted protons[40].

The form of proton momentum dependence presented in Fig.15.and Fig.16. may be treated as a "point of reference" to look for other particular effects. The relationship, shown in Fig.16., is similar to that observed by J.Bartke for pions[17] but the projectile mass has been replaced by the target one for protons.

The following picture seems to emerge from the data analysis performed above. Fast protons are emitted successively in the first stage of the collisions. Their space-time characteristics reflect the collision geometry. The decrease of the proton momenta is related to the expansion of the emission zone over the whole target nucleus. Corresponding time intervals increase as well. Slow protons are emitted according to the typical thermalized process.

Summarizing, the factors considered can explain, in principle, our experimental results although some comments are in order.

1. Taking target mass as a reference variable, the tacit assumption has been made that registered protons are mainly target fragments, which is not a case when a massive projectile with not too high energy per nucleon is used. The influence of the projectile mass and energy on the overall pair energy dependence was observed for the reactions in a relatively low energy region[38]. Since for a higher energies the projectile mass dependence has not been observed,[28] it seems natural to suppose that, if the incident energy is insufficient for colliding nuclei to penetrate each other, the measured sizes reflect the limits of the reaction zone.

2. Correlations between the space and time characteristics are possible during the emission process. Assuming sequential proton emission from the target nucleus along the path of the projectile one obtains the following relation which join the distance r^* "seen" in the rest frame of the pair and the space interval $r_\parallel$ between the consecutive emission points:[42]

$$(r^*)^2 = r_\parallel^2 \; [\; \sin^2\theta + \gamma^2 (v/u - \cos\theta)^2] \qquad (16)$$

where u is the projectile velocity, θ, v, γ - emission angle, velocity and Lorentz factor of the proton pair. This effect is consistent with the assumption of a small perpendicular size, fast protons are ejected from. Angular dependence of the two-proton correlations has been observed experimentally.[40,45]

3. Different emission angles can reflect different mechanisms of the particle emission and consequently, different space-time characteristics of the emission process. Backward emitted protons are often considered as "essentially emitted from a thermalized system"[45].

4. Experimental data show the influence of several interrelated factors upon the correlations observed. Some of them are listed above. A detailed analysis is still necessary for separated proton momentum and angular intervals.

The experimental material presented here represents the work of many people. I am grateful to my collegues from the Chamber Division at the Laboratory of High Energies JINR for nice and fruitful cooperation. I am indebted to R.Lednicky for making accessible his numerical program for theoretical computations of the two-proton correlation function. Useful discussions with M.I.Podgoretsky, V.G.Grishin, R.Lednicky, V.L.Lyuboshitz, B.Lörstad, D.Ardouin, and Yu.Sinyukov are gratefully acknowledged. The investigations were supported in part by CPBP Program 01.09 in Poland.

REFERENCES

1. G. Goldhaber et al., *Phys.Rev.* **120**, 300 (1960).
2. T. Siemiarczuk, P. Zielinski, *Phys.Lett.*, **24B**, 675 (1967).
3. K. Eskreys *Acta Phys.Polon.*, **XXXVI**, 237 (1969).
4. G.I. Kopylov, M.I. Podgoretsky, *Yad.Fiz.* **15**, 392 (1972) [*Sov.J. Nucl.Phys.* **15**, 219, (1972)], *Yad.Fiz.* **18**, 656 (1973) [*Sov.J.Nucl. Phys.* **18**, 336 (1974)].
5. G.I. Kopylov, *Phys.Lett.* **50B**, 472 (1974).
6. V.G. Grishin, *IV Int. Sem. on High Energy Physics Problems*, Dubna, June 5÷11 (1975), JINR, D1, 2-9224, p.11.
7. N. Angelov et al., *Yad.Fiz.* **26**, 796 (1977).
8. N. Angelov et al., *Yad.Fiz.* **27**, 675 (1978)
9. N. Angelov et al., *Yad.Fiz.* **31**, 411 (1980).
10. N. Angelov et al., *Yad.Fiz.* **33**, 1257 (1981).
11. N.S. Angelov et al., *Dokl.BAN* **35**, 1061 (1982)

12 M.I. Podgoretsky, *Yad.Fiz.* **37**, 455 (1983)
13. N. Akhababian, et al., *Z.Phys.C*, **18**, 97 (1983)
14. G.N. Agakishiev et al., *Yad.Fiz.* **39**, 543 (1984)
15. J. Pluta, *Proc.Int.Nucl.Phys.Conf. Harrogate U.K. 25÷30 August 1986*, **1**, 463 (1986)
16. V.G. Grishin et al., *Yad.Fiz.* **47**, 439 (1988)
17. J. Bartke, *Phys.Lett*, **B174**, 32 (1986)
18.R. Lednicky, V.L. Lyuboshitz, JINR, E2-81-453, Dubna (1981), *Yad.Fiz*, **35**, 1316 (1982).
19. N. Angelov et al., *Yad.Fiz.* **32**, 1357 (1980).
20. N. Akhababian et al., *Z.Phys.C*, **26**, 245 (1984).
21. N. Akhababian et al., *Yad.Fiz.* **44**, 434 (1986).
22. J. Bartke et al., *Z.Phys.A*, **324**, 471 (1986).
23. G.N. Agakishiev al., *Z.Phys.A*, **327**, 443 (1987).
24. V.G. Grishin et al., *Yad.Fiz.* **46**, 832 (1987)
25. G.N. Agakishiev et al., *Yad.Fiz.* **47**, 1292 (1988).
26. J. Pluta, JINR, P1-89-247, Dubna (1989)
27. V.G. Grishin et al., JINR, P1-89-639, Dubna (1989)
28. H. Aihara et al., *Phys.Rev.* **D31**, 996 (1985).
29. M. Althoff et al., *Z.Phys.* **C30**, 355 (1986).
30. I. Juricic et al., *Phys.Rev.* **D39**, 1 (1989).
31. T. Åkesson et al., *Z.Phys.C* **36**, 517 (1987)
32. A.M. Baldin et al., *Z.Phys.* **C33**, 363 (1987); JINR, P1-87-552 (1987); JINR, P1-88-331, Dubna (1988)
33. A.M.Baldin et al., JINR, E1-84-317, Dubna (1984); *Yad.Fiz*, **39**, 1215 (1984)
34. A. Bamberger et al., *Phys.Lett.***B203**, 320 (1988).
35. J. Pluta, JINR, E1-88-745, Dubna, (1988)
36. J. Pochodzalla et al., *Phys.Lett.* **B174**, 36 (1986).
37. J. Pochodzalla et al.,*Phys.Rev.* **C35**, 1695 (1987).
38. Z. Chen et al., *Phys.Rev.* **C36**, 2297 (1987).
39. Yu.D. Bayukov et al. *Phys.Lett.* **B189**, 291 (1987).
40. Yu.D. Bayukov et al. ITEP 88-2, Moscow, (1988); *Yad.Fiz.* **34**, 95 (1981).
41. B.M. Bobchenko et al., *Yad.Fiz.* **30**, 1553 (1979).
42. M.L. Allaberdin et al., *Yad.Fiz.* **46**, 1785 (1987).
43. W.G. Lynch et al., *Phys.Rev.Lett.* **51**, 1850 (1983).
44. S.A. Azimov et al., *Phys.Rev.* **D29**, 1304 (1984).
45. D. Ardouin INTERNE LPN-88-04, Univ. de Nantes, (1988).

PROTON-PROTON CORRELATIONS AT SMALL RELATIVE MOMENTUM IN NEON AND ARGON COLLISION WITH NUCLEI AT SATURNE

J.P. ALARD, P. DUPIEUX, P. CHARMENSAT, J. AUGERAT, N. BASTID, F. BIAGI, L. FRAYSSE, J. MARRONCLE, G. MONTAROU, P. MOREL, M.J. PARIZET, D. QASSOUD, A. RAHMANI,

LPC Clermont-Ferrand, IN2P3/CNRS, Université Blaise Pascal
F-63177 AUBIERE CEDEX FRANCE

and

R. BABINET, C. CAVATA, M. DEMOULINS, J. GOSSET, D. L'HOTE, B. LUCAS, J. POITOU, Y. TERRIEN, O. VALETTE,

DPHN, CEN Saclay, F-91191 GIF-SUR-YVETTE CEDEX

and

M.C. LEMAIRE

LNS, F-91191 GIF-SUR-YVETTE CEDEX

and

F. BROCHARD, P. GORODETZKY, C. RACCA

CRN, BP 20CR, F-67037 STRASBOURG CEDEX

ABSTRACT

Proton proton correlations at small relative momentum have been studied using the Diogene $4\pi-$ Facility at Saturne. The incident energies vary from 200 to 1000 A.MeV for Ne and Ar projectiles on various targets. Experimental correlation functions are compared to theoretical calculations of S. Koonin, corrected for experimental biases. We give the values of the source size as a function of incident energy, target-projectile system and centrality of the collisions. Extracted values of the density of nuclear matter show that the method is mainly sensitive to the final stage of the interaction ; nevertheless, we have some indications of a highly compressed phase at the beginning of the collision.

Introduction

Two particle correlations at small relative momentum can be an useful tool for the extraction of the size and lifetime of the source in relativistic nuclear collisions. If we can measure the number of participant nucleons in the source, we also can extract the value of the density of nuclear matter ; such an information is of fundamental interest for our knowledge of the equation of state of hadronic matter.

Correlations between two particles with nearly the same momentum are sensitive to the space time structure of the source, through Coulomb and nuclear interactions and statistical effects. Interesting results have been obtained showing evidence for a great difference between low incident energy collisions[1,2] (hot spot) and very high energy ones[3] (possible signature of QGP). In the range of Saturne energies (200 – 1 000 A.MeV) many results[4,5] confirm the validity of the participant spectator model and the existence of a highly excited fireball. Such studies are important both for our knowledge of the properties of hadronic matter (Equation of state) and for an approach of in medium effects.

In such analyses, 4π detectors allow to study the correlation function versus other observables (centrality for instance). But due to the complexity of such detectors, it is necessary to make a lot of corrections in order to take into account the efficiency of the experimental set up.

We present in the part I of the paper a brief description of the Diogene experimental set up ; Part II present some results concerning (pp) correlations at small relative momentum[7]. We give some values of source sizes and of extracted densities of the nuclear matter.

I – The Diogene 4π facility at Saturne.

Diogene[6] is made of a Pictorial drift chamber (PDC) placed in a magnetic field, which measures light charged particles (π^-, π^+, p, d ...) emitted from 20° to 132° and above energy thesholds of $\sim$ 25 MeV for the pions and $\sim$ 45 MeV for baryons (participants) and a plastic wall which measures particles below 6° Lab. Fig 1 presents a general view of the experimental set up. A longitudinal magnetic field permits the determination of the magnetic rigidity of particles inside the chamber. Such a measurement, combined with the value of the energy loss dE/dx in the gas of the PDC, permits the identification of the emitted light fragments.

A detailed description of the acceptance cuts is given in ref [6]. In what follows, data are extracted only from the PDC.

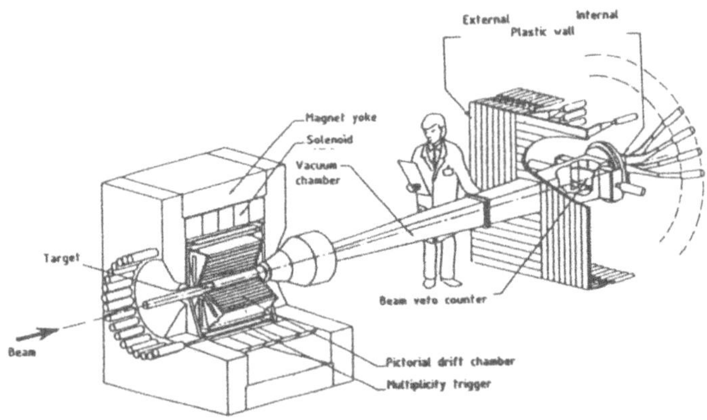

Fig. 1 : Exploded view of the Diogene detector.

100 to 300 000 events were stored for each projectile-target combination. The trigger selected non peripheral reactions. The centrality of the events was determined from their proton like multiplicity $M(\overline{p})$. We estimated the mean squared reduced impact parameter $\overline{b^2} = b^2/(R_1 + R_2)^2$ corresponding to a given $M(\overline{p})$ interval from the total cross section in this interval ($R_1 = 1.12\ A_P^{1/3}$ and $R_2 = 1.12\ A_T^{1/3}$ are the projectile and target radii in Fm).

The plastic wall, which is made of plastic scintillators, permits the measurement of velocity and charge of fragments by the correlation between energy loss in the scintillators and time of flight. The plastic wall is not used in our analysis of (pp) correlations.

II – Proton-proton correlations.

The correlation function C_2 is calculated as the ratio of the spectra of "correlated protons" (belonging to the same event) to "uncorrelated protons" (belonging to different events) for each channel of relative momentum Δp [$\vec{\Delta_p} = (\vec{p_1} - \vec{p_2})/2$]. Experimental correlation functions are then compared to Koonin calculations[8] distorted by experimental biases in order to extract the proton source radius r_0. Experimental and theoretical curves are normalized to each other for the three points $\Delta p = 80 - 90 - 100$ MeV/c.

II.1 – Experimental biases.

We correct directly the theoretical correlation functions for three main biases (in the following order) :

• Double track resolution due to electronics inefficiencies. The effect of this bias is evaluated using a simulation program of the whole Diogene apparatus, called TONO. For each Δp channel up to 150 MeV/c, we simulate couples of protons with various individual angles and momenta, having this Δp value. We look then at the remaining number of couples, after running TONO. The results are given in Table I for 1 000 initial pairs, for each Δp channel up to 50 MeV/c. We loose ~ 28% of the pairs in the channel [20-30] MeV/c, which corresponds to the peak of the correlation function (let us recall that the value of the correlation function at this peak is sensitive to the radius of the source).

Table 1 : effect of the double track resolution versus Δ_p

Δ_p (MeV/c)	0 – 10	10 – 20	20 – 30	30 – 40	40 – 50
Remaining pairs of protons (1 000 initially)	274	572	725	815	856

• Resolution on the relative momentum Δp. It is evaluated by a simulation program in which each Δp channel of the theoretical curve is replaced by a distorted and normalised spectrum taking this experimental resolution into account. The effect of this correction is mainly to flatten the peak near Δp = 20 MeV/c. The sharper the peak, the more it is flattened (see ref [7]).

• Creation of spurious pairs due mainly to multiple scattering which can enough change the trajectory of an individual particle so that the reconstitution program interpret an individual track as two tracks. In this case, the relative momentum of the pair is always small. From TONO, we evaluate that about one pair of proton over two in the first Δp channel [0-10] MeV/c is a spurious pair. This effect is negligible in other bins.

II.2 – Results for the source size r_0

(i) – *Discussion about the lifetime of the source*

When extracting r_0 values from the comparison of experimental and theoretical curves, we have to take into account the lifetime τ of the source. It is possible, in principle, to evaluate the lifetime of the source too from p-p correlations. In practice, and looking at the case $\tau \neq 0$ in the Koonin calculations[8], it becomes very difficult, due to necessary drastic experimental selections.

We have evaluated lifetime values using the intra-nuclear cascade code of J. Cugnon[9]. For each event, we define time origin at the first collision of any particle of the system and we draw the histogram of the last collision time (emission) for each proton inside the Diogene acceptance. The lifetime of the source is simply defined as the rms parameter of this spectrum, following the Koonin assumption. Lifetime values from 4 – 7 fm/c are obtained for the various analysed systems and energies. Running the Koonin code with these values, we found very small deviation $\Delta r_0/r_0$ of the order of 3% for the NaF target and 5% for the Pb one. Thus, in the following, we neglect the lifetime of the source.

(ii) – r_0 *values without any selection of the data*

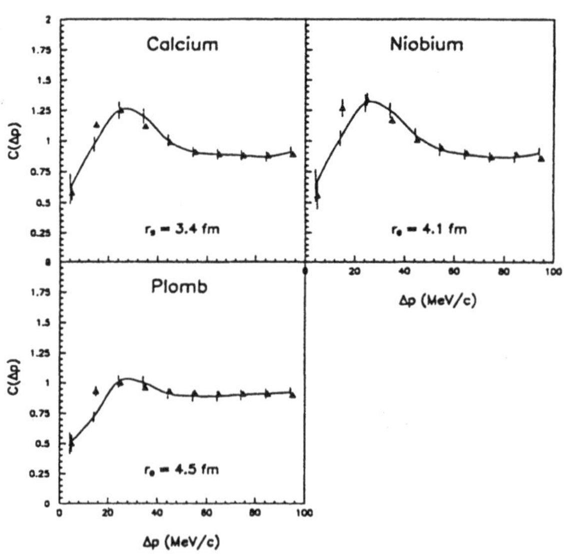

Fig. 2 : Proton–Proton correlation functions for Argon projectiles at 400 A.MeV.

We show on Fig. 2 some example of experimental correlation functions, with the associated theoretical ones, concerning the systems Ar + (Ca – Nb – Pb) at 400

MeV per nucleon. All the r_0 values are summarized on Figs. 3 and 4, respectively for Ne + (NaF − Nb − Pb) and Ar + (Ca − Nb − Pb) versus incident energy. It appears clearly that the obtained values are independent of incident energy, for a given system. It is also clear that the heavier the system, the greater r_0. Such results agree with a geometrical description of the collision, like in participant–spectator concept, although all obtained r_0 values correspond to equivalent spherical radii ($R_s = \sqrt{\frac{5}{2}} * r_0$) bigger than the classical hard sphere radii of the colliding system.

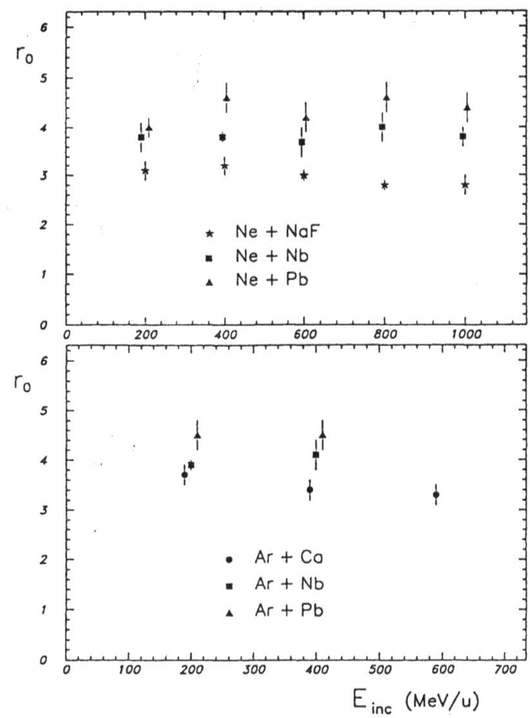

Fig. 3–4 : Extracted r_0 values vs incident energy.

In order to go further in these observations we have compared R_s to the radius R_g ("geometrical radius") corresponding to the clean cut geometry when considering a staight–line collision. R_g is calculated by a simulation program taking the Diogene trigger into account and is weighted on the impact parameter. We show on Fig. 5 the values of R_s/R_g for various systems in case of the 400 MeV per nucleon incident energy. We see that we measure always the same fraction of the geometrical radius R_g. No compressionnal effects are pointed out. An absolute value of $R_s/R_g \simeq 2$ is in good agreement with freeze-out values.

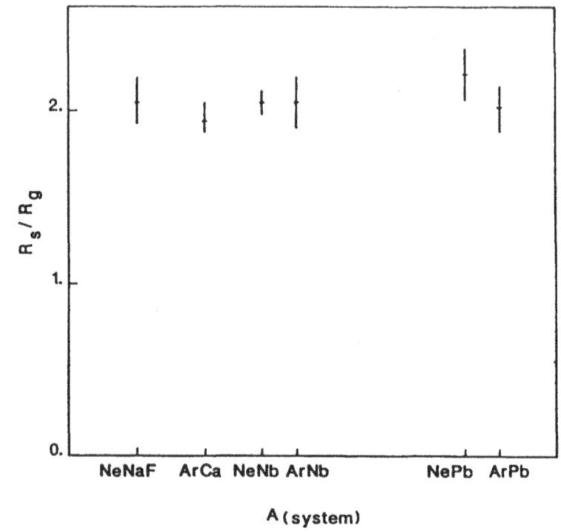

Fig. 5 : values of the R_s/R_g ratio for different projectile-target systems.

(iii) – r_0 *values in case of impact parameter selections*

The impact parameter is evaluated from the proton-like multiplicity assuming a monotonous correlation between these two quantities[10]. For each system, the ratio of the proton-like cross section to the geometrical one is considered as a measurement of the square of the reduced impact parameter $(b/b_{max})^2$.

For all systems, we obtain greater r_0 values in case of more central collisions. We give on Fig. 6 r_0 values corresponding to four impact parameter classes for the system Ar + Ca and various incident energies. These results are also in good agreement with a geometrical description of the collision.

We have calculated the ration R_s/R_g, always in case of four classes of impact parameter, for the system Ar + Ca at 600 MeV per nucleon incident energy. The results are given on Fig. 7. In contrary to what we observe when no impact parameter selections are done, R_s/R_g is not constant. It increases when going to more peripheral collisions. It can be interpreted as a signature of a bigger compression in the earlier stages of the interaction, for the more central collisions.

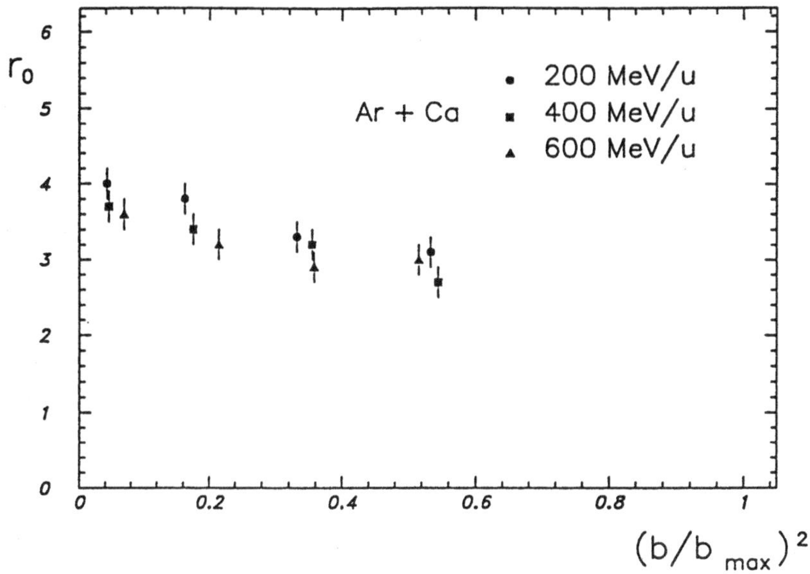

Fig. 6 : Extracted r_0 values vs impact parameter.

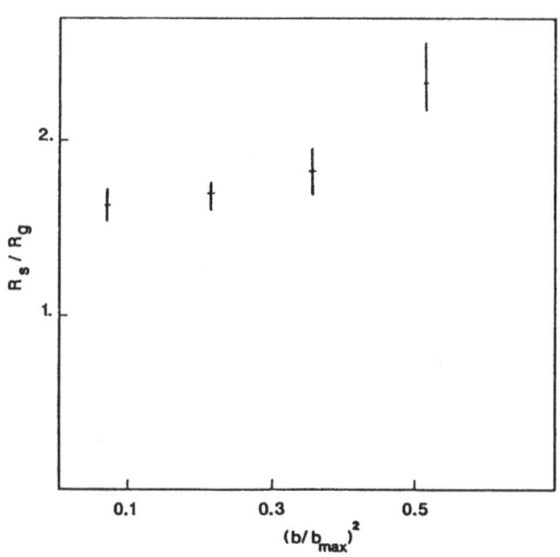

Fig. 7 : values of the R_s/R_g ratio vs impact parameter for Ar + Ca at 600 A.MeV.

(iv) – r_0 *values in case of high momentum protons selection*

Results from Dubna[11] have shown a very significant decrease of the measured source radii when considering high energy protons. In our case, we see only a very slight decrease of the r_0 values when doing such selections (see Table II), probably due to less sensitivity of high energy protons to final state interactions. We have to notice however that Dubna results concern lighter projectiles at higher energy and lighter targets. For instance, they don't see any increase of r_0 with increasing the mass of the colliding system, like in other groups[4].

Table 2 : r_0 values in case of high momentum protons selection, for a Ne projectile.

Target	Incident Energy A.MeV	Momentum selection MeV/c	r_0 with selet. fm	r_0 no select fm
NaF	200	≥ 500	$3.3^{+.3}_{-.3}$	$3.1^{+.2}_{-.2}$
Nb	200	≥ 550	$3.4^{+.3}_{-.4}$	$3.8^{+.2}_{-.3}$
Pb	200	≥ 550	$3.3^{+.5}_{-.1}$	$4.0^{+.2}_{-.2}$
NaF	600	≥ 500	$3.2^{+.3}_{-.2}$	$3.0^{+.1}_{-.1}$
Nb	600	≥ 550	$3.6^{+.3}_{-.3}$	$3.7^{+.3}_{-.3}$
Pb	600	≥ 550	$3.8^{+.6}_{-.2}$	$4.2^{+.2}_{-.3}$
NaF	1000	≥ 550	$2.9^{+.3}_{-.2}$	$2.8^{+.1}_{-.2}$
Nb	1000	≥ 600	$3.8^{+.5}_{-.4}$	$3.8^{+.2}_{-.2}$
Pb	1000	≥ 650	$4.0^{+.2}_{-.4}$	$4.4^{+.4}_{-.2}$

II.3 – The density of nuclear matter

We can evaluate the density of the nuclear matter if we know both the value of the size r_0 and the number of participant protons. This last quantity can be obtained from the experimental multiplicity of proton–like $M_{\bar{p}}$. Nevertheless we loose a part of the multiplicity due to the Diogene acceptance. We have made a correction on $M_{\bar{p}}$, using the intranuclear cascade code of J. Cugon[9]. Finally, the effective proton like multiplicity has been given by :

$$M_{\bar{p}} = F \,.\, M_{\bar{p}}\,(\mathrm{exp}).$$

where F is an acceptance factor which has been determined for all projectile-target systems at different energies and integrating on all impact parameters.

On Fig. 8 we present the extracted values of the relative density ρ/ρ_0 as a function of incident energy, for Ne and Ar projectile. Fig. 9 presents the ρ/ρ_0 values versus the centrality of the collision.

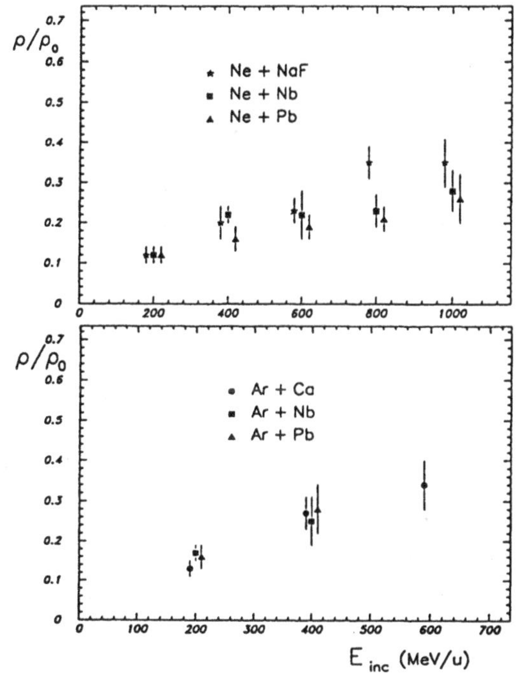

Fig. 8 : Extracted values of the reduced density for Ne and Ar projectiles.

In all cases, the ρ/ρ_0 values are less than 1. Such a result confirms that the method is sensitive to final state interaction and to the final stage of the collision.

Nevertheless, ρ/ρ_0 increases as the incident energy increases and as the centrality increases ; such observations are in agreement with a greater compression effect for high incident energies and more central collisions.

III – Conclusions.

The results of (pp) correlation studies at Saturne are in all cases representative of the freeze out stage of the interaction. That can be seen both from the size of the sources and the extracted densities of the nuclear matter.

The size of the source increases when the mass of the projectile target system increases. It also increases when the centrality of the collision increases. Such observation are in good qualitative agreement with the participant spectator concept. On the other hand, we observe no clear variation of the size with the incident energy.

The extracted values of R_s/R_g can show that the system keep some information from the early stage of the collision, as the compression of the nuclear matter seems to be greater for central collisions.

Finally we note that our results for (pp) correlation studies are in good agreement with those obtained with Diogene for $(\pi^-\pi^-)$ and $(\pi^+\pi^+)$ interferometry. They are also in good agreement with the Plastic ball results. It would be interesting in future experimental studies at these energies, to focus on large angle correlations, fragment fragment analyses, and measure correlations between particles which are less sensitive to final state interactions, in order to obtain more information on the initial stage of the collision.

References

1. W.G. Lynch, C.B. Chitwood, M.B. Tsang, D.J. Fields, D.R. Klesch, C.K. Gelbke, G.R. Young, T.C. Awes, R.L. Ferguson, F.E. Obenshain, F. Plasil, R.L. Robinson, A.D. Panagiotou, *Phys. Rev. Lett.* **51**, 1850 (1983).
2. J. Pochodzalla, C.K. Gelbke, W.G. Lynch, M. Maier, D. Ardouin, H, Delagrange, H. Doubre, C. Grégoire, A. Kyanowski, W. Mittig, A. Péghaire, J. Péter, F. Saint–Laurent, B. Zwieglinski, G. Bizard, F. Lefèbvres, B. Tamain, J. Québert, Y.P. Viyogi, W.A. Friedman, D.H. Boal, *Phys. Rev.* **C35**, 1695 (1987).
3. A. Bamberger et al, *Phys. Lett.* **203**, 320 (1988).
4. H.A. Gustafsson et al, *Phys. Rev. Lett.* **53**, 544 (1984).
5. J. Bartke, *Phys. Lett.* **174**, 32 (1986).
6. J.P. Alard, J. Arnold, J. Augerat, R. Rabinet, N. Bastid, F. Brochard, M. Crouau, N. de Marco, M. Drouet, P. Dupieux, H. Fanet, Z. Fodor, L. Fraysse, J. Girard, P. Gorodetzky, J. Gosset, C. Laspalles, M.C. Lemaire, D. L'Hôte, B. Lucas, G. Montarou, A. Papineau, M.J. Parizet, J. Poitou, C. Racca, W. Schimmerling, J.C. Tamain, Y. Terrien, J. Valero, O. Valette, *N.I.M.* **A261**, 379 (1987).
7. P. Dupieux, J.P. Alard, J. Augerat, R. Rabinet, N. Bastid, F. Brochard, P. Charmensat, N. de Marco, H. Fanet, Z. Fodor, L. Fraysse, J. Girard, P. Gorodetzky, J. Gosset, C. Laspalles, M.C. Lemaire, D. L'Hôte, B. Lucas, J. Marroncle, G. Montarou, M.J. Parizet, J. Poitou, D. Qassoud, C. Racca, A. Rahmani, W. Schimmerling, O. Valette, *Phys. Lett.* **B200**, 17 (1988).
8. S.E. Koonin, *Phys. Lett.* **B70**, 43 (1977).
9. J. Cugnon, *Phys. Rev.* **C22**, 1885 (1980).
10. C. Cavata. M. Demoulins, J. Gosset, M.C. Lemaire, D. L'Hôte, J. Poitou, O. Valette, *"Determination of the impact parameter in relativistic nucleus-nucleus collisions", to be published in Phys. Rev.* **C.**
11. G.N. Agakishiyev, R.R. Mekhteev, V. Boldea, S. Ditza, J. Bartke, M. Kowalski, D. Armutliyski, J. Bogdanowicz, A.P. Cheplakov, L.A. Didenko, A.P. Gasparian, V.G. Grishin, I.A. Ivanovskaya, T. Kanarek, E.N. Kladnitskaya, D.K. Kopylova, V.B. Lyubimov. K. Miller, V.F. Nikitina, J. Pluta, M.I. Soloviev, R. Togoo, G.P. Toneeva, M. Chubarian, R.N. Bekmirzayev, U.D. Sherkulov, Z.V. Metreveli, Z. Strugalski, *Z. Phys.* **A327**, 443 (1987).

EXCLUSIVE STUDY OF LIGHT PARTICLE CORRELATIONS IN THE INTERMEDIATE ENERGY DOMAIN

François SAINT-LAURENT, Alain FERRAGUT

GANIL, BP 5027 14021 Caen-Cedex, France

ABSTRACT

Two-particle correlations at small relative momenta were measured for ^{40}Ar-induced reaction on ^{197}Au at 60 MeV per nucleon and for ^{16}O-induced reactions at E/A = 94 MeV, in coincidence with a plastic multidetector system. The light charged particle multiplicity observed in the multidetector was found to be a rather good filter on the violence of the collision. The correlations decrease monotonously with increasing observed multiplicity. An interpretation of the correlation functions in terms of final-state interaction model and/or unstable cluster emission is discussed.

INTRODUCTION

Intermediate-energy heavy-ion induced reactions are of considerable interest since they provide the opportunity to study unknown features of highly excited nuclear matter. Although the final disintegration stages of these reactions may be approximated in terms of statistical emission, a large amount of pre-equilibrium particles and clusters can be emitted in the first stage of the collisions. This emission could originate from a localized zone, as at relativistic energies where the interaction zone is restricted to the geometric overlap of the two nuclei [1]. Due to the very short collision times, dynamical treatment of the collisions taken into account both the nuclear mean field and the individual nucleon-nucleon collisions must be done, like in the Landau-Vlasov or BUU calculations [2-4]. An overall feature of these collisions is a very strong dependence of the dissipated energy with the impact parameter.

At relativistic energies, p-p and π-π correlation functions are extensively used to extract the space-time extent of the interaction zone [5]. In the Fermi energy domain, light particle correlations have also been used to provide informations on the size of the emitting source [6]. Until now, investigations in this energy domain have been limited mainly to correlation functions derived from two-particle inclusive data [7-9]. While these measurements can provide insights into the average properties of the emitting system, more detailled information must be obtained from more exclusive measurements in which specific classes of reactions can be suppressed or enhanced.

Very few exclusive measurements have already been reported in the litterature. A pionner experiment have been performed by Chen et al [10]. They have studied the ^{14}N+Au system at 35 MeV per nucleon. The centrality of the collision was extracted from the measurement of the linear momentum transfered (LMT) by the projectile to the target : large momentum transfers can be associated with more central collisions. The LMT was obtained by measuring the correlation angle between the two fission fragments emitted in the collision. The light particle coincident pairs were detected at 20° lab.

They found that the selection of events with large momentum transfers produces enhanced maxima in correlation functions as compared to that with low momentum transfers. Interpreting these correlation functions using the final-state interaction model [5,6], larger source radii were extracted for more peripheral collisions, and a significantly smaller radius than the entire system was found for the more central collisions.The authors interpreted these puzzling results by some important temporal effects, which must not be neglected at this incident energy. But, even with a time scale dependence, the data are understandable with difficulty. A lifetime of 25 fm/c, the typical collision time at this energy, will destroy the peak at $q \approx 20$ MeV/c in the correlation function. An other explanation could be that the coincident pairs were detected at 20°, an angle not too far from the grazing angle. The production rate of protons and light clusters could strongly be affected by a pre-equilibrium emission and by the sequential decay of the projectile-like fragment.

EXPERIMENTAL SET-UP

To go further in the exclusive study of correlation function, we have performed exclusive experiments at Ganil, using a completely different reaction filter. Two different systems have been studied : $^{40}Ar+^{197}Au$ at 60 MeV per nucleon and $^{16}O+^{197}Au$ at 94 MeV per nucleon. Light particles ($Z\leq3$) were detected in a close-packed hexagonal array of ΔE-E telescopes. The hodoscope allowed us to detect coincident pairs in order to built the correlation functions. For the ^{40}Ar (respectively ^{16}O) experiment, the hodoscope were mounted at an average angle of 30° (resp. 45°), far away from the grazing angle (less than 5° for the 2 systems). More experimental details can be found in references

In order to select on event type, light charged particles associated to the coincident pairs were detected in a multidetector array. For the ^{40}Ar experiment, it consisted of 96 plastic scintillators (plastic wall [11]), covering the whole angular range between 5° and 30° in the laboratory frame. For the ^{16}O experiment, this plastic array was extended toward smaller angle (3°), and completed in the 30°-90° angular range by the forward part of the "Tonneau" [12]. This multidetector, covering the full azimuthal angle ϕ, consisted of 36 plastic scintillator half-staves. The charge and the velocity of each light charged particle detected in the two plastic arrays were determined by a ΔE-time-of-flight measurement.

The data taken with the hodoscope are presented in terms of the correlation function R(q), defined for a particle pair (1,2) by

$$Y_{1,2}(p_1,p_2) = C.Y_1(p_1).Y_2(p_2).[1 + R(q)]$$

where $Y_{1,2}$ denotes the coincidence yield, Y_1 and Y_2 the single yields. The laboratory momentum of particles 1 and 2 are p_1 and p_2 respectively, and q is the momentum of relative motion. The mormalization constant C is determined by the requirement that R(q) = 0 at large relative momenta.

For the two studied systems, inclusive data have already been reported [7,8], as well as some exclusive results [13,14].

RESULTS AND DISCUSSION

The multidetector data are useful to give more information on the origine of the coincident pairs detected in the hodoscope. The figure 1 shows the observed multiplicity spectra of charged particle detected in the plastic wall for $^{40}Ar+^{197}Au$, under two triggering conditions : the left hand part of the figure corresponds to the minimum bias trigger, and the right hand side when two light particles are required in the correlator.

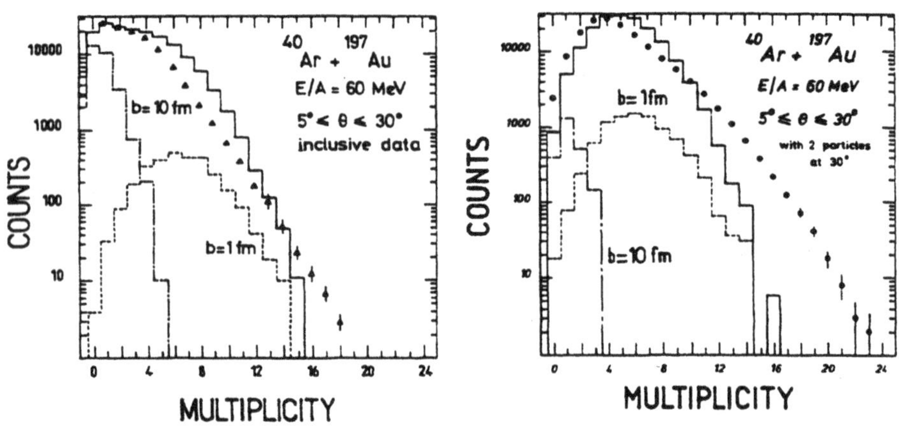

Figure 1 : *Charged particle multiplicity distributions taken with a minimum bias condition (left side) and when triggering by two particles detected at 30° (right side). The experimental data are compared to the prediction (full line) of a Monte-Carlo simulation (see text), normalized to the data at M = 4. Predicted spectra for two given impact parameters, b =1 fm (dashed line) and b = 10 fm (dotted-dashed line) are also displayed with the same normalization.*

One sees clearly that the coincident pairs are likely associated to the larger multiplicities. To facilitate the interpretation of these data and because of the restricted solid angle of the plastic array for this system which can bias the interpretation [15], we have performed a Monte-Carlo simulation in order to predict the dependence of the observed multiplicity on the impact parameter.The reaction model used for this calculation is the abrasion-ablation model as extended by Dayras et al [16]. The predicted multiplicity spectra are compared to the data in figure 1. For the two triggering conditions the data are fairly well reproduced, given some confidence in this simple model, especially for the respective contribution of different impact parameter. High multiplicities are only achieved in central collisions, while low multiplicities are essentially fed by peripheral collisions. When triggering by two particles, the contribution of large impact parameter reactions is strongly reduced.

4π-neutron multiplicity measurements [17] performed for the same system give an other information : light particles and clusters emitted at large angle (θ>20°) are also likely associated to larger neutron multiplicities. These results indicate clearly that coincident pairs detected in the correlator are producted mainly in highly dissipative collisions.

On the figure 2 the velocity spectra of proton (upper part) and alpha particle (lower part) are reported.

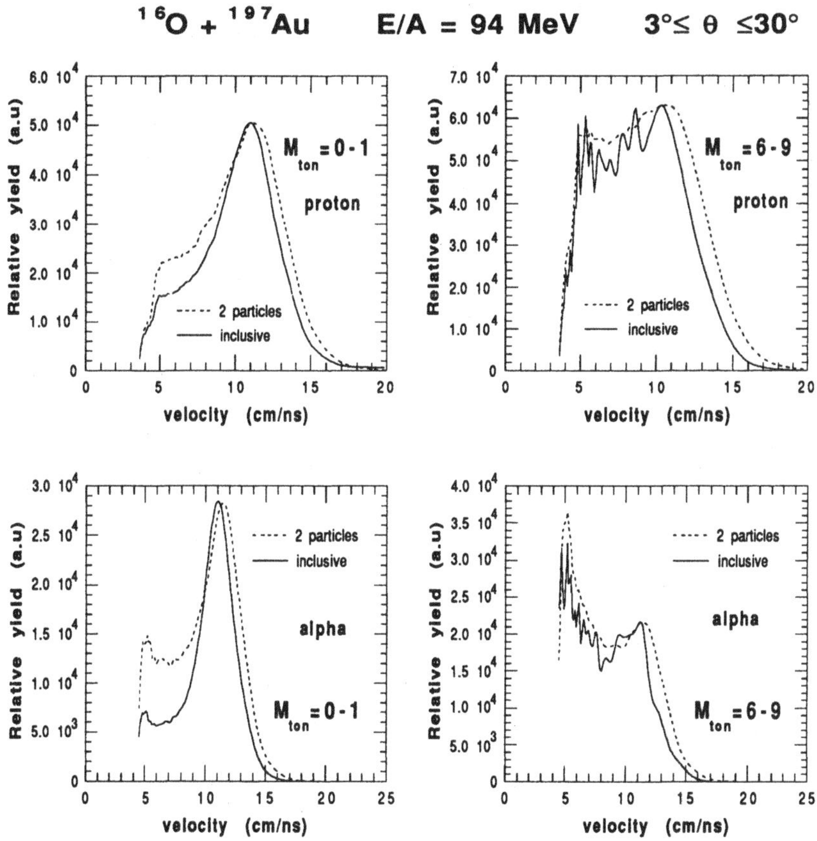

Figure 2 : *Velocity spectra of proton (upper part) and alpha particle (lower part) integrated over the 3°-30° angular range for the $^{16}O+^{197}Au$ system. The spectra taken with a no external condition are compared to those obtained when a coincident pair is detected at 45°, under the further condition of a low (left part) or a high (right part) multiplicity gate of light charged particles detected between 30° and 90°.*

These spectra are integrated over the 3°-30° angular range for the $^{16}O+^{197}Au$ system. The full curves are obtained from a inclusive trigger condition, the dashed curves are spectra associated with two light particles detected at 45° in the correlator. The left hand parts correspond to the further condition of a low observed multiplicity in the tonneau (30°-90° angular range), the right hand side to a high multiplicity gate. In the forward direction, light particle velocity spectrum exhibits two component :

- particles emitted from the target-like fragment give a low velocity component. The greater this component is, the more dissipative the collision is.

- the high velocity component, centered at about 80% of the beam velocity, originates from the projectile-like fragment desexcitation and from preequilibrium emission in the first stage of the collision. This component will dominate for peripheral reactions.

In the figure 2, the low velocity component enhances when trigerring by two particles in the correlator and for low multiplicity events. The coincident pairs are associated with violent collisions even for this system. For a high multiplicity gate, the inclusive and triggered spectra exhibit the same shape. The further condition of two particles detected at 45° do not select any special class of events, among the most violent ones. Even for these events, a high velocity component remains, given some indication that a preequilibrium emission already takes place for central collisions at 94 MeV per nucleon

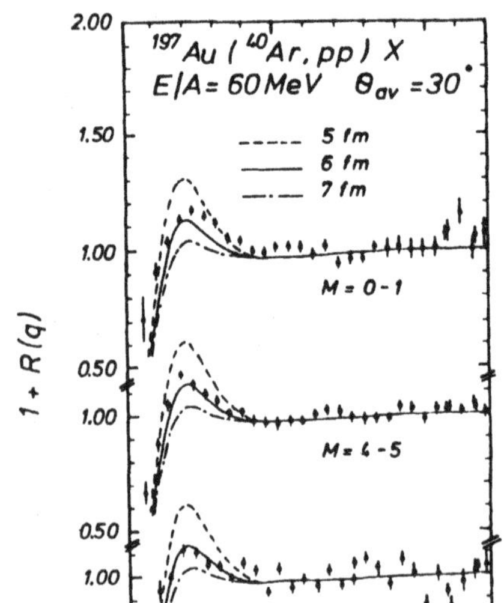

From the low to high multiplicity events, the huge component near the beam velocity is strongly suppressed, indicating that these events originate from more and more central collisions. The observed multiplicity of light charged particles detected between 30° and 90° lab. can be used to filter on the violence or the centrality of the collision.

Figure 3 : *Two-proton correlation functions under different constraints on the multiplicity. of charged particles detected between 3° and 30° for the $^{40}Ar+^{197}Au$ system. The curves show the calculations with the final-state interaction model [5].*

Using the observed multiplicity as a filter, we have studied the correlation functions of several light particles pairs : p-p, p-d, p-t, p-α, d-d, d-^{3}He, d-α, α-α, p-^{7}Li...

For the $^{40}Ar+^{197}Au$ system at 60 MeV per nucleon, the p-p correlation function obtains for different multiplicity gates is shown in the figure 3.

The maximum at q≈20 MeV/c is caused by the attractive singlet S-wave interaction between the two protons, or in other words, to the emission of a particle instable ^{2}He cluster. This maximum decreases smoothly with increasing observed multiplicity. The curves in the figure 3 correspond to theoretical correlation functions predicted by the final state interaction model [5]. The extracted radii r_o, assuming a source of negligible lifetime and Gaussian density distribution, range from 5.6 fm to 6.1 fm. These values are larger than the size of the target nucleus, $r_o(Au) \approx 5$ fm. Including a finite lifetime of the source will reduce these values.For the $^{16}O+^{197}Au$ system, the extracted r_o values range from 4.5 fm to 4.9 fm, from peripheral to central collisions. These values, smaller by 1 fm than for $^{40}Ar+^{197}Au$, could indicate a shorter lifetime of the emitting source at the higher incident energy.

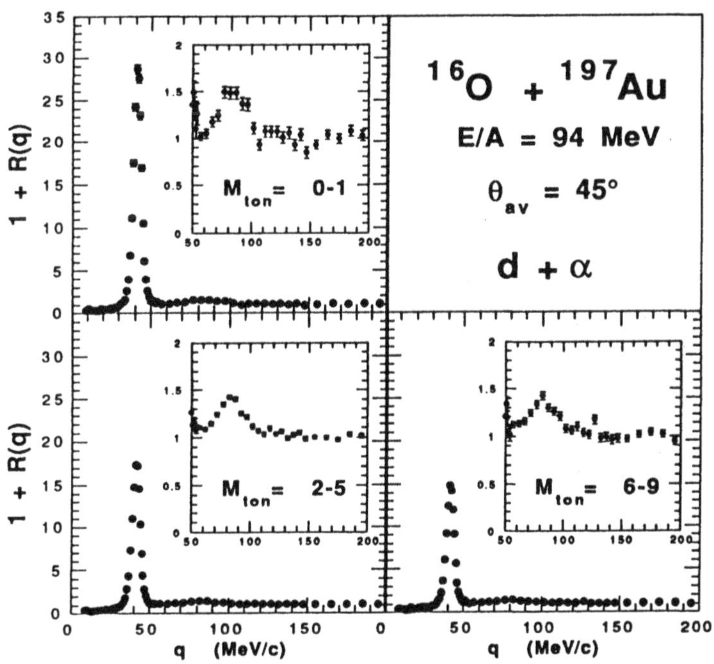

Figure 4 : *d-α correlation functions under different constraints on the multiplicity. of charged particles detected between 30° and 90° for the* $^{16}O+^{197}Au$ *system.*

In the figure 4, the d-α correlation function for the $^{16}O+^{197}Au$ system is shown, under different multiplicity gates, as a function of the relative momentum q. The sharp peak at q = 42 MeV/c is associated to the 2.186 MeV state in ^{6}Li, and the broad peak at q ≈ 85 MeV/c is due to the overlapping states at 4.31 MeV and 5.65 MeV in ^{6}Li.

A stronger dependence to the light particle multiplicity than for the two-proton correlation is observed. The maximum of the peak at q ≈ 42 MeV/c enhances by a factor 2 from a high to a low multiplicity gate. Such a strong dependence has already been reported for the $^{40}Ar+^{197}Au$ system [13]. Calculations [6] of the d-α correlation function, based on the final state interaction model and corrected for the finite resolution of the hodoscope, give extracted radii r_0 ranging from 2.95 fm to 4.1 fm when neglecting the source lifetime, from peripheral to central reactions. These values are definitively smaller than the target size, even for central collisions, and will be lowered when including a finite lifetime of the emitting source. For the low multiplicity gate, the extracted radius is comparable to the size of the projectile nucleus.

From the analysis of the different two-particle correlation functions, we found that the extracted radii are larger for correlations involving particle pairs that do not have resonances (p-d, d-d and t-t), than for those which have resonances. This pattern have been seen before [7-9]. A new piece of information to these puzzling results is that the dependence remains when selecting a given class of events, either central or peripheral reactions. It has been suggested that these results may indicate the sequential freeze-out of different particle species at different stages of the reaction [6].

An alternative to the interpretation of the light particle correlation functions in terms of final-state interaction is the break-up of a particle unstable cluster emitted by the source. Statistical calculations to reproduce the p-p correlation function has already been reported [18]. Such a break-up leads to a resonance in the correlation function at the same relative momentum that in the final-state interaction approach. As far as one considers the two-particle system, these two descriptions are exactly the same. But, in the final-state interaction model, the two particles are supposed to be independantly emitted. The final-state interaction originates only from the space-time proximity of the two particles. No prescription is used to determine the production rate of the two-particle species. Furthermore, the same single particle source functions contribute to the resonant and to the uncorrelated background parts in the correlation function.

On the other hand, only one fragment is emitted in a cluster emission interpretation. The correlation function is the result of the superposition of two contributions : the resonant decay of a instable particle cluster and the uncorrelated background of two sequentially emitted particles. In the intermediate energy domain, the time scales characteristic of the first steps of the statistical emission become of the same order of magnitude (≈10^{-22}s) that the relative emission time needed to observed a resonant contribution in the correlation function using the final-state interaction model. The final observed correlation function could be the result of the combination of the two processes [9].

In a thermal model approach, the authors of reference [19] have shown that, when the interactions between the two particles are must more important than the interactions with the rest of the system, the predicted correlation function is equivalent to the prediction of the final state interaction model assuming a source of zero lifetime. The production rate of uncorrelated pairs is given by the product of the single particle yields. The inclusive yields depend on the temperature of the source, but particle emissions do not affect subsequent decays, so the size and the temperature of the source remain unchanged. A consequence is the non dependence of the predicted correlation functions to the temperature of the system.

But the thermal model is an oversimplificated approach of the source deexcitation, especially at low incident energies. At these energies, the source decay would be more likely described as a statistical emission of particles and clusters from a highly excited nucleus. Even in the intermediate energy domain, the overall feature of the nuclei deexcitation can be estimated using a Hauser-Feschbach formalism. In such a description, a competition between the whole open channels takes place at each step of the cascade. Moreover, the choosen channel affects the condition (size, temperature, spin) of the source for the next step. So, even if the sequentially emitted particles are independant - it is the fundamental hypothesis of the statistical approach -, it exists like a correlation between the production rates of the different open channels, especially between the emission of a cluster and the sequential emission of two particles. An other consequence is that the production rate of two sequentially emitted particles cannot be directly related to the single particle yields. So, the correlation function must not necessary be normalized to the unity at large relative momentum.

Of course, because the two particles are emitted independantly, the corresponding correlation function will be flat, like the uncorrelated background, and the cluster decay will contribute to the resonant peak.So the peak to background ratio will depend upon a much more conplex way to the source conditions.

Such an effect is well known at low excitation energy where the Q values govern the partial cross-sections of each open channel. This feature remains at the high excitation energy reaches in the intermediate energy heavy-ion collisions for light systems ($A \leq 20$) [20], but is questionnable for heavy systems. Detailed statistical calculations are needed to answer this point. Anyway, from the data we can try to estimate the contribution of a cluster emission to the correlation function.

In such an interpretation, the observed evolution of the d-α correlation function with the violence of the collision would be the result of a balance between excited ^6Li yield and d-α uncorrelated production rate : either the $^6Li^*$ yield increases or the d-α uncorrelated yield decreases for low multiplicity events. Because the first excited state in ^6Li at 2.18 MeV is not too far from the ground state, we expect a near constant relative production yield between the two ^6Li states. In the left hand side of the figure 5, the yield of ^6Li ground state, integrated over the whole solid angle of the hodoscope, is compared to the production rate of alpha particle and deuteron, for several multiplicity gates. We found a slight decrease of ^6Li production for peripheral reactions.

The relative ^6Li single particle cross-section to the d-α uncorrelated pairs detected in the correlator is shown in the right hand side of the figure 5. The number of uncorrelated pairs is taken by integrating the measured d-α coincident pair at large relative momentum. This result is not bias by the experimental set-up. Of course, the efficiency to detect coincident pairs in the hodoscope will not be the same than to detect single particle, but the data are presented as a function of the observed multiplicity between 30° and 90°. The geometric efficiency of the hodoscope remains the same for each multiplicity gate. The total efficiency for a coincident pair detection depends also on the single particle energy spectra. These spectra have been found to be insensitive to the multiplicity selection.

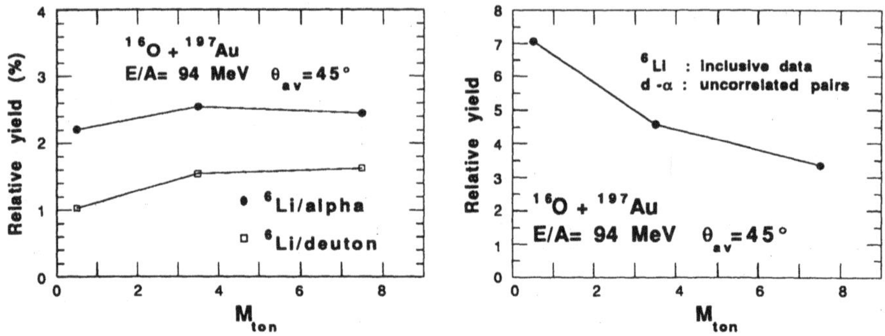

Figure 5 : *Relative production rates, integrated over the whole solid angle of the hodoscope, as a function of the charged particle multiplicity measured between 30° and 90°. Left hand part : relative 6Li single particle cross-section with respect to alpha particle yield and to deuteron yield. Right hand part : ^{6}Li single particle yield to d-α uncorrelated pairs ratio.*

The ^{6}Li ground-state yield to d-α uncorrelated pairs ratio decreases by a factor two from the low to high multiplicity gate, the same factor observed for the evolution of the d-α correlation function (fig 4). Combining this result with the rather insensitivity of the relative single particle yields to the event multiplicity, we conclude that the relative d-α uncorrelated background to either alpha particle or deuteron yields increases with the increasing multiplicity.

CONCLUSION

The ^{40}Ar+^{197}Au at 60 MeV per nucleon and ^{16}O+^{197}Au at 94 MeV per nucleon systems have been investigated at Ganil. The light particle coincident pairs were detected far away from the grazing angle, and the associated charged particles were detected in a plastic array mounted in the forward hemispher.

We have studied the multiplicity distribution and velocity spectra of the light particles detected in the plastic multidetector for minimum bias events and for events given a detected coincident pair in the correlator. The comparison of the data obtained with these two triggering conditions shows that the coincident pairs are mainly associated with the most violent and dissipative collisions, and they definitively do not originate from a projectile-like decay.

The observed light particle multiplicity has been found to be a rather good filter on the violence of the collision. The inclusive energy spectra of the light particles in the correlator are insensitive to this observed multiplicity. On the other hand, the light particle correlation functions evolve with the multiplicity gate : enhanced correlation functions are associated with low multiplicity events. Interpreting the correlation functions within the framework of the Final-State interaction model developped by Koonin [5] and extended by Boal et al [6], a reduction of the space-time localization of the coincident particle emission is observed at larger impact parameters where fewer nucleons are violently interacting. This evolution is much more pronounced for highly resonating pairs, like d-α pair.

For the $^{40}Ar+^{197}Au$ system at 60MeV per nucleon, the extracted source radii, assuming a zero-lifetime, are always larger (about 1 fm) than for the $^{16}O+^{197}Au$ system at 94 MeV per nucleon, even for the more central collisions. This could be an indication of a shorter source lifetime when increasing the incident energy.

The dependence of the extracted source radii to the particle pairs, already observed from inclusive data [7-9], remains when selecting a given class of events, either peripheral or central collisions.

A tentative reinterpretation of the two-particle correlation functions in terms of a cluster emission rather than an in-flight final-state interaction has been done. Detailed statistical calculations are needed to support such a interpretation. Nevertheless, experimental evidence have been presented that the thermal model approach is oversimplificated : the two-particle uncorrelated background yield does not evolve as the single particle yields product.

Acknowledgments.

We thank our colleagues which have participate to the data taking :
D. Ardouin, H. Delagrange, H. Doubre, C. Grégoire, A. Kyanowski, W. Mittig, A. Peghaire, J. Peter, Y.P. Viyogi, B. Zwieglinski, J. Kasagi, E. Rosato, Z. Chen, C.K. Gelbke, W.G. Gon g, Y.D. Kim, W.G. Lynch, M.R. Maier, J. Pochodzalla, M.B. Tsang, G. Bizard, F. Lefebvres, B. Tamain, and J. Québert.

References

[1] G.D. Westfall et al, *Phys. Rev. Lett.* **37** (1976) 1202.

[2] J. Aichelin and G. Bertsch, *Phys. Rev.* **C31** (1985) 1730.

[3] J. Aichelin, *Phys. Rev.* **C33** (1986) 537.

[4] C. Grégoire, B. Remaud, F. Sébille, L. Vinet, Y. Raffray, *Nucl. Phys.* **A465** (1987) 317.

[5] S.E. Koonin, *Phys. Lett.* **B70** (1977) 43.

[6] D.H. Boal and J.C. Schillcock, *Phys. Rev.* **C33** (1986) 549.

[7] J. Pochodzalla et al, *Phys. Rev.* **C35** (1987) 1695.

[8] Z. Chen et al, *Phys. Rev.* **C36** (1987) 2297.

[9] D. Fox et al, *Phys. Rev.* **C38** (1988) 146.

[10] Z. Chen et al, *Phys. Lett.* **B186** (1987) 280.

[11] G. Bizard el al, *Nucl. Instr. Methods* **A244** (1986) 549.

[12] A. Peghaire et al, Submitted to Nucl. Instr. Methods, GANIL P89 24.

[13] A. Kyanowski et al, *Phys. Lett.* **B181** (1986) 43.

[14] F. Saint-Laurent et al, *Phys. Lett.* **B202** (1988) 190.

[15] M.B. Tsang et al, *Phys. Lett.* **B220** (1989) 492.

[16] R. Dayras et al, *Nucl. Phys.* **A460** (1986) 299.

[17] B. Lott et al, Proceedings of XXVII Inter. Winter Meeting on Nucl. Phys. (Bormio 90) Ed I. Iori., and D. Guerreau, private communication.

[18] M.A. Bernstein et al, *Phys. Rev. Lett.* **54** (1985) 402.

[19] B.K. Jennings et al, *Phys. Rev.* **C33** (1986) 1303.

[20] J. Pouliot et al, *Phys. Lett.* **B223** (1989) 16.

Source Size Measurements for Heavy Ions Collisions at Intermediate Energies from Particle Correlations : Results and Questions

D. Rebreyend [1], Serge Kox [1], J.C. Gondrand [1], B. Khelfaoui [1], F. Merchez [1], B. Norén [1,2], C. Perrin [1]

The data have been obtained in experiments performed in collaboration with :
for the SARA experiments : H.A. Gustafsson [2], B. Jakobsson [2], A. Kristiansson [2], H. Ryde [2], A. Oskarsson [2], M. Westenius [2], M. Cronquist [3], S. Mattson [3], O. Skeppsted [3], M. Rydehell [3], G. Lovhoiden [4], T.F. Thorsteinsen [4], L. Westerberg [5], J.P. Bondorf [6]
for the AMPHORA experiment : P. Desesquelle [1], H. El Haage [1], A. Giorni [1], D. Heuer [1], A. Lleres [1], C. Morand [1], B. Chambon [7], B. Cheynis [7], D. Drain [7], T. Awes [8]
for the Saturne experiment : T. Motobayashi [1,9], J. Arvieux [10], J. Yonnet [10], G. Gaillard [11], G. Guillaume [12], A. Boudard [13], M. Garcon [13]

1) Institut des Sciences Nucléaires de Grenoble, UJF-IN2P3, 38026 Grenoble Cedex, France
2) Division of Cosmic and Subatomic Physics, University of Lund, Sweden
3) Department of Physics, Chalmers Institute of Technology, Göteborg, Sweden
4) Department of Physics, University of Bergen, Norway
5) The Svedberg Institute, University of Uppsala, Sweden
6) The Niels Bohr Institute, University of Copenhagen, Denmark
7) Institut de Physique Nucléaire de Lyon, 69622 Villeurbanne Cedex, France
8) Oak-Ridge National Laboratory, Tennessee 37831, USA
9) Rikkyo University, 3 Nishi-Ikebukoro, Toshima, Tokyo 171, Japan
10) Laboratoire National Saturne, 91191 Gif sur Yvette Cedex, France
11) DPNC, Université de Genève, CH-1211 Geneva, Switzerland
12) Centre de Recherches Nucléaires, 67037 Strasbourg, France
13) DPhN/ME-Saclay, 91191 Gif sur Yvette Cedex, France

Abstract:

Nuclear source sizes can be determined by measuring the correlation of light charged particles at small relative momentum. This is achieved by constructing a correlation function and analyzing it in the framework of final state interaction models. If numerous results are now published, questions still remained addressed. This contribution will discuss some of these problems using results of measurements performed on light heavy ion reactions at intermediate energies.

1. INTRODUCTION

Particle correlation measurements at small relative momenta are widely used to determine the extent of emitting sources formed in nuclear reactions by the mean of a correlation function. At high energies, meson correlations [1-2] have provided a systematic of results for emission times and source sizes [3] in the framework of models considering quantum statistics correlations of two identical bosons. In the intermediate energy range one has used pairs of baryons, but the correlation is then dominated by final state interactions (coulomb and nuclear). A large set of results has been published on proton-proton correlation measurements over a wide range of incident energy [4-12] but several questions remain still addressed. First of all authors desagree on the way of constructing the correlation function which role is to remove all the effect induced by the set-up. In the intermediate energy range one also faces the difficulty to isolate coincident particles arising from a single nuclear source and experiments have tried to constraint the data with conditions set on the energies of the detected protons [6,7] and on the multiplicity of charged particles emitted at forward angles [13]. However the relation between such quantities and the impact parameter or the time of the collision are not yet well established [14]. Finally most of the source sizes have been deduced under the assumption of zero lifetime and are then only upper limits of the spatial extent of these sources. A recent article even questioned which quantity is really extracted from these measurements : source size or simply the characteristics of the final state interaction of the two protons [15].

All these observations show that new experiments and results are still needed. In this contribution we would like to discuss the various problems of this kind of measurements in the intermediate energy range and will illustrate it with results recorded in recent experiments [12].

2. EXPERIMENTS

2.1 Experimental set-up

The EMRIC set-up used in our experiments has been detailed in a previous publication [16]. EMRIC is composed of an array of CsI crystals (4x4x10 cm^3) used in conjunction with a multiwire proportional chamber (MWPC) which provides precise localization of the impact point of the detected particles. The CsI crystals, optically coupled to phototubes, allow particle identification (in mass and charge up to $Z< 4$) by using a pulse shape analysis technique [17]. The energies of the particles are measured in these crystals with a precision of the order of 2 %. Gain shifts must however be carefully controlled to retain such a resolution. Dead area between contiguous detectors has been minimized to less than 3 mm allowing the detection of particles at very small relative momenta. The detection of the charged particles is made with high efficiency in the MWPC whereas neutral ones are undetected. Also pile-ups in the same CsI detector can be detected in the MWPC and rejected. The performance of this set-up is very interesting because one is able to measure relative momenta less than 5 MeV/c with typical precision of 1 MeV/c and for its wide dynamic range for particle detection (5 to 200 MeV proton).

2.2 Experiments

In a first set of experiments [12] two particles inclusive measurements have been made. Several targets (of ≈ 1 mg/cm^2 thick) ranging from ^{12}C to ^{197}Au have been bombarded with a ^{20}Ne beam of 30 MeV/u delivered by the SARA facility. EMRIC was placed at 80 cm from the target and at several angles in the forward hemisphere. The energy threshold was set to 8 MeV for protons, accounting for the identification limitations at very low light signal. Energy calibration was achieved by means of a time of flight technique [16] with a precision of about 5 % (mainly due to gain shift problems).

In a second very exlusive experiment we have tried to distinguish the various sources produced in heavy ion collisions in the intermediate energy range by setting EMRIC in combination with the AMPHORA 4π detector [17]. Events were recorded whenever two coincident particles were detected in EMRIC, placed at 50°. AMPHORA provides the information multiplicity of charged particles over almost 4π solid angle. From these information we plan to select impact parameter ranges and isolate the various produced nuclear sources.

Finally p-p correlations were measured for the reaction H(d,2p)n. These results, primarely measured to determine tensor analyzing powers [18], are presented only to illustrate some points of our discussion as they constitute a very peculiar set of data. Measurements were performed at 200 and 350 MeV by bombarding a liquid Hydrogen target with polarized deuterons beams delivered at the Laboratoire National Saturne. EMRIC was placed at 70 cm of the target and moved at 3 different mean angles (0°, 7.5° and 20°).

3. FINAL STATE INTERACTION MODEL

The source sizes are extracted in the framework of models whose predictions depend on the extent of the emitting source. Let us briefly recall the main physical ingredients of one of these models (Koonin [19]) which calculates the correlation function in terms of the final state interaction of the two protons. The particles are emitted independently from an equilibrated source parameterized as a gaussian in space and time extensions, with respectively r_0 and τ as free parameters (note that r_0 is a reduced radii and that the corresponding value of rms radii is larger). The mean-field effect of the emitter on the two protons is neglected and the final state interaction of the two protons takes into account the coulomb and nuclear interactions as well as the correct symmetrization (quantum statistics) of the relative wave function.

The left part of Figure 1 presents the predictions of Koonin's model for a zero lifetime (τ= 0) and r_0 value of 5 fm. The figure also details the coulomb and statistical contributions to the final correlation function. Note that the quantum statistics effect does not contribute significantly to the correlation function in that case, as compared to final state interaction. The correlation function shows an anticorrelation at q = 0, related to the coulomb repulsion, a correlation peak around 20 MeV/c, due to the nuclear attractive interaction in $^1S_0$ state, and no correlation at large relative momenta. The height of the correlation increases

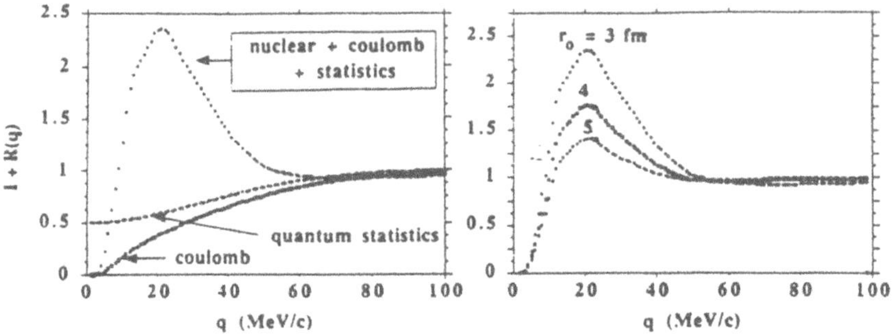

Figure 1 : Predictions of Koonin's model for zero lifetime. Left part of the figure details the different contributions to the correlation function while right part shows the evolution with the source size parameter r_0.

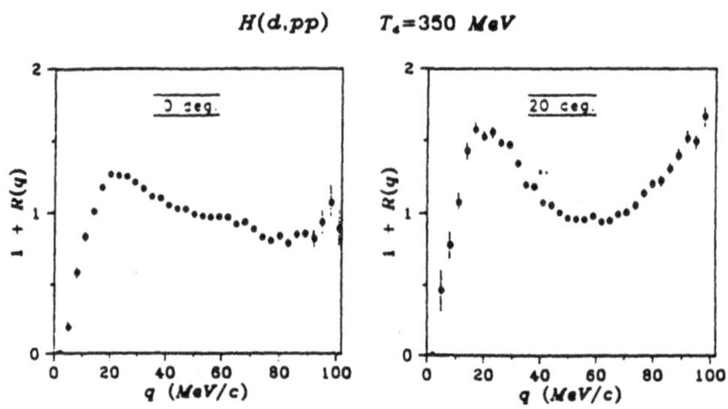

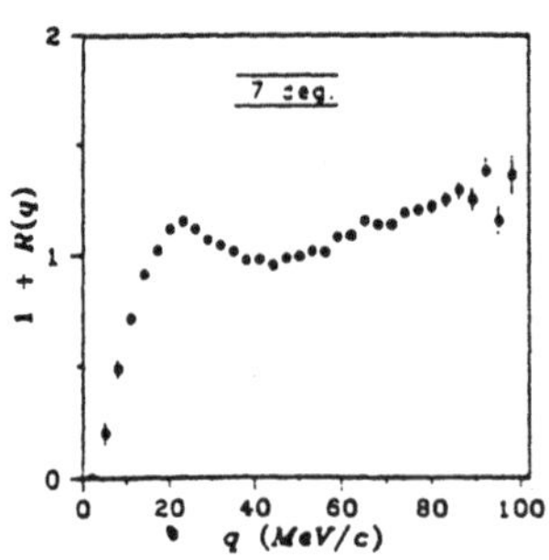

Figure 2 : Experimental correlation functions for the H(d,pp) reaction , measured at 3 angles (corresponding to 3 transfer momenta of the undetected neutron). These results illustrate the influence of kinematical constraints on the shape of the correlation function.

when the source size decreases (see right part of figure 1) expressing, in the framework of this model, that the proportion of proton pairs which are emitted at small relative distance (thus sensitive to the nuclear interaction), is larger when the source size is smaller.

4. THE CORRELATION FUNCTION

We will compare the results of our experiments and the predictions of the Koonin model in the following sections. Let us first discuss some of the problems linked to the correlation function.

Intensity interferometry was first used in nuclear physics by Goldhaber et al. [1] for p-p anhilation. This experiment allowed the determination of the volume of interaction by looking at the angular dependence of identical created pions. For experiments with a large number of particles in the exit channel, the integration over the phase space of undetected particles becomes too difficult. Then Kopylov and Potgoretzkii [20] have proposed to use a two particles correlation function to remove the experimental biases (correlations induced by the set-up) from the data :

$$C(p_1,p_2) = N * [d^6\sigma(p_1,p_2) / d^3(p_1)*d^3(p_2)] / [d^3\sigma(p_1)/d^3p_1 * d^3\sigma(p_2)/d^3p_2] \quad (1)$$

- with N being a normalization factor taking into account the different normalization for the single and double cross sections

Indeed the denominator takes into account the geometrical and detection characteristics of the device. One has however to mention that this expression is correct only if the phase space of the two coincident proton is large enough. As an example we have constructed the correlation function for the coincident protons of the H(d,2p)n reaction in which only three particles are produced in the final channel. As one can see in figure 2, the shape of the correlation function at large relative momenta depends strongly on the observation angle which is related to the momentum of the undetected neutron. This condition of large multiplicity or loose phase space constraint is however usually achieved in heavy ion reactions. Authors using this expression commonly construct the denominator by making simply the product of single cross sections measured with the set-up during the run [4,6,7,11]. **This method is however not taking into account the fact that the detection of particles in coincidence put a strong constraint on the reaction mechanism.** In fact the single cross sections of expression (1) should be derived by integrating the doubly differential cross section of the coincident particles over 4π for one particle. This is almost impossible to perform and Kopylov [21] proposed later to use the following expression :

$$1+R(q) = C * \frac{N_c\ (q)}{N_{nc}\ (q)} \quad (2)$$

- $q = (p_2 - p_1)/2$, is the modulus of the relative momentum of the two particles
- $N_c(q)$ denotes the yield of coincident pairs at relative momentum q
- N_{nc} corresponds to the yield of "non-correlated" pairs. These are generated by mixing the coincident event using particles from two diferent coincident events.
- C is a constant normalizing the correlation function to unity at large q, where no correlation is expected

The drawback of this technique is that correlation can be retained especially for small acceptance detectors. One can remove it using an iterative method [2]. An alternate method is to simulate this effect and apply it on the theoretical predictions to be compared with unchanged data. This is the procedure we have followed performing a realistic simulation of our set-up using a monte-carlo code to generate simulated correlated events for a given source radius which are written in a file, **in order to be further analyzed in exactly the same way than the experimental data.**

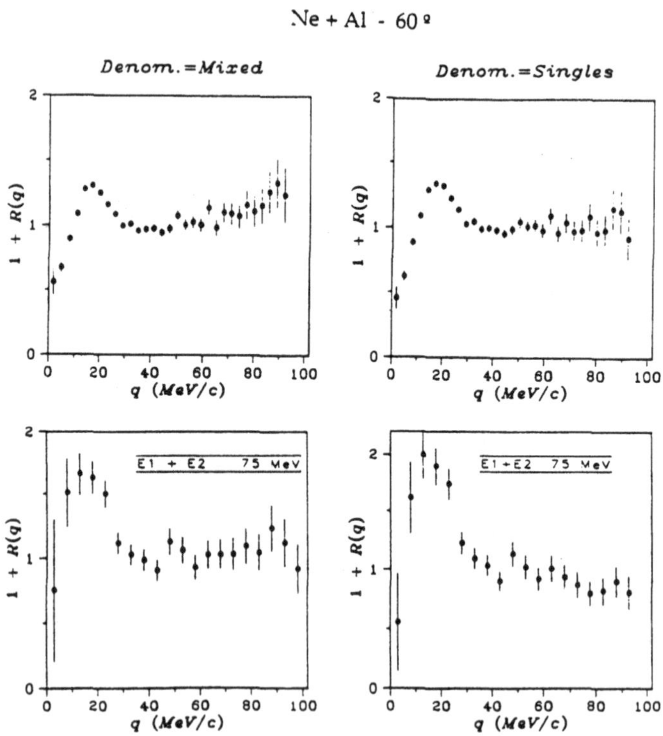

Figure 3 : Comparison of the 2 methods (singles/mixed, see text) used to contruct the denominator.

These two different expressions (1) and (2) are used by authors. We have constructed the correlation function using singles data and compared it in figure 3 with the one constructed using mixed coincidences. We see that the correlation functions differ when conditions are set on the energy of the two detected protons and that a larger enhancement with energies is observed for singles. In previous experiments performed in the intermediate energy domain, such conditions have been set in order to select the first emission steps of the collision [4,6,7]. The enhancement of the correlation function with increasing energies was interpreted as the evidence of the emission of the high-energy protons on the shortest time scale. We have not yet calculated the difference in radii between the two methods but even if it remains small one has really to be cautious in interpreting this increase. In particular if cuts are made in the phase space of the coincident protons, one has to fully simulate if these changes are partly due the response of the set-up. Indeed with the help of our simulation program, we found that despite the good angular and energy resolutions, the simulated correlation functions exhibit deviations from the initial theoretical curves showing that the correlation function does not cancel all the experimental biases mainly due to the limited acceptance of our device [12].

5. EXPERIMENTAL RESULTS

5.1 Source radii

Results measured up to now were obtained for heavy target and source radii extracted found to be of the order of the size of the whole system. Cebra et al [15] performed an experiment with proton projectile and found similar source size. We have chosen to perform experiments with rather light heavy ion systems in order to investigate smaller sources.

Figure 4 displays the results for three different systems at nearly the same angle (25°). If we consider all the results, neither the mass (i.e. the target radius) nor the detection angle [12], essentially affects the correlation function. The ^{12}C target gives slighty smaller correlation. This might be due to the fact that one is in inverse kinematics and that preequilibrium particles are emitted in the backward hemisphere. On figure 4, we have also displayed the simulated correlation functions associated to the best fit performed with a zero lifetime assumption. As one can see, the simulated correlation function reproduces not only the height but also the overall shape of the experimental one except for the region of very low relative momenta. The extracted source radii for the different systems are nearly identical but larger than the dimension of the target radius ($r_0 = 5$ fm corresponds to a ^{197}Au nuclei !!). Our results agree with those obtained by other groups in the intermediate energy domain (20-100 MeV/u) [15].

5.2 Lifetime effect

As neither the way of calculating the correlation function nor a set-up effect can be invoked to account for the large observed radii, the most natural physical explanation is to consider the lifetime of the

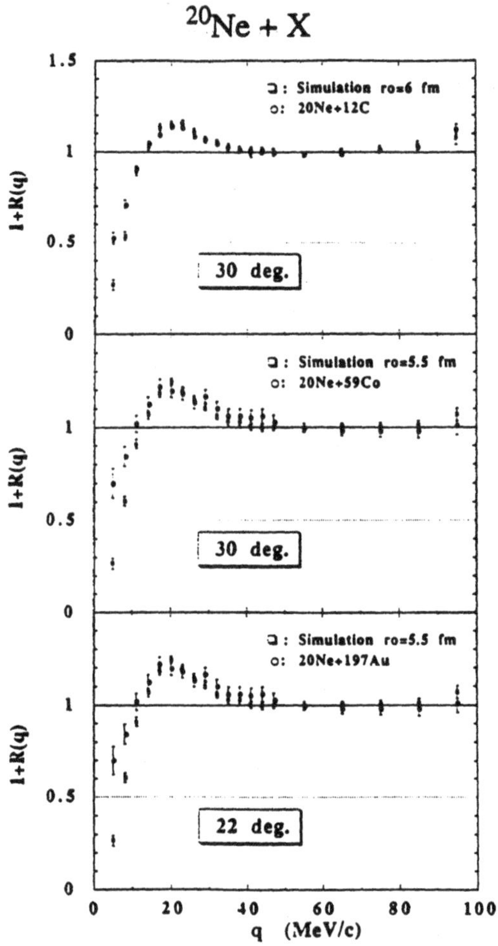

Figure 4 : Experimental correlation functions for 3 systems (Ne + C, Co, Au) (black dots). The theoritical predictions, corrected for the set-up effects, and corresponding to the best fit for a zero lifetime are also.displayed (open circles).

emitting source. This point is illustrated by figure 5 which shows that a long lifetime would result in a larger apparent size and a non spherical shape of the emitting source.

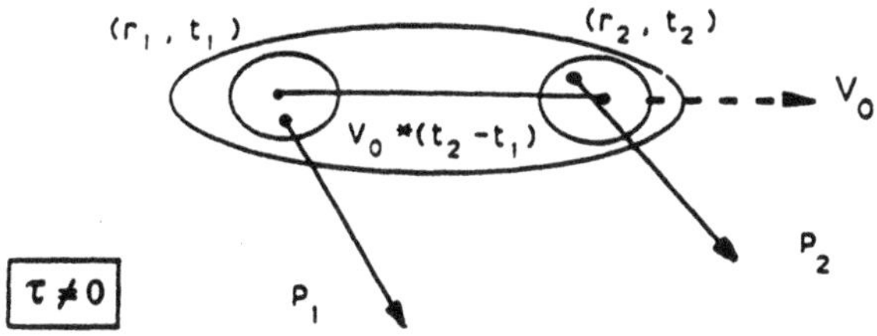

Figure 5 : Apparent shape for a long-lived source.

Koonin has first pointed out the possibility to extract lifetime information from the data by selecting different directions of the relative momentum **q** [19]. However, the reference direction proposed by Koonin made it necessary to know the source velocity. For that reason and in order to avoid an additional assumption, other methods were recently proposed to seek for a time effect. The basic idea is still to select different directions of **q** but relative to a wellknown direction. If the lifetime becomes large enough, then the correlation depends upon the angle between **q** and the reference direction **D**, the two limiting cases being **q** **D** (transverse case), and **q** // **D** (longitudinal case). A first alternative method was proposed by S. Pratt et al. [22] who takes as reference direction the total momentum P_T of the detected pair. This prescription has been applied by T. Awes et al. [23] for the system $^{32}S + ^{nat}Ag$ at 23 MeV/u, and no difference was observed between the transverse and longitudinal correlation functions ruling out the lifetime argument. More recently the same treatment was applied to data for the $^{40}Ar + ^{nat}Ag$ reaction at 44 MeV/u by D. Ardouin et al. [24] and, at variance from the former experiment, they observed a small enhancement of the longitudinal correlation function (15-20 %), as expected in case of a non negligible lifetime. Unfortunetaly, in our case given the limited acceptance of our set-up, we could not use the same method (For the transverse case, we cannot obtain values of q larger than 30 MeV/c, making the normalization impossible). We rather used an extension of Koonin's model performed by S. Adachi et al. [25], in which the spherical shape of the source associated with a lifetime is replaced by a source of ellipsoidal shape with no lifetime. In the framework of this model, the reference direction is the symetry axis z of the ellipsoid, that we considered to be the beam axis. As an example, figure 6 shows the predictions of this model for an ellipsoidal shape (R_L = 3 fm, R_T = 10 fm) for two directions of **q** (right side of the figure), compared to the results of Koonin's model for a radius of 5 fm and a zero lifetime (left side of the figure). We have performed this analysis for the $^{20}Ne + ^{12}C$ (30º) and $^{20}Ne + ^{27}Al$ (60º) reactions, whose results are displayed on figure 7. Let us first focus on the $^{20}Ne + ^{27}Al$ system. A clear enhancement of the correlation function is observed in the longitudinal case as compared to the transverse case indicating, in the framework of the model, a non spherical source with a longitudinal axis (parallel to the beam) larger than the transverse one. This corresponds to what is expected from a long-lived source, and could be considered as an evidence of a lifetime effect. However, if we turn now to the ^{12}C case, the situation is inverted. That is, we

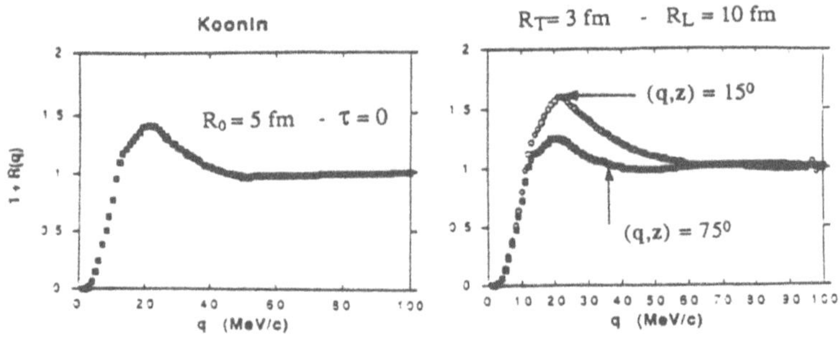

Figure 6 : Predictions of extended Koonin's model for an ellipsoidal shape and for 2 directions of q relative to the symmetry axis, compared to a spherical shape with zero lifetime.

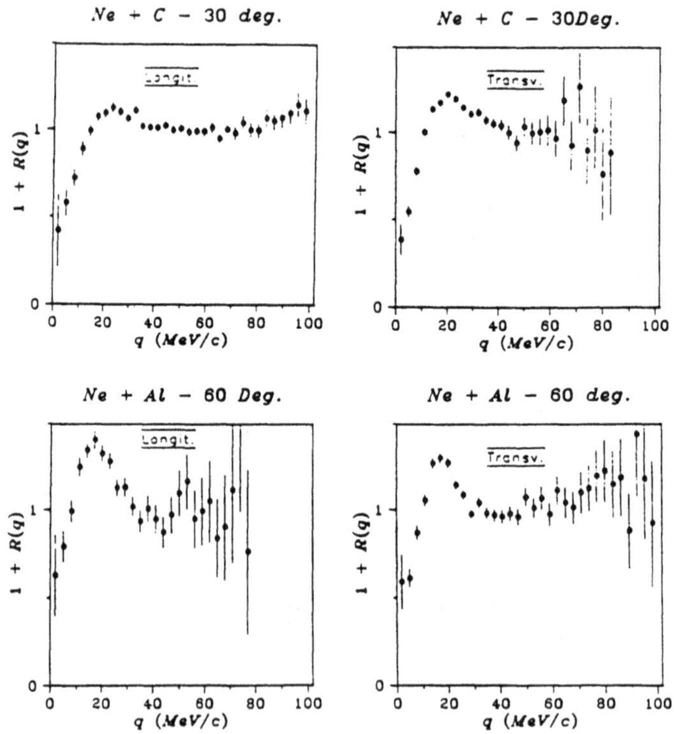

Figure 7 : Longitudinal ($0^0 < (q,z) < 30^0$ and $150^0 < (q,z) < 180^0$) and transverse($60^0 < (q,z) < 120^0$) correlation functions for ($^{20}Ne + {}^{12}C$, 30^0) and ($^{20}Ne + {}^{27}Al$, 60^0) systems.

observe an enhancement of the transverse correlation function, corresponding to a source with a pancake shape! To test an eventual set-up effect we set the same kind of conditions to simulated data. But, no difference has been observed between the longitudinal and transverse correlation functions with simulated data. The approximate values of the extracted longitudinal and transverse dimensions are respectively $R_T = 8$ fm, $R_L = 4$ fm and $R_T = 4$ fm, $R_L = 8$ fm for the $^{20}Ne + ^{12}C$ (30º) and $^{20}Ne + ^{27}Al$ (60º) systems. Independently of the conflicting behaviour of the two systems, it has to be noticed that this observed effect do not allow to recover small radii. Thus, no clear conclusion can be drawn from this attempt to seek for a lifetime effect in our data, letting the problem open.

5.3 Selection of a source

In two-particle inclusive experiments performed at intermediate energies and forward angles, the problem of the identification of the observed source(s) is always underlying. It is clear that when the source is well identified, results are more easily interpretable. Two very different sets of measurements support this idea. For experiments performed at higher energies (200-1000 MeV/u) [8] lifetime can be neglected and one is more confident to measure a pure spatial extent. Actually, a clear evolution of the extracted radii with target mass has been observed. Secondly for experiments performed at low energies [10], backward angles in direct kinematics [11,24], or forward angles in inverse kinematics [25], lifetime is the driving quantity. Indeed, these experimental conditions select an equilibrated longlived source and they have provided emission times of the source. In the other conditions, we are dealing with a mixture of different reaction mechanisms, occuring on different time scales, and the extracted radius is an average value with poor physical meaning. However, having a selection on the impact parameter is not an easy task. The commonly used selection consists of setting gates on the sum of the kinetic energies of the two detected particles. If a clear change was observed its physical meaning is not really clear. Some attempts were also undertaken [13], using the multiplicity of charged particles at forward angles, but it has been shown that this may be not a relevant quantity [14]. Therefore, it is worthwhile to perform very exclusive experiments in order to obtain reliable selection of the reaction mechanism. We have recently performed such an experiment for the system ($^{32}S + {}^{nat}Ag$) at 30 MeV/u and will present some preliminary results. On figure 8, we have displayed the total charged particle multiplicity measured with the AMPHORA detector for two different triggering conditions. The right part of figure 8 corresponds to the case of AMPHORA triggered in single. Two well separated peaks appear : the first one at very low multiplicity is associated to peripheral collisions, the second one corresponds to semi-peripheral and central collisions. The left part of the figure 8 corresponds to data recorded using EMRIC (detecting two particles) as the trigger of the experiment. The striking change is the disappearence of the low multiplicity peak. This clearly shows that coincident experiments impose a selection of the reaction mechanism which is different from the single experiments. Therefore, it confirms the choice made to use mixed events to calculate the denominator. Figure 9 shows the correlation function for three different gates on the total multiplicity of charged particles measured in AMPHORA. For the intermediate multiplicities (lower part of the figure), we observe the usual shape of the correlation function for similar sytems, with a small peak at 20 MeV/c, corresponding to a radius of 5-6 fm.

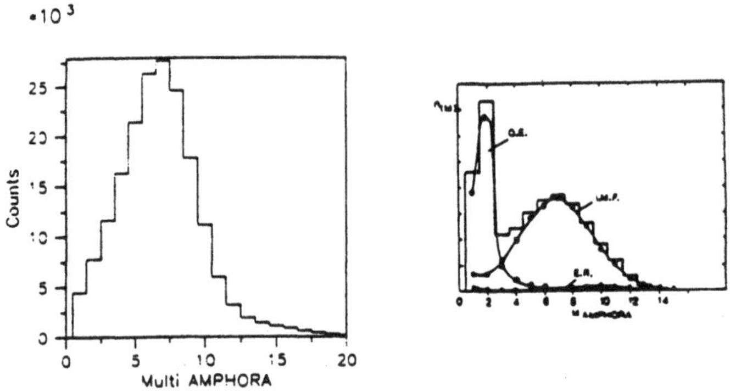

Figure 8 : Total multiplicity of charged particles in the 4π detector AMPHORA for 2 different triggering conditions (see text).

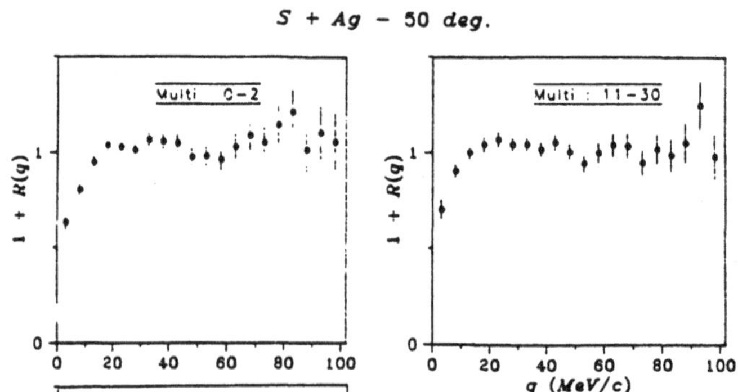

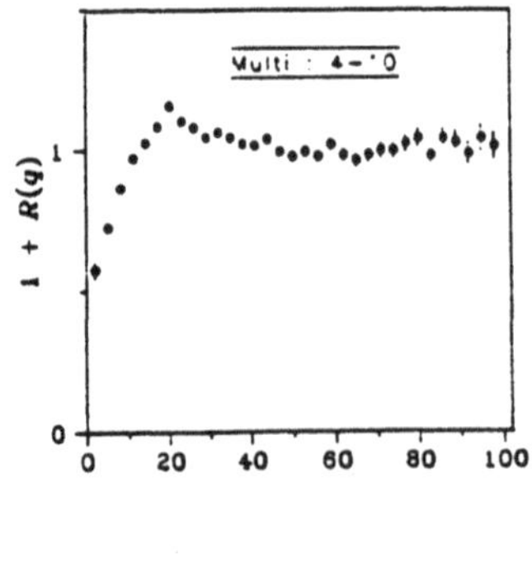

Figure 9 : Evolution of the correlation function measured in the EMRIC set-up as a function of the total multiplicity of charged particles measured in coincidence by AMPHORA device.

For higher multiplicities (right upper part), where we expect more central collisions and therefore long-lived source, the correlation function clearly becomes flatter, in agreement with results mentionned above. More surprisingly the correlation function does exhibit the same trend for the low multiplicity case. At this stage we need to analyze the data in more details to draw conclusions on these changes. In particular, the total multiplicity of charged particles is not believed to be a very reliable criterion for impact parameter selection and we have still to define more realistic ones. However, these results are rather promising.

A second way to select a source is to deal with low multiplicity reactions. In an expriment performed at Saturne on the H(d,2p)n reaction we have been able to exclusively select the elementary process of charge exchange schematized in left upper part of figure 10 using kinematical arguments. In this experiment one is then sensitive to the distribution of the neutron and the proton in the deuteron at the time of the charge exchange reaction. One is then expecting r_0 values of the order of 1.5 fm. For these results however, there is a strong kinematical constraint that will not be removed by the denominator (see section 3). At this stage we have not yet fully simulated this effect and extracted source radii. Using the polarization information one can also investigate the variation of the correlation as a function of the spin projection (right upper part of figure 10). Indeed the shape of the deuteron and the mean distance of the proton and neutron in the deuteron is related to Mz and to the tensor polarization (t_{20}) of the beam. We show in figure 10 the correlation functions measured at 7° for four different "spin state" of the beam delivered by the source. We can see that the shapes for spin states 5 and 8 are different than those for spin 6 and 7. This maybe a signature for different mean distances of the nucleons in the deuteron but one has still to analyze further this result to be sure that these changes are really due to a different distribution of the nucleons in the deuteron nuclei. Moreover we can notice that for this case where the source is perfectly selected and that the final state interaction is not disturbed by the field of the emitterat, the correlation function tends to zero at very small relative momenta. This is at variance from the other correlation functions measured for heavy ion systems with the same set-up.

6. CONCLUSION

We have presented in this contribution a set of new results measured with the EMRIC set-up in three experiments. We have used these results to discuss in a critical way some of the problems inherent of the interferometry technique when used in the intermediate energy range. From the very technical point of the construction of the correlation function up to the problem of selecting a source one can see that this physics has still to be explored. We found that the way of constructing the correlation function can give differences for limited solid angles detectors and when cuts are applied to the phase space of the detected particles. Secondly from the simulation of our data we found that a careful simulation of the effect of the set-up and data handling has to be performed to fully account for the effect of the set-up as the correlation function does not completely cancel all the experimental biases.

We have measured source size for different light heavy ion systems in the forward hemisphere and obtained values rather insensitive to the studied system and large (5-6 fm) compared to the radii of the

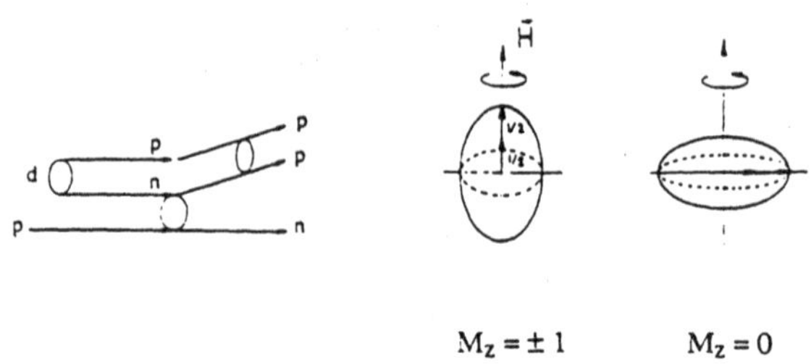

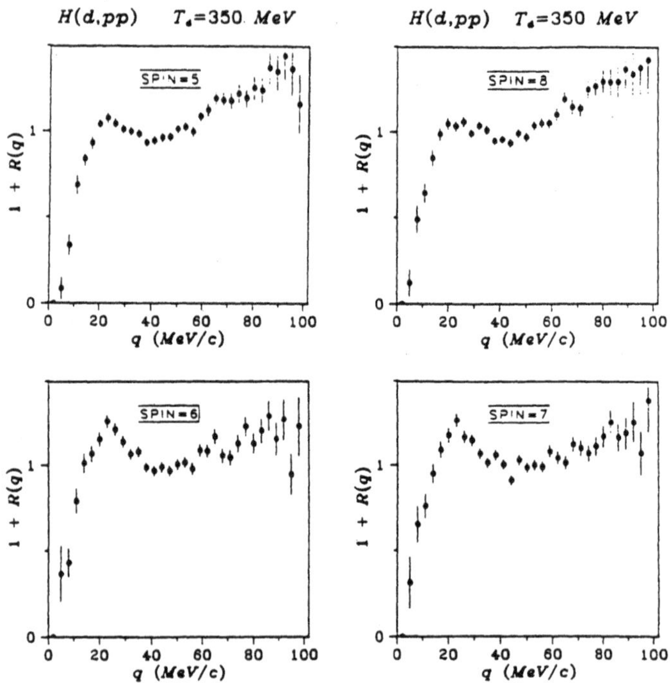

Figure 10 : Correlation function for the system H(d,pp), for 4 different "spin states" of the polarized beam of deuterons.

studied nuclei confirming the observations of D. Cebra [15]. The selection of high energy protons did not yield the small radii as might be expected for our light systems. The increase of the correlation function for the higher energies appeared to result party from a set-up effect. We tried to interpret these large values by considering the lifetime of the emitting source and indeed observed differences in the longitudinal and transverse case. However the geometry chosen for this comparison such as the effect observed have still to be investigated before drawing definite conclusions.

Finally attempts were made aiming at selecting a nuclear source. Two very different experiments were performed. First a very exlusive experiment using a 4π detector and second an experiment selecting a pseudo-deuteron as the source for the two protons. The preliminary results show a dependence of the results with conditions applied on these sets of data. In our mind this is certainly the only way to measure physical radii from experiments. Indeed one can always determine source size or lifetime emission but they will remain of small interest as long they cannot be associated to a well defined nuclear object.

References

1 G. Goldhaber et al., Phys. Rev. **120**, 300 (1960)
2 W. A. Zajc et al., Phys. Rev. **C29**, 2173 (1984)
3 J. Bartke, Phys. Lett. **B174**, 32 (1986)
4 W. G. Lynch et al., Phys. Rev. Lett. **51**, 1850 (1983)
5 H. A. Gustaffson et al., Phys. Rev. Lett. **53**, 544 (1984)
6 J. Pochodzalla et al., Phys. Rev. **C35**, 1695 (1987)
7 Z. Chen et al., Nucl. Phys. **A473**, 564 (1987)
8 P. Dupieux et al., Phys. Lett. **B200**, 17 (1988)
9 D. Fox et al., Phys. Rev. **C38**, 146 (1988)
10 P.A. DeYoung et al., Phys. Rev. **C39**, 128 (1989)
11 J. Quebert et al., Proc. of the XXVII Meeting on Nucl. Physics, Bormio, Italy (1989), p. 220
12 D. Rebreyend et al., Proc. of the XXVIII Meeting on Nucl. Physics, Bormio, Italy (1990), p. 161
13 A. Kyanowski et al., Phys. Lett. **B181**, 43 (1986)
14 M. B. Tsang et al., Phys. Lett. **B220**, 492 (1989)
15 D. A. Cebra et al., Phys. Lett. **B227**, 336 (1989)
16 F. Merchez et al., Nucl. Instr. and Meth. **A275**, 133 (1989)
17 D. Drain et al., Nucl. Instr. and Methods **A281**, 528 (1989)
18 T. Motobayashi et al., Phys. Lett. **B233**, 69 (1989)
19 S. E. Koonin, Phys. Lett. **70B**, 43 (1977)
20 G. I. Kopylov et al., Sov. J. Nucl. Phys. **19**, 215 (1974)
21 G. I. Kopylov, Phys. Lett. **50B**, 411 (1974)
22 S. Pratt et al., Phys. Rev. **C36**, 2390 (1987)
23 T. Awes et al., Phys. Rev. Lett. **20**, 1850 (1988)
24 D. Ardouin et al., Proc. of the International School Enrico Fermi, Varenna, Italy, July 1989
25 S. Adachi et al., ISN Report and private communication
26 W. G. Wong et al., MSUCL-718, 1990 and C.K. Gelbke, this conference

HARD PHOTON CORRELATIONS IN HEAVY-ION COLLISIONS

Reint OSTENDORF[1], Yves SCHUTZ

GANIL, BP 5027 14021 Caen-Cedex, France

ABSTRACT

All experimental data agree that the most likely mechanism for the production of high energy photons ($E_\gamma \geq 20$ MeV) in heavy ion collisions is bremsstrahlung radiation emitted in first chance nucleon-nucleon collisions. In order to measure the space-time extent of the source of the hard photons, we propose measuring their correlation. We present here a simulation demonstrating the feasibility of the experiment using TAPS.

1. PRODUCTION OF HIGH ENERGY PHOTONS

The observation of hard photons of energy greater than 20 MeV produced in heavy ion collisions at intermediate bombarding energies (20 MeV $\leq E_B \leq$ 100 MeV/u) has triggered a lot of work in order to understand the origin of these photons. Experimentally, the basic characteristics of the emission of hard photons have been determined. The photons are emitted from a source moving with half the beam velocity, which coincides with the nucleon-nucleon center of mass. The angular distribution in the source frame is nearly isotropic with a small dipole component. It has thus been suggested that the photons are associated with bremsstrahlung radiation mainly produced in individual nucleon-nucleon collisions. Systematic studies of hard photon production enabled its variation with bombarding energy and with the mass of the projectile and target to be deduced. By assuming that the photons are produced exclusively in first chance neutron-proton collisions, the variation of the production of hard photons in heavy ion collisions can be described by a single parameter: the probability to produce a photon per elementary neutron-proton collision [1]. The variation of this probability with energy is compared to the same probability for free neutron-proton

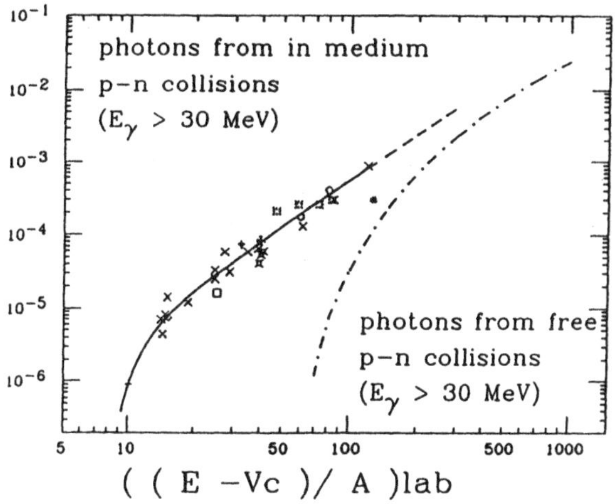

Figure 1: *Emission probability for photons with $E_\gamma \geq 30$ MeV per in-medium proton-neutron collision. For comparison the proton yield from free proton-neutron collisions is shown. From reference [1].*

collisions in figure 1. The relative displacement of the two curves reflects the Fermi motion of the nucleons which adds to the relative motion of the colliding nucleons. A complete review of the experimental status of hard photon production can be found in references [1,2].

Several theoretical models have been proposed to explain the origin of the experimentally observed hard photons. One approach suggests that coherent nucleus-nucleus bremsstrahlung is a possible production mechanism. In this model the high energy photon is produced in the sudden deceleration of both the projectile and target nucleus. However, based on the measured angular distribution and the source velocity of the hard photons, this mechanism does not make the dominant contribution to the hard photon production. A different approach suggests that incoherent nucleon-nucleon bremsstrahlung is the main production mechanism of high energy photons. In this picture high energy photons are produced either cooperatively, with several proton- neutron collisions adding up incoherently to produce a high energy photons, or as bremsstrahlung from a single proton-neutron collision, with the Fermi-motion of the nucleons providing the extra boost needed to produce the high energy photon. This approach is supported by essentially all available experimental data. Calculations of the BUU type indicate that hard photons are produced mostly in first chance collisions and that subsequent collisions do not produce hard photons efficiently. A detailed review of the various theoretical approaches is found in references [1,2].

[1]supported by the Instituut voor Fundamenteel Onderzoek der Materie, the Netherlands

It seems very likely that the hard photons emitted in heavy ion collisions at intermediate energy are emitted in a very early stage of the collision. Their yield is related to the number of individual nucleon-nucleon collisions taking place in the collision and could thus be a measure of the impact parameter. Finally their energy distribution reflects the Fermi energy distribution of the colliding nucleons and thus is a photograph of the phase space distribution of the nucleons within the nucleus. What makes the photons particularly interesting is that they only interact weakly (through the electromagnetic force) with the surrounding nuclear medium and thus provide an undistorted probe suffering from neither absorption nor final state interactions.

So far, no experimental evidence exists to support the first chance collisions hypothesis. Hard photons could as well be emitted at a later stage of the collision when the participant zone is already thermalized. We therefore propose to measure the space-time extent of the hard photon source using the Bose-Einstein correlation method.

2. PHOTON-PHOTON CORRELATION

Since we are dealing with two identical bosons we can apply the technique of second order interferometry to correlated photons emitted in heavy ion collisions at intermediate energies. Compared to other particles commonly used in nuclear physics to construct the correlation function, photon-photon correlation is less complicated because corrections for final state interactions or reabsorption in the nuclear medium are not needed. Nevertheless a difficulty arises from the fact that photons have a polarization which has a large influence upon the correlation function.

In general the correlation function is defined as:

$$C_2\left(\tilde{k}_1, \tilde{k}_2\right) = \frac{P\left(\tilde{k}_1, \tilde{k}_2\right)}{P\left(\tilde{k}_1\right) P\left(\tilde{k}_2\right)} \tag{1}$$

where $\tilde{k}_1$ and $\tilde{k}_2$ are the four-momentum vectors of the two photons and $P\left(\tilde{k}\right)$ is the one photon inclusive distribution and $P\left(\tilde{k}_1, \tilde{k}_2\right)$ is the two photon inclusive distribution.

Taking into account the polarization of the photon and under the following three assumptions:

- coherent electromagnetic current can be neglected,
- all collisions produce the same magnitude of electromagnetic current,
- currents from different collisions have uncorrelated directions,

D. Neuhauser has shown [3] that the two photon correlation is:

$$C_2\left(\tilde{k_1},\tilde{k_2}\right) = 1 + \frac{1+\left(\hat{\mathbf{k_1}}\cdot\hat{\mathbf{k_2}}\right)^2}{2}\frac{\left|\rho\left(\tilde{k_1}+\tilde{k_2}\right)\right|^2+\left|\rho\left(\tilde{k_1}-\tilde{k_2}\right)\right|^2}{2} \tag{2}$$

where $\hat{\mathbf{k_1}}$ and $\hat{\mathbf{k_2}}$ are unit vectors and $\rho\left(\tilde{k}\right)$ is the Fourier transform of the normalized space-time density $\rho(\vec{r},t)$ of the photon source. As usual, we assumed a source of symmetric Gaussian shape:

$$\rho(\vec{r},t) = \frac{1}{\pi^2 R^3 \tau}\exp\left(-\frac{r^2}{R^2}-\frac{t^2}{\tau^2}\right) \tag{3}$$

where R and τ characterize the radius and lifetime of the source. The maximum of the correlation function is, for $\tilde{k_1}=\tilde{k_2}=\tilde{k}$:

$$C_2\left(\tilde{k},\tilde{k}\right) = 1.5 + \frac{\left|\rho\left(2\tilde{k}\right)\right|^2}{2} \tag{4}$$

For photons of energy larger than 40 MeV, the second term in equation 4 can be neglected and the maximum of the correlation function at zero relative momentum is equal to 1.5 [3]. The reason why the maximum of the correlation function is not equal to 2, as commonly expected for interference of identical bosons, is due to the fact that photons with orthogonal polarization do not interfere.

3. EXPERIMENTAL SETUP AND SIMULATIONS

In March 1990 we ran an experiment at GANIL to measure the correlation between hard photons emitted in the heavy ion collisions of ^{129}Xe at 45 MeV/u on ^{197}Au. The photons were measured with the TAPS [2] multidetector. For this experiment, the TAPS detectors were completed by a set of BaF_2 detectors from the Oak Ridge National Laboratory, giving a total number of 247 hexagonal shaped BaF_2 detectors. The detectors were assembled in 13 blocks of 19 detectors, the central detector looking at the target from a distance of 42.4 cm. They covered a solid angle of nearly 30% of 4π and spanned angles between 60° and 160° in θ for all ϕ with respect to the beam direction. Each individual detector covered 8° in θ and ϕ. During the experiment a hardware trigger was set up to accept only the events where two high energy photons ($E_\gamma \geq 6$ MeV) are detected in a single block or in two different blocks. Single photon events have been recorded with a scaled down rate.

For the simulation, we have described the exact geometry of the detectors as used in the experiment. The detector response was generated by the code

[2] TAPS is a collaboration between GANIL, the University of Gießen, the KVI of Groningen, the GSI of Darmstadt and the University of Münster

GEANT3 [4]. The measured and calculated response function of BaF_2 detectors have been checked to match [5].

The two photon events were generated assuming that, in the half beam-velocity frame, their energy spectrum has an exponential shape with a slope of 12.1 MeV and that their angular distribution is isotropic with a 30% dipole component [2]. Only photons with energies larger than $E_\gamma \geq 30$ MeV were considered. Furthermore, events where the two photons hit the same 19-pack were rejected. The detected photon energy was taken as the total energy deposited in the whole 19-pack, and the direction of the photon has been taken from the position of the detector where the largest amount of energy is deposited. Correlation between photons is taken into account using equation 2. The denominator of equation 1 was calculated by generating events with two photons having the same energy and angular distribution but having a flat correlation function. Experimentally, correlated photons have been measured with sufficient statistics for energies between 20 MeV and 60 MeV. At lower energies the contribution from the decay of giant resonances and of thermal photons is dominant. At higher energies the photon spectrum is contaminated by photons from the decay of π^0. We have estimated that, within this energy range and by requiring that the correlation function be measured up to at least twice its width, the smallest source size we can measure is R = 5 fm or τ = 5 fm/c. On the other hand, because of the energy resolution of our detectors (FWHM ~ 13% $\cdot E_\gamma$), the largest source size we can measure is R = 25 fm or τ = 25 fm/c.

Figure 2 a shows the generated correlation function (about 30,000 events) as a function of the relative four-momentum for R = 8 fm and τ = 8 fm/c (i.e. a source lifetime of 16 fm/c). The spatial extent represents the size when one reaches maximum overlap in the ^{129}Xe + ^{197}Au system and the time extent is the lifetime of the source as deduced from calculations in reference [1]. Figure 2 b shows the same correlation function as in figure 2 a with the generated events passed through our detection filter. We have then fitted the correlation function with the following expression:

$$f(x) = 1 + N \cdot \exp\left(-\frac{x^2}{x_0^2}\right) \tag{5}$$

The values of N and x_0 obtained for the "raw" spectrum and for the "detectors filtered" spectrum are listed in table 1. It should be noticed that the point for zero relative momentum is spurious and has been excluded from the fit. Qualitatively the correlation pattern is conserved in the filtered spectrum. But we observe a rather strong change in the width of the correlation function. This is due to the bias introduced by the limited coverage of the solid angle of our detector system and to the energy resolution of the BaF_2 detectors. At this stage we have not yet estimated the relative importance of the two effects. It is thus too early to say if an optimum detector geometry would give a width closer to the real width of the correlation function. At least, we have learnt that the measured width should be

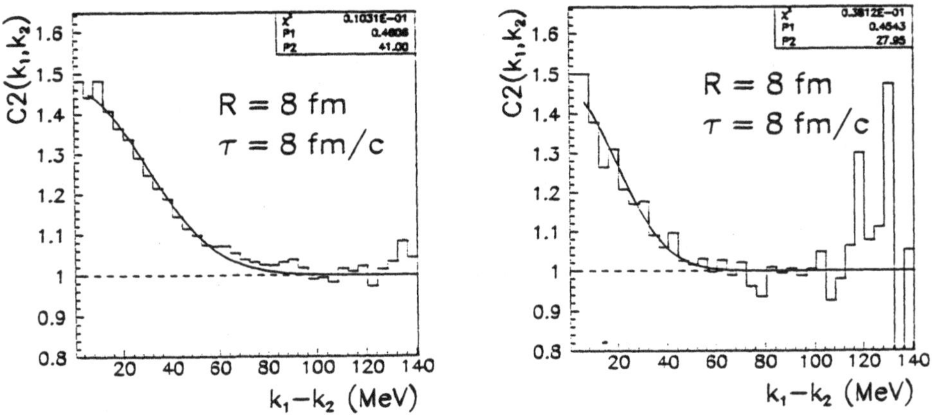

a) b)

Figure 2: *Simulated correlation function as a function of the relative four-momentum for $R = 8$ fm and $\tau = 8$ fm/c, a) from raw events b) from events measured with TAPS.*

	R		F	
	N	x_0	N	x_0
R=8fm,τ=8fm/c	0.46	40.9	0.45	28.0
R=2fm,τ=8fm/c	0.33	96.8	0.38	86.3
R=8fm,τ=15fm/c	0.51	36.1	0.39	28.5

Table 1: *Fit parameters of* **R***aw and detector* **F***iltered correlation functions for various source sizes*

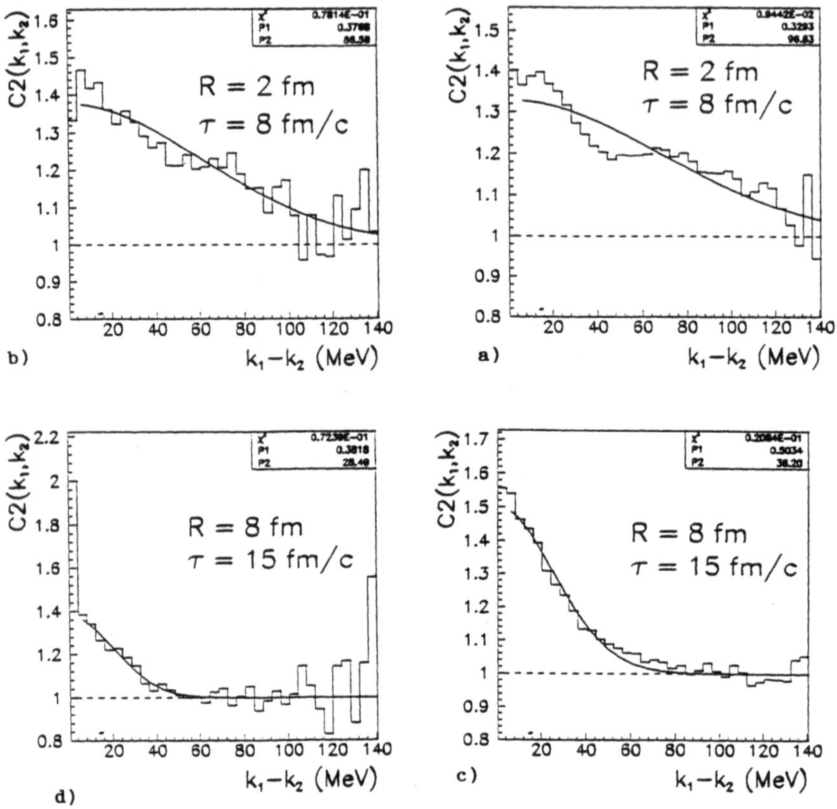

Figure 3: *Correlation functions constructed from detector filtered events for R = 2 fm, τ = 8 fm/c a) raw, b) filtered and R = 8 fm, τ = 15 fm/c c) raw, b) filtered.*

taken with great care and that it is necessary to fold the correlation function with the response function of the detector system before one can extract values for the source size.

To check the sensitivity of the method we have performed the same calculation assuming very different source sizes: R = 2 fm, τ = 8 fm/c (figure 3 a,b) and R = 8 fm, τ = 15 fm/c, i.e. a source lifetime of 30 fm/, (figure 3 b,c). The first set of parameters would correspond to a large impact parameters and to the first chance nucleon-nucleon collision hypothesis and the second set to a central collision and emission from a thermalized source. The result of the fit to expression 5 are listed in table 1. We observe that our experimental setup will be sensitive enough to distinguish only different values for R. To distinguish different values of τ it would be necessary to measure the correlation as a function of relative energy at fixed relative angle. Nevertheless, at this stage of the data analysis we are not yet sure if the statistics will be sufficient to measure separately the correlation function for relative energy and for relative three-momentum. The bump which is observed from 60 MeV relative momentum on the raw correlation function in figure 3 c can be easily understood by examining the expression for the correlation function in equation 2. Since we only take photons with energies larger than 30 MeV, anti-parallel photons contribute only above 60 MeV relative momentum and they contribute strongly through the $\left(1+\left(\hat{\mathbf{k}}_1\cdot\hat{\mathbf{k}}_2\right)^2\right)$ factor. This bump would vanish if we would consider all photon energies. The bump is also present in the spectra calculated for larger values of R, although it is much weaker. This is because for these source sizes the factor $\left|\rho\left(\tilde{k}_1-\tilde{k}_2\right)\right|$ is dominant. In the filtered spectra the bump is reduced because our detector setup is relatively less efficient for the detection of photons emitted back to back.

4. CONCLUSION

The hard photons emitted in heavy ion collisions at intermediate energies are expected to be a unique probe which carries undistorted information on the dynamics of the collision. It is furthermore assumed that they are emitted incoherently from individual nucleon-nucleon collisions at a very early stage of the collision. They should therefore show Bose-Einstein correlation effects. We have shown that it is reasonable to measure these effects and that our detector system does not distort dramatically the shape of the correlation function. Nevertheless, to interpret the data a good knowledge of the detector system response function is crucial. The simulation shows that our measurement should be sensitive enough to distinguish between extreme source sizes and should therefore be able to provide direct evidence for the origin of the hard photons. Nevertheless, we should notice that the simulations were performed in an ideal environnement. We have reasonably assumed that we will be able to remove from the photon spectrum the contamination due to neutrons, charged particles and π^0.

References

[1] W. Cassing, V. Metag, U. Mosel, and K. Niita, Physics Reports 188 (1990) 363.

[2] H. Nifenecker, and J.A. Pinston, Prog. Part. Nucl. Phys. 23 (1989) 271.

[3] D. Neuhauser, Phys Lett B182 (1986) 289

[4] R. Brun, F. Bruyant, A.C. McPherson, and P. Zanirini, GEANT3, CERN Report DD/EE/84-1 (1986), unpublished.

[5] R. Novotny, S. Riess, R. Hingmann, H. Ströher, R.D. Fischer, G. Koch, W. Kühn, V. Metag, R. Mülhans, U. Kneissl, W. Wilke, B. Haas, J.P. Vivien, R. Beck, B. Schoch and Y. Schutz, Nucl. Instrum. Methods Phys. Res. A 262 (1987) 340.

MEASUREMENTS OF SMALL ANGLE PROTON-PROTON CORRELATIONS IN INTERNAL TARGET EXPERIMENTS AT DUBNA SYNCHROPHASOTRON

V.A.Budilov, A.Filipkowski, A.Golembiewski, V.I.Ilyuschenko, A.Korejwo, P.Kozma, A.P.Laritcheva, V.A.Nikitin, P.V.Nomokonov, M.Traikova, N.K.Zhidkov
Joint Institute for Nuclear Reserch - Dubna - U.S.S.R

A.Kotus, A.Nawrot, M.Szawlowski, I.P.Zielinski
Institute for Nuclear Studies - Warsaw - Poland

G. Sidor, M.Surala, A.Turowiecki, Z.Wilhelmi, J.Zlomanczuk
Institute of Experimental Physics of Warsaw University
Warsaw - Poland

Abstract

Small angle correlations of protons emitted in collisions of protons, deuterons and ^{4}He ions with carbon and gold nuclei in the energy range of several GeV were measured using plastic scintillator hodoscopes and NaI(Tl) detectors. The detectors were placed at large scattering angles ranging from 60 to 115 degrees. It was found that for both targets the two-proton correlation function depends on the emission angle and that this dependence is stronger for the carbon target than for the gold one. The extracted source radii are in the limits of errors constant with respect to the average momentum of emitted pairs.

The study of correlations between identical particles emitted in nuclear interactions enables one to determine the spatial and temporal sizes of emitting regions[1-3]. Since these quantities play a substantial role in testing various reaction models they have been determined in a number of experiments for different energies and projectile-target combinations[4-6]. At medium and high energies mostly bubble chambers were used. Since it is rather difficult to obtain good statistics with this technique, several attempts were made to carry out electronics experiments to overcome this drawback[5,6].

Starting from 1986 our collaboration has built a detection system consisting of two sets of plastic scintillator and NaI(Tl) detectors capable of registering pairs of light particles. The apparatus, which is schematically shown in Fig.1, was placed at the internal target of the Dubna synchrophasotron. Due to the geometry of the accelerator linear section we had to place the detectors at fairly large distances from the target which resulted in a small solid angle. It is illustrated by the data of Table 1 containing the energy and angle acceptance of the apparatus[7]. The whole system was tested in the 1-st half of 1987. After necessary corrections two 3-week runs were carried out in 1987 and 1988 devoted to measurements of small and large angle proton-proton correlations in the a+b $\rightarrow$ 2p+X reactions where a stands for protons, deuterons and alphas and b for carbon and gold. The internal targets were made of 2-4 μm thick polyethylene (CH_2) and gold foils of 0.5 cm^2 area. The energy resolution of the detectors was good enough to resolve the hydrogen isotopes.

Small solid angle of the apparatus forced us to work at high

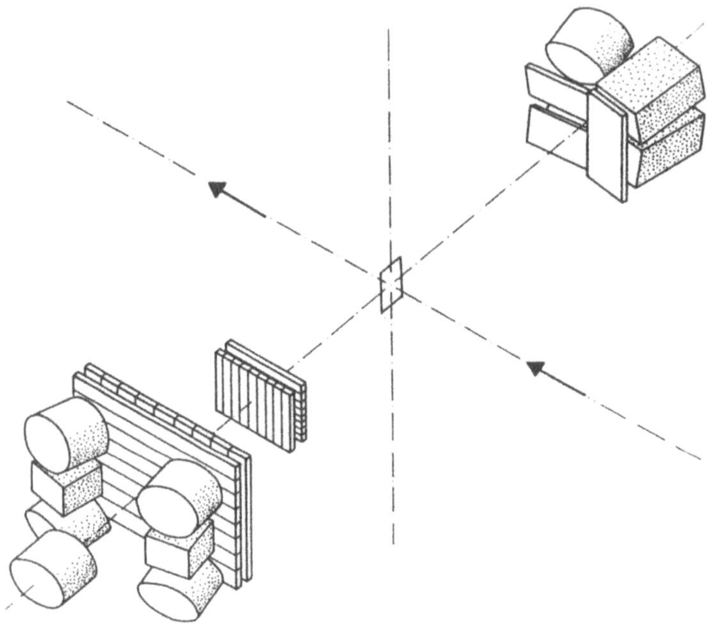

Fig.1. Schematic layout of the experimental setup.

luminosities of the order of $10^{33} cm^{-2} s^{-1}$ in order to achieve a reasonable counting rate. At such luminosities there is a problem of mixing particles from different projectile-target interactions or, in other words, accidental coincidences. This is illustrated by Fig.2 showing the distribution of the difference of emission times Δt of two registered protons. Protons emitted in the same interaction give the same time difference which manifests itself as a clear peak. From the analysis of the distribution presented in Fig.2 follows that the

Table 1. Energy and angle acceptance of the experimental setup.

Parameter	Set 1	Set 2
angular acceptance in θ	97.5°-104.5°	69.5°-74.5°
angular acceptance in ϕ	0°-7°	180°-185°
angular resolution	0.2°	0.75°
minimum energy of registered protons	57 MeV	42 MeV
maximum energy of registered protons	240 MeV	185 MeV

number of accidental coincidences is quite high but in the peak area does not exceed 15%. This admixture of uncorrelated background in the data was carefully controlled during the off-line analysis and the final distributions were corrected for it.

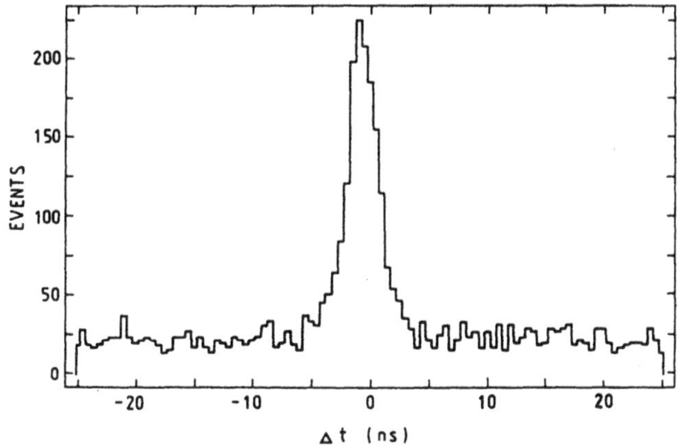

Fig.2. Distribution of the difference of emission times Δt for a pair of registered protons.

As it is seen from Table 1 the observation angles for both sets of detectors were well defined and did not overlap. In consequence, we were able to look for the angular dependence of the two-proton correlation functions[7].

The experimental two-proton correlation functions were obtained using the following definition[8]:

$$C_2(k^*) = F\, N_c(k^*)/N_{nc}(k^*),$$

where $k^* = |\bar{p}_1 - \bar{p}_2|/2$ is a half of the two proton momenta difference calculated in their center of mass, $N_c(k^*)$ is the k^* distribution for both protons belonging to the same event (correlated), $N_{nc}(k^*)$ is the same distribution but for protons belonging to different events (non-correlated) and F is the normalization constant obtained under the condition that the average value of C_2 for k^* in the 80 - 120 MeV/c range is equal to 1.

The correlation functions obtained for the $^4He+C \longrightarrow 2p+X$ and $^4He+Au \longrightarrow 2p+X$ reactions at 2 GeV/c/nucleon momentum are shown in Fig.3 taken from Ref.7.

As it has been shown in Ref. 2 and 3 the Coulomb repulsion and the anti-symmetry of the p-p wave function in the most probable spin triplet state tend to reduce the p-p correlation at small angles, while the short range nuclear attraction in the spin singlet state increases it. The resulting p-p correlation curve shows a minimum at small values of k^* and a broad peak at about 20 MeV/c. The height of this peak strongly depends on the size of the source and, if the emission time is large, or the source is not isotropic, or both, the height can also depend on the relative orientation of k^* and the p-p CMS velocity.

The discussed gross features of the p-p correlation function are clearly seen in the experimental distributions of fig. 3 despite of

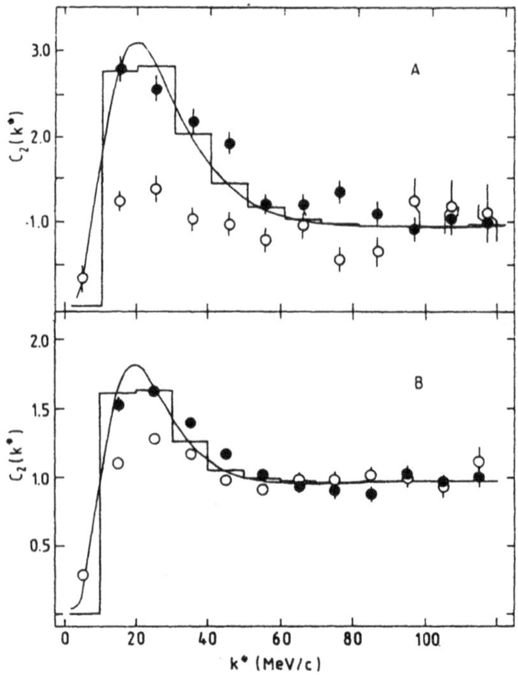

Fig. 3. The two-proton correlation functions for the ^{4}He+C (A) and ^{4}He+Au (B) reactions. Circles represent experimental data for the 72° (full) and 101° (open) emission angles. The histograms represent the simulated p-p correlation functions corresponding to the Lednicky-Lyuboshitz curves calculated for τ=1 fm/c, r_o = 1.56 fm (A) and 2.53 fm (B) and shown as solid lines.

the distortion caused by the apparatus acceptance and response. The interesting aspect of the data is a large difference between experimental distributions corresponding to 72° and 101° emission angles. Since these distributions cannot be directly compared because of the different detector granularities at both angles we have used a two-step procedure to interpret the data. In the 1-st step the theoretical two-proton correlation functions were computed using the Lednicky-Lyuboshitz model[3] for the emission points distributed in space and time according to the four-dimensional Gaussian law with the space and time parameters r_o and τ. The calculations were done for the proton CMS velocity v = 0.4 c, τ = 1 fm/c and the relative proton momentum parallel to v. This choice of parameters is compatible with our experimental data. The requirement that k^* is parallel to v follows from the apparatus acceptance given in Table 1.

In the next step a careful Monte Carlo study of the influence of the detector acceptance and the energy and angle resolution on the measured p-p correlation functions was done. In this study we have used the same sets of data for which the experimental correlation functions were found to ensure the proper proton momentum distributions.

As a result of the Monte Carlo study the theoretical curves obtained in the 1-st step were converted into histograms representing simulated p-p correlation functions with the apparatus features taken into account. The simulated distributions concerning the r_o values which represent the best fits to the experimental data for 72° are shown as histograms in fig. 3. The original curves corresponding to the same values of r_o are represented by solid lines.

The values of r_o obtained by comparison of experimental and simulated distributions are given in Table 2. The errors are purely statistical. Since the solid angle of the apparatus was small the statistics in some cases was poor which resulted in large errors.

Table 2. The space parameters r_o obtained by comparison of experimental and theoretical two-proton correlation functions

Reaction	Projectile momentum [GeV/c/nucleon]	Average proton momentum [GeV/c]	Average observation angle	r_o [fm]
p+C	5.1	0.441	72°	$1.66^{+0.27}_{-0.28}$
d+C	2.0	0.419	72°	$1.68^{+0.16}_{-0.14}$
^{4}He+C	2.0	0.431	72°	$1.56^{+0.10}_{-0.10}$
p+C	5.1	0.474	101°	$2.35^{+0.44}_{-0.41}$
d+C	2.0	0.458	101°	$2.55^{+0.62}_{-0.63}$
^{4}He+C	2.0	0.465	101°	$3.21^{+1.37}_{-0.37}$
p+Au	5.1	0.445	72°	$2.55^{+0.27}_{-0.20}$
d+Au	2.0	0.428	72°	$1.99^{+0.17}_{-0.21}$
^{4}He+Au	2.0	0.442	72°	$2.53^{+0.06}_{-0.06}$
p+Au	5.1	0.480	101°	$2.97^{+0.30}_{-0.18}$
^{4}He+Au	2.0	0.469	101°	$3.23^{+0.35}_{-0.22}$

For ^{4}He - induced reactions the registered number of proton pairs was sufficiently large to observe the r_o dependence on the average proton momentum (defined as a result of averaging of $|p_1+p_2|/2$ over the set of events for which the correlation function has been built). In the case of the gold target it turned out to be possible for both observation angles but for the carbon target only for the angle of 72°. The results obtained for both targets are shown in Fig.4. It is seen that for the gold target r_o is constant in the limits of errors for both angles but the average value corresponding to 101° is nearly 30% greater than that corresponding to 72° emission angle. For the carbon target and the emission angle of 72° r_o slightly decreases when proton momenta are increased. The average value of r_o corresponding to 101° is twice as large as that for the 72° emission angle.

Since it was relatively easy for us to get the synchrophasotron beam time, and initial analysis proved that it was possible to measure

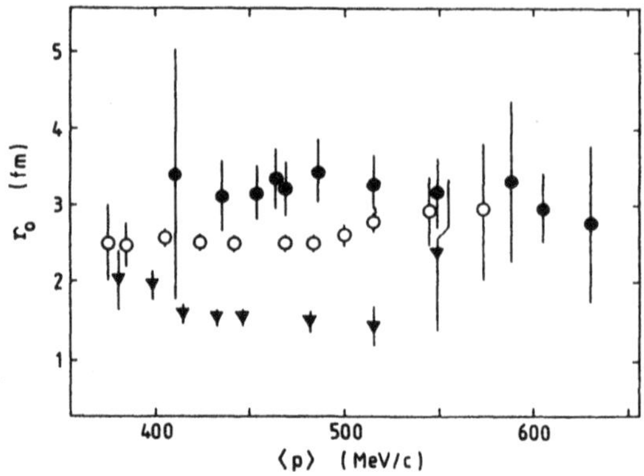

Fig.4. Dependence of r_o on the average proton momentum. The circles represent the ^4He+Au reaction for 72° (full) and 101° (open) emission angles and the triangles the ^4He+C reaction for 72° emission angle.

the p-p correlations in internal target experiments, we decided to upgrade our experimental setup by placing the detectors closer to the target. The schematic layout of the new configuration is shown in Fig.5. In the 2-nd half of 1989 a 5 weeks long run was carried out. The goal was to measure the two-proton correlation functions in interactions of proton, ^4He, and ^6Li beams with carbon and gold targets with statistics large enough to study the dependence of the correlation function on the relative orientation of k^* and p-p CMS velocity. One of the immediate effects of changing the detector configuration was the reduction of the luminosity by a factor of 20. This in turn reduced the rate of accidental coincidences what can be seen from Fig.6 showing the distribution of the difference of emission times of two registerd protons.

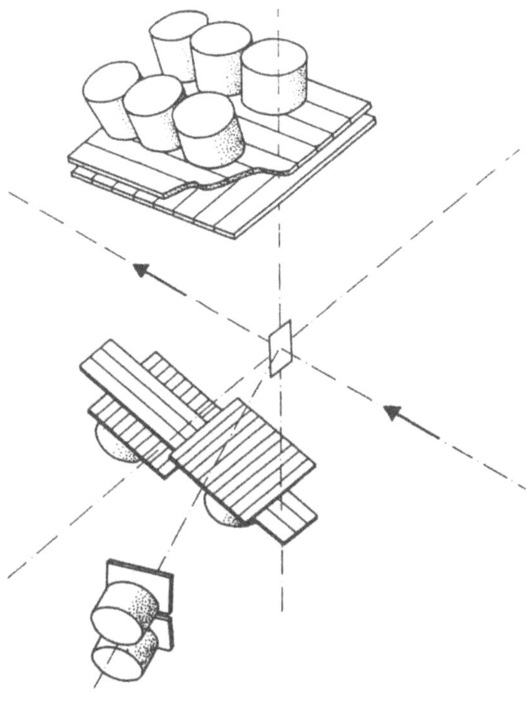

Fig.5. Schematic layout of the new experimental setup.

As it was originally pointed out by Koonin[2] and later by Pratt[9] the study of the dependence of the correlation function on the relative

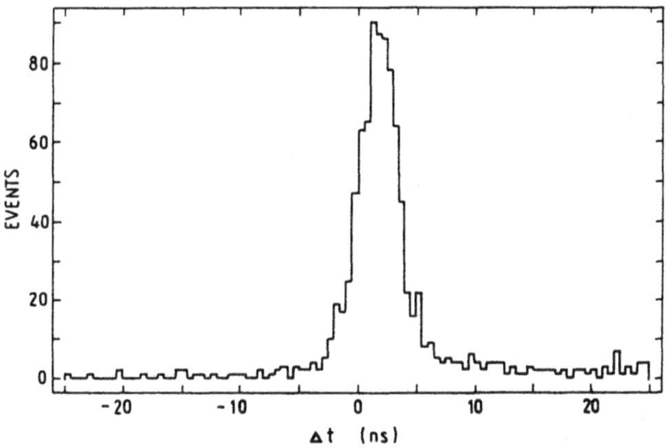

Fig.6. Distribution of the difference of emission times Δt for a pair of registered protons.

orientation of k^* and v (p-p CMS velocity) could provide information on the source shape and the emission time. In order to illustrate how strong is this dependence, we have calculated the two-proton correlation functions for k^* parallel and perpendicular to v. The calculations were done for a spherically symmetric source characterized by r_o = 1.45 fm (which corresponds to the carbon nucleus), v = 0.4 c and two emission times: τ = 1 and 10 fm/c. The calculated functions are shown in Fig.7. One can see that for τ = 1 fm/c both functions are practically the same (curves A and C in Fig.7) while for τ = 10 fm/c they are differrent (curves B and D) and it seems that the difference is large enough to be measured in a high statistcs experiment.

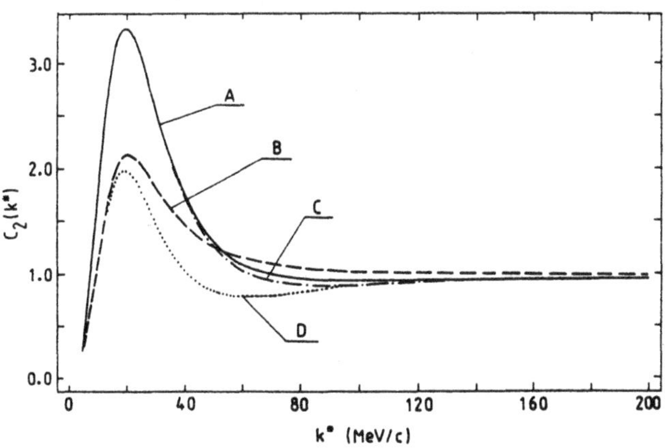

Fig.7. The two-proton correlation functions calculated for r_o=1.45 fm and v=0.4 c. Curves A,C correspond to τ=1 fm/c and B,D to τ=10 fm/c. Relative momentum k^* is parallel to v for curves A,B and perpendicular for C,D.

The experimental two-proton correlation functions measured in the new configuration for p+C interactions are shown in Fig.8. The 3 sets of points correspond to 3 conditions imposed on the relative orientation of k^* and v: full diamonds - all orientations , empty diamonds - angle between k^* and v smaller than 30°, full triangles - angle between k^* and v greater than 60°. The data are very preliminary and contain only about 15% of the total amount of events collected for this reaction so they cannot be used to draw any conclusions about the difference between the correlation functions. However they illustrate the apparatus acceptance and it looks that it is sufficiently large in both cases i.e. for k^* parallel and perpendicular to v.

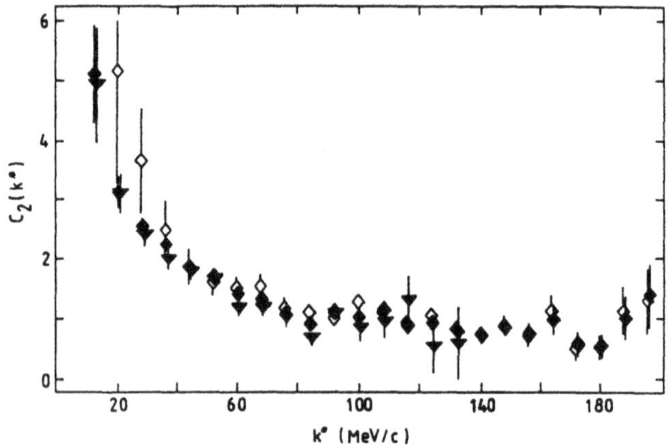

Fig.8. Experimental two-proton correlation functions.

To summarize, we have observed large differences of the two-proton correlation functions and corresponding source sizes for the 72° and 101° emission angles. For fixed angles the source sizes are constant (in the limits of errors) with respect to the average momentum of emitted protons. The above observation could be helpful in the interpretation of the bubble chamber data showing an increase of r_o with decreasing of the proton momenta since in these experiments the emission angle was not fixed and, on average, faster protons are emitted at smaller angles.

Preliminary analysis of the experimental data concerning the dependence of the p-p correlation functions on the relative orientation of k^* and v shows that both the apparatus acceptance and the statistics will probably allow us to draw some conclusions about the emission times of the sources.

We would like to express our gratitude to dr.R.Lednicky for his computer code to calculate the two-proton correlation functions and to dr. J. Pluta for helpful discussions.

This work was partly supported by the Polish Central Program for Fundamental Research CPBP 01.06.

References:

1. G. I. Kopylov, M. I. Podgoretsky, *Yad. Fiz.* **18**, 656 (1974).
2. S. E. Koonin, *Phys. Lett.* **70B**, 43 (1977).
3. R. Lednicky, V.L.Lyuboshitz, *Yad. Fiz.* **35**, 1316 (1982).
4. J. Pluta, JINR E1-88-754, Dubna, 1988 and references therein.
5. F. Zarbakhsh et al. *Phys. Rev. Lett.* **46**, 1268 (1981).
6. H. A. Gustafsson et al. *Phys. Rev. Lett.* **53**, 544 (1984).
7. V. A. Budilov et al. to be published in *Phys. Lett.* **B**.
8. G. I. Kopylov, *Phys. Lett.* **B50**, 472 (1974).
9. S. Pratt,M.B.Tsang, *Phys. Rev.* **C36**, 2390 (1987).

Proton Pictures of $^{40}Ar+^{197}Au$ Reactions at E/A=200 MeV

G.J. Kunde[a], J. Pochodzalla[b], E. Berdermann[a], B. Berthier[c], C. Cerruti[d], C.K. Gelbke[e], J. Hubele[a], P. Kreutz[b], S. Leray[d], R. Lucas[c], U. Lynen[a], U. Milkau[a], C. Ngô[d], C.H. Pinkenburg[b], R. Raciti[f], H. Sann[a], W. Trautmann[a],

[a] *Gesellschaft für Schwerionenforschung, D-6100 Darmstadt, F.R.G.*
[b] *Institut für Kernphysik, Univ. Frankfurt, D-6000 Frankfurt, F.R.G.*
[c] *DPhN, CEN Saclay, F-91191 Gif-sur-Yvette Cedex, France*
[d] *Laboratoire national Saturne, CEN Saclay, F-91191 Gif-sur-Yvette Cedex, France*
[e] *National Superconducting Cyclotron Laboratory, Michigan State University, East Lansing, MI 48824, USA*
[f] *Department of Physics, University of Catania, Catania, Italy*

ABSTRACT

Proton correlations observed in $^{40}Ar + ^{197}Au$ reactions at E/A = 200 MeV are presented. The observed two - proton correlation functions are consistent with a fast emission process from a source which does not exceed the size of the target nucleus. Some indication for a true three - proton correlation is found.

The size of the emission zone and its temporal evolution is of fundamental interest for our understanding of the fragment production mechanism. Whereas fast light particles are believed to have their origin in small subsystems at a rather early stage of the reaction [1], slow particles with energies close to the Coulomb barrier are associated with the decay of a more equilibrated compound system. In evaporative models [2,3] the spatial extent of the emitting region corresponds to the size of the system at normal nuclear density. In addition, these models predict a non-zero lifetime which will stretch the phase space distribution along the direction of the particle's velocities [4]. On the other hand, a simultaneous breakup of an (equilibrated) expanded nuclear system (e.g. refs. [5,6]) may manifest itself in increased longitudinal *and* transverse source radii.

Pair-correlations between identical [7,8,9] and non-identical [10,11,12] light particles have widely been used to explore the space-time characteristics of the emission zone in the Fermi energy domain, E/A < 100 MeV. From the analysis of the directional dependence of the two-proton correlation function an extended spherical shape for the emission of low-energy protons has been inferred [9]. For more energetic nucleus - nucleus collisions (E/A > 100 MeV) the analysis has always been based on the assumption of a spherical source [13,14,15]. Combining the information on the associated light (charged) particle multiplicity indications for low freeze out densities of the participant zone were found [14,15].

However, the interpretation of the apparent source dimensions may be influenced by lifetime effects [7,16] and the reaction dynamics [4,17]: even two - particle correlation functions measured in exclusive experiments are a replica of the whole history of the emission zone. Additional complications might result from sensitivities to fluctuations of the Coulomb field of the residual nuclear system [18,19] and theoretical uncertainties of the interaction between the two nucleons in a nuclear medium. Finally, all two - particle correlation studies face the problem that an unambiguous relation between the observed correlation function(s) and the source distribution (see e.g. ref. [20]) does not exist. However, each reaction theory makes definite prediction about the source dimensions and, therefore, the correlation function. Thus, these measurements can at least help to *exclude* certain reaction scenarios.

In this letter, we report correlation functions of protons emitted in collisions between ^{40}Ar and ^{197}Au at E/A = 200 MeV. The experiment was performed at the Laboratoire National Saturne in Saclay. A gold target of 87 mg/cm^2 areal density was irradiated by ^{40}Ar ions with an incident energy of E/A = 200 MeV. Light particles were detected by a closely packed quadratic array of 64 ΔE - E telescopes, each consisting of a 300 μm thick silicon detector and a 6 cm long CsI(Tl) detector read out via a photodiode [21]. Each telescope had an active area of 25 $\times$ 25 mm^2. The center of the hodoscope was positioned at a laboratory angle of $\Theta_{av} = 40°$ at a distance of about one meter from the target. The calibration of the silicon detectors was achieved by using radioactive sources of α - particles and assuming linearity in the detector response. The calibration of the CsI(Tl) detectors was obtained from particles whose energy loss was measured with the silicon detectors and by using tabulated energy loss data. The energy calibration of individual telescopes is accurate to within 3%. In the off-line analysis, a low-energy threshold of 10 MeV was placed by software on the kinetic energy of the protons.

We investigate the energy and directional dependence of two - proton coincidences. As a first step to unravel the temporal evolution of the emission zone, we address the question of sequential decays of particle unstable fragments. We define the experimental correlation functions $R_{12}(q)$ in terms of the number of true coincidences $N_{12}(q)$ for a given momentum of relative motion, q, and uncorrelated two-particle events $M_{1,2'}(q)$ which were obtained by mixing individual particles from different coincidences:

$$R_{12}(q) + 1 = C_{12} \cdot N_{12}(q)/M_{1,2'}(q). \tag{1}$$

The normalization constant C_{12} is chosen in such a way that $\langle R_{12}(q) \rangle = 0$ in a q - interval at sufficiently large relative momenta. In order to explore the energy and

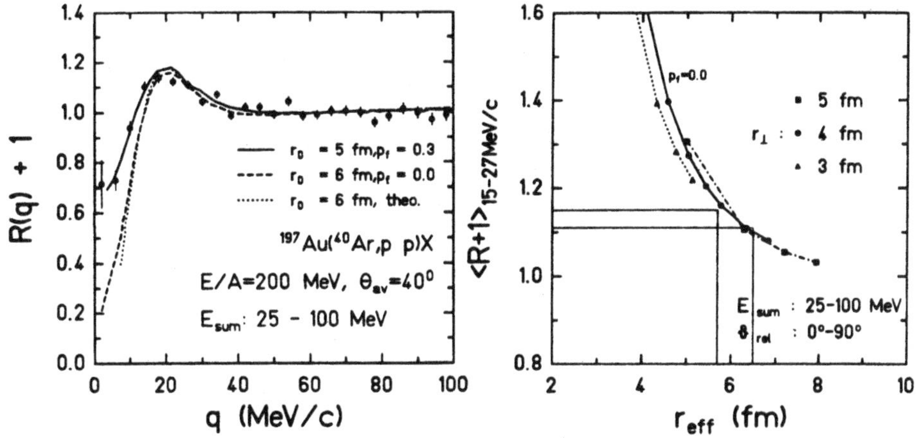

Figure 1:
Two - proton correlation function measured for ^{40}Ar induced reactions on ^{197}Au at E/A = 200 MeV, and an average emission angle of 40°. The total kinetic energy of the proton pairs has been constrained to values between 25 and 100 MeV.

directional dependence of the correlation functions, the correlation functions were evaluated for different constraints on the sum energy of the two particles $E_{sum} = E_1 + E_2$ and the relative angle θ_{rel} between the relative momentum and the total momentum. For each constraint on E_{sum} the normalization constant was adjusted individually. For a quantitative comparison, all theoretical correlation functions were normalized in the same interval.

As an example, the p - p correlation function for a gate on the sum energy of 25 MeV $\leq E_{sum} <$ 100 MeV and normalized in the range 50 MeV/c $\leq q \leq$ 80 MeV/c is shown in Fig. 1 (solid points). The dotted line in Fig. 1 represents the theoretical correlation function for a spherical Gaussian - shaped source with a radius parameter $r_{||} = r_\perp = 6$ fm [9]. These calculations, however, can not be directly compared to the data, since the experimental cuts on angles and energies and the distortions due to the angular and energy resolution will modify the correlation function [15]. Moreover, corrections due to spurious correlation in the background events have to be applied [22,23].

In order to evaluate these effects, a distribution $n_{12}(q)$ of primary relative momenta q was folded by means of a Monte Carlo calculation with the efficiency function of the experimental setup, $\epsilon(q,\hat{q})$. Here, $\hat{q}$ denotes the "measured" relative momentum. The distorted correlation function $\hat{R}(\hat{q})$ was then generated by weighting each event with the value of the primary theoretical correlation function at the initial relative momentum q:

$$\hat{R}_{12}(\hat{q}) + 1 = \frac{\sum_q (R_{12}(q) + 1) \cdot \epsilon(q, \hat{q}) \cdot n_{12}(q)}{m_{1,2'}(\hat{q})}. \tag{2}$$

The background spectrum $m_{1,2'}(\hat{q})$ was again generated by event mixing. Analogous to the experimental conditions, each fake coincidence was given a weight of $(R_{12}(q)+1)\cdot(R_{1'2'}(q')+1)$. The primary spectrum $n_{12}(q)$ was chosen in such a way that the denominator in Eq. 2 reproduces approximately the corresponding experimental distribution of mixed coincidences $M_{1,2'}(q)$ in Eq. 1. It turned out, that the filtered correlation functions are rather insensitive to the detailed shape of $n_{12}(q)$.

The dashed line in Fig. 1 shows the filtered theoretical correlation function for a spherical source with a radius parameter of 6 fm. For the present detector system, distortions due to the detector properties are only important in regions where the correlation function varies strongly, e.g. in the vicinity of sharp resonances (d-α correlation) or at small relative momenta $q < 10$ MeV/c in case of the p-p correlation. Modifications of the correlation function due to spurious correlations in the mixed background spectra are negligible at small relative momenta. It was verified that the width of sharp resonances which have been observed in correlation functions of complex particles (e.g. 2.186 MeV state in ^{6}Li or the ground state of ^{8}Be) are rather well reproduced by the calculations. This may be taken as evidence, that the effects due to the detector properties are well accounted for in the Monte Carlo calculations. Note, that in figure 1 there still is a remarkable difference between the data and the calculations at relative momenta $q < 10$ MeV/c.

In order to characterize the p - p correlation functions we evaluated the average of the correlation function $<R + 1>_{15-27}$ in the range 15 MeV/c $\leq q \leq$ 27 MeV/c. As expected [4], this quantity is in good approximation only dependent on the effective radius $r_{eff} = (r_{\|} \cdot r_{\perp}^2)^{1/3}$ and rather insensitive to the shape of the source, i.e $r_{\|}/r_{\perp}$. This is demonstrated in the right part of figure 1, where the (distorted) theoretical value for $<R + 1>_{15-27}$ is plotted as a function of r_{eff}. The various symbols correspond to different values for $r_{\perp}$ between 3 and 6 fm. The range of $<R + 1>_{15-27}$ which is consistent with the experimental correlation function is indicated by the dashed region and corresponds to an effective radius of 6.1 $\pm$ 0.4 fm. No significant dependence of the effective radius on the summed kinetic energy E_{sum} was observed (see table 1).

An example for the directional dependence of the correlation function is shown in figure 2 for the gate on E_{sum} from 25 to 100 MeV. The correlation function were accumulated for the relative momenta either parallel (θ_{rel} = 0°- 30°, or 150°- 180°) or transverse (θ_{rel} = 90°$\pm$ 30°) to the direction of the emitted proton pair. (Note, that for different cuts on θ_{rel} a common normalization is used.) Within statistics, no difference between these two directions is observed.

For a more quantitative analysis of the two cuts on θ_{rel} we again use the average theoretical correlation $<R + 1>_{15-27}$. It turns out, that the ratio of this value for

Figure 2:
Directional dependence of the p-p correlation function for E_{sum} = 25 - 100 MeV.

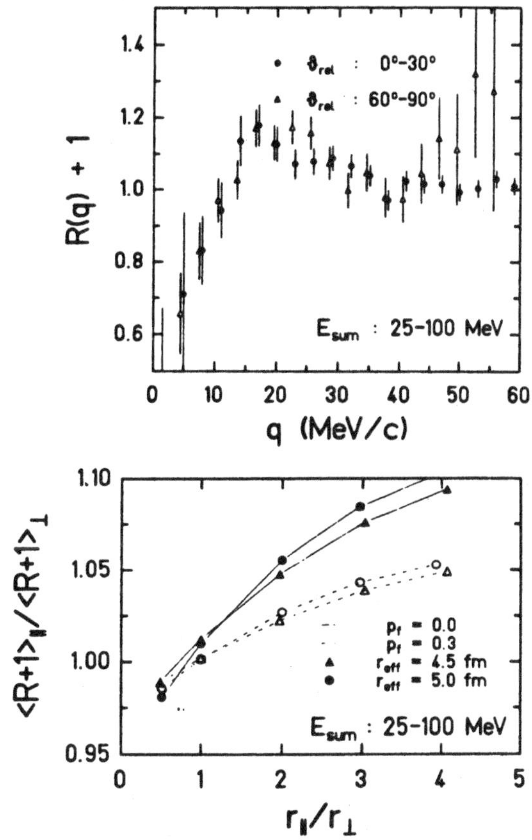

E_{sum} (MeV)	r_{eff} (fm)	$r_{\parallel}/r_{\perp}$
25-100	6.1±0.4	≤2.0
100-160	5.4±0.5	≥1.0
160-300	5.6±0.6	≤7

Table 1:
Effective source radii, r_{eff}, and deformations, $r_{\parallel}/r_{\perp}$ of the two - proton sources in ^{40}Ar + ^{197}Au reactions at E/A = 200 MeV. E_{sum} denotes the total kinetic energy of the coincident protons.

the two directions $<R + 1>^{\parallel}_{15-27}/<R + 1>^{\perp}_{15-27}$ increases monotonically with $r_{\parallel}/r_{\perp}$, but is rather insensitive with respect to r_{eff}. This is illustrated in the bottom part in figure 2. Because of the limited statistics, the data shown in the top part of figure 2 do not permit to extract a definite value for $r_{\parallel}/r_{\perp}$. Nonetheless, the experimental correlation functions exclude source deformations with $r_{\parallel}/r_{\perp} > 2$. Similarly, a lower bound for $r_{\parallel}/r_{\perp}$ of 1.0 could be extracted for the range 100 MeV $\leq E_{sum} <$ 160 MeV. Table 1 summarizes the present results.

The lines in figure 3 mark the transverse and longitudinal source radii which are consistent with the observed correlation functions for the constraints on the summed kinetic energy of 25 to 100 MeV and 100 to 160 MeV. Interpreting these apparent radii, however, one has to keep in mind that a significant fraction of the observed protons does not result directly from the reaction zone but is a result of sequential decays of particle unstable primary fragments. Most of these fragments are rather long - lived. Therefore, these secondary protons will produce a nearly uncorrelated

Figure 3:
Apparent transverse $r_\perp$ and longitudinal $r_\parallel$ dimension of the two - proton sources. The lines mark the values which are consistent with the observed correlation functions.

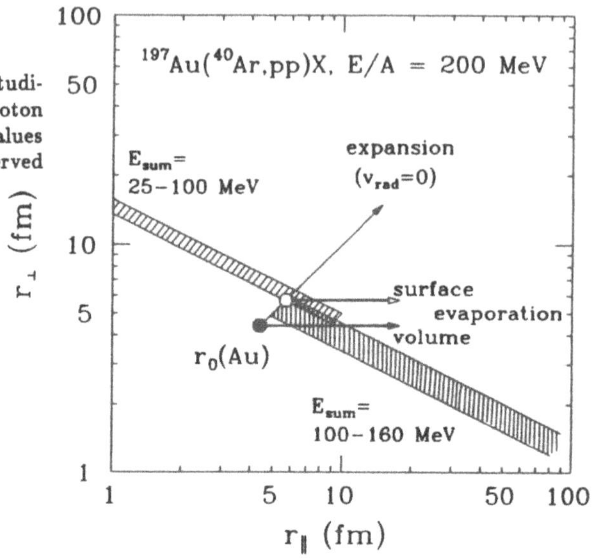

flat background in the correlation function. Since the correlation R_{12} is approximately proportional to $1/r^3_{eff}$ a fraction p_f of sequentially produced protons will cause an apparent increase of the emission radius according to

$$r_{eff} = r'_{eff} \cdot (1 - p_f)^{-2/3}. \tag{3}$$

Here, r'_{eff} denotes the primary effective emission radius. Emission temperatures observed in the present reaction are of the order of 5 - 7 MeV. At this temperature approximately 30 % of all observed protons are due to sequential decays. Of course, protons which result from the decay of ^{2}He itself are not included in this value. Thus the primary effective emission radius is of the order of 5 fm instead of the apparent 6 fm.

Indeed, an uncorrelated contribution is also supported by the shape of the correlation function. The solid curve in figure 1 shows for illustration a calculation assuming a spherical source with r_0 = 5 fm and a flat background due to a contamination of 30 % by sequentially produced protons. It has been checked that the values quoted for $r_\parallel/r_\perp$ in table 1 are not significantly affected by the sequential decays.

The observations are consistent with an evaporative process from a targetlike nucleus provided the temporal extent of the emission zone does not exceed a value of $\tau \approx 40$ fm/c for protons with 25 MeV $\leq E_{sum} <$ 100 MeV and $\tau \leq 30$ fm/c for 100 MeV $\leq E_{sum} <$ 160 MeV. Taking into account the contaminations of the sequential decays, the data exclude an expanded targetlike system being the dominant source of the observed proton - proton coincidences.

Figure 4:
Lower part: Three - proton correlation function which includes effects from two - proton correlations. Upper part: True three - proton correlation function, where the influence of two - proton correlations has been removed.

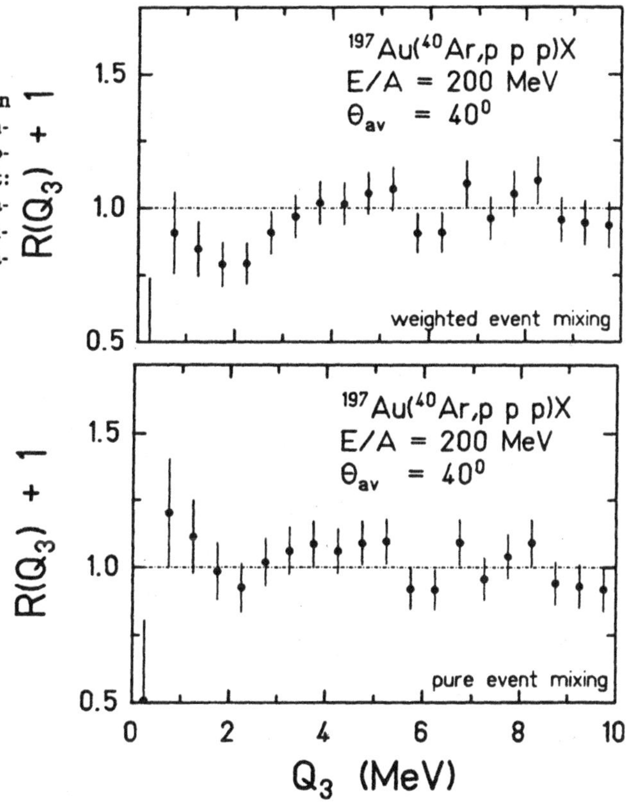

The high efficiency of the hodoscope used in the present experiment enabled us to investigate also the three - proton correlation function. Figure 4 shows this correlation function as a function of the effective mass of the three-proton system $Q_3 = M(3p) - 3m_p$. In the lower part, the uncorrelated background was again obtained by randomly mixing protons from *three* different events with the same number of protons. This correlation function, however, is not the real three - proton correlation function, since it still shows the influence of two - proton correlations. In order to eliminate this effect, the true three - proton correlation function was generated by weighting the uncorrelated background with a value [24]:

$$R_3(Q_3) = 1 + \sum_{i \neq j} R_2(q_{ij}) + 2 \cdot Re\left[\sqrt{R_2(q_{12})} \cdot \sqrt{R_2(q_{13})} \cdot \sqrt{R_2(q_{23})}\right]. \qquad (4)$$

The resulting correlation function is shown in the upper part of figure 4. Here, R_2 denotes the two - proton correlation and $q_{i,j}$ is the relative momentum between particle i and j. The data indicate a weak anti - correlation at low values of Q_3. Experiments with higher statistics, however, are clearly needed to confirm this observation.

References

[1] G.D. Westfall *et al.*, Phys. Rev. Lett. **37**, 1202 (1976).

[2] W.A. Friedman and W.G. Lynch, Phys. Rev. C **28**,16 (1983).

[3] R.J. Charity *et al.*, Nucl. Phys. **A483** (1988) 371.

[4] S. Pratt and M.B. Tsang, Phys. Rev. C **36**, 2390 (1987).

[5] D.H.E. Gross, Zhang Xiao-ze, and Xu Shu-yan, Phys. Rev. Lett. **56** (1986) 1544.

[6] J.P. Bondorf *et al.*, Nucl. Phys. **A443**, 321 (1985).

[7] S.E. Koonin, Phys. Lett. **70B**, 43 (1977).

[8] W.G. Lynch *et al.*, Phys. Rev. Lett. **51**, 1850 (1983).

[9] T.C. Awes *et al.*, Phys. Rev. Lett. **23**, 2665 (1988).

[10] D.H. Boal and J.C. Shillcock, Phys. Rev. C **33**, 549 (1986).

[11] J. Pochodzalla *et al.*, Phys. Lett. B **174**, 36 (1986).

[12] D.A. Cebra *et al.*, Phys. Lett. B **227**, 336 (1989).

[13] F. Zarbakhsh *et al.*, Phys. Rev. Lett. **46**, 1268 (1981).

[14] H.A. Gustafsson *et al.*, Phys. Rev. Lett. **53**, 544 (1984).

[15] P. Dupieux *et al.*, Phys. Lett. B **200**, 17 (1988).

[16] A. Kyanowski *et al.*, Phys. Lett. B **181**, 43 (1986).

[17] S.E. Koonin, W. Bauer, and A. Schäfer, Phys. Rev. Lett. **62**, 1247 (1989).

[18] J. Pochodzalla *et al.*, Phys. Lett. B **161**, 256 (1985).

[19] J. Pochodzalla *et al.*, Phys. Lett. B **175**, 275 (1986).

[20] J. Bartke and M. Kowalski, Phys. Rev. C **30**, 1341 (1984).

[21] C.H. Pinkenburg, diploma thesis, Universität Frankfurt (1988).

[22] D. Drijard, H.G. Fischer, and T. Nakada, Nucl. Instr. and Methods **225**, 367 (1984).

[23] W.A. Zajc *et al.*, Phys. Rev. C **29**, 2173 (1984).

[24] T. Akesson *et al.*, Z. Phys. C **36**, 517 (1987).

II. Correlations in Nuclear Matter at Equilibrium and Out-of-Equilibrium — Experimental and Theoretical Approaches

Particle interferometry for an evaporation process at intermediate energy

J. Québert, R. Boisgard, P. Lautridou
CENBG-IN2P3-CNRS and University of Bordeaux I
Le Haut-Vigneau, F 33175 Gradignan Cedex

D. Ardouin, H. Dabrowski, D. Durand, B. Erasmus, P. Eudes,
D. Goujdami, F. Guilbault, C. Lebrun, R. Tamisier
LPN-IN2P3-CNRS and University of Nantes
2. Rue de la Houssinière,F 44072 Nantes Cedex

A. Péghaire, F. Saint-Laurent
GANIL-IN2P3-CNRS,IRF-CEA Caen, F 14021 Caen Cedex

ABSTRACT :

The p-p correlation function at backward angles displays a strong reduction of the well-known ^{2}He final state interaction at q = 20 MeV/c. This effect is attributed to a sequential process where time ordering is estimated to be in the range of 10^{-21} s.

1. Introduction

The conventional methods like gamma spectroscopy, elastic or inelastic scattering measurements, which are used to study the intrinsic properties of nuclei near stability are not appropriate for probing very excited nuclei or complex transient systems. In this range of excitation energies and of short lifetimes, methods such as multiparticle correlations can only be used, as an indirect way, to learn about the size of the system and its timescale. More specifically, two-particle correlations are a useful method to estimate such quantities. Moreover, they are a means to eliminate many dynamical or experimental corrections since the absolute normalisation is not needed in the method.

Two-particle correlations for close angular emissions have been used in domains in which data processing is based upon two different concepts :

- The first field, which has been developed for many years in particle physics, may be called Boson interferometry[1,2]. This method uses the fact that two detected particles are identical (e.g. $\pi^- \pi^-$; $K^- K^-$) to obtain information on the size of the source from interference patterns occuring at small relative momenta (the GGLP effect[1]) . This method remains an important tool in ultrarelativistic heavy ion collisions for tracking down superdensities[3] and volumes of small sizes. The interferometric method, so-called intensity interferometry, appears to have an analogy with correlations defined in chaotic optics[4].

- The other field in which two-particle correlation functions have been extensively used is heavy ion collisions. Because of the final state interactions always present among heavy particles, it is difficult to apply the afore-mentioned method. Occurence of nuclear resonances between two identical[5,6] or non-identical particles has also been successfully exploited to estimate the size of the source[7] or its temperature[8].

In this work, we present some results concerning an evaporation process in which a progressive reduction of the mutual final state interactions is noticed. This effect seems to be due to a time ordering of emission of the two detected particles. We will try to show that the conventional analysis of the mutual final state interactions[5]) cannot account for the large apparent sources which have to be used to reduce the resonance; therefore new developments are needed to understand the final state of particle emission. One of the possibilities is the competition of other final state interactions which occur between one of the detected particles and the residual source, due in fact to the sequential decay in the deexcitation of the source.

2. Specificity of this work

We present several correlation functions which have been obtained in backward angle measurements for three collisions : ^{16}O + ^{197}Au(94 MeV/A), ^{40}Ar + ^{nat}Ag(44 MeV/A), ^{40}Ar + ^{197}Au(44MeV/A). As shown in Fig. 1 (lower right part) the singles spectra, obtained at backward angles (115°), exhibit a shape due to a clear evaporation process, contrary to the complexity of the mechanisms usually observed in forward angle measurements.

The multidetector, shown in the same figure, was able to provide invariant plots in the backward hemisphere. This multidetector is made of small CsI scintillators, grouped in a close-packed geometry (24 elements). Each element of the multidetector consists of a small cylinder of CsI (12.7 mmr of diameter, 10 cm long) placed on the photocathode of a small Hamamatsu photomultiplier (R647

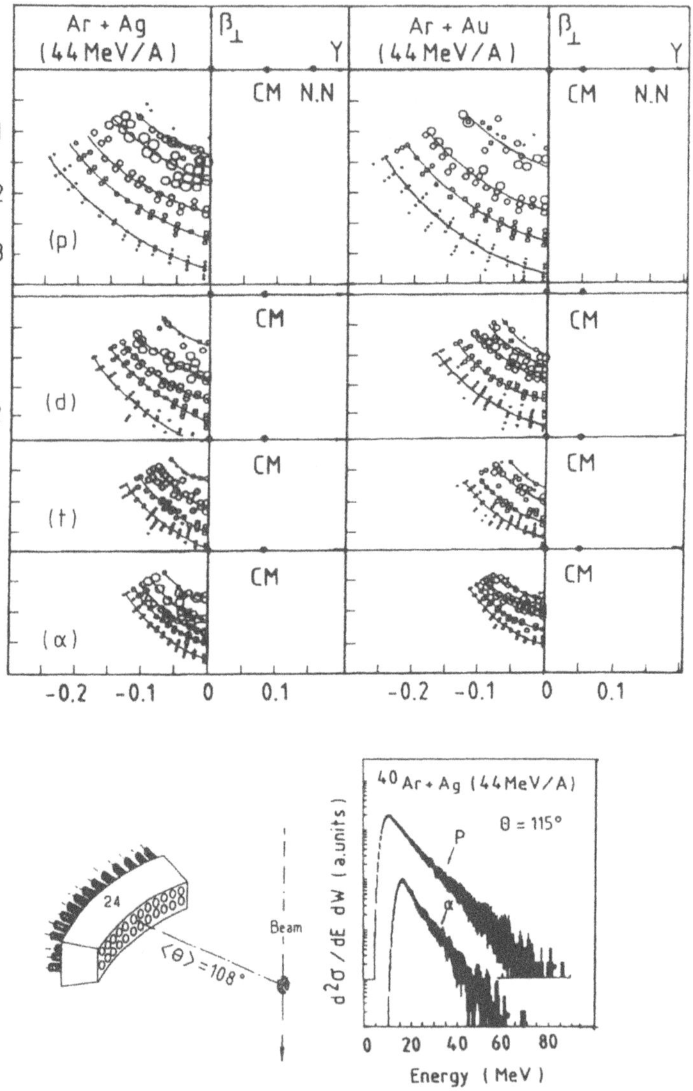

Fig. 1 : Detection at backward angles and invariant plots obtained with a multidetector of 24 CsI counters recording singles spectra of typical evaporative shapes.

01). Particle discrimination is achieved by the double sampling method of the delivered pulses for which the relative integrated charges are coded. At the end of the experiment a calibration in energy is performed by recording in each element a yield of secondary light particles which are of known energy (such particles are selected by beam-like transport).

As it can be noticed in Fig. 1, the invariant plots $d^2\sigma/\beta_\perp d\beta_\perp dY \propto E^{-1/2}$ $d^2\sigma/dEd\Omega$ clearly display a source, the average rapidity of which is closer to the target rapidity than to the C.M. rapidity.

The exact location of the source rapidity certainly depends on the nature of the emitted particle and on its energy. The highest energies locate the source at a rapidity which does not exceed 50% of the C.M. rapidity. The low energies roughly locate the source at rest.

Therefore, if we perform a measurement of two-particle correlations at large angles, we avoid the mixing of different mechanism contributions and we situate the correlation measurements for a source which, on average, can be considered roughly at rest in the interpretation of the correlation functions.

The studies which will be presented come from two series of experiments carried out at the Ganil facility.

3. Kinematic location of the source

The plots in Fig. 1 concern the collisions induced by an ^{40}Ar beam. In this case, the C.M. rapidity is distinct from that of the target. The location in energy of the maxima of the different inclusive spectra is in agreement with Coulomb repulsion due to the charge of the quasi-target (Ag, Au). The maximum energy measured for a single particle, or a pair of particles in coincidence, does not exceed 100 $\pm$ 20 MeV. The fact that the source remains almost at rest for the numerous particles of the lowest energies means that in the total C.M. system we chiefly deal with an inelastic scattering in which the target and projectile trajectories are not strongly deviated.

The question of the total energy left in the real initial source is difficult to clearly elucidate since we know that in the same collisions neutron multiplicities are very high[9]. Compared to the large values which are estimated for the energy deposit ($\simeq$ 600 MeV), charged particles and light excited clusters detected at backward angles seem not to carry a so large total energy (however, their multiplicity is not known in this measurement).

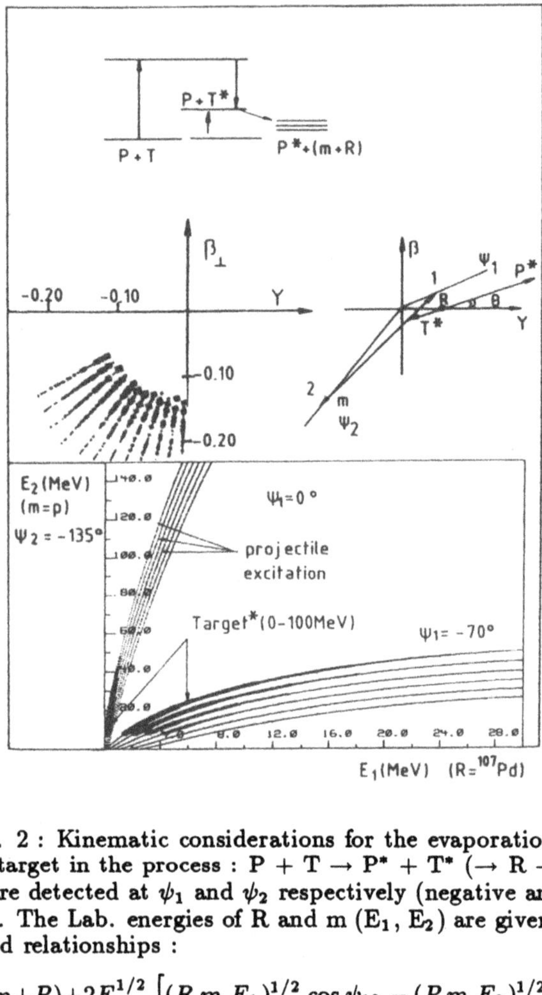

Fig. 2 : Kinematic considerations for the evaporation of the excited target in the process : $P + T \to P^* + T^*$ ($\to R + m$). R and m are detected at ψ_1 and ψ_2 respectively (negative angles for $\beta_\perp < 0$). The Lab. energies of R and m (E_1, E_2) are given by the combined relationships :

$$1)\quad E_2(m+P)+2E_2^{1/2}\left[(R.m.E_1)^{1/2}\cos\psi_{12}-(P.m.E_0)^{1/2}\cos\psi_2\right] + \left\{E_1(R+P)-2(P.R.E_0.E_1)^{1/2}\cos\psi_1 - P.Q\right\} = 0$$

$$2)\quad E_{12}^* + Q_{12} - \mu_{12}\left[\frac{E_1}{R} + \frac{E_2}{m} - 2\left(\frac{E_1.E_2}{mR}\right)^{1/2}\cos\psi_{12}\right] = 0$$

(Q, Q_{12} are the Q values in the different channels ; E_o is the incident energy , μ_{12} the reduced mass and ψ_{12} the relative angle). Kinematic locii are shown at the bottom of the figure with evaporative densities due to the target excitation E_{12}^*. Different projectile excitations (10,20,30 ..MeV) are also considered and give the slightly shifted locii.

To analyse the kinematic location of the source more precisely and get a simulation of the invariants (Fig. 2), we have used three-body kinematics for the two-step process: P + T $\rightarrow$ P* + T* (T*evap $\rightarrow$ R + m), in which a light fragment: m (^{8}Be,Li,α,..p) is evaporated from an excited target T*. The energy distribution in this excited quasi-target was assumed to have a Maxwellian shape with a temperature θ. The light fragment m (or m*, which gives rise to two-particle correlations) is detected at an arbitrary angle ψ_2 and its associated residue R at angle ψ_1. For each pair of m and R directions we deduce the target excitation E*(eq. 2 in caption of Fig. 2) and a distribution in $e^{-E^*/\theta}$. Two sets of loci are displayed in the lower part of the figure for two directions of R ; the distributions of target excitations are shown by heavy lines in the plots (also computed for different projectile excitations). A projection on the E_2 axis would give the single spectrum for m. A typical proton evaporation process from the target (m = p ; with projectile excitation) is shown in this example.

4. Complementary information given by the coincidence energy plots

The Dalitz plots relative to the two particles detected in coincidence (for instance from a resonant state m* afore-defined) give valuable information and indicate trends in the analysis of correlation functions.

In Fig. 3 we show three typical plots for p - p, $\alpha - \alpha$ and α - d data obtained with a chosen group of closely spaced detectors ($\theta_{12} \simeq 5°$) in the multidetector. We immediatly notice whether a mutual final state interaction is present or not by comparing data accumulations to expected Breit-Wigner distributions. Such plots can be considered as describing the final sequential process T* $\rightarrow$R + m*$_{\rightarrow m_1, m_2}$ where T* is the evaporating target (defined in Fig. 2) and m* the invariant mass of the two detected particles. In this example, levels such as ^{8}Be(g.s.), or ^{6}Li(2.18) are clearly observed. In contrast, the ^{2}He interaction (m* $\simeq$ 0.4 MeV) is more difficult to locate.

Such final state interactions have also to be combined with the total available kinetic energy (or the apparent source excitation defined by the dotted lines). We notice that the full available energy carried by the decaying nuclei Be(0^+) and ^{6}Li(3^+) does not exceed 100 $\pm$ 20 MeV.

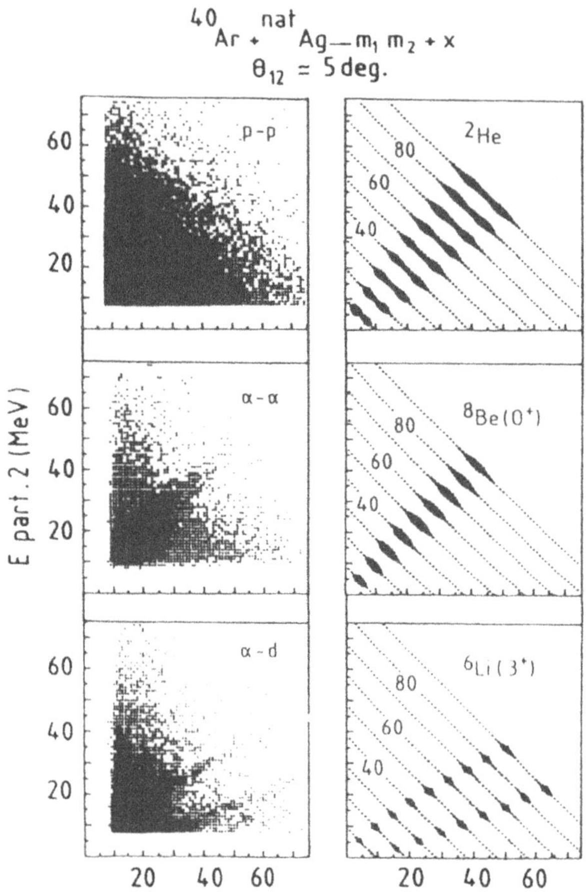

Fig. 3 : Partial coincidence plots for a detection at backward angles. As it can be noticed, the particle energies are relatively low; the experimental sums of energies indicate that the recoiling nucleus is heavy (the dotted lines due to different excitations of the source, should be curved for a light residue). On the right, are shown the expected loci for some mutual final state interactions.

5. Correlations as a function of dynamical variables

A confirmation of the reduction (or disappearance) of the ^{2}He state is displayed in the typical p-p correlation function at backward angles compared to $\alpha - \alpha$ correlations (Fig. 4).

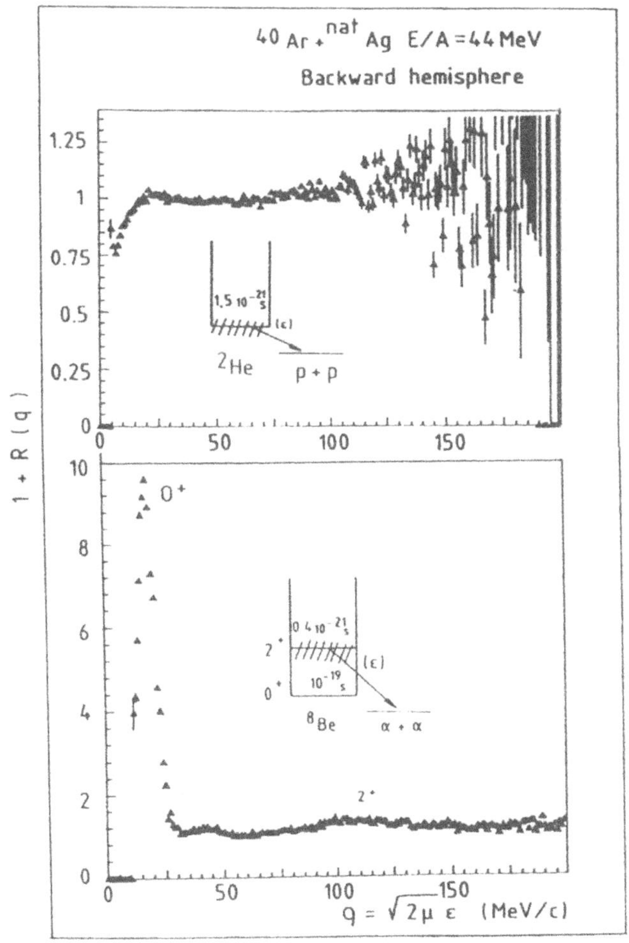

Fig. 4 : Two typical correlation functions obtained with a multidetector of 24 elements, set up in the backward hemisphere. We notice a persisting yield of ^{8}Be(0^+), when the other resonances, easily detectable at forward angles, are strongly reduced (^{8}Be(2^+), ^{2}He).

This strong reduction of the final state interaction of ^{2}He is observed in many collisions either at intermediate energy and backward angles (this work; Fig. 5), or at low energy[10).

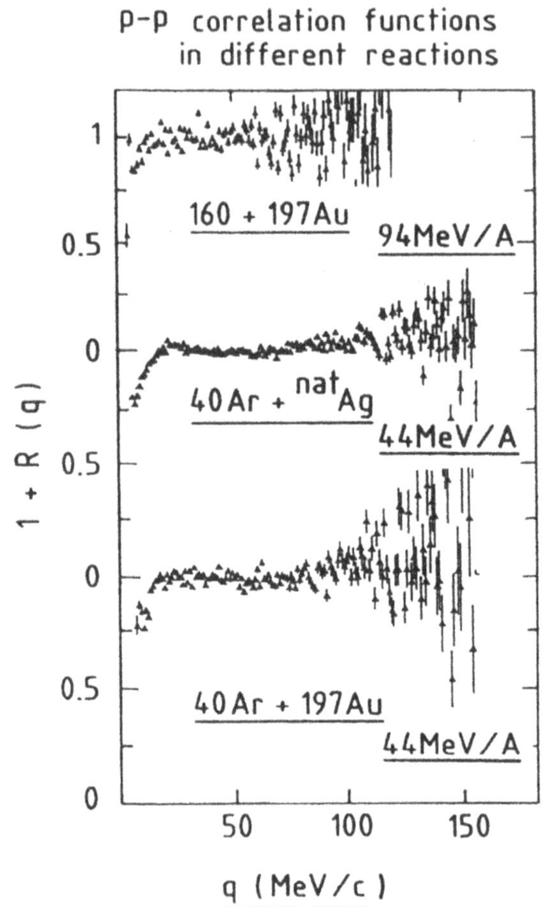

Fig. 5 : p-p correlation functions measured at backward angles in three different collisions. We notice a unique typical flat shape.

As shown in Fig. 6, this reduction is in fact depending on the energy of the detected pair: We notice a clear signature of the ^{2}He state for protons of high energy and its disappearance for protons of low energies.

The absolute values of the normalisations ($R(q \rightarrow \infty)=1$) are much lower when we gate on high energies than when we gate on low energies. This means that, on average, the final state interaction strength is very weak and that the major part of the p-p correlation function displaying a flat plateau near q =20 MeV/c in Fig. 4, is chiefly caused by low energy protons.

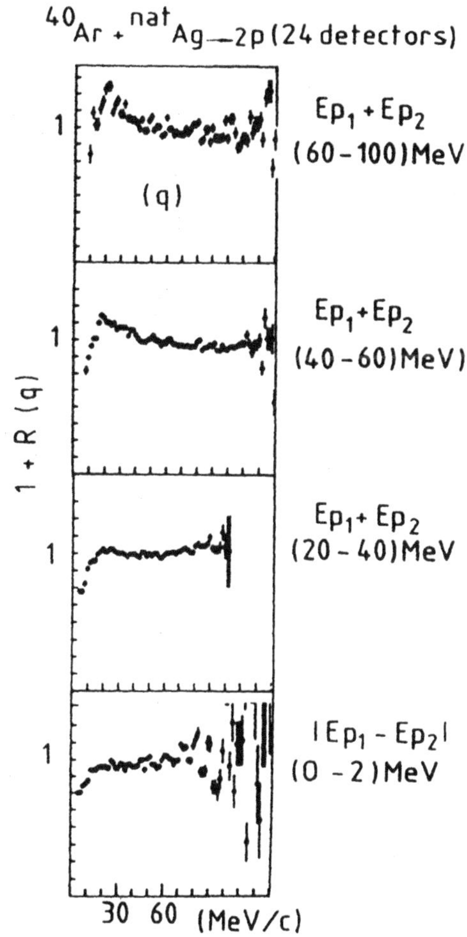

Fig 6 : The backward p-p correlation function versus different gates. For high energy protons the mutual final state interaction is apparent; this interaction vanishes at lower energies. At the bottom, the correlation function for $q_o \simeq 0$ is displayed; this gating does not change the shape significantly.

In Fig. 7 are also shown different representations of the p-p correlation function versus typical variables used in particle correlations, i.e. q, Q_T(transverse to the momentum pair), Q_L(longitudinal to the momentum pair), $Q_\perp$(perpendicular to the beam), $Q_\parallel$(parallel to the beam). All representations are compared to simulations in which an isotropic evaporation process is detected by the multidetector.

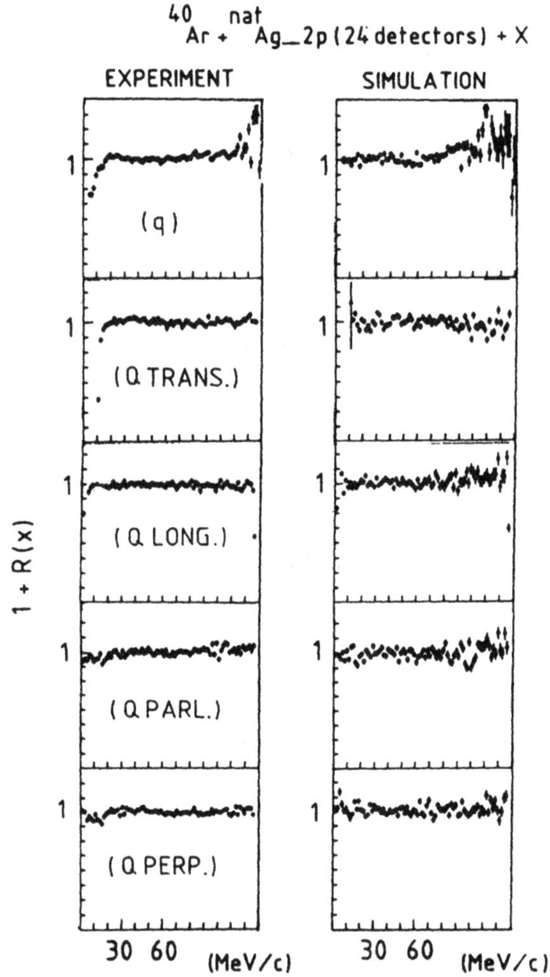

Fig. 7 : The different representations of the p-p correlation function compared to a Monte Carlo simulation. The q distribution only differs clearly from the accidental coincidences due to the evaporation process.

The data display a relative insensitivity to such directional choices. The only significant correlation to be detected, is a depletion near q=0.

These features indicate that two protons of low energy have a tendency to ignore each other in the emission as this can occur in a sequential decay, or in a time-ordering process, where large effective ranges reduce all mutual correlations. This feature is also noticed in other correlation functions which are not discussed in this study.

From our experimental results at backward angles we draw the conclusion that in a collision, the correlation function depends on dynamical variables. In particular, all correlation functions vary with the average scattering angle[11] at which they are measured, or on the excitation of the source, as discussed here.

Therefore, this effect appears as an interesting signature for the dynamics of nuclear disassembly.

6. Other kinds of final state interactions

In the preceeding discussions, only mutual nuclear interactions have been considered. Two other possibilities of interactions exist in the emission of three particles, as shown in the lower plot of Fig. 8. Such interactions correspond to the last two possibilities which are encountered in the decays:

1) $Source \rightarrow (p_1 + p_2)^* + X$

2) $\rightarrow p_1 + (p_2 + X)^*$

3) $\rightarrow p_2 + (p_1 + X)^*$

These three processes are illustrated by the hatched areas in the Dalitz plot. The upper plot corresponds to the first process, usually considered in the study of particle correlations at small relative angles. For instance, the bump found at q = 20 MeV/c in the p-p correlation function, corresponds to this process when the others can be assigned to sequential decays which are usually neglected. These last two processes in which detection cannot distinguish between particle 1 and particle 2 have the following effects:

- reduce the usual nuclear resonant interaction between p_1 and p_2
- reduce the Coulomb interaction between p_1 and p_2
- decorrelate the two particles on a more general point of view.

This kind of competing emission, which certainly occurs in an evaporation process, is probably signed by the features of the p-p correlation functions shown in Fig. 5. The delay in time emission allows to consider the emission as a two-step process where particle indistinguishability only has to be considered.

Final state interactions are well known[12–15] in many three-body emissions. It is also known that, due to indistinguishability, interference effects occur in the overlap area of the Dalitz plot where two or three resonance bands are crossing.

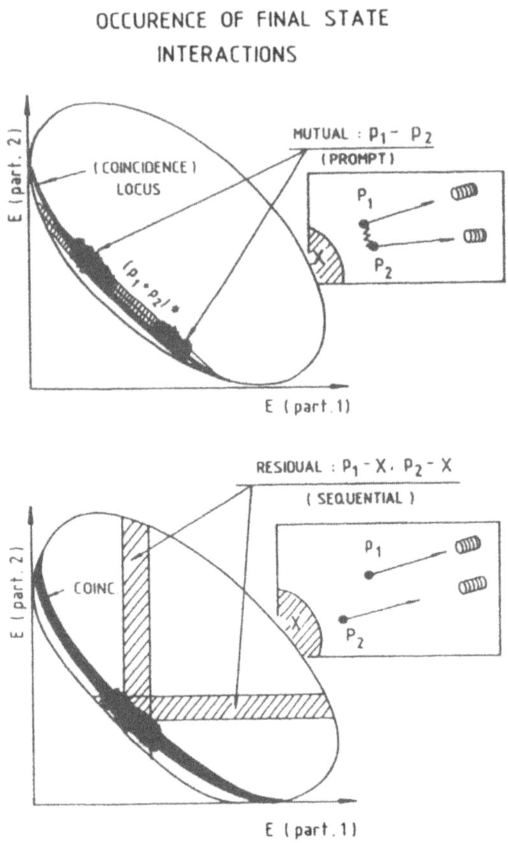

Fig. 8 : Final state locations in an ideal Dalitz plot. The upper plot shows that a mutual final state interaction is only observable in a finite range of θ_{12} values and cannot be observed at large angles (this is the case of interferometry). The lower part shows that a sequential emission has another location in the Dalitz plot and can be observed anywhere in θ_{12}. The overlap of the two alternate ways due to the detection of like particles can be observed along the diagonal of the plot and more specifically for q=0.

Such effects are explained by a symmetrization or an antisymmetrization of the total matrix element describing the deexcitation process with respect to energies or momenta. According to this, when detection cannot distinguish the elementary process, two or three amplitudes can contribute to give rise to such interference patterns.

An example of this effect is given in the reaction leading to three alpha particles where anomalous departures from simple sums of intensities are noticed. This effect occurs when kinematics do not allow a distinction between first emitted and second emitted particles (Fig. 9).

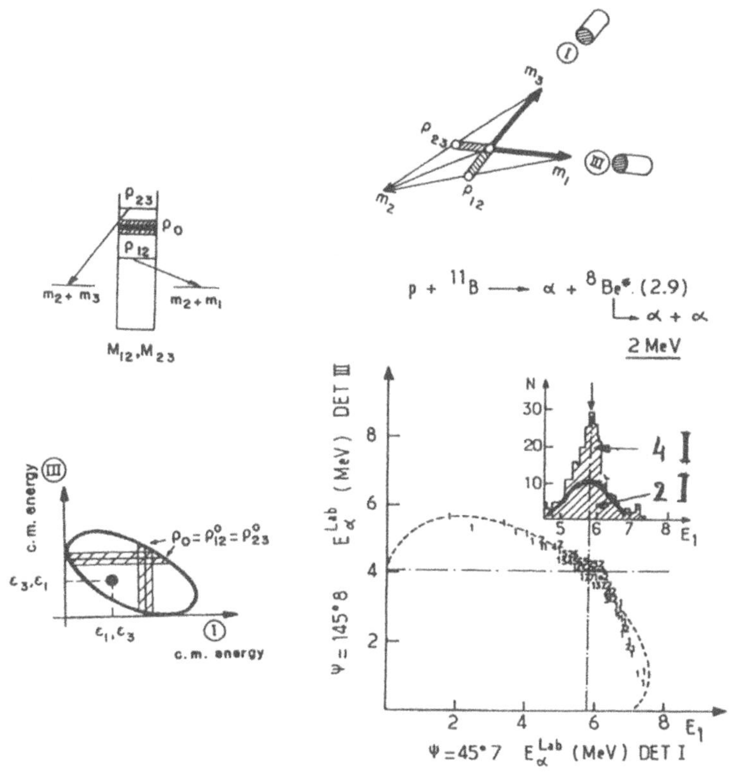

Fig. 9 : Occurence of interference effects in a three-body reaction when detection cannot distinguish between m_1 and m_3 and their origin in the complex decay. At the crossing of two resonance bands of intensity I, the yield can be 4 I instead of 2 I.

If we observe in detail the coincidence plot for the two protons detected in the evaporation process (Fig.10), we notice that the coincidence events are lying, with a larger probability, on loci parallel to the energy axes, in similarity with the resonance bands in the lower plot of Fig. 8.

The question is thus to know whether the partly depleted zone near q $\simeq$ 0 displays destructive interferences like the patterns of two overlapping resonance bands.

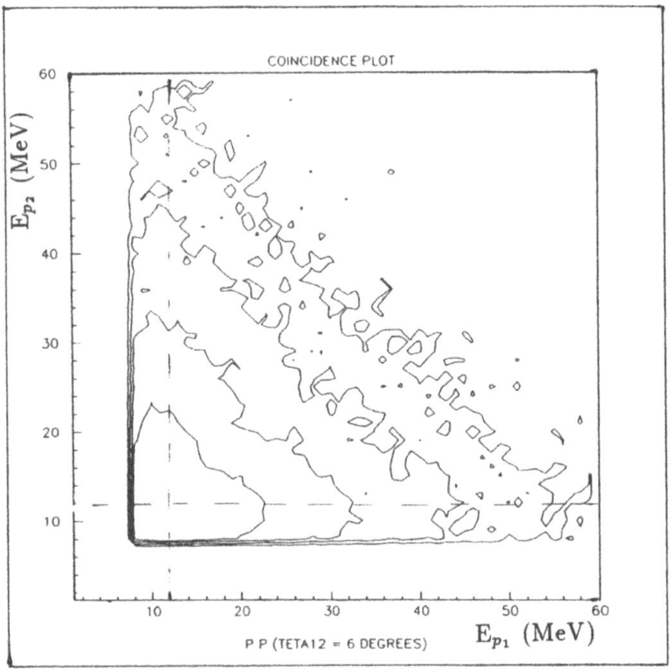

Fig. 10 : p-p coincidence plot. The data are lying in areas which are closer to the ones defined by two resonance bands in Fig.8 than to the one which defines a mutual interaction in the same figure.

7. Analysis of the "background" p-p correlation function

Some difficulties to fit the "background" p-p correlation function are encountered when using the conventional Koonin's analysis[5)]:

- first, large Gaussian source sizes are necessary to reduce the resonance strength ($R_G \geq 10$ fm).

- second, according to the same analysis[5]), it should be possible to detect any lifetime effect through sensitivity of directional distributions. We have checked that our data do not have a great sensitivity with respect to the different representations and thus, to the lifetime effect so defined. This result, which would lead to the conclusion of a negligible lifetime, is in contradiction with the real and clear effect due to the velocity of the pair, observed in Fig.6 (in the Koonin's formula, this effect is taken into account by the term $V\tau$, when $\tau \neq 0$).

- finally, there is no possibility with this formula to explain the partly-depleted data yield when q→0 (the experimental limit is not zero).

One of the possibilities to account for such correlation functions is the one developed in ref.(10), where three-body Coulomb trajectories are computed. The major problem of such calculations is that they depend very much on initial conditions and that many sets of initial conditions can provide the same asymptotic effect.

The other question which can be raised, is the limit of the correlation function near q=0. This limit seems to be close to 0.5, but this has to be confirmed by a specific measurement. This value is expected by the Fermionic nature of the two protons (Pauli blocking, spins weighting, etc...)

In the sense of interference effects described above for the Dalitz plots we find an analogy with interferometry in the work of ref.(2). Of course, other views of the relevant matrix elements can be considered. One of these can be found in ref.(16), where correlations, due to sequential decay in the compound nucleus, are also computed.

A fit of our p-p correlation function has been obtained by a simple formula, obtained in the framework of interferometry[2]). In the calculations, the two particles are considered as free particles (for two protons, this is certainly a rough approximation, only valid when Coulomb effects are considerably reduced by sequential emission ; two plane waves ψ_i, ψ_j and symmetric/antisymmetric considerations are used in ψ_{ij}. The same effect as the one due to the overlapping of two resonance bands in the Dalitz plot is expected by these latter considerations.

This is included in the expression :

$$\frac{\sigma_{ij}}{\sigma_i \sigma_j} = \frac{< \mid \psi_{ij} \mid^2 >}{< \mid \psi_i \mid^2 > < \mid \psi_j \mid^2 >} \tag{1}$$

Usually the mean values are computed over the same source volume of uniform density (a Gaussian distribution is also used). Using this simple calculations we get :

$$1 + R(Q) = 1 - 0.5 \left[\frac{3j_1(QR)}{QR} \right]^2 \tag{2}$$

(j_1, spherical Bessel function).

Introduction of the lifetime τ of the source leads to multipling the last term of Eq. (2) by :

$$F(q_o, \tau) = (1 + q_o^2 \tau^2)^{-1} \tag{3}$$

Assuming first $\tau = 0$ we tried to fit the correlations functions presented in Fig. 5 with Eq. (2): we found a good fit, but only with a value of R which does not seem realistic : i.e. R = 25 fm (a Gaussian equivalent source would be $R_G = \sqrt{5/2}R \simeq 16fm$).

We then introduced the lifetime effect. In this case, the correlation function has to be considered as a projection on the Q axis of all the events defined by Q and q_o (Q and q_o are not independant variables).

To take the data depending on q_o into account, we have computed the average value $< F(q_o, \tau) >$ over the experimental distribution of events which have different q_0 values for each bin of Q ; this expression finally depends on τ and Q and enables a fit versus Q only.

We also analysed [17] the two-dimensional plot (Q,q_0) by a least-square fitting procedure to search for a unique solution in R and τ. With a conventional size of the source R $\simeq$ 6.5 fm, we find a lifetime $\tau = (1.1 \pm 0.2)10^{-21}s$, but alternatively, we cannot exclude other combinations in space and time in the study of the two-dimensional plot.

A test to check the validity of the combined Eqs. (2),(3) and the sensitivity with respect to τ, is gating the correlation function on small q_o values (this is obtained by selecting equal energies). The result of this procedure is presented at the bottom of Fig. 6 . No significant change has been observed for the experimental correlation function which , for R =6.5 fm , should be shifted to the location of curve 1 in Fig. 11.

This test shows that the above expressions cannot explain the data with a conventional nuclear size and that the apparent size of the system is actually large. This is also the conclusion we need to draw by studying the effect of the source size to reduce the resonance in the framework of the study of ref.(5). This conclusion is also in some coherence with a relative insensitivity (on average and without particular gates) of directional effects.

The way Eq. (2) is obtained, shows in fact that there is no need to compute the average values of the wave functions, as well as the cross terms, over the same density distributions. In the scenario in which a particle is emitted before the other, Eq. (2) can be folded over two different spheres of radii R_S (source) and R_1 (active shell of the first particle around the source) without using time considerations directly.

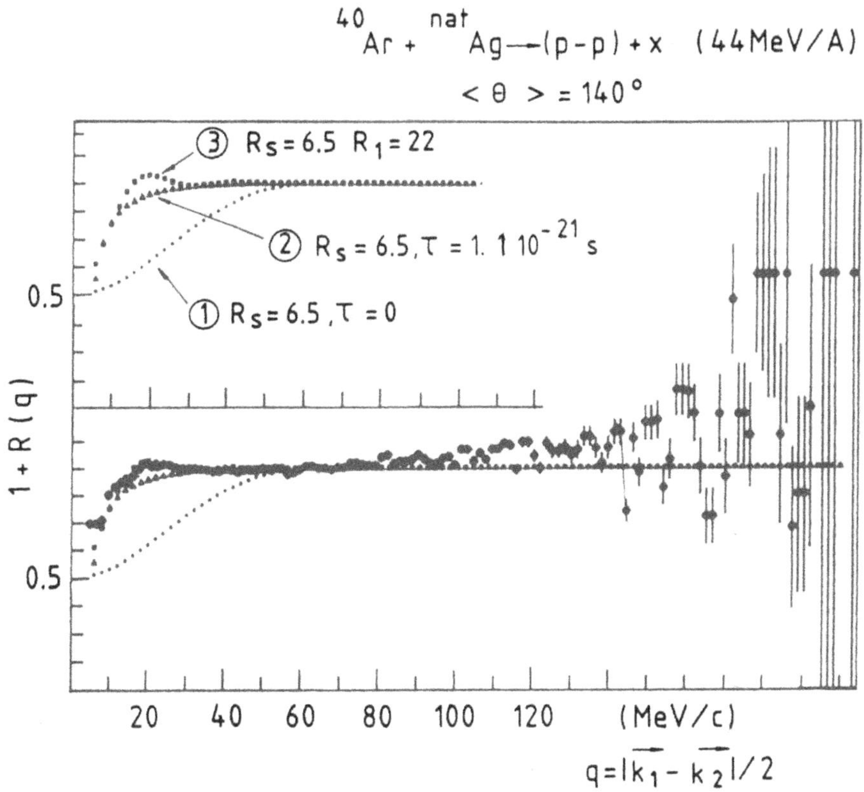

Fig. 11 : Different fits of the p-p correlation function assuming no mutual interactions between the particles for different space-time distributions of the source (R values are in fm.).

We then get:

$$1 + R(Q) = 1 - 0.5 \left[\frac{3j_1(QR_1)}{QR_1}\right] \left[\frac{3j_1(QR_S)}{QR_S}\right] \tag{4}$$

We have derived this expression in order to fit the p-p correlation function; It gives a good interpretation of the distribution, assuming that one of the two protons is present in a cloudy volume of radius R_1=22 ± 2 fm around the source of size R_S (6.5 fm). This may be pictured in terms of time ordering but not in terms of the same lifetime governing the emission of two protons. For instance, the first emitted proton of 11 MeV (maximum of the singles) has a statistical advance time of $\simeq 0.4\ 10^{-21}$s with respect to the second one.

8. Conclusion

In this work we have tried to show that a coexistence of almost independent and correlated emissions may originate from the same source (the excited quasi-target). The progressive change from one process to the other is related to the excitation of the source. We can assign this change to time ordering of emission. This emission can be considered either as a prompt emission of two particles (short delay, high excitation of the source, occurence of final state interactions), or as a sequential decay (larger delay at low excitation and loss of nuclear interactions). The range of times involved is estimated qualitatively in the framework of interferometry, applied to the p-p correlations observed at backward angles. This simple approximation allows to state that the space-time extent of the source is certainly large for an evaporative source observed on its whole range of excitation.

9. Acknowledgements

We are indebted to MM Guiard, Balauge, Brut, Laugenie, Barbas, Rio, Ropert for their efficient contributions during the different phases of the experimental set up. We also thank the Ganil crew for excellent beam transport.

10. References

1) G. Goldhaber, S. Goldhaber, W. Lee, A. Pais, *Phys. Rev.* **120** (1960) 300

2) G.I. Kopylov, *Phys. Lett.* **50B** (1974) 472 and references therein

3) T.J. Humanic et al, *Z. Phys. C - Particles and Fields* **38** (1988) 79

4) L. Mandel; E. Wolf, *Rev. Mod. Physics* **37**(1965) 231

5) S. Koonin,*Phys. Letters* **70B**(1977) 43

6) W.G. Lynch, C.B. Chitwood, M.B. Tsang, D.I. Fields, D.R. Klesch, C.K. Gelbke, G.R. Young, T.C. Awes, R.L. Ferguson, F.E. Obenshain, F. Plasil, R.L. Robinson, A.D. Panagiotou, *Phys. Rev. Lett.* **51**(1983) 1850

7) D.H. Boal, H. De Guise, *Phys.Rev.Lett.* **57**(1986) 2901

8) J. Pochodzalla, C.K. Gelbke, W.G. Lynch, M. Maier, D. Ardouin, H. Delagrange, H. Doubre, C. Gregoire, A. Kyanowski, W. Wittig, A. Peghaire, J. Peter, F. Saint-Laurent, B. Zwieglinski, G. Bizard, F. Lefebvres, B. Tamain, J. Québert, Y.P. Viyogi, W.A. Friedman, D.H. Boal, *Phys. Rev.* **C35**(1987) 1695

9) Jiang et al, *Nucl. Physics* **A503**(1989) 560.

10) P.A. De Young, M.S. Gordon, Xiu qin Lu, R.L. MCcGrath, J.M. Alexander, D.M. de Castro Rizzo, L.C. Vaz, *Phys. Rev.* **C39**(1989) 128

11) J. Québert, *Proceedings of the XXVII Winter Meeting on Nuclear Physics* (Bormio) 1989-p.220

12) R. H. Dalitz, *Rev. of Modern Physics* **33**(1961) 333

13) K. M. Watson, *Phys. Rev.* **88**(1952) 1163

14) G.C. Phillips, *Rev. of Modern Physics* **37**(1965) 409

15) R.J. de Meijer, R. Kamermans, *Rev. Mod. Phys.* **57**(1985) 1

16) S. E. Koonin, W. Bauer, A. Schaäfer, *Phys. Rev. Letters* **62** (1989) 1247

17) D. Ardouin, *Proceedings of the E. Fermi Physics School* (Varenna) 1989

Particle-Particle Correlations for Both Large and Small Angle Separations: Characterizations of Composite Nuclei and Probes of Their Lifetimes

A. Elmaani,[(1)] *J. M. Alexander,*[(1)] *R. Lacey,*[(1,2)] *S. Kox,*[(3)] *E.Liatard,*[(3)] *F. Merchez,*[(3)] *T. Motobayashi,*[(3)] *B. Noren,*[(3)] *C. Perrin,*[(3)] *D. Rebreyend,*[(3)] *Tsan Ung Chan,*[(3)] *G. Auger,*[(4)] *and S.Groult*[(4)]

[(1)]*Department of Chemistry, State University of New York at Stony Brook, Stony Brook, NY 11794-3400*

[(2)]*Centre d'Etudes Nucléaires de Saclay, F-91191, Gif-sur-Yvette, France*

[(3)]*Institut des Sciences Nucléaires de Grenoble, 53 avenue des Martyrs, 38026 Grenoble, CEDEX, France*

[(4)]*Grand Accélérateur National d'Ions Lourds, Boîte Postale 5027, 14021 Caen, CEDEX, France*

ABSTRACT

For evaporative particle emission the statistical model framework gives a clear route for interpretation of spectral shapes and angular anisotropies in terms of emitter temperature, barrier and rms spin. We review such interpretations for the reactions (E/A = 6-8.5 MeV) ^{40}Ar + Ag. The same methods are used to characterize the composite nuclei formed by 680 MeV ^{40}Ar + Ag; we find ≈ 4.5 MeV for the initial temperature and ≈ 60 $\hbar$ for the rms spin. Small-angle correlations between p-p and d-d pairs are measured to give a test of the emitter lifetimes. These measured correlations are compared to reaction simulations made by classical trajectory calculations. The results are consistent with lifetimes predicted by the phase-space, statistical model. This result implies that energy mixing and evaporative particle emission steps occur with intervals of ≈ 10^{-22} s, a time period comparable to the traversal time for a projectile velocity of 17 MeV/nucleon.

In this report we want to present some new experimental data on particle-particle correlations from the reactions 312 and 680 MeV ^{40}Ar + ^{nat}Ag.[1] Our major interest is to characterize the properties of a very highly excited composite nucleus namely its temperature, spin zone, radial extent and mean lifetime. For this purpose one must make comparisons to reaction model simulations, and I will describe a new code that we use, COULBOMB.[2] The approach starts by examining data for 6-8.5 MeV/nucleon ^{40}Ar reactions where the dominant process is evaporation after essentially complete fusion.[3] Let us first review some experimental data for 247 and 337 MeV ^{40}Ar [3] and show some relationships of those data to our new results for 312 MeV ^{40}Ar.[1] Then we will examine some relationships between the new results for 312 MeV and 680 MeV ^{40}Ar, and finally, some model calculations in comparison to the data.

Fig. 1 shows the experimental setup used in Ref. 3 and relates it to the geometry of "in-plane" and "out-of-plane" angular correlations. The experimental points reveal an almost flat in-plane angular correlation in contrast to a rather strong out-of-plane anisotropy. The strongest anisotropy is found for a sweeper detector at $\theta_{c.m.} \approx 90°$ in coincidence with in-plane versus out-of-plane triggers.

The first question to ask concerns the extent of thermalization in these reactions and the momentum transfer to the emitter. Fig. 2 shows invariant cross section maps for α-α coincidences from the in-plane correlations. Circles have been drawn centered on the c.m. velocity; since they fit the data, one can conclude that the emitter velocity is consistent with the c.m. velocity. Also, one can conclude that since the spectra are relaxed, the thermalization is essentially complete.

Next we want to examine the details of the angular correlations and their information content concerning the emitter spin. Fig. 3 shows angular distributions for different α-energy cuts for both in-plane and out-of-plane geometries. The calculated curves show a good fit for evaporative emission from an emitter with spin zone $35 \leq J \leq 70\hbar$, just the higher part of the zone for evaporation residues (0-70$\hbar$). These curves show an impressive amount of detail for the fits; however, the configuration of maximum anisotropy really provides the major constraint on the spin. The rest of the points give only consistency checks. Fig. 4 shows another view of these data that actually carries more information, but in a more compact display. Here we have observed energy spectra for a detector at $\theta_{c.m.} \approx 90°$ in coincidence with in-plane and out-of-plane triggers. The bin-by-bin ratios provide those most constraining anisotropies from Fig. 3. In addition, of course, the spectral shapes give constraints on the effective emission barriers and temperatures. Therefore, we can obtain all the available information on the emitter from the shapes of these spectra and the associated anisotropy ratios.

Now let's turn to the experimental setup used recently at Grenoble; a simplified drawing is shown in Fig. 5. To obtain the same kind of information described above, we placed two trigger detectors at $\theta_{lab} = 68°$ or $\theta_{c.m.} \approx 90°$, one in-plane and one out-of-plane. Energy spectra were recorded in EMRIC, also at $\theta_{c.m.} \approx 90°$, in coincidence with these two triggers. All detectors were mounted in the Magnolia chamber of the SARA double cyclotron facility. Beam intensities were 2-20 na, and the Ag target had an areal density of $\approx$ 500 μg/cm^2. Light charged particles ^{1,2,3}H and ^{4}He were observed in rectangular CsI detectors (4 x 4 x 10 cm^3), sixteen of which constituted the EMRIC array[4] for small-angle correlations. A number of other CsI units were located to obtain large-angle correlations; Fig. 5 shows only the two most important ones for measuring angular anisotropies. In addition to the CsI detectors, four photovoltaic detectors [5] were used to record fission-like fragments in coincidence with the light particles. In this report we present some initial results from the simplified configuration shown in Fig. 5.

For most runs we enabled the data acquisition by a coincidence between any two detectors; singles data were also obtained in several runs for each incident energy. Time of flight measurements for continuous ^{1}H and ^{4}He spectra gave energy calibrations for some of the detectors. However, the primary method of energy calibration was to require conformity of proton and alpha spectra from the 312 MeV ^{40}Ar reaction to the results from Ref. 3. In that work multi-element Si detector telescopes were carefully calibrated by the 6.05 and 8.78 MeV α-particles from a Th-C source. The observed spectra were fitted in great detail by a simulation program[6], for which the parameters are published.[3] In this work we have assumed a linear response for energy versus the fast light output signal of the CsI detectors.[4] For each detector, two separate linear relationships faithfully reproduced the spectral shapes[3] for ^{4}He and ^{1}H respectively from ≈ 14-32 MeV and from ≈ 6-19 MeV. Thresholds were ≈ 9 MeV and ≈ 4 MeV for ^{4}He and ^{1}H. Average energies of the ^{1}H and ^{4}He particles were reproducible to ≈ 0.3 MeV. The energy response functions for ^{1}H were also used for $^{2,3}H$; this assumption gave ^{2}H and ^{3}H spectra for 312 MeV ^{40}Ar that are also consistent with those from Ref. 3.

In Fig. 6 we show four pairs of energy spectra, each recorded in the sixteen-detector EMRIC array. Those on the left (right) result from ^{4}He - ^{4}He (^{1}H-^{1}H) coincidences; the solid (open) points are from the reaction with ^{40}Ar of 312 MeV (680 MeV); the circles (squares) are from in-plane IP (out-of-plane OOP) coincidences with the single detector shown in Fig. 5 on the right of (above) the beam axis. The data in Fig. 6 give a lot of information about the reactions. Let us begin by examining the solid points on the right for ^{1}H from 312 MeV ^{40}Ar + Ag. They show classic Maxwellian-like shapes for proton evaporation: a sharp rise at low energies from ≈ 4-8 MeV, a maximum at ≈ 9 MeV, and a straight exponential decline for energies ≥ 12 MeV. The maximum for the IP data is normalized to 6000 per MeV; bin by bin ratios of the IP to OOP spectra are shown by triangles on the lower scale. Here these ratios are essentially unity for all proton energies, i.e. there is no significant anisotropy. In Table I we list three quantities to characterize these spectra (a) $\tilde{T}$ the spectral temperature or slope parameter $[(1/\tilde{T}) = -(d \log \sigma(\epsilon)/d\epsilon)]$; (b) $<\epsilon>_{IP}$ the average energy, and (c) the energy integrated anisotropy ratio IP/OOP.

Let us now compare the data for protons from 312 and 680 MeV ^{40}Ar; we see the same classic evaporation shapes. The high energy spectral slope has changed from $\tilde{T}$ of 2.4 to 4.2 MeV, and the average proton energy has increased from 10.5 to 11.4 MeV. This rather small increase in $<\epsilon>$ results because the spectrum has also broadened at low energies to give more particles of < 10 MeV. Finally, the angular anisotropy has increased from 0.99 to 1.09 in favor of the in-plane over the out-of-plane configuration. The same pattern is also exhibited by the ^{4}He-^{4}He data, but each quantity is larger than that for ^{1}H - ^{1}H. Spectral slopes change from 3.6 to 6.7 MeV, and the average energies from 20.0 to 23.2 MeV. The angular anisotropy increases with ^{4}He energy (Fig. 6) from almost unity for alphas of $\epsilon \leq 10$ MeV to IP/OOP ≥ 2 for those alphas of $\epsilon \geq 30$ MeV. Energy integrated anisotropies for the ^{4}He-^{4}He pairs (Table I) increase with beam energy from 1.82 to 1.95. Table I also gives data for several other particle pairs.

Before interpreting these results, let us give a brief overview of the classical picture of statistical evaporation.[7,8] An intuitive feeling can be obtained from the schematic diagram in Fig. 7. Particle emission from a spinning nucleus can be pictured as emission of a charged, but sticky granule from the surface of a charged rotating sphere. Emission probability $d\Gamma_{E,J}$ in a given direction $\vec{n}$ (or angle ϕ) is anisotropic; the average energy $<\epsilon>_{\phi}$ also depends on ϕ. The strength of these anisotropies is determined by β_2 the ratio of the rotational energy of the granule on the emission surface $[(\hbar^2(J + 1/2)^2/2\, \Im_d)\, (\mu R^2/(\Im_d + \mu R^2))]$ to its temperature T (symbols defined in Fig. 7 caption). Therefore, the anisotropy increases with rotational energy of the emitter and with mass and radial coordinate of the granule, but decreases with increasing temperature. Of course, the average emission energy $<\epsilon>_{\phi}$ increases with increasing temperature, but it decreases with radial expansion due to a concomitant decrease in the Coulomb barrier B.

These features of the mechanical analogue model also result from a more complete statistical model.[7] Therefore, they allow us to make a simple qualitative interpretation of the observations given in Fig. 6 and in Table I. First, the spectral slopes $\tilde{T}$ increase by factors of 1.6-2.1, a reasonable change for the expected increase in energy deposition by fusion reactions[10] of 7.8 to 17 MeV/nucleon ^{40}Ar. However, there are rather modest increases in the average energies $<\epsilon>_{IP}$ due to a growth in the abundance of low as well as high energy particles. This increase in low energy particles signifies a decreased Coulomb barrier to particle emission. Finally, the slight increase in angular anisotropies indicates somewhat larger anisotropy parameters (β_2). Since the temperature has also increased, this change in β_2 points toward a significantly larger emitter spin.

These observations have been interpreted more quantitatively by use of the statistical-model simulation code GANES[3,6,7]; we have made fits to the ^{1}H and ^{4}He data in order to obtain parameterizations of the average emitter characteristics. These parameters are listed in Table II. As discussed above, the root-mean-square emitter spin J_{rms} has increased along with β_2, the anisotropy parameter. The effective temperatures have also increased, and the emission barriers have decreased. Even at the relatively low average temperature of ≈ 2.2 MeV for 312 MeV ^{40}Ar, these emission barriers indicate an of increase of ≈ 1.5 fm in the average surface radius for protons compared to cold nuclei.[3] Similar evidence for radial distention has been reported in Ref. 9. The major point here is that the qualitative role of the Coulomb barriers is quite evident in the low energy cutoffs of the spectral shapes (Fig. 6). Hence the notion still obtains of particles evaporating from a rather compact source, held together by its mean field.

In fact, all of these observations seem to be consistent with a reasonable picture for the properties of a thermalized, but very hot and somewhat swollen compound nucleus. Furthermore, the statistical evaporation model provides a formula, simply from phase space considerations, to estimate the emission lifetimes of such hot compound nuclei.[8] For these estimations we have assumed an average momentum transfer of 80% for incomplete fusion of 680 MeV ^{40}Ar (i.e. complete fusion of 544 MeV ^{32}S + Ag).[10] Calculated values of τ are shown in Fig. 8 along with values for the average emission temperatures. These lifetime estimates are very short indeed, but they can be tested by means of the small-angle correlation measurements recorded in EMRIC.

Data analysis procedures followed those described in Refs. 11 and 12. In brief, two-particle coincidence events from EMRIC were summed over energy and angle and binned in relative momentum P_{rel} to form a spectrum $A(P_{rel})$. A two-particle correlation function is defined as the ratio $A(P_{rel})/B(P_{rel})$ where $B(P_{rel})$ is a reference spectrum for the same class of particles, but from separate events. We have chosen to use event mixing for the construction of $B(P_{rel})$ as described in Refs. 11-13. There is no arbitrary normalization of the correlation function in this method; one simply uses the same number of events for both $A(P_{rel})$ and $B(P_{rel})$.

In Fig. 9 we show measured correlation functions for three particle pairs: ^1H-^1H (a and b), ^1H-^2H (c and d), and ^2H-^2H (e and f). Data points are compared to curves obtained by reaction simulations described below. The major feature of the data from 312 MeV ^{40}Ar is that all the correlations are very weak. In fact, only for ^1H-^1H is there any significant deviation from unity. However, for the data from 680 MeV ^{40}Ar + Ag there are noteworthy anticorrelations for both ^1H-^1H and for ^2H-^2H for P_{rel} values ≤ 20 MeV/c. The flat correlation functions for 312 MeV can be said to result from rather large time intervals (hence spacial distances) between the particles; the anticorrelations for 680 MeV presumably arise from considerably smaller time intervals between the first few emitted particles. The reaction simulations described below translate this idea into a quantitative form.

To obtain predictions of these correlation functions from the statistical model, one must first have a description of the stepwise decay of the emitter. This involves estimation of the lifetimes, spectral shapes and relative yields as discussed above. With this information as input, we have made reaction simulations that follow the trajectories of each particle. The simulation program we use is an extension of the Monte Carlo code KALLIOPI described elsewhere.[13] The new simulation program [2] proceeds as follows: (a) Input parameters are read that describe the physical processes as well as the acceptance geometry and energy thresholds of the detectors. (b) Individual emission steps, angles, and energies are chosen for each particle of a pair by random selections. (c) The relative orbital angular momenta are also chosen by random selections. (d) Finally the cummulative time delay between two particle emissions is chosen by random selections. For example, if the first particle is emitted at step i and the second at step j, then t_{delay} = t (i + 1) + ...t (j) where in each step t(k) is chosen from the distribution t(k) = exp $(-t/\tau_k)$. (e) For the time interval between emissions a two-body Coulomb trajectory is followed for the first particle plus the recoil nucleus. (f) Then the second particle is born and a three-body trajectory is followed. (A nuclear force can be used between particles as well as the Coulomb force, here its effect is minimal.) (g) The final laboratory directions and energies are checked to test for a valid coincidence. (h) Spectra are binned and correlation functions are built in exactly the same way as in the data analysis.

The smooth curves in Fig. 9 represent simulation calculations as just described. For the 312 MeV ^{40}Ar reaction, the calculated correlation functions are quite weak, in agreement with the data (Figs. 9a, c,e). These weak correlations result from accumulations of time intervals over randomly selected steps. Although the initial mean lifetime is calculated to be only 6.6. x 10^{-22} s, it is relatively rare that one selects particles from steps 1 and 2. More often, at least one of the particles is chosen later in the chain, thus giving a much greater time interval.

For the 680 MeV ^{40}Ar reaction, the calculated anticorrelations are much stronger, especially for ^{2}H-^{2}H pairs. The reason for this is clear from Fig. 8; we see that the average statistical-model lifetimes are $\leq 5 \times 10^{-22}$ s for each of the first seven steps. The stronger predicted anticorrelation for ^{2}H-^{2}H compared to ^{1}H-^{1}H arises from their respective relative yields; for ^{1}H the emission yield is rather uniformly spread over fourteen steps, whereas the ^{2}H yields are heavily biased toward the first half of the chain where the time delays are short.[14] These predictions for ^{1}H-^{1}H and ^{2}H-^{2}H are in agreement with the data. However, the data for ^{1}H-^{2}H pairs seem to indicate longer time delays than predicted. Nevertheless, the overall agreement between the simulations and the data is rather good, and we conclude that in general the predictions of the evaporation model seem to be acceptable.

It is quite striking that such short lived nuclei, $\tau \approx 10^{-22}$ s, can be essentially completely thermalized. One normally expects that the thermalization time is much greater than the traversal time. A traversal time (radius/projectile velocity) for these collisions at 17 MeV/nucleon is only $\approx 10^{-22}$ s. Nevertheless, the implication of these results is that essentially complete thermalization has occurred, and that relaxed particle evaporation is also ocurring on a timescale of 10^{-22} s. If this is correct, then the timescales for fusion-evaporation reactions can be of the same order as the collision durations in peripheral collisions. A commonly cited distinction between evaporative and direct reactions is based on their timescales; here that distinction seems to be essentially nil.

In summary, we have measured particle-particle correlations from the reaction systems 312 and 680 MeV ^{40}Ar + ^{nat}Ag. Energy spectra and angular anisotropies are consistent with an evaporation model and allow for a characterization of thermalized composite nuclei with initial temperatures of ≈ 4.5 MeV and root mean square spins of ≈ 60ℏ. The observed small-angle correlations are consistent with statistical evaporation on a time scale of ≈ 10^{-22} s. This implies energy mixing among the nucleons in times comparable to the traversal time.

References

1. A. Elmaani, J. M. Alexander, R. Lacey, S. Kox, E. Liatard, F. Merchez, T. Motobayashi, B. Noren, C. Perrin, D. Rebreyend, Tsan Ung Chan, G. Auger and S. Groult, (unpublished preprint 1990.

2. A. Elmaani, N. N. Ajitanand, J. M. Alexander, T. Ethvignot and R. L. McGrath, (unpublished preprint 1990).

3. R. Lacey, N. N. Ajitanand, J. M. Alexander, D. M. de Castro Rizzo, G. F. Peaslee, L. C. Vaz, M. Kaplan, M. Kildir, G. La Rana D. J. Moses, W. E. Parker, D. Logan, M. S. Zisman, P. DeYoung and L. Kowalski, *Phys. Rev. C* **37**, 2540 and 2561 (1988).

4. F. Merchez, S. Kox, C. Perrin, J. Mistretta, J. C. Gondrand and L. N. Imouk, *Nucl. Inst. and Meth. in Physics Research* **A275**, 133 (1989).

5. E. Liatard, S. Akrouf, J. C. Bruandet, A. Fontenille, F. Glasser, P. Stassi and Tsan Ung Chan, *Nucl. Inst. & Meth. in Physics Research*, **A267**, 231 (1988).

6. N. N. Ajitanand, R. Lacey, G. F. Peaslee, E. Duek and J. M. Alexander, *Nucl. Inst. and Meth. in Physics Research* **A243**, 111 (1986).

7. N. N. Ajitanand, G. LaRana, R. Lacey, D. J. Moses, L. C. Vaz, G. F. Peaslee, D. M. de Castro Rizzo, M. Kaplan, and J. M. Alexander, *Phys. Rev.* **C34**, 877 (1986).

8. T. Ericson, *Adv. Phys.* **9**, 423 (1960); T. Dossing, *Licentiat Thesis,* University of Copenhagen (1979).

9. G. Nebbia, K. Hagel, D. Fabris, Z. Majka, J. B. Natowitz, R. P. Schmitt, B. Sterling, G. Mouchaty, G. Berkowitz, K. Strozewski, G. Viesti, P. L. Gonthier, B. Wilkins, M. N. Namboodiri and H. Ho, *Phys. Lett.* **176B**, 20 (1986); K. Hagel, D. Fabris, P. Gonthier, H. Ho, Y. Lou, Z. Majka, G. Mouchaty, M. N. Namboodiri, J. B. Natowitz, G. Nebbia, R. P. Schmitt, G. Viesti, R. Wada and B. Wilkins, *Nucl. Phys.* **A486**, 429 (1988).

10. D. Jacquet, G. F. Peaslee, J. M. Alexander, B. Borderie, E. Duek, J. Galin, D. Gardes, C. Gregoire, H. Fuchs, M. Lefort, M. F. Rivet and X. Tarrago, *Nucl. Phys.* A511, 195-220 (1990), and references therein.

11. P. A. DeYoung, M. S. Gordon, Xiu qin Lu, R. L. McGrath, J. M. Alexander, D. M. de Castro Rizzo, L C. Vaz, Phys. Rev. C **39**, 128-131 (1989).

12. P. A. DeYoung, C. J. Gelderloos, D. Kortering, J. Sarafa, K. Zienert, M. S. Gordon, B. J. Fineman, G. P. Gilfoyle, X. Lu, R. L. McGrath, D. M. de Castro Rizzo, J. M. Alexander, G.

Auger, S. Kox, L. C. Vaz, C. Beck, D. J. Henderson, D. G. Kovar, and M.F. Vineyard, *Phys. Rev.* C41, R1885 (1990).

13. R. L. McGrath, A. Elmaani, J. M. Alexander, P. A. DeYoung, T. Ethvignot, M. S. Gordon, and E. Renshaw, (*Computer Physics Communications*, in press, 1990)

14. N. N. Ajitanand and J. M. Alexander, (unpublished preprint, 1990).

Table I: Spectral slopes $\tilde{T}$ (MeV), average energies $<\varepsilon>_{IP}$(MeV), and anisotropies IP/OOP [a]

	312 MeV $^{40}Ar + ^{nat}Ag$			680 MeV $^{40}Ar + ^{nat}Ag$			
Particles	$\tilde{T}$	$<\varepsilon>_{IP}$	IP/OOP	$\tilde{T}$	$<\varepsilon>_{IP}$	IP/OOP	$\tilde{T}_{680}/\tilde{T}_{312}$
PROTONS							
$^1H-^1H$	2.4	10.5	0.99	4.2	11.4	1.09	1.8
$^1H-^2H$	2.3	10.3	1.04	4.3	11.4	1.13	1.9
DEUTERONS							
$^2H-^1H$	2.7	13.7	1.04	4.3	14.7	1.13	1.6
$^2H-^2H$				4.3	14.6	1.25	
ALPHAS							
$^4He-^1H$	3.4	19.6	1.17	7.0	22.4	1.27	2.1
$^4He-^4He$	3.6	20.0	1.82	6.7	23.2	1.95	1.9

[a]Relative uncertainties are ≈ 5% for $\tilde{T}$, and ≈ 2% for $<\varepsilon>_{IP}$ and IP/OOP

Table II: Statistical properties of the composite nucleus emitters.

	312 MeV ^{40}Ar		680 MeV $^{40}Ar^{a}$	
	^{1}H - ^{1}H	^{4}He-^{4}He	^{1}H - ^{1}H	^{4}He-^{4}He
$J_{rms}(\hbar)$	50	50^{b}	53	64
β_2	0.43	1.7	0.52	1.8
B (MeV)	6.0^{b}	13.2^{b}	5.5	11.0
B_{fus} (MeV)	8.49^{b}	15.9^{b}		
<FEL> [c]	0.33^{b}	0.33^{b}	0.5	0.25
<T> (MeV)	2.2	2.2	2.95	4.0

[a] Values taken from fits to the data in Fig. 6. and Table: I, Due to the dominance of incomplete fusion we make calculations for complete fusion of ^{32}S also at E/A= 17 MeV (i.e. 544 MeV ^{32}S + ^{nat}Ag)

[b] Values taken from Ref. 3.

[c] FEL denotes fractional energy loss to evaporation and deformation for the average emission process

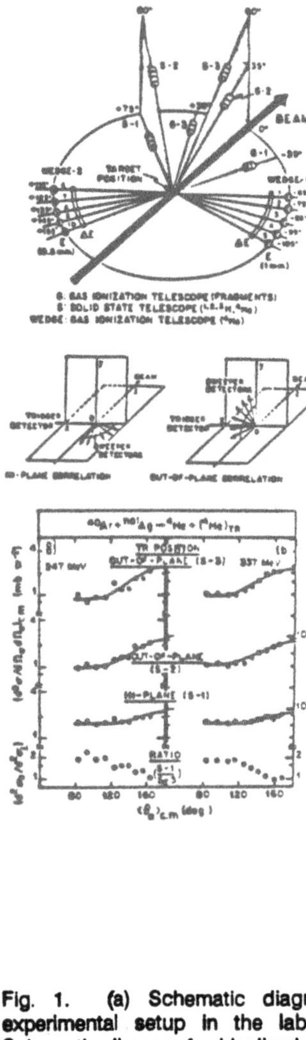

Fig. 1. (a) Schematic diagram of the experimental setup in the laboratory. (b) Schematic diagram for idealized in-plane and out-of-plane correlations in the c.m. For particle-particle correlations the former employs S-1 as trigger, while the latter employs S-3 as trigger. (c) Measured angular correlations for ^{4}He in coincidence with the ^{4}He trigger. Solid curves were calculated for spherical emitters with initial spin zone $35 < J < \ell_{ER}$ and mean excitation energy 2/3 of that calculated for the compound nucleus (E_{CN}). After Ref. 3.

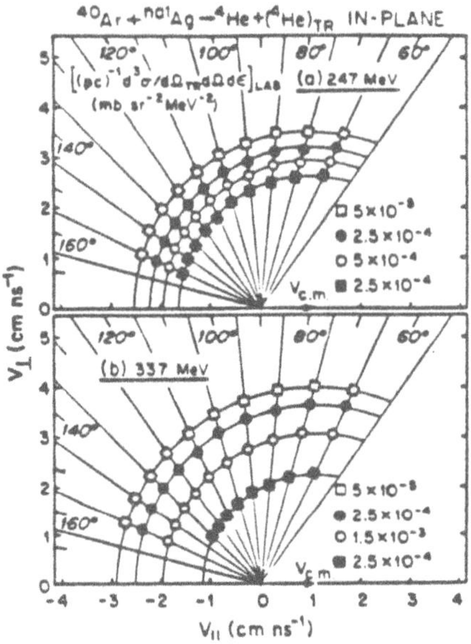

Fig. 2. Countour maps of the invariant cross section for ^{4}He triggered for ^{4}He (in-plane, S-1) for ^{40}Ar beams of (a) 247 MeV and (b) 337 MeV. The circular arcs are centered on $V_{c.m.}$, the velocity of the center of mass. Straight lines are drawn along the laboratory angles as indicated. After Ref. 3.

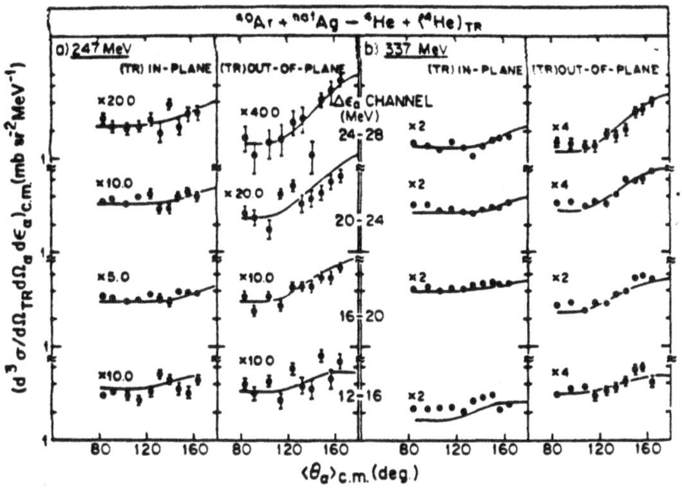

Fig. 3. Measured ^{4}He-^{4}He correlation distributions (in the c.m.) as a function of the energy bin $\Delta\epsilon_a$. Solid curves were calculated as described in Fig. 1(c). After Ref. 3.

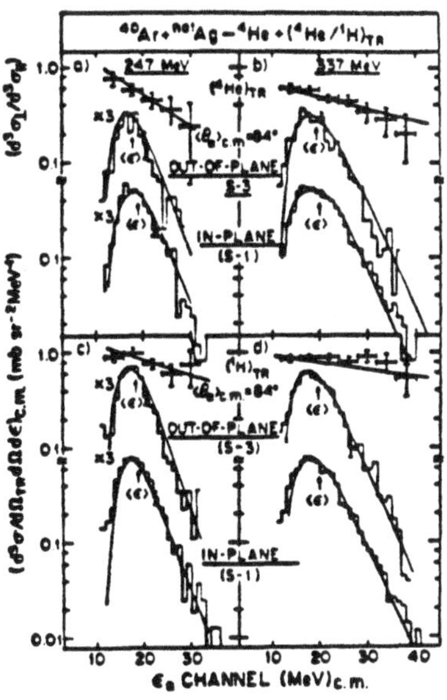

Fig. 4. Energy spectra (c.m.) for ^{4}He detected at $<\theta>$ = 84° in coincidence with ^{4}He [(a) and (b)] or ^{1}H [(c) and (d)]. The trigger detector was S-3 (out-of-plane) or S-1 (in-plane) as indicated. The ^{40}Ar beam energy was 247 MeV [(a) and (c)] or 337 MeV [(b) and (d)]. The ratio of out-of-plane to in-plane cross sections is indicated by the points at the top of each frame. The smooth curves were calculated for a spherical emitter with s-wave emission barrier B = 13.2 MeV, fractional energy loss of 30% and the same spin zones indicated in Fig. 1. After Ref. 3.

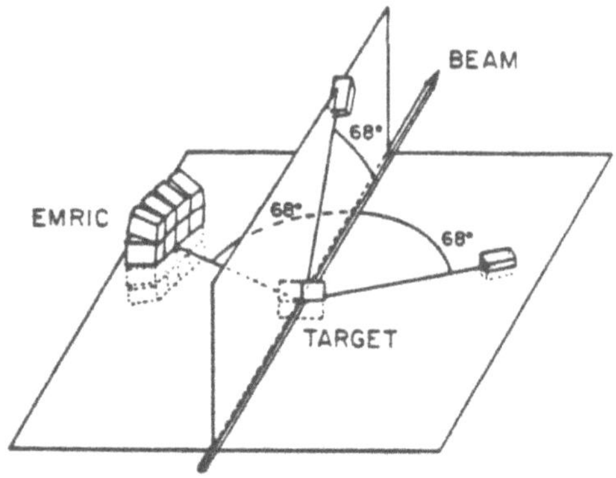

Fig. 5. Schematic diagram of the experimental setup.

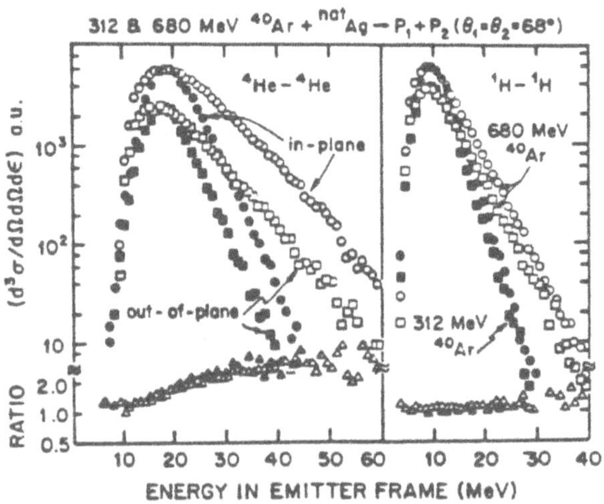

Fig. 6. Energy spectra recorded in EMRIC for ^{1}H-^{1}H and ^{4}He-^{4}He pairs; solid points for 312 MeV ^{40}Ar; open points for 680 MeV; circles for in-plane (IP) coincidences; squares for out-of-plane (OOP). Triangles show the anisotropy ratios IP/OOP. Emitter frame transformations assumed 100% and 80% momentum transfer respectively for 312 and 680 MeV ^{40}Ar.

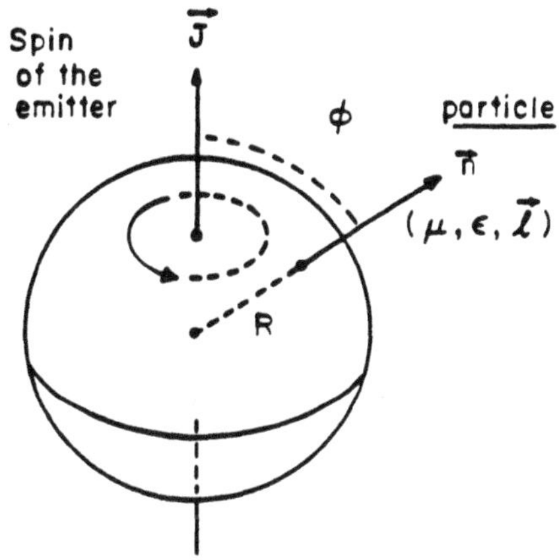

$$d\Gamma_{E,J}(\vec{n}) \propto \exp(\beta_2 \sin^2\phi)$$

$$\langle \epsilon \rangle_\phi = B + T + T\left[\frac{\mathcal{J}_d}{\mathcal{J}_d + \mu R^2}\right](1 + \beta_2 \sin^2\phi)$$

$$\boxed{\beta_2 = \frac{\hbar^2 (J+1/2)^2}{2\mathcal{J}_d T}\left[\frac{\mu R^2}{\mathcal{J}_d + \mu R^2}\right]}$$

Fig. 7. Schematic diagram for emission of a particle (reduced mass μ, channel energy ϵ, orbital angular momentum $\vec{\ell}$) in the direction $\vec{n}$ at angle ϕ to the spin $\vec{J}$ of a spherical emitter of excitation energy E. The emission radius is R, emission barrier is B, and daughter moment of inertia is $\mathcal{J}_d$. Emission width is $d\Gamma$ and anisotropy parameter is β_2.

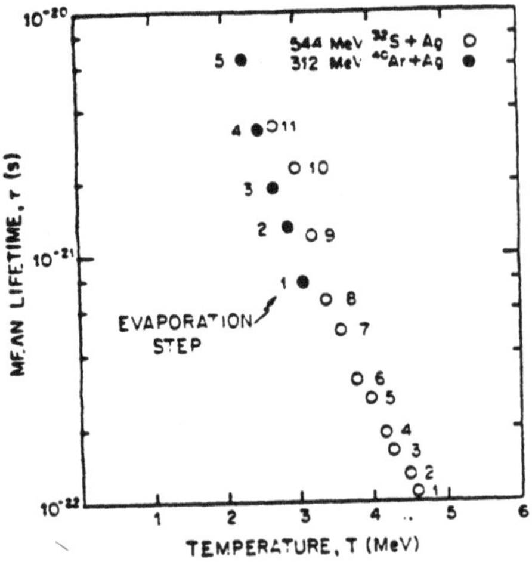

Fig. 8. Calculated mean lifetimes[2,8] versus temperature and evaporation step for the two reactions of interest.

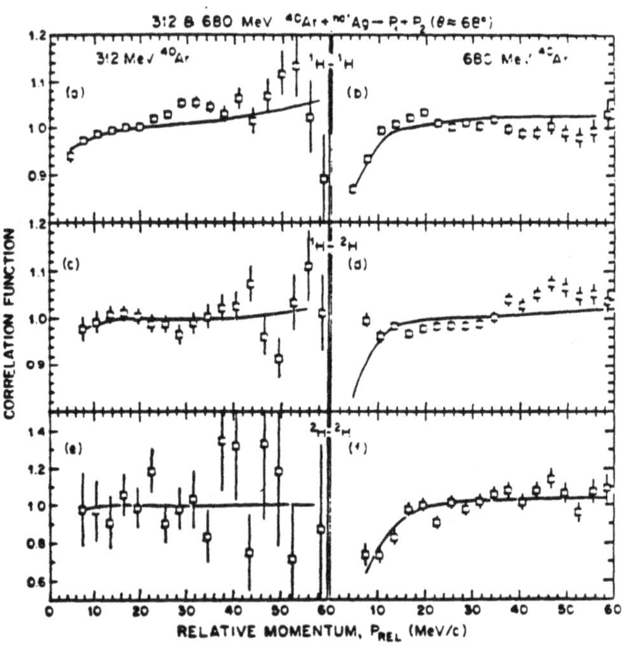

Fig. 9. Correlation functions vs relative momentum P_{rel} for ^1H-^1H, ^1H-^2H, and ^2H-^2H. Points are for small-angle correlation data taken in EMRIC; smooth curves are from reaction simulations. The results from 312 MeV ^{40}Ar are on the left and 680 MeV ^{40}Ar on the right. These simulations use only Coulomb trajectories, but inclusion of a reasonable nuclear force makes very little difference; statistical uncertainties are comparable to or better than shown for the data.

PROBING THE LIFETIME OF EXCITED COMPOSITE NUCLEI WITH PARTICLE-PARTICLE CORRELATIONS

P.A. DeYoung[1], J.M. Alexander[3], J.J. Kolata[4], D. Kortering[1], R.A. Kryger[4], C.J. Gelderloos[1], M.S. Gordon[2], R.L. McGrath[2], J. Sarafa[1], and R. Sedlar[1]

[1]Physics Department, Hope College, Holland, MI 49423 USA
[2]Department of Physics, State University of New York at Stony Brook, Stony Brook, NY 11794-3400 USA
[3]Department of Chemistry, State University of New York at Stony Brook, Stony Brook, NY 11794-3400 USA
[4]Physics Department, University of Notre Dame, Notre Dame IN 46556 USA

ABSTRACT

Particle-particle correlation techniques can be applied to small angle coincidence measurements where the primary source of the detected particles is thought to be a statistically equilibrated composite nucleus. Such application allows one to investigate the lifetimes in such systems and probe the limits of the statistical model for high excitations. In this report data will be shown for the small-angle particle-particle correlations from the reactions 80, 140, 215, and 250 Mev $^{16}O + ^{27}Al$ giving p-p, p-d, p-α, d-α, and α-α pairs. In addition the results from n-p, n-d, and n-α measurements at 215 MeV will be presented. Finally, the results from $^{32}S + ^{27}Al$ at 105 and 215 MeV will be shown. The correlations are seen to be weak at the lowest energies and become progressively stronger with increasing excitation energy. Simulations of the correlation functions which include: lifetimes and decay probabilities based on the statistical model, particle trajectories, and cooling of the source, will be compared to the measured results. The general trends of the p-p, p-d and n-d data are well reproduced by the simulations. However, the correlations of other particle pairs are complicated by the sequential breakup of particle unbound states such as ^8Be or ^6Li. Some preliminary results where this process is included will be shown.

1. Introduction

The characteristics of highly-excited composite nuclei are often modeled within the context of a statistical approach. While one cannot predict the specific particle, directions, or energies of the emitted particles, it is felt that since the composite nucleus is (in some sense) a soup of nucleons that an average or statistical approach to predicting the decay on average is valid. The validity of this approach is limited to excitation energies where the time for particle emission (evaporation) is greater than the thermal equilibration time. It is important to establish the limits of this approach in order to gain some guidance in understanding both the statistical model and the newer and more novel physics which sets in at the higher energies.

One way to gain insight into the space-time properties of emitting sources is with small angle correlation techniques. In recent years these correlation techniques have been widely applied to a wide variety of questions at both relativistic, intermediate, and low energies.[1-8] The correlations may come from a wide variety of interactions (Coulomb forces, quantum statistics, nuclear force etc.) but in all cases the size of the measured correlations (or anticorrelations) is thought to be related to the space-time extent of the source of particles. This work, which concentrates on evaporation from a composite source, is mainly sensitive to Coulomb interactions and lifetime effects.

2. Experimental Configuration and Data Analysis

The results presented in this report come from six different experiments performed with the superconducting linac at the State University of New York at Stony Brook and with the ATLAS facility at Argonne National Laboratories. The numerous specific details of each experiment can be found elsewhere[7,8]. They were all similar enough that only an overview will be given here. The $^{16}O + ^{27}Al$ system was measured at the four beam energies of 80, 140, 215, and 250 MeV and the $^{32}S + ^{27}Al$ system was done at 105 and 215 MeV. In these works light charged particles (LCP) (p, d, t, and α) were detected in NaI(Tl) crystals. Particle identification was based on pulse shape and time-of-flight.[9] The detectors were placed between 70 and 124 cm (depending on experiment) from the target in a hexagonal close-packed configuration involving seven detectors. The mean laboratory scattering angle (the angle of the center detector) ranged from 45° to 53.5°. The angle between adjacent detectors varied from 4.62° to 3.00° and the solid angle of each detector was on the order of 1 msr. Only data from small angle coincidences (both particles detected in the same hexagonal array) has been included in this report. Sufficient statistics were obtained in the p-p, p-d, p-α, d-α, and α-α channels for all cases except the lowest energy S+AL work. Here the low number of emitted deuterons prevented measurement of the p-d or d-α channels.

A standard procedure for forming the correlation function was followed for all the cases. The relative momentum (P_{rel}), defined as

$$P_{rel} = \mu \left(v_1^2 + v_2^2 - 2v_1v_2\cos\theta\right)^{1/2}$$

where μ is the reduced mass, v is the velocity, and θ is the opening angle of the two detectors, was calculated for each coincidence event from within a single cluster of seven detectors. The resulting values were then summed over energy and angle to form a spectrum $A(P_{rel})$. The correlation function is the ratio $A(P_{rel})/B(P_{rel})$ (bin by bin) where $B(P_{rel})$ is a reference spectrum which is free from the correlations of interest. Because $A(P_{rel})$ was formed by summing over different detector pairs (with differing energy thresholds and angles) the $B(P_{rel})$ must also reflect these biases in order for the correlation function to be accurate. For all the results presented here the $B(P_{rel})$ spectrum was formed by event mixing[10]. In this process the energies for two events which have the same characteristics (same pair of detectors, same particle types in each detector, same real or random classification, etc.) are interchanged. That is, a new event is formed from the energy of the first particle from event i and the energy of the second particle from event $i+1$. This guarantees that the biases of the A are reflected in the B. As discussed in Reference 10 this may underestimate the strength of the correlations and anticorrelations and distort the shape of the correlation function when strong correlations and anticorrelations are present. As will be discussed below, this technical difficulty is overcome by applying event mixing to the simulated data.

The particle energies were calibrated with scattering from lithium hydride and and lithium deuteride targets, the $^{12}C(^{16}O,\alpha)$ reaction at 53 MeV, and from the kinematics of the breakup of ^{6}Li. Because of the large pulse height defect in NaI it was necessary to separately calibrate the Z=1 and Z=2 particles. The breakup information serves as a check of the consistency between the two calibrations.

Neutron-LCP correlations from the $^{16}O+^{27}Al$ system at 215 MeV were also measured. The neutron-LCP measurement was made in similar fashion to the LCP-LCP measurements described above. The differences were in three areas. The first is that in the hexagonal close-packed array the two locations above the plane and the two locations below the plane were occupied by neutron detectors and the NaI detectors were placed in the three in-plane locations. The second difference from the above experimental description is that neutrons were discriminated from γ rays based on the signal pulse shape. Finally, the neutron energy was determined from the neutron time-of-flight.

3. Overview and Orientation

The systems which are covered in this report are distinguished from most other correlation studies by the fact that they are relatively low energy studies. For these systems and the particular scattering angles chosen, the majority of the light particles are thought to be evaporated from an equilibrated composite nucleus. The energies of these studies are such that contributions from other processes, besides the formation of an equilibrated compound nucleus, especially for the 250 MeV measurements, are occurring. With this limitation in mind we have proceeded as though all particles come from an equilbrated compound nucleus (CN). Table 1 shows some of the gross characteristics for the various reactions. We see that the S+Al reaction at 105 MeV is matched to the O+Al system at 80 MeV and the 215 MeV S+Al reaction has the same excitation as the 140 MeV O+Al case. Of course in each case the energy is spread over more nucleons (59 for S+Al versus 43 for O+Al) so that the temperature of the CN will be lower in the S+Al cases than the corresponding O+Al compound nucleus. Based on the spins and excitation energies the assumption that an equilibrated compound nucleus is formed may not be unreasonable. This assumption is also confirmed by the measured angular

Table 1: A summary of the major characteristics of the systems studied in this report. The compound nucleus spin is the average value based on a statistical model calculation. Similarly, the chain length is an approximate value for the average number of particles emitted during the decay process as predicted by the statistical model.

System	Beam Energy	C.M. Energy	CN Excitation	CN Spin	Steps
	MeV	MeV	MeV	$\hbar$	
$^{16}O + ^{27}Al$	80	50.2	64.5	19.6	4.5
	140	87.9	102.2	22.8	6.5
	215	135.0	149.2	22.7	8.0
	250	157.0	171.2	23.0	9.0
$^{32}S + ^{27}Al$	105	48.1	61.2	18.8	5.0
	215	98.4	111.5	30.3	6.5

distributions for p[7] and α at 140 and 215 MeV. These indicate that the cross sections at the angles chosen for the small angle correlation measurements agree with predictions based on the cross sections at more backward angles (which are more certain to originate from an equilbrated compound nucleus).

Since the small angle correlation techniques are applied here in an effort to learn about the lifetimes of the decaying CN, Table 2 is provided to give an orientation to the anticipated average (or typical) lifetimes for the early stages of the decay process. As one would expect, the lifetimes for the stages become progressively longer as the nucleus cools. The shorter lifetimes in the higher energy cases result in particles having a higher probability of being close together so that the forces acting between them with have a larger effect and produce stronger correlations. Of course these crude systematics need to be tempered by the specific probabilities

Table 2: The average lifetime for the first four decay stages in units of 10^{-22} seconds for p-p events. The values are based on the results of a Monte Carlo version of the statistical model code CASCADE[11].

System	Beam Energy (MeV)	τ_1	τ_2	τ_3	τ_4
$^{16}O + ^{27}Al$	80	6.7	12.7	395.	39000.
	140	3.1	4.5	6.6	12.1
	215	1.7	2.1	2.8	6.0
	250	1.4	1.7	2.0	2.8
$^{32}S + ^{27}Al$	105	17.2	50.5	408.	8300.
	215	4.9	7.2	14.3	35.0

of particle emission at each step and the length of the decay chain. That is, long chains tend to have short first step lifetimes (resulting in stronger correlations) but may also have more particles in the latter stages where the lifetime has become long (weaker correlations). These effects are included in the model discussed below.

While the results presented in Figure 1 and Table 3 are based on the model presented below, their implications fit best in this section which seeks to provide an overview and general understanding of the physics at work in these systems. In principle there are many effects that determine the specific shape of the correlation functions. These include Coulomb repulsion, fermion and boson spin statistics, and the nuclear force. For the systems studied in this work only the first is important. This is simply due to the fact that the lifetimes are still relatively long (even for the first steps) and the particle energies are relatively low so that the particles get close enough to experience these other forces and interactions only infrequently. Figure 1 shows the distribution of the minimum distance between the two particles

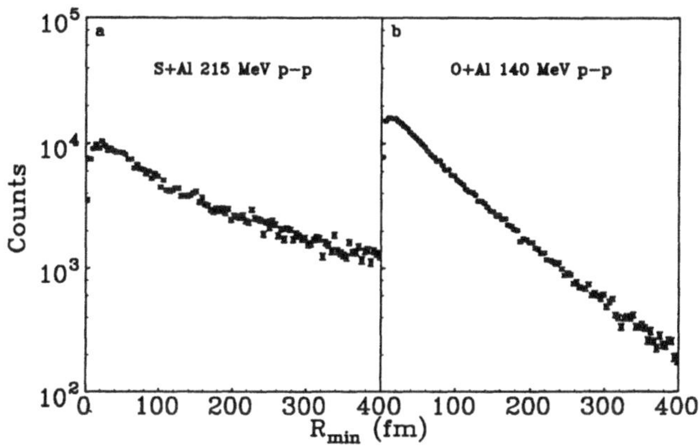

Figure 1. Two typical distributions of the distance-of-closest-approach or R_{min}. The left hand plot (a) is from simulations of p-p events from S+Al at 215 MeV and the right hand plot (b) is from p-p events from O+Al at 215 MeV. The small area below R_{min} of 5 fm may be influenced by the nuclear force.

or distance-of-closest-approach (DOCA) for p-p simulations for O+Al and S+Al at 215 MeV. Careful examination shows that the distributions have turned over at a distance of 10-20 fm and that for most of the events the particles are so far apart that only the Coulomb force can act. The results of integrating that portion of these curves which lies below 5 fm (a conservative cutoff for nuclear force effects) are shown in Table 3. Events where the two particles are close enough to experience the nuclear force do occur. However, since the numbers involved are small, a few

Table 3: The fraction of events in p-p simulations which had minimum separation less than 5 fm (at any time during their flight to the detectors). The 5 fm value was chosen as a conservative cutoff for estimating the effects of nuclear forces in these systems.

System	Beam Energy (MeV)	Percent $< 5\ fm$
$^{16}O + ^{27}Al$	80	1.6%
	140	4.1%
	215	5.1%
	250	6.5%
$^{32}S + ^{27}Al$	105	0.4%
	215	1.0%

percent at most, the nuclear force plays a minor role in determining the shape of the correlation function and will be ignored in favor of the long range Coulomb forces.

In summary, at the energies under consideration here the particles will be assumed to be statistically emitted from a thermally equilibrated compound nucleus. Futhermore, it is expected that the Coulomb force plays an overwhelming role in determining the correlation function and that the nuclear force and statistics can be ignored.

4. Experimental Results

The measured correlation functions from the different beam/target/energy configurations are shown in Figures 2–6. For purposes of discussion, the results have been divided into two sections: the first dealing with those correlations which do not exhibit resonant behavior in the two particle system (no particle-unstable breakup) and the second section dealing with those correlation functions where the breakup of particle-unstable states ($^2H^*$, ^5He, ^5Li, $^6Li^*$, and ^8Be) can influence the results. For the systems covered by this work it is believed that the first group is useful for understanding the lifetime behavior of the compound nucleus while the second set is not. The lines seen in the various correlation functions are the results of simulations described below.

4.1 Results Free from the Influence of Particle-Unbound States.

Figure 2 shows all the p-p and p-d results from the $^{16}O + ^{27}Al$ reaction. Several general features can be seen in these results. The most striking feature is the anticorrelation which is present in all the correlation functions for values of P_{rel} below 25 MeV/c. Futhermore, this anticorrelation is seen to become stronger with increasing energy (or excitation energy). In principle, the p-p results do not belong in this section due to the possibility of emitted ^2H perturbing the results. Indeed a broad peak ($\approx$ 25 MeV) may be present in the p-p results at 215 MeV which could be due to the influence of ^2H, but the results at 250 MeV are even less conclusive.

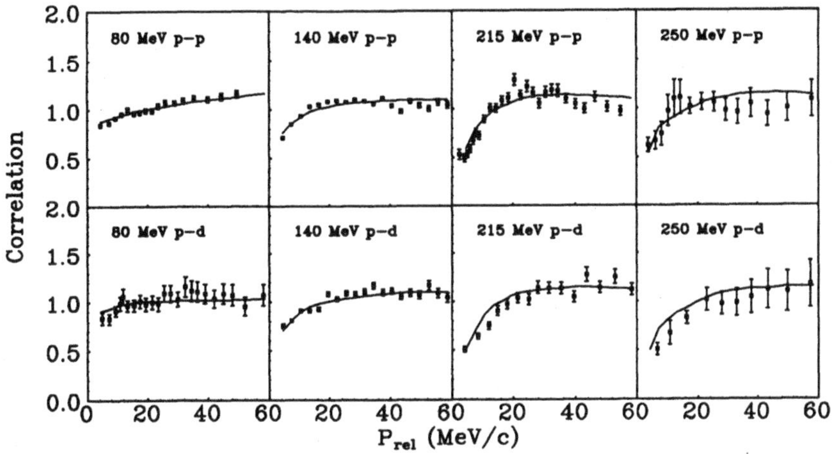

Figure 2. The p-p and p-d correlations for O+Al at 80, 140, 215, and 250 MeV incident energy. For both the data and the calculations there is no arbitrary normalization.

However, these data have been treated as being free from such effects for purposes of comparison with simulations based on the evaporation model.

Results from the $^{32}S + ^{27}Al$ system for p-p and p-d correlations are shown in Figure 3. The p-p and p-d correlation functions exhibit less pronounced correlations than the O+Al results. (No p-d is shown for 105 MeV because of insufficient statistics.) This is not unexpected and is due to the lower excitation energies and the fact that the excitation energy must be divided among the larger number of nucleons present in the compound system. Indeed careful comparison of the S+Al and O+Al results shows that the strength of the anticorrelation tracks with the excitation-energy/number-of-nucleons.

All the neutron-LCP correlation functions (both those affected by particle-unstable states and those unaffected by such states) from O+AL at 215 MeV are shown in Figure 4. Here we will only focus on the n-d result. The unique feature in this case is that there is no Coulomb force acting between the two particles. Indeed, the correlation function is seen to be flat. (The single point at the lowest value of P_{rel} which does not overlap a value of 1.0 is interesting but not understood at this time.) For LCP-LCP coincidences the nuclear forces can be ignored in comparison to the Coulomb effects, but here the attractive nuclear force (even though the potential contains no resonances) should be the dominant force determining the shape of the correlation. If enough particle pairs have a minimum distance within the range of the nuclear force (which will depend on the particular lifetimes) then a positive correlation should be observed. This question will be addressed during the discussion of the simulations and interpretation of the data in light of the

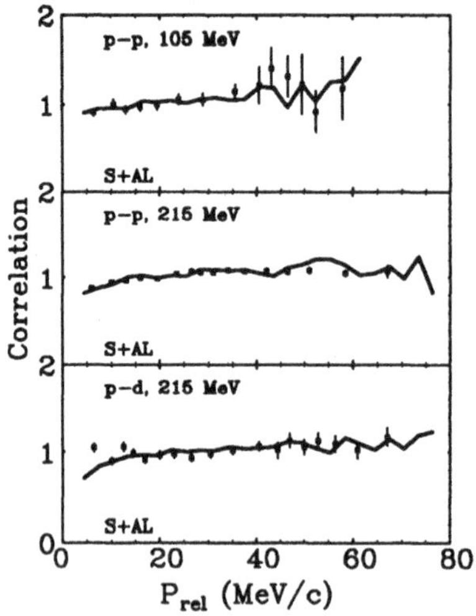

Figure 3. The p-p and p-d correlations for S+Al at 105 and 215 MeV incident energy. For both the data and the calculations there is no arbitrary normalization. (The jaggedness of the line is due to statistics on the simulation.)

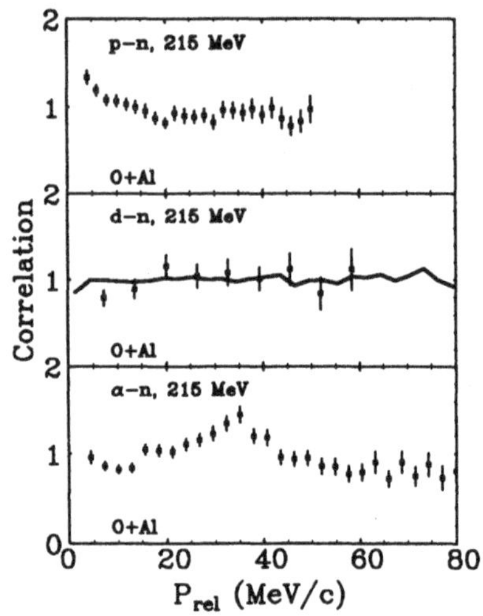

Figure 4. The p-n, d-n, and α-n correlations for O+Al at 215 MeV incident energy. (The jaggedness of the solid line is due to statistics of the simulation.)

simulations.

4.2 Result Affected by Particle-Unbound States

The remainder of the correlation functions, those with contributions from the break up of particle-unbound states, are shown in Figures 4–6. The main feature of of these correlation functions is the large positive correlations due to the breakup processes. Again, the correlations are observed to become stronger as the energy is increased but probably because the production of the unstable particles is increasing. The particular detector configuration for these measurements is very efficient at measuring these breakup events. To first order the efficiency for detecting independent particles goes as the square of the efficiency (determined mainly by the solid angle of a single detector) of detecting one particle. The detection efficiency for the breakup events has a stronger energy dependence but is much larger than the efficiency for independent particles. Basically, if one of the daughter particles hits a detector the other is directed in the general vicinity of the cluster rather than being directed into 4π. In principle, the interaction of independent particles which approach close enough for the nuclear force to play a role would also give such positive correlations. As discussed above, there are very few such events (due

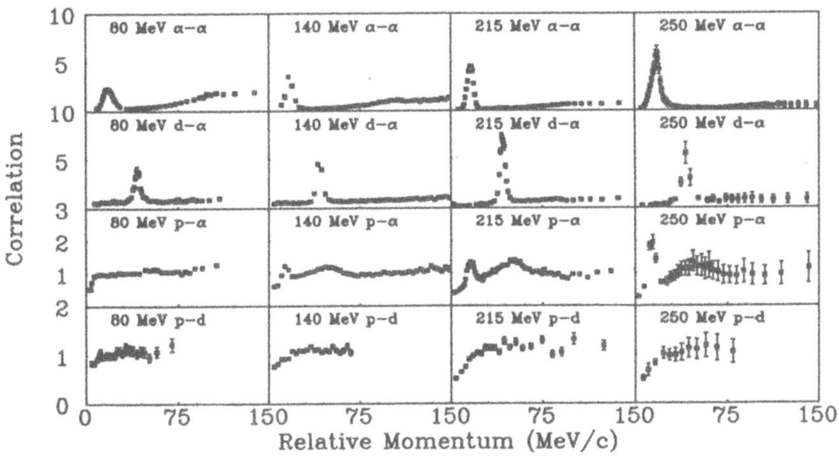

Figure 5. The p-α, d-α and α-α correlations for O+Al at 80, 140, 215, and 250 MeV incident energy. The p-d correlations are include for purposes of comparison.

to the relatively long lifetimes) and certainly not enough to produce the prominent correlations observed here.

The ^{32}S $+^{27}$ Al results, Figure 6, deserve one additional comment. While the p-p and p-d results seemed to scale according to the excitation-energy/nucleons, the S+Al correlations involving unstable states are more arbitrary. The correlations (p-α, d-α, and α-α) from S+AL at 215 MeV seem more consistent with the O+Al results from 80 MeV. (Again there is no d-α correlation at 105 Mev for S+Al because of low statistics.)

5. Statistical Model Simulations

Because the majority of particles emitted in these studies are believed to be emitted from an equilibrated compound nucleus, a simulation based on the statistical model was developed[12]. Basically, the statistical evaporation model, which forms the basis of our simulation is based on the principle of detail balance[13]:

$$\rho_A w_{ab} = \rho_B w^*_{ba}$$

Where the number of distinguishable states for the initial (final) system is denoted by ρ_A (ρ_B) and the transition rate from a to b (b to a) is w_{ab} (w^*_{ba}). From this the total transition rate from a to b is P_ν which is given by the integral

$$P_\nu = \int w_{ab} = \int \rho_B w^*_{ba}/\rho_A$$

where the integral is over all possible paths from a to b. This equation has been implemented in a number of computer codes. Often these codes have calculated

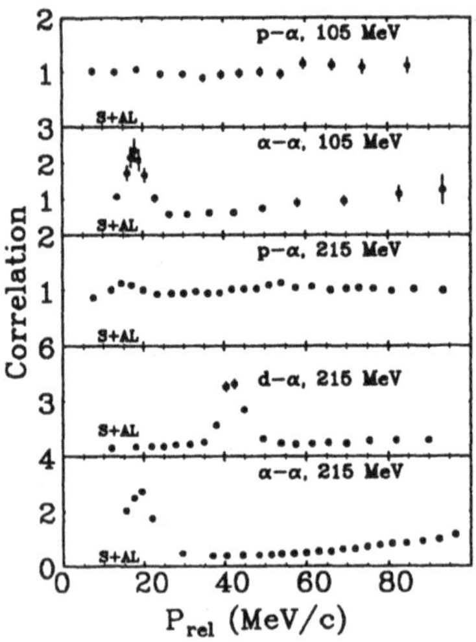

Figure 6. The p-α, d-α, and α-α correlations from S+Al at 105 and 215 MeV incident energy.

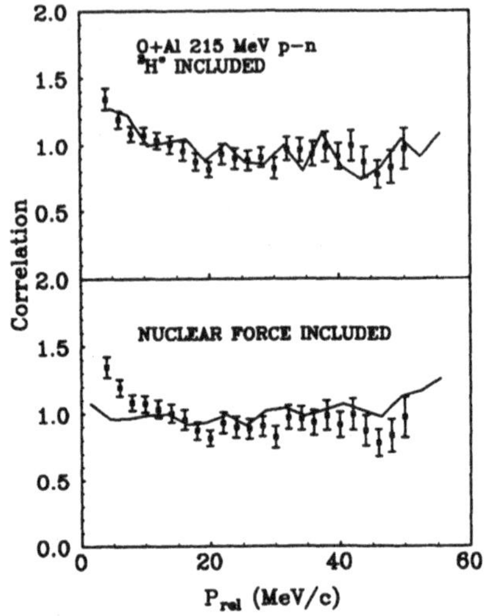

Figure 7. The p-n correlation function from 215 MeV O+Al. The solid line is a preliminary result of a COULBOMB simulation in which 1.5% of the trial events are $^2H^*$. The dashed line is a preliminary COULBOMB simulation in which the particles interact via both the Coulomb force and a classical implementation of the nuclear force. (The jaggedness is due to the statistics of the simulation.)

only the relative decay rates rather than the absolute rates (which can be inverted to find lifetimes) but if the level densities of both the emitter and the residual nucleus (ρ_A and ρ_B are generally taken from the Fermi-gas model) are determined and some model in used for the inverse reaction probabilities w^*_{ba}, the computer code can predict the absolute decay rate or lifetime.

All the simulations shown (solid lines in the figures) are based on the computer code CASCADE[11] for the character of the decay process. CASCADE was run in a Monte Carlo mode to generate a large number of decay cascades. These cascades were sorted to determine the decay probabilities and energy spectra at each stage of the decay. These stage-by-stage energy spectra were then fit with the function

$$P(E) = (E - V_i)^2 \exp(-E/T_i)$$

with individual V_i and T_i for each stage. (These values are purely empirical and have no role in the interpretations.) A typical lifetime was determined for each stage from the distributions of lifetimes at each stage (see Table 2).[14] This information is then combined with a separate simulation program, COULBOMB[15], which is based on a classical idea of evaporative emission. The program has a large number of options, but for the calculations shown in this report the following four elements form the heart of the simulation. First, particles are evaporated radially from the surface of the composite system in randomly chosen directions. Secondly, the particle energies are randomly chosen from the distributions given by CASCADE. Thirdly, the particular stages for the two particles are randomly chosen according to CASCADE. Finally, the distribution of time delays between two particles is given by an exponential decay with characteristic time, τ_i (see Table 2). It is important to realize that, in fact, the simulation takes into account intermediate times should the particles not be emitted in consecutive stages. That is, for one particle from stage i and one from stage j the time delay between the emission of the first and second particle will be $t_{delay} = t_{i+1} + \cdots + t_j$ where each t_k is chosen from an exponential distribution. Once angles, energies, and the time delay is chosen, the program calculates the Coulomb trajectories and then tests to see if both particles hit individual detectors in the cluster. COULBOMB allows for detector thresholds, detector resolution, and detector geometry. It also takes great care to treat the simulated data in the same fashion as the real data. The relative momentum is always calculated based on the center angle of the detector (even though COULBOMB knows exactly where the particle hit) and the same event mixing prescription is applied to the simulated data that is applied to the real data. In general, both quantum mechanics and nuclear forces are neglected but the program does contain an option to turn on a classical short ranged nuclear force in addition to the Coulomb force. One other option allows the simulation to include the random inclusion of unstable particle emission in addition to the independent emission.

This classical emission approach is somewhat free of arbitrary parameters. The correlation functions calculated in this fashion can be directly compared to the experimental results without any adjustment to compensate for asymptotic behavior which differs from unity. However, it is sensitive to the stage lifetimes, probabilities, and energy distribution, which in turn, depend on the CASCADE input. For example, the CN cross section can have a large effect on the correlation function because the CN cross section determines the CN spin cut off, which plays a large role in determining the lifetimes and cooling rate.

6. Comparison with the Data

In Figures 2-4 the solid lines shown in the p-p and p-d correlations are the results of the simulations described in the preceding section. (Some of the simulations show a certain jaggedness. This is due to the fact that sufficient statistics have not yet been accumulated.) No calculations have been superimposed on the correlation functions involving particle-unbound states in Figures 4-6. The COULBOMB simulations including the the unbound state emission are still in progress for most of these cases. (The n-p case is shown below.) The Koonin-Boal[16] formalism would give positive correlations at the appropriate values of P_{rel} but this model is simply not appropriate for these systems. Because of the particle-unbound state contribution a naive application of the Koonin-Boal approach to these correlations gives a ridiculously small space-time extent which is inconsistent with the statistical model expectations and the p-p and p-d correlations.

On examination, one sees agreement of the simulated and measured correlation functions which is remarkable. Especially so when one is reminded that there is no arbitrary normalization of the simulated results to make them fit the data and that the default values were adopted for the CASCADE calculations. However, the agreement is not perfect. For the O+Al correlations, the simulations tend to slightly underestimate the measured functions, while in the S+Al system the agreement is better, at least for the p-p correlations. Also at the higher energies for O+Al one can begin to see what may be the effects of unstable ^{2}He production. The discrepancy in the O+Al cases may be simply a matter of slightly adjusting the parameters of the statistical model calculation but may also indicate that additional refinements need to be made to COULBOMB. It is also remarkable that the simulations can reproduce the general trends of the data (implying that the assumption of thermal equilibrium still holds) for O+Al at 250 MeV where the lifetimes are approximately the same as the transit time for a beam nucleus to cross a target nucleus.

At the time of this writing, two extended COULBOMB simulations have accumulated marginally sufficient statistics to compare them with the data. The results of a simulation where a small fraction (1.5% of the initial emissions) of the events were from the breakup of ^{2}H* and the results of a simulation where a classical nuclear force plus Coulomb force acted on the particles are shown in Figure 7 for the p-n results from O+Al at 215 MeV. The breakup simulation agrees nicely with the measured results. The nuclear force simulation does show a very slight tendency toward a positive correlation at low P_{rel} but does not reproduce the data. At this time it is not clear whether this is confirms that the nuclear force is not important or that the implementation is too crude.

7. Conclusion

We have measured a number of correlation functions to study the properties excited composite nucleus and test the statistical model. The general trends of the p-p and p-d correlations are reproduced within the context of the statistical model, even at energies where we expected breakdown of the model would be obvious. The other correlations, which involve particle-unstable states, do not give lifetime information.

Several questions remain open and need further study. The program has advanced to the point that one must consider some of the more subtle effects and include more physics in the model. Also, while the correlations involving unstable states may not give information on lifetime, measurements of these states, along with measurements of the stable counterparts, could place additional constraints on the statistical model. One additional topic involves the question of separating correlation functions into a part due to independent emission (which may have positive peak) and a part due to unstable states (the main source of positive correlations in this work.) Work on all these topics is underway.

8. Acknowledgements

We thank the many people who helped carry out the particular experiments, provided discussion, and provided assistance with the calculations. These include: C. Copi and K. Zienert from Hope College, D.M. de Castro Rizzo, B.J. Fineman and X. Lu from SUNY-Physics, A. Elmaani, G. Auger, S. Kox, L.C. Vaz from SUNY-Chemistry, G.P. Gilfoyle and S.K. Sigworth from the University of Richmond, C. Beck, D.J. Henderson, D.G. Kovar, and M.F. Vineyard from Argonne National Laboratories, and W. Chung, S. Dixit, R.J. Tighe, and J.J. Vega from Notre Dame. The accelerator staffs at both Stony Brook and at Argonne were very helpful and efficient. This research was supported by the United States National Science Foundation (NSF) and Department of Energy (DOE), the William and Flora Hewlett Foundation of the Research Corporation, and the Centre National de la Recerche Scientifique (CNRS) of France.

9. References

1. Conference Proceedings of the Third International Conference on Nucleus-Nucleus Collisions, Saint Malo, France, June 1988; *Nucl. Phys. A* **488** (1988).

2. Conference Proceedings of the Texas A&M Symposium on Hot Nuclei, College Station, Texas, Dec. 1987; (*World Scientific*, edited by S. Shlomo, R.P. Schmitt, and J.B. Natowitz).

3. Proceedings of the Symposium on Central Collisions and Fragmentation Processes, 193rd National Meeting of the American Chemical Society, Denver Colorado, April 1987; *Nucl. Instr. and Meth.* **A471** 1987.

4. T.C. Awes *et al., Phys. Rev. Lett.* **61**, 2665 (1988), and references therein.

5. Z. Chen *et al., Phys. Rev. C* **36**, 2297 (1987), and references therein.

6. D. Ardouin *et al., Nucl. Phys. A* **495**, 57 (1989).

7. P.A. DeYoung, M.S. Gordon, Xiu qin Lu, R.L. McGrath, J.M. Alexander, D.M. de Castro Rizzo, and L.C. Vaz, *Phys. Rev. C* **39**, 128 (1989).

8. P.A. DeYoung, C.J. Gelderloos, D. Kortering, J. Sarafa, K. Zienert, M.S. Gordon, B.J. Fineman, G.P. Gilfoyle, X. Lu, R.L. McGrath, D.M. de Castro Rizzo, J.M. Alexander, G. Auger, S. Kox, L.C. Vaz, C. Beck, D.J. Henderson, D.G. Kovar, and M.F. Vineyard, *Phys. Rev. C* **41**, R1885 (1990).

9. P.A. DeYoung, R.L. McGrath, and W.F. Piel, Jr., *Nucl. Inst. and Methods* **226**, 555 (1984).

10. W. Zajc *et al., Phys. Rev. C* 29, 2173 (1984).

11. M. Herman, U. of Rochester *NSRL Report* UR-NSRL-318 (unpublished), and private communication; F. Puhlhofer, *Nucl. Phys.* **A280**, 267 (1977).

12. R.L. McGrath, A. Caraley, B. Fineman, M.S. Gordon, J.M. Alexander, A. Elmaani and P.A. DeYoung, *Annual Report of the Nuclear Structure Laboratory,* SUNY at Stony Brook, p. 89 (1989); R.L. McGrath, A. Elmaani, J.M. Alexander, P.A. DeYoung, T. Ethvignot, M.S. Gordon and E. Renshaw, *Computer Physics Communications,* (in press, 1990).

13. T. Ericson, *Adv. Phys.* **9**, 423 (160) and references therein.

14. One simulation has been performed (by RLM) where each Monte Carlo CASCADE event is followed to the detectors rather than picking from average distribution. There is a small discernable difference between the results from the two methods.

15. A. Elmaani, N.N. Ajltanand, J.M. Alexander, T. Ethvignot and R.L. McGrath, (unpublished preprint (1990).

16. S.E. Koonin, Phys. Lett. **70B**, 43 (1977), and D.H. Boal and J.C. Shillcock, Phys. Rev. **C33**, 549 (1986).

PROTON CORRELATIONS FROM EVAPORATING AND EXPLODING SOURCES

Scott Pratt
Department of Physics, University of Wisconsin
Madison, WI 53706, USA

Wolfgang Bauer
National Superconducting Cyclotron Laboratory and Department of Physics,
Michigan State University, East Lansing MI 48824-1321, USA

ABSTRACT

The proton correlation function is calculated from both an evaporative model and from a Boltzmann simulation. We predict a qualitatively different behavior for lower energy collisions which should be well described by the compound nucleus picture and the higher energy region (E = 75 MeV/A) where the Boltzmann simulation should be appropriate.

1. Introduction

When infinite nuclear matter is heated to the point where the excitation energy approaches the binding energy, the state of the matter should change from that of a liquid to that of a gas[1]. For heavy-ion collisions we expect a corresponding change in reaction mechanisms when the center-of-mass beam energy per particle approaches that of the nuclear binding energy. At low energies the nuclear matter fuses and the hot liquid drop slowly evaporates particles until it reaches zero temperature. At high excitation the binding energy is overcome and the matter attains a finite pressure. The hot gaseous matter then suddenly expands, as it attempts to fill the available infinite volume. In this talk we show how correlation measurements yield tremendous insight into understanding the spatial and temporal development of a nuclear reaction. As examples we show predictions for both an evaporative source at low excitation and an explosive reaction modelled by the Boltzmann equation.

In the second section we demonstrate how coulomb repulsion, strong interactions and the fermi nature of the protons contribute to the correlation function. We then explain what information about a collision's development can be extracted from the correlation function. The third and fourth sections present results from the evaporative and Boltzmann model respectively. We discuss prospects for future research in the conclusion.

2. The Power of Correlation Measurements

The correlation function is the ratio of the two-particle probability to the product of the single-particle probabilities.

$$C(\boldsymbol{K},\boldsymbol{k}) = \frac{P(p_1,p_2)}{P(p_1)P(p_2)} \tag{2.1}$$

Here $\boldsymbol{K}$ is the total momentum, (p_1+p_2), and $\boldsymbol{k}$ is the relative momentum, $(p_1-p_2)/2$. The correlation function $C(\boldsymbol{K},\boldsymbol{k})$ can be calculated if one knows the final phase-space or Wigner distribution of the particles with one half the total momentum, $\boldsymbol{K}/2$.

$$C(\boldsymbol{K},\boldsymbol{k}) = \int d^3r\, F_K(\boldsymbol{r})\, |\phi(\boldsymbol{k},x_2-x_1)|^2 \tag{2.2}$$

where ϕ is the relative wave function of the two protons and $F_K(\boldsymbol{r})$ is defined by,

$$F_K(\boldsymbol{r}) = \frac{\int d^3\frac{r_1+r_2}{2} f(K/2,r_1,t_>) f(K/2,r_2,t_>)}{|\int d^3r_1\, f(K/2,r,t_>)|^2}$$

The Wigner function $f(\boldsymbol{p},\boldsymbol{x},t_>)$ is the phase-space distribution of particles of momentum $\boldsymbol{p}$ at position $\boldsymbol{x}$ at some time after both particles have been emitted $t_>$. It can also be extracted from the emission probability $g(\boldsymbol{p},\boldsymbol{x},t)$.

$$f(p,r,t_>) = \int dt\, g(p,r-v_p\cdot(t_>-t),t) \tag{2.3}$$

Thus any dynamical model that predicts the final positions and velocities of the emitted particles can be used to generate the correlation function.

For a given total momentum $\boldsymbol{K}$ the correlation has three degrees of freedom, $\boldsymbol{k}$ which are a function of the three dimensional function $F_K(\boldsymbol{r})$. The most we can hope to extract from the correlation function is $F_K(\boldsymbol{r})$ which is the normalized probability of two protons with the same momentum $\boldsymbol{K}/2$ being separated by $\boldsymbol{r}$. To calculate the correlation function one need only know the final phase space distributions or the emission probability $g(\boldsymbol{p},\boldsymbol{r},t)$. In principle two different emission probabilities $g(\boldsymbol{p},\boldsymbol{r},t)$ could lead to the same Wigner functions $f(\boldsymbol{p},\boldsymbol{r},t_>)$ and therefore yield identical correlation functions.

This function $F_K(\boldsymbol{r})$ carries a great deal of information about the dynamics of the collision. For instance a long lived source will lead to an extended separation when $\boldsymbol{r}$ is parallel to the velocity $\boldsymbol{K}/2m$. Thus in addition to the spatial extent of $F_K(\boldsymbol{r})$ one gains insight into the lifetime of the collision. This is discussed in detail in reference (1).

The $\boldsymbol{K}$ dependence yields enormous insight into the dynamics of the collision. Both the spatial size and the lifetime can depend on the total momentum. Cooling is signified by increasingly large lifetime for particles with smaller energies. We discuss this later in the section on the compound nucleus. Collective explosive flow is signaled by short lifetimes and shrinking apparent size for increasing energy. There should be a transition in reaction mechanisms, from evaporative to exploding, at excitation energies near the nuclear binding energy. This is the dynamical equivalent of the liquid-gas phase transition. At low excitation the nuclear matter slowly evaporates particles, cooling like a hot liquid drop. While at sufficient excitation the nuclear matter explodes, expanding to fill the available volume, like a gas. This is also discussed in reference (1).

The relative wave function ϕ is influenced by three different effects: identical particle interference, short range hadronic interaction and the coulomb repulsion of the protons. Here we present a brief discussion of the contributions to the correlation function for each of these three effects, and how the resulting correlation can be used to determine $F_K(r)$.

The squared wave function for identical non-interacting particles is $1 \pm cos(2k \cdot r)$. Thus if there were no interactions one could do an inverse Fourier transform on the correlation function and obtain complete three-dimensional information about $F_K(r)$. For spin-half particles the correlation function is reduced to one half at $k=0$ and returns to unity with a width of $k_x \approx 1/R_x$. Experiments that gate on the direction of the relative momentum can then determine all three spatial dimensions of $F_K(r)$.

Coulomb interactions yield a dip in the correlation function which goes to zero as k approaches zero. As long as the characteristic sizes of $F_K(r)$ are much less than the two-proton Bohr radius of 58 fm, the shape of the dip does not depend on the shape or size of the source but only on the Gamov factor. For source sizes larger than 58 fermi the coulomb interaction begins to lend information about the source size. The coulomb dip in the correlation function will weaken for large sources. The most unfortunate aspect of the coulomb interaction is that the coulomb dip lessens the number of available pairs at small relative momentum, making it more difficult to see the effects of identical particle interference.

Strong interactions provide excellent gauges of the size of small sources. The d^* appears as a bump in the proton-proton correlation function at a relative momentum of 20 MeV/c. The size of the bump is proportional to the percentage of pairs whose relative position is within the size of the nearly bound state. Thus the height goes roughly as $1/R^3$. This provides a very sensitive test of the size but not of the shape.

Choosing a distribution $F_K(r)$ that fits the correlation function requires more than just the appropriate size as measured by a single parameter. One must have $F_K(r)$ correct for large r in order to fit the correlation function at low k in the coulomb dip and one must have $F_K(r)$ correct at small r in order to fit the height of the correlation function. If the source is not so large that the coulomb dip erases the effect of identical particle interference, the shape must be chosen correctly as well to fit the correlation function for different directions of k. Thus all the physical characteristics of $F_K(r)$ can be thoroughly tested with correlation measurements. It is extremely unlikely one can fit the six-dimensional correlation function unless one really has all the correct dynamical information in $F_K(r)$. Correlation functions provide a powerful method for reconstructing a dynamical physical picture of a collision.

Figure (1) shows the correlation functions for two spherically symmetric gaussian sources, characterized by R=5 fermi and R=20 fermi. The full correlation is shown for both cases and is depicted by a solid line. The dashed line represents the correlation function without the strong interaction and the dotted line shows the correlation function for non-identical particles interacting only through the coulomb potential. By comparing the full line to the dashed and dotted lines one sees that both the strong and coulomb interactions play important roles for the small source and a negligible role for the large source. By performing this same procedure with phase-space distributions from the evaporative model of the next section, we have seen that the strong interaction still contributes less than 5% unless the excitation energy is greater than 4.0 MeV/Nucleon.

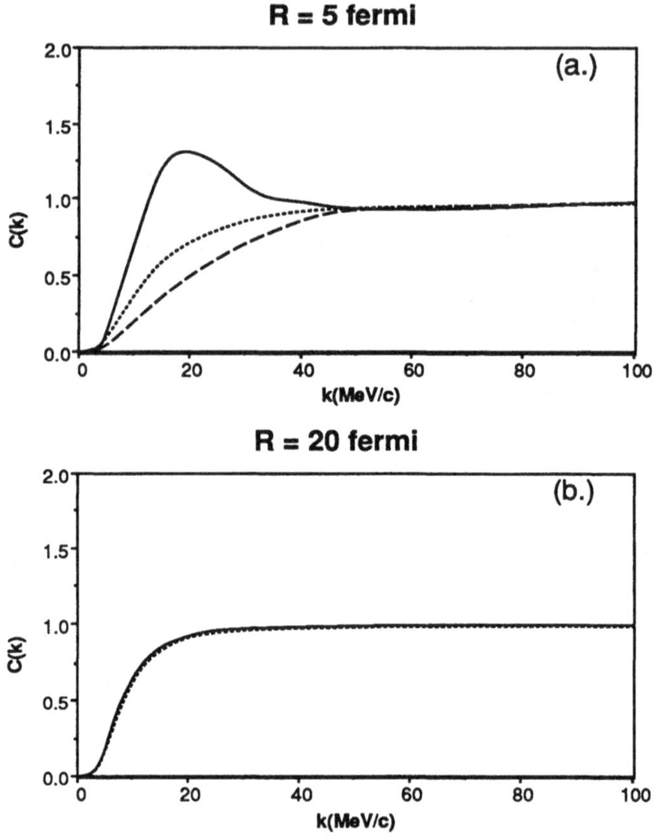

Figure 1. The correlation functions (full lines) are shown for gaussian sources of size 5 and 20 fermi. The dashed line is the prediction for when there is no strong interaction but the fermi nature of the protons is taken into account. The dotted line shows the prediction for non-identical particles interacting solely through the coulomb interaction.

3. Predictions of the Evaporative Model

The compound nucleus is the prototypic picture of a hot-liquid drop. The probability for the emission of particles can be determined by time-reversal arguments.

We calculate the emission probability according to the prescription of Friedman and Lynch[2], including fragments up to the size of Lithium. Most of the emitted particles are protons and neutrons. The evolution of a compound nucleus of size A=122, Z=52 and initial excitation energy of 2.5 MeV/A (T_0=6 MeV) is shown in figure 2. The emission occurs rapidly at first with half of the particles being emitted in the first 250 fm/c. The emission rate then slows to zero which is characteristic of a cooling source.

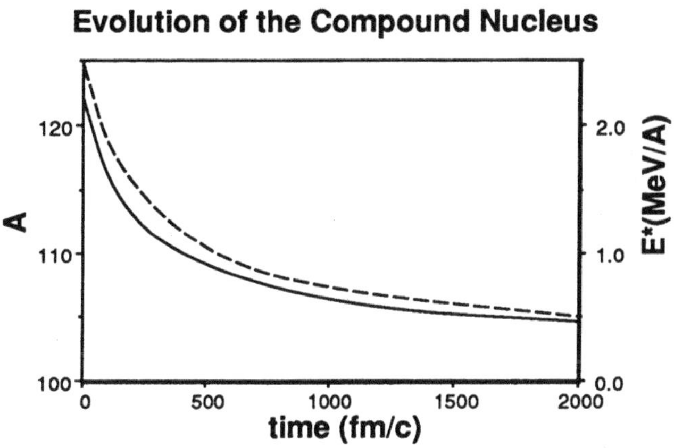

Figure 2. The evolution of a compound nucleus with an initial temperature of 6 MeV and an initial size of A = 122. The baryon number (full line) and the excitation energy (dashed line) are shown as a function of time.

The correlation functions for this example are shown in figure (3). We show the correlation function for two angle cuts and two values of the total momentum $\boldsymbol{K}$. In figure (3a) the average momentum of the pair is set to 175 MeV/c. The full line shows the correlation function for when the direction of the relative momentum $\boldsymbol{k}$ is within 30 degrees of the direction of the total momentum $\boldsymbol{K}$. The dashed line represents the result for when $\boldsymbol{k}$ is within 30 degrees of being perpendicular to $\boldsymbol{K}$. Experiments have not yet had the statistical accuracy to distinguish the difference between the two curves[3]. Figure (3b) shows the same calculation except that the average momentum of the pair is set to 300 MeV/c. This momentum cut yields a somewhat higher correlation function indicating a smaller apparent source.

The correlation is weak due to the long lifetime of an evaporating source. The shape of the correlation functions in figure (3) can not be reproduced with gaussian sources. Since the compound nucleus emitted a fair fraction of its particles in a short amount of time one would need to choose a gaussian source of about 10 fermi in order to fit the height of the bumps in figure (3). However the correlation function at very low momentum is sensitive to the many protons separated by more than the two-proton

Bohr radius of 58 fermi. In order to fit this portion of the correlation functions in figure (2) a gaussian source needs to be of the order of 50 fermi.

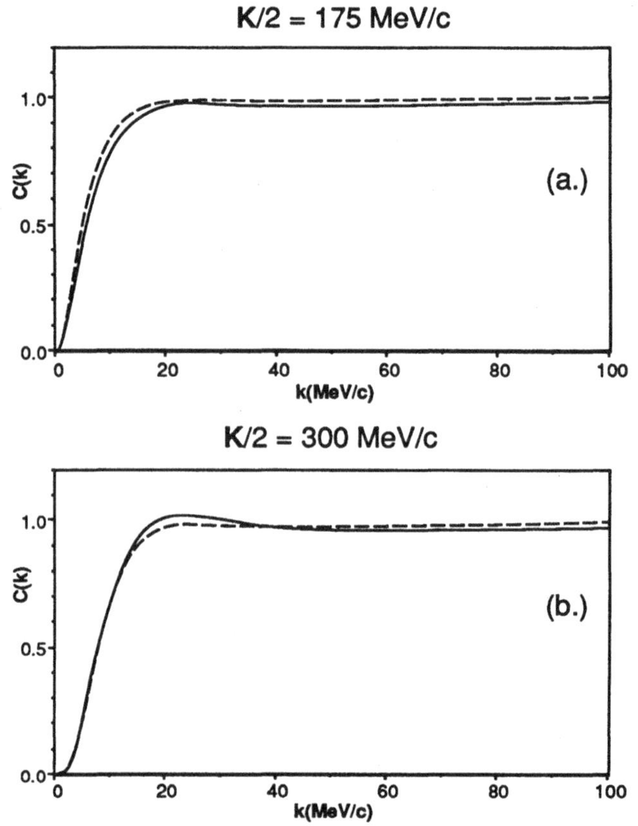

Figure 3. Correlation functions for the evaporative model for two different total momenta. In (a.) where the total momentum of the pair is $K/2$ = 175 MeV/c and in (b.) the total momenta is $K/2$ = 300 MeV/c. The solid lines are for where the relative momentum $\boldsymbol{k}$ has the same direction as the total momentum $\boldsymbol{K}$ whereas the dashed lines are for when $\boldsymbol{k}$ and $\boldsymbol{K}$ are perpendicular to within 30 degrees.

In the last year experimental results have appeared which test the evaporative picture. Goujdami[4] and collaborators have presented results for ^{40}Ar on ^{108}Ag collisions at 44 MeV/A. The detectors were located at backward angles, allowing the experiment to neglect the forward spray of the projectile particles due to pre-equilibrium emission. Within the accuracy of the experiment the results confirm the calculations of figure (3a). In a similar experiment Gong and collaborators[5] made measurements for ^{129}Xe on ^{27}Al and ^{129}Xe on ^{122}Sn at 31 MeV/A. The detectors were located in the forward plane but the reverse kinematics minimized the contamination due to pre-equilibrium emission. The same behavior as the backward angles experiment was observed. Both experiments saw very weak correlations, which hint very strongly that the sources were long lived. In both cases the predictions of the evaporative model fit the experimental results better than simple gaussian parameterizations of the source which do not include both prompt and slow emission.

One remaining question is whether the small difference in the correlation function for the two angular cuts shown in figure (3) will be able to be measured experimentally. This would give the most dramatic verification of both the methods and the evaporative model.

4. The Boltzmann Model

The evolution of a nuclear collision should be reasonably well described by a Boltzmann simulation when the excitation energies are high enough to disintegrate the nuclear matter. Here we present calculations for central collisions of ^{14}N on ^{27}Al at 75 MeV/A. The correlation functions show strong peaks that increase when gating on higher transverse momentum. Three energy cuts are shown in figure (4). The total momenta are confined to within 45 degrees of the beam axis. In a forthcoming paper we will compare the Boltzmann predictions averaged over all impact parameters with experimental results from the same system. We expect this to supply a stringent test of the dynamical picture generated from the Boltzmann simulation.

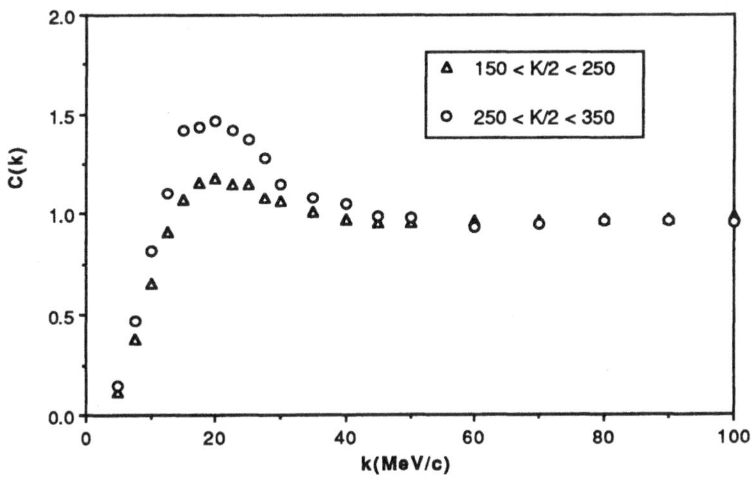

Figure 4. Correlations for forward angles as predicted by a Boltzmann simulation of a central collision of a 75 MeV/A ^{14}N projectile on a ^{27}Al target. The triangles correspond to the magnitude of one half the total momentum being between 150 and 250 MeV/c. The circles represent a momentum cut of 250 to 350 MeV/c. The statistical uncertainty for each point is about 5%.

There is certainly a dramatic difference between the correlation functions calculated in the evaporative picture and the ones calculated from the Boltzmann code. This difference stems from the dramatic difference in the dynamical behavior of a hot liquid versus that of a gas. Careful exploration of the entire range of excitation should clarify the abruptness with which the change of reaction mechanisms occurs with respect to increasing the beam energy. Hopefully, the Boltzmann approach will be useful for lower energies as well. Then one theoretical model could describe the transition of reaction mechanisms, which would certainly provide greater insight.

5. Conclusions

The correlation function is a six-dimensional function. Originally only the magnitude of the relative momentum was varied. This meant that the data could be summarized with one-parameter fits. However we now know that the shape of the correlation function contains much more information than can be obtained by fitting with a single parameter. In addition to varying the magnitude of $\boldsymbol{k}$, the direction of the relative momentum can be varied in an attempt to gain information about the three-dimensional shape. These dependences are of the order of a few percent. In the near future experiments will be able to measure to this accuracy. The correlation function depends very strongly on the magnitude of the pair's total momentum. This dependence is connected to the most physical aspects of the dynamical behavior such as cooling and expansion. The direction of the pairs total momentum is another critical parameter. The peak in the correlation function for system with projectiles at 50 MeV/A can be as high as 2.0 when the pair is moving in the forward direction and then disappear entirely in the backward direction. Thus only the azimuthal angle of the total momentum can be neglected due to symmetry. The remaining five dimensions all yield fascinating insight into the description of an energetic reaction.

The prospect for future study in the next few years is tremendously bright. Since predicting correlation functions means matching a five-dimensional measurement, theoretical models will be tested much more stringently than ever. This is especially promising as a test for three-dimensional numerical simulations. The statistics and accuracy of experiments are becoming sufficient to test all these proposed five-dimensional variations. With the theoretical apparatus maturing simultaneously we expect a very promising future.

This paper has covered only proton correlations at intermediate energy. Neutron-neutron and proton-neutron experiments are also expected to yield results in the near future. The neutron-neutron experiment would be clean of all the coulomb effects and thus is an extremely good candidate for study. Proton and pion correlations at higher energies have not been studied with nearly the diligence as at intermediate energies. At energies in the range of 1.0 GeV/nucleon, correlations might provide a valuable measurement of the compression in the collision and perhaps contribute to the determination of the nuclear equation of state at high compression. Pion correlations from ultrarelativistic nuclear collisions has received tremendous attention in the last few years. If a dedicated π-π experiment is built that can match the statistics of the proton measurements at intermediate energy, many questions about the evolution of the proposed quark-gluon plasma can be thoroughly investigated[6]. The experience gained at intermediate energy should prove invaluable as we are interested in basically the same question at RHIC energies, how a first order phase transition manifests itself in the dynamics of a collision.

Acknowledgments

This work was supported by the National Science Foundation under grant NSF-PHY-881439 and by the Wisconsin Alumni Research Foundation.

---

[1]S. Pratt and M.B. Tsang, *Phys. Rev.* **C36**, 2390 (1988).
[2]W.A. Friedman and W.G. Lynch, *Phys. Rev.* **C28** , 16 (1983).
[3]T.C. Awes et. al., *Physical Review Letters* **61**, 2665 (1988)
[4]D. Goujdami et.al., Lab. de Phys. Nucléaire-Univ. de Nantes Preprint,"Lifetime Effects of an Evaporative Source via Two-Proton Correlations" (1990).
[5]W.G. Gong et.al., Mich. State Univ. Preprint, "Two-Proton Correlation Functions for Equilibrium and Non-equilibrium Emission" (1990).
[6]G. Bertsch and G.E. Brown, *Phys. Rev.* **C40**, 1830 (1989).
S. Pratt, *Phys. Rev.* **D33**, 1314 (1986).

CLASSICAL APPROACH TO LIGHT CHARGED PARTICLE CORRELATIONS

N. CARJAN
Centre d'Etudes Nucléaires de Bordeaux-Gradignan
33175 Gradignan Cedex, France

B. ERAZMUS, D. ARDOUIN
Laboratoire de Physique Nucléaire de Nantes
IN2P3/CNRS et Université de Nantes
2, rue de la Houssinière 44072 Nantes

Abstract

The Coulomb interaction between two particles evaporated from the same compound nucleus is studied as a possible origin for their measured correlations. This is done in the frame of a three-body trajectory calculation with both Coulomb and nuclear forces. The temperature and mass of the emitting nucleus can be determined from the measured energy spectra of single particles. The correlation function was found to be very sensitive to the mean time delay between two particle emissions. A value $\tau_0 = (1 \pm 0.5)\ 10^{-21}$s was obtained by comparison of our calculations with two-proton correlation function measured at backward angles in the reaction $^{40}Ar + ^{108}Ag$ at 44 MeV/u.

I-Introduction

It is at present well established [1,2] that the shape of the correlation function for the emission of two like-particles with small relative momenta (q < 30 MeV/c) during nuclear reactions contains valuable information about the spatial and temporal extension of their emission source. This is in analogy with photon interferometry used to measure the size of the stars [3] or with like-pion correlation used to measure the fireball in high energy proton-proton collisions [4].

Unfortunatelly this information is difficult to extract since the shape of the correlation function is usually due to several, difficult to separate, effects :

a) The quantum statistical effect : when we deal with identical bosons (fermions) it is necessary to (anti)symmetrize their wave function and this leads to a constructive (destructive) interference between particles with small relative momenta. The deviation of the correlation function from unity is of the order

$$R(q) = \genfrac{}{}{0pt}{}{+}{(-)} \frac{1}{2} \cos \left(\frac{p_0}{\hbar} d_0 \theta \right)$$

where θ is the angle between the emission directions of the two particles, d_o and p_o their separation distance and average momentum upon emission.

b) The final state interaction effects : unlike photons the outgoing charged particles interact through repulsive Coulomb forces and, when they are close to each other, also through attractive nuclear forces. The Coulomb interaction for instance causes a complete depletion of events for

$$q < \left(\frac{Z\mu e^2 \; Z_1 \; Z_2}{d_0} \right)^{1/2}$$

where μ is the reduced mass. For $Z_1 = Z_2 = 1$ and $d_0 = 6$ fm (an average nuclear radius) one obtains $q < 14.2$ MeV/c. On the contrary the nuclear interaction produces an enhancement of events around $q = 20$ MeV/c.

It is therefore useful to look for a situation in which one of this effects can be isolated and this is the purpose of the present paper. Two protons sequentially evaporated from a thermalized system will be mainly affected by Coulomb interaction. At the moment of emission of the second proton, the first proton is far away so that one can neglect both the nuclear interaction and the effect of the antisymmetrization of their wave function (no spatial overlap anymore). Indeed, for a nuclear temperature of 4 MeV and 10^{-21}s between the two emissions $\langle d_0 \rangle$ = 43 fm, $\langle q_0 \rangle$ = 87 MeV/c and one obtains R = 0 for $\theta > 3^\circ$ (the effect of non-orthogonality of d_0 and q_0 was neglected in this estimate).

It is very probable that the conditions for such sequential evaporation have been experimentally fulfilled in a recent measurement of p-p correlation at backward angles in the reaction ^{40}Ar + ^{108}Ag at 44 MeV/u [5]. These data were used for comparison with the present calculations and in this way an average lifetime of the emitting system was deduced. Lifetimes for particle emission at this intermediate bombarding energy are of particular interest since they help the understanding of the thermalization process and set the applicability limits for the statistical description of the very hot nuclei formed in this type of nuclear reactions.

In high energy heavy-ion collisions the interaction energy between the emitting nucleus and the protons can be neglected since it is much smaller than the energies of the emitted protons. This is no more justified in lower or intermediate energy regimes where a three-body problem has to be solved.

The correlation function was therefore calculated based on a classical simulation for the dynamical evolution of the two protons and of the residual nucleus in their mutual field. The formalism used is described in Sec. II. The initial conditions were selected, using the Monte Carlo method, from distributions compatible with the sequential evaporation hypotheses (Sec.III). The way of constructing experimental and theoretical correlation functions is explained in Sec. IV. In Sec. V are presented the numerical results and Sec. VI contains the conclusions.

II - Formalism

To describe the separation of the reaction products for the sequential evaporation of two light charged particles from a residual recoiling nucleus, three-body bidimensional trajectory calculations have been performed. The two light particles were approximated by point charges and the emitting nucleus was supposed spherical.

This motion of the residual nucleus (index N) and of the light particles (index 1 and 2) in their mutual field was calculated using the Lagrangian :

$$L(\langle x\rangle,\langle y\rangle,\langle\dot{x}\rangle,\langle\dot{y}\rangle, t) = \begin{cases} \sum\limits_{i=1,N} \frac{m_i}{2} \dot{r}_i^2 - V_{1N} & \text{for } t < T \\ \sum\limits_{i=1,2,N} \frac{m_i}{2} \dot{r}_i^2 - \sum\limits_{\substack{i,j=1,2,N \\ (i<j)}} V_{ij} & \text{for } t > T \end{cases}$$

with the following notations :

$$\dot{r}_i^2 = \dot{x}_i^2 + \dot{y}_i^2 \qquad i = 1,2,N \quad ;$$

$$V_{12} = \frac{Z_1 Z_2}{r_{12}} \quad ;$$

$$V_{iN} = \frac{Z_i Z_N e^2}{r_{iN}} - \frac{V_0}{1 + \exp\left(\frac{r_{iN} - r_0 A^{1/3}}{\alpha}\right)} \qquad i = 1,2 \quad ;$$

and

$$r_{ij} = [(x_i - x_j)^2 + (y_i - y_j)^2]^{1/2} \qquad i = 1,2,N \, .$$

τ is the time interval between first and second emission. An attractive Woods-Saxon nuclear interaction V_0 was included only between the residual nucleus and each of the light particles but not between particles themselves since at the time τ (of the order of 10^{-21}s) the first particle is much farther from the nuclear surface than the range of the nuclear interaction. Fig. 1 shows the three-body problem which we are interested in and the geometry of the detection.

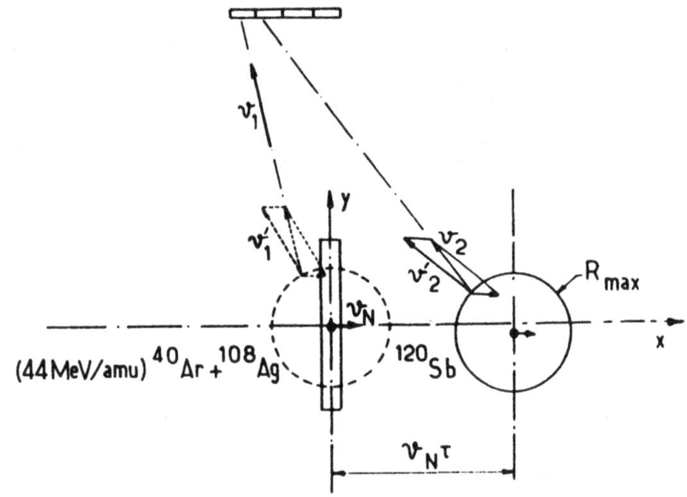

Fig. 1 : Two particles are sequentially emitted from a residual nucleus (see text). The set of four detectors covering the angles from 87.5° to 107.5° with respect to the beam axis is schematically presented.

III - Initial conditions

The light particle was initially placed at R_{max}, the position of V_{max} (the maximum of its interaction with the emitting nucleus) at any angle θ with respect to the beam direction. The emission was supposed isotropic and radial with a maxwellian distribution for the initial particle energies (evaporation spectrum) :

$$f(E_{1(2)}) = \frac{E_{1(2)} - V_{max}}{T^2} \exp\left(-\frac{E_{1(2)} - V_{max}}{T}\right)$$

where T is the nuclear temperature. The initial energies E were selected randomly in the interval [V_{max}, 60 MeV] from the above distribution using the Von Neuman algorithm. The magnitude of the initial velocities $v_{1(2)}$ were determined from their initial kinetic energies ($E_{1(2)} - V_{max}$). Finally the initial angles $\theta_{1(2)}$ were selected randomly in the restricted (by the geometry of the detection) interval [80°, 150°]. All above light-particle characteristics have been defined in the system of the emitter (residual nucleus).

As far as the characteristics of the residual nuclei are concerned we have assumed a gaussian distribution for their masses :

$$\omega(A) = \frac{1}{(2\pi\ \sigma_A^2)^{1/2}} \exp\ [-(A-A_0)/2\ \sigma_A^2]$$

and a maxwellian distribution for their recoil velocities :

$$g(v_N) = (v_N - a)\ \exp(-v_N/b).$$

The coefficients in $g(v_N)$, a = 0.05 cm/ns and b =0.6 cm/ns, were fixed by reproducing approximatively experimental velocity distributions of residual nuclei. We have tried to interpolate the data for two reactions : ^{40}Ar + Ag at 39 MeV/u and 50 MeV/u [6].

Some of the parameters defining the initial conditions are related like its charge Z and its mass A : Z = 0.425 A from the assumption that, on the average, the number of preequilibrium neutrons and protons are equal.

At t= 0 the residual nucleus was placed in the center of the target (x_N = 0). Calculations were also done for x_N = 3 fm and x_N = 10 fm but no influence on the calculated correlation function was noticed.

The distribution of the time differences between the two emissions was taken in accordance with the radioactive decay low :

$$\rho(\tau) = \exp(-\tau/\tau_0)$$

The parameters V_0 and r_0 of the Woods-Saxon nuclear interaction between the emitter and the particle were taken from the analyses of the angular distribution in elastic scattering [7,8] while the parameter α was modified to reproduce the empirical fusion barriers [9] :

$$V_0 = 58.75 \text{ MeV} \quad r_0 = 1.17 \text{ fm} \quad \alpha = 0.65 \text{ fm} \quad .$$

IV - Construction of correlation function

Once the initial conditions of the particle emission have been chosen, the classical three-body trajectories are calculated. Then the particle is checked to see if it hit the set of 4 detectors covering the angles from 87.5° to 107.5° with respect to the beam axis. There are two types of events :

a) singles : at least one particle has hit a detector,

b) coincidences : two particles have hit the same or two separate detectors.

For each coincidence pair a value of relative momentum q_c is calculated and a spectrum $C(q_c)$ is constructed. The single events provide a reference spectrum $S(q_s)$ without final state interaction. A relative momentum q_s is calculated for two particles from two successive single events. A two-particle correlation function COR(q) is defined as the ratio :

$$COR\ (q) = C(q)/S(q)\ .$$

For both coincidence and single events the relative momentum q can be calculated taking the exact final particle positions. This ideal correlation function may contain an interesting information but it can not be nevertheless directly compared to the experimental data. In fact, the experimental correlation function is generated in the exactly same way as the calculated one but, because of instrumental contraints, the final particle position is the position of the detector center (90°, 95°, 100° or 105°). Moreover, two particles from the same pair (true coincidence or two single successive events) cannot hit the same detector. The last condition imposes a threshold for the minimum value of the relative momentum.

Therefore, the calculated correlation functions are always constructed in two manners :

a) using the exact final particle positions (to provide the ideal correlation function),

b) taking the positions of the detector centers and rejecting two particles hiting the same detector (to be compared with the experimental data).

V - Results

I . Description of particle emitting system.

Information about the residual nucleus is certainly contained not only in the correlation function, but also in the energy spectrum of the single protons. The experimental data have been used (Fig.2) to determine the nuclear temperature T and parameters of the gaussian mass distribution of the emitter : the mean value A and the width σ_A.

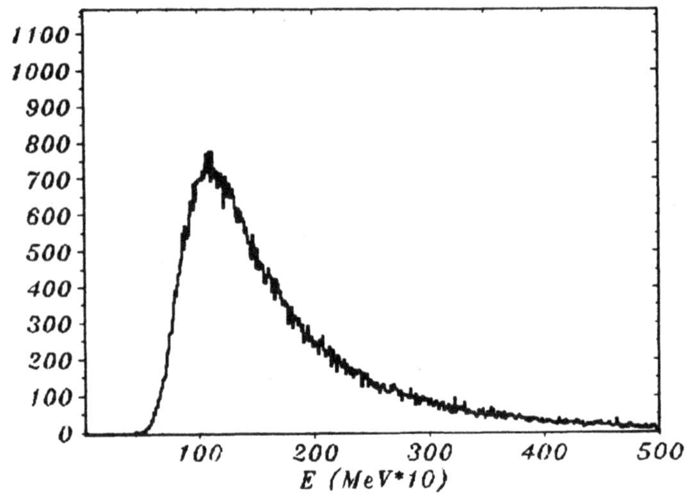

Fig. 2 : Energy spectrum of single protons measured in the reaction ^{40}Ar + ^{108}Ag at 44 MeV/u.

The calculated energy spectra of single protons are, in fact, very sensitive to the characteristics of the emitting system (Fig.3). A good reproduction of the experimental data has been obtained by choising T = 4 MeV, A_O = 120 amu (the mass after emission of the first proton) and σ_A = 10amu. It should be emphasized that we cannot reproduce the high energy tail of the experimental spectrum with our single source assumption.

Fig. 3 : Energy spectra of single protons calculated with different characteristics of the residual nucleus :

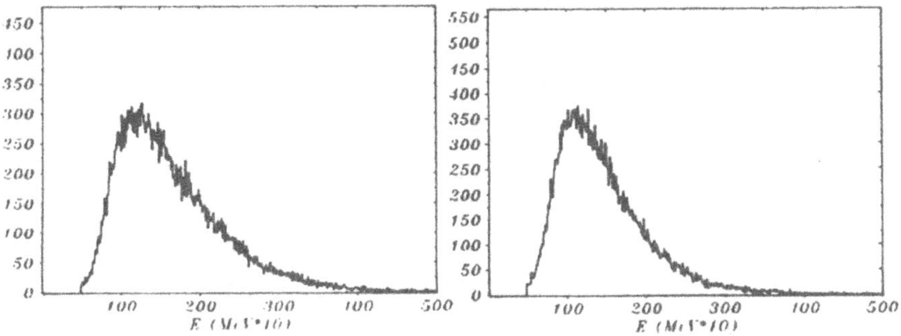

a) T = 5 MeV (left side) and T = 4 MeV (right side) with the same parameters of the gaussian mass distribution : A_0 = 120 amu, σ_A = 10amu ;

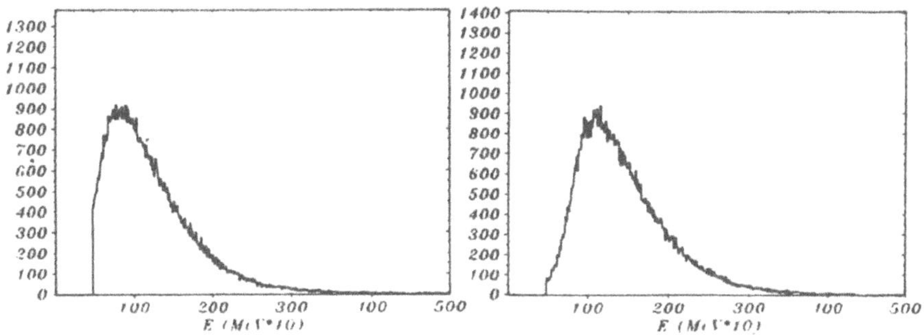

b) A_0 = 90 amu (left side) and A_0 = 120 amu (right side) with the same temperature T = 4 MeV and the same width of the mass distribution σ_A = 10amu ;

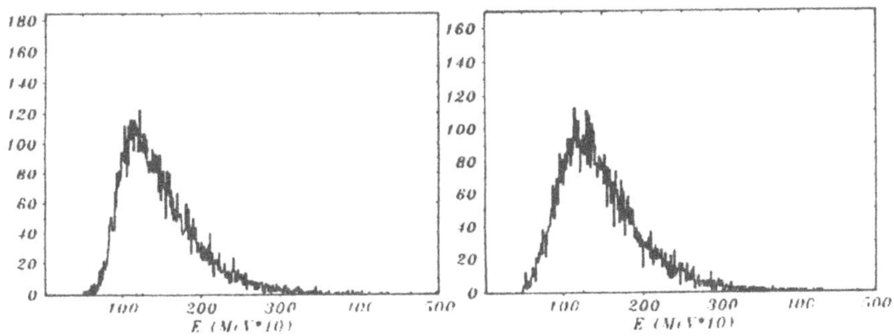

c) σ_A = 0amu (left side) and σ_A = 20amu (right side) with the same temperature T = 4 MeV and the same A_0 = 120 amu.

II . Correlation function.

Once the main characteristics of the residual nucleus have been found we have analyzed the sensitivity of the calculated correlation function to the value of τ_0, the mean time between the two emissions of protons. Correlation functions shown in Fig.4 were calculated for four different values of τ_0, using the exact final particle positions. We can notice a significant anticorrelation for $\tau_0 = 5\ 10^{-22}$s and almost the loss of correlation for $\tau_0 = 10^{-20}$s.

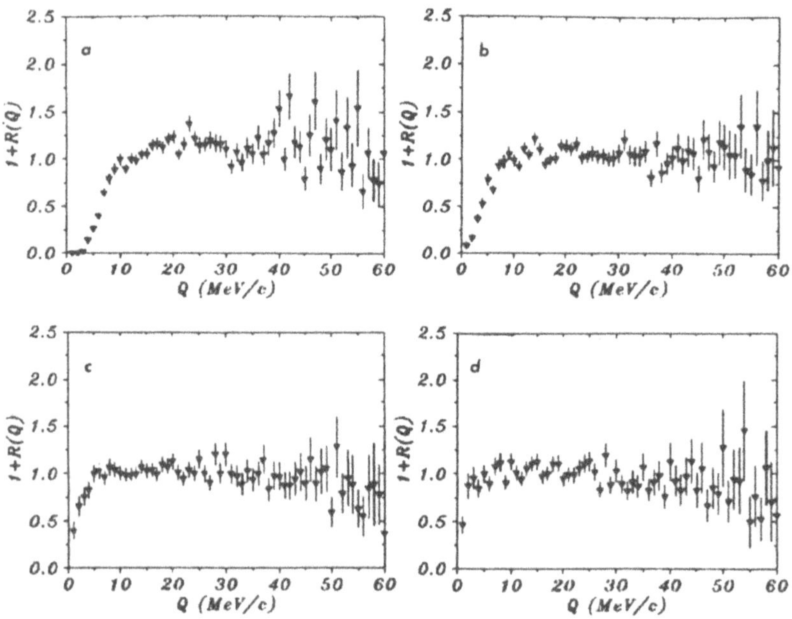

Fig.4 : Correlation functions calculated, using the exact final particle positions, for four different values of the time delay τ_0 between the two particle emissions from the same residual nucleus (T = 4 MeV, A_0 = 120 amu, σ_A = 10 amu) :
a) $\tau_0 = 5\ 10^{-22}$s, b) $\tau_0 = 1.5\ 10^{-21}$s, c) $\tau_0 = 5\ 10^{-21}$s, d) $\tau_0 = 10^{-20}$s

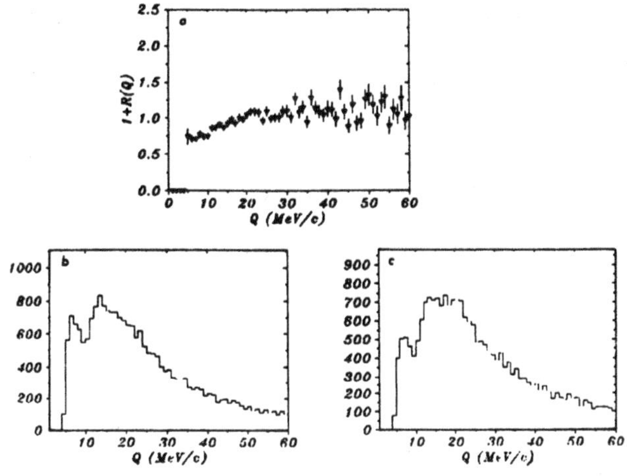

Fig. 5 Correlation function measured in the reaction $^{40}Ar + ^{108}Ag$ at 44 MeV/u (a) with corresponding spectra of single S(q) (b) and coincidence C(q) (c) events (see text).

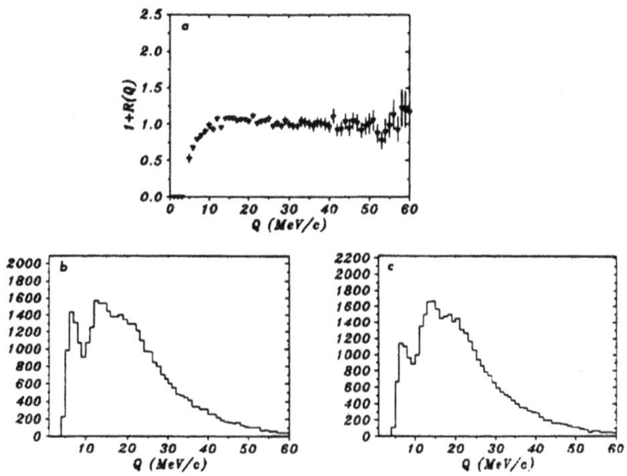

Fig. 6 : Correlation function calculated, taking the positions of detector centers as the final particle positions, with the mean time $\tau_0 = 10^{-21}$s (a). Corresponding single (b) and coincidence (c) events spectra are also represented.

The experimental correlation function with corresponding spectra of coincidence C(q) and single S(q) events (see IV) are presented in Fig.5. The best reproduction of the data is given with the mean time τ_0 equal $(1 \pm 0.5)\ 10^{-21}$s. Resultant correlation function as also coincidence and single events spectra -all of them calculated taking the positions of detector centers- are shown in Fig.6.In so far as only proton emission is concerned, our estimates of the mean time τ_0 give an upper limit for the lifetime of the residual nucleus. In such a sense the value 10^{-21}s is consistent with lifetimes predicted by the statistical model [10].

VI - **Conclusions**

Light charged particles sequentially evaporated from a thermalized system are affected mainly by Coulomb interaction. In order to study this effect as a possible origin of measured correlation, three-body trajectory calculations with both Coulomb and nuclear forces have been performed. The data from recent measurements of two proton correlations at backward angles in the reaction ^{40}Ar + ^{108}Ag at 44 MeV/u have been used for comparison with the present calculations. The main characteristics of the residual nucleus i.e. its temperature T = 4 MeV as also the mean value A_0 = 120amu and the with σ_A = 10 amu of the gaussian mass distribution were determined by reproducing experimental energy spectra of single protons. The calculated correlation function was found to be very sensitive to the mean time delay between two proton emission. The best reproduction of experimental data is obtained with τ_0 $(1 \pm 0.5)\ 10^{-21}$s, a value which, as far as considered as an upper limit for the emitting nucleus lifetime, is consistent with predictions of the statistical model.

Acknowledgments

We very much appreciated numerous and stimulating discussions with our colleagues. We would like to thank especially Ph. Eudes, D. Goujdami, F. Guilbault, C. Lebrun and B. Remaud. We thank Mrs. E. Gerbaud for a careful typing of this manuscript.

REFERENCES

1) G. Kopylov and P. Podgorestkii, Sov. J. Nucl. Phys. 18 (1974) 336
2) S. Koonin, Phys. Lett. 70 B (1977) 43
3) R. Hanbury-Brown and R. Twiss, Nature 178 (1956) 1046
4) G. Goldhaber, S. Goldhaber, W. Lee and A. Pais, Phys. Rev. 120 (1960) 300
5) P. Lautridou, Thesis, Université de Bordeaux (1989)
D. Ardouin, Nucl. Phys. A 495 (1989) 57c
J. Quebert, Proceedings of the XXVII Winter Meeting on Nuclear Physics, Bormio, Italy (1989) 220
6) Ph. Eudes, Thesis, Université de Caen (1988)
B. Tamain, Proceedings of the International School Ernico Fermi, Varenna, Italy, July 1989
7) C.M. Perey and F.G. Perey, At. Data Nucl. Data Tables 17 (1976) 1
8) L. McFadden and G.R. Satchler, Nucl. Phys. 84 (1966) 177
9) L.C. Vaz and J.M. Alexander, Z. Phys. A 318 (1984) 231
10) T. Ericson, Adv. Phys. 9 (1960) 423

FERMIONIC SECOND-ORDER INTERFERENCE IN COMPOUND NUCLEAR NEUTRON-NEUTRON CORRELATIONS

W.Dünnweber, W.Lippich, D.Otten, W.Assmann, K.Hartmann[a], W.Hering, D.Konnerth and W.Trombik[b]

Sektion Physik, Universität München, D-8046 Garching

ABSTRACT

In coincidence experiments with time-of-flight spectrometers, an anti-correlation at small relative momentum is observed for neutrons from the $^{18}O+^{26}Mg \rightarrow {}^{44}Ca$ (E_X = 60-70 MeV) compound nuclear reaction, giving evidence for quantum-statistical second-order interference (Hanbury-Brown and Twiss effect). An average decay width of 60 keV is extracted from the measured energy correlations. The observed insignificance of the attractive nn final-state interaction is attributed to the large extension of the neutron wave packets, exceeding by far the nn scattering length.

1. Motivation

Coincidences of compound-nuclear neutrons are an ideal case to look for correlations of quantum-statistical character. The absence of Coulomb interaction and the strong suppression of nucleonic final-state interaction at typical distances in the evaporation chains are distinctive prerequisites to let the correlations due to the antisymmetrization of the overlapping wave packets become visible. In contrast, previous interferometric experiments with protons or charged nucleonic clusters, as presented at this workshop, have revealed correlations dominated by final-state interactions. In those cases the shape of the correlation function is essentially determined by the mutual scattering in the exit channel.The contribution from the nucleon-nucleon strong interaction is found to decrease with respect to that from Coulomb interaction with increasing lifetime of the source and hence increasing separation of the successively emitted particles. This tendency suggests that in the case of neutrons from a source of sufficiently large lifetime, such that the average separation is large compared to the scattering length a_{nn} = 16 fm, the effects of the attractive nn final-state interaction should practically vanish, allowing purely quantum-statistical interference to come out in the experimental correlations.

2. Experimental Method to Study Neutron-Neutron Correlations

In our experiments[1,2] four time-of-flight neutron spectrometers are arranged in pairs symmetrical to the beam axis (Figs.1,2). Neutrons are detected by their scintillation pulses in large-area NE213 counters[3]. The time delay with respect to the beam pulse reference signal yields the time-of-flight for the 2 m distance from the target (Fig.3). The neutron energy is determined by the time-of-flight with an overall resolution varying between 200 keV and 1100 keV full width at half maximum (FWHM) in the accepted range between 2 MeV and 10 MeV, respectively. Gamma rays from the target are discriminated by their different pulse shapes. Cross talk, due to the scattering of neutrons from one detector into another one was suppressed by means of borated polyethylene shieldings and by imposing thresholds on the scintillation pulse heights. Neutron-induced γ-ray cross talk was suppressed by the pulse-shape discrimination. Measurements with a $^{241}Am/^{9}Be$ source in the target position showed that the shielding efficiency was sufficient.

Coincidences were recorded simultaneously for two pairs of detectors in close proximity at angles θ_1,θ_2 and $-\theta_1,-\theta_2$ and for two pairs of distant detectors at angles $\theta_1,-\theta_2$ and $-\theta_1,\theta_2$ (Figs. 1,2). Since each detector takes part in a close and a far combination of the symmetrical set-up, all detection efficiencies cancel in the double ratio of coincidence yields,

$$\frac{N(\theta_1,\theta_2)\cdot N(-\theta_1,-\theta_2)}{N(\theta_1,-\theta_2)\cdot N(-\theta_1,\theta_2)} = \left[\frac{\sigma_{lab}(\theta_1,\theta_2)}{\sigma_{lab}(\theta_1,-\theta_2)}\right]^2 \equiv \left[\frac{\sigma^{close}}{\sigma^{far}}\right]^2 \quad . \qquad (1)$$

Fig.1.: Apparatus consisting of two pairs of scintillation counters placed symmetrically on both sides of the beam axis at a distance of 2 m from the target mounted in an aluminum vacuum box. The scintillation counters of 1 m height and 8 cm diameter are contained in long vertical aluminum boxes that carry photomultipliers at both ends[3]. Shieldings of borated polyethylene are placed between the counters and, in combination with lead bricks, around the beam stop.

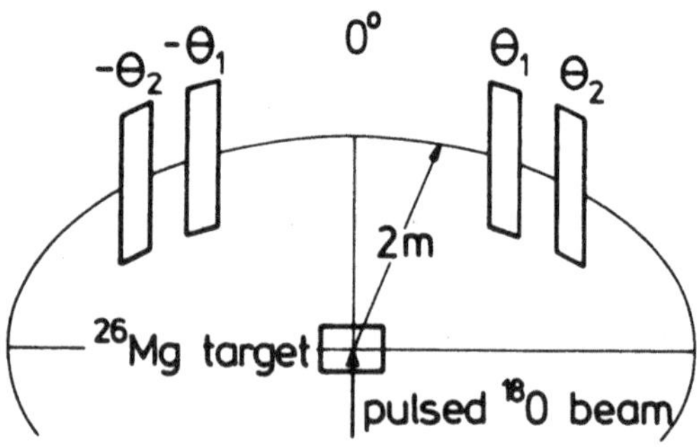

Fig.2: Schematic view of the experimental set-up. The scattering angles were $\theta_1 = 25°$, $\theta_2 = 38°$ and $\theta_1' = 38°$, $\theta_2' = 51°$ in two separate experiments.

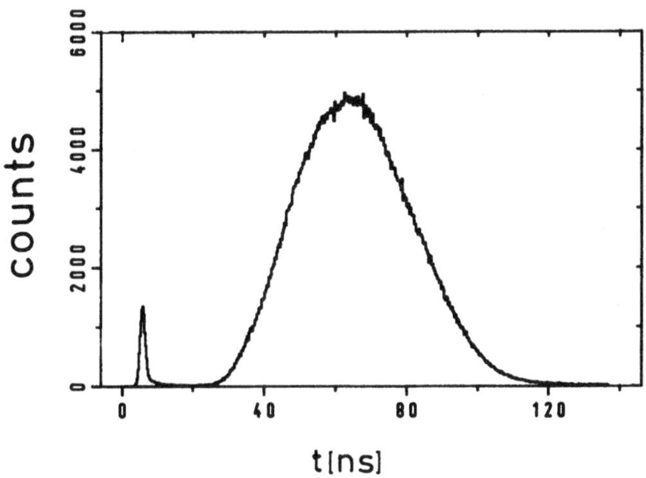

Fig.3.: Spectrum of flight times, $t = t_{scintillator} - t_{beam\ pulse}$ at 25° for 88 MeV $^{18}O + ^{26}Mg$. The sharp peak (FWHM = 2 ns) at t = 6 ns is due to γ-rays, the broad bump is due to neutrons.

3. Quantum-Statistical Interference versus Final-State Interaction

For the neutron-neutron correlation function,

$$C(\vec{p}_1,\vec{p}_2) = \text{Constant} \times \sigma_{coinc}(\vec{p}_1,\vec{p}_2)/(\sigma_{single}(\vec{p}_1)\sigma_{single}(\vec{p}_2)), \qquad (2)$$

two different characteristics (Fig.4) are expected from approximative theoretical treatments which are valid for two different regimes of source lifetimes:

(i) For simultaneous emission, the shape of $C(\vec{p}_1,\vec{p}_2)$ is essentially determined by the attractive final-state interaction, since nuclear source radii are generally smaller than the singlet s-wave scattering length a_{nn} = 16 fm. The height of the resulting peak at momentum difference $\vec{p}_1-\vec{p}_2$ = 0 is determined by the source radius. Its half width is $2\ \hbar/a_{nn}$ = 25 MeV/c in both the longitudinal and the transversal momentum difference of the two neutrons (see Refs. 4–7).

(ii) For large source lifetimes, leading to average distances of successive neutrons much larger than a_{nn}, it is expected that the final-state interaction becomes insignificant[5,6]. Quantum-statistical antisymmetrization (which is of course also obeyed in case (i)) leads to a dip at $\vec{p}_1-\vec{p}_2 = 0$ with source independent depth (1/2) $C(\vec{p}_1-\vec{p}_2 \rightarrow \infty)$, due to the larger weight of the triplet configuration for random spin orientation. The transverse-momentum half width of this dip is about $\pi/(2r)$, corresponding to an angular half width $\lambda_n/(4r)$, for an emitting disk of radius r (see Ref.8). The half width in the longitudinal momentum difference amounts to $\langle\Gamma/v_n\rangle$, corresponding to an energy difference Γ, where $\Gamma = \hbar$/(source lifetime) is the decay width of the source. Hence, unlike case (i), the transversal and the longitudinal momentum widths are different. These features arise merely from the antisymmetrization of the overlapping neutron wave packets, in close analogy to optical second-order interference (Hanbury-Brown and Twiss effect)[8,9].

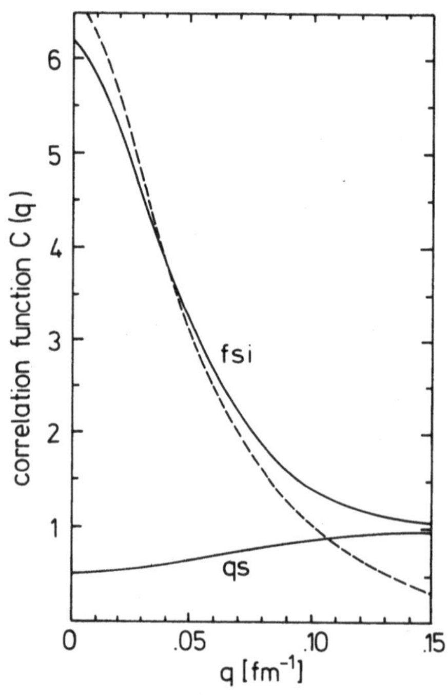

Fig.4: Examples of calculated nn correlation functions dominated by final-state interactions (fsi) or quantum-statistics (qs), as a function of the relative momentum $q=(p_1-p_2)/2$. Simultaneous emission, Gaussian spatial distribution (r_o=6 fm) and random spin orientation are assumed. The full fsi curve is taken from Ref.5. The dashed curve shows an approximation by means of an elementary potential well phase-shift calculation, assuming two point sources separated by 7 fm. These correlations, which include proper antisymmetrization, are found to be dominated by the attractive singlet s-wave interaction. To simulate the purely quantum-statistical correlation as a function of the transverse q, the final-state interaction has been switched off.

4. Experimental Results

The chosen reaction was $^{18}O+^{26}Mg$ at E_l = 65 and 88 MeV in two separate experiments[1,2], producing ^{44}Ca compound nuclei at E_x = 60–70 MeV. Measured neutron singles spectra are in agreement with a superposition of calculated Maxwellian spectra for the various evaporation stages, with a temperature parameter varying between 2 and 3 MeV. The mean energy of the detected neutrons is 5.2 MeV, corresponding to 3.1 MeV in the c.m. system.

Coincidences were recorded for neutrons from the same and from successive pulses of the beam (prompt and accidental coincidences, respectively, as shown in Fig.5). The results for true coincidences have been corrected for the contributions from accidentals to the prompt peak.

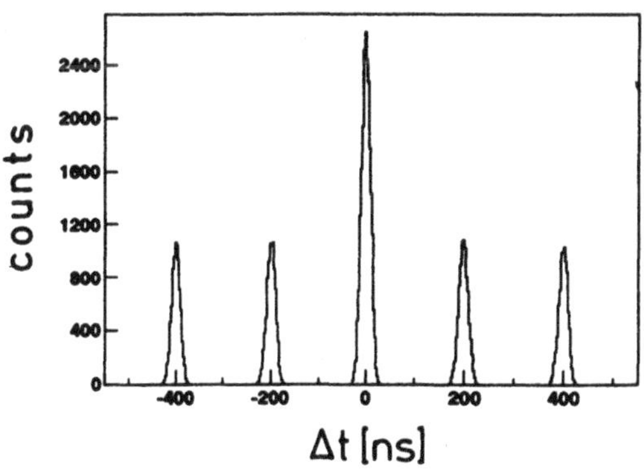

Fig.5.: Spectrum of the time-of-flight differences for one pair of neutron detectors. Gamma rays were discriminated by their different scintillation pulse shapes. The ^{18}O beam impinging on the ^{26}Mg target is pulsed with a spacing of 200 ns. Neutrons from the same pulse fall in the central peak, neutrons produced accidentally in successive pulses fall in the other peaks.

The coincidence data are represented in Fig.6 in the form of double ratios (Eq.(1)) as a function of the difference Δt of the times of flight. In the two separate experiments different angles were chosen, with the same angular distance of coincident neutrons in the close geometry, $\langle\Delta\theta_{cm}^{close}\rangle$ = 20°, and with $\langle\theta_{cm}^{far}\rangle \gtrsim$ 100° in the far geometries. These angular distances are substantially smaller and larger, respectively, than the half width $\lambda_n/4r \approx 55°$ of the expected effect. In accordance, constant true-to-accidental ratios were found in the far geometry. Hence,

$$C(\Delta E) = C_0 \times \sigma_{lab}(\Delta\theta^{close})/\sigma_{lab}(\Delta\theta^{far}) \quad , \qquad (3)$$

which is derived from the double ratio (Eq.(1) after converting times of flight into energies, represents the ΔE-dependent small-angle correlation, normalized to unity at large momentum difference (Fig.7).

The observation of double-ratios consistent with unity for neutrons from different pulses of the beam, i.e. for uncorrelated neutrons from different events (see Fig.5), confirms the reliability and the high sensitivity of the double-ratio technique as a means of detecting correlations (Fig.6a). The levels below unity for the true coincidences (dashed lines in Figs. 6b and c) are in agreement with a purely kinematical correlation due to the recoil imparted to the intermediate compound nucleus by emission of the first neutron.

Evidence for the quantum-statistical interference effect is provided by the dip appearing in Figs.6 and 7 at those values of Δt and ΔE in the lab system that correspond to $E_{cm} = 0$. The width of about 1.5 MeV (FWHM) is due to the experimental resolution and to the summing over different evaporation steps with different E_{cm} to E_{lab} conversions.

The area of the dip in Fig.7 is not affected by the experimental broadening and can be compared to that of the theoretical energy correlation[8,9]

$$C(\Delta\theta^{close},\Delta E) = 1 - C(\Delta\theta^{close},0)/[1+(\Delta E/\Gamma)^2] \quad . \qquad (4)$$

Instead of $C(\Delta\theta^{close},0) = 0.5$, valid for vanishing $\Delta\theta^{close}$ and random relative spin orientation, we have used a value of 0.55 which takes into account the finite value of $\Delta\theta^{close}$ in our experiment and the statistical-model spin polarization correlation. This analysis of our experimental result yields a decay width $\langle\Gamma\rangle = (60 \pm 15)$ keV which is to be understood as the average over all nn-decays of the primary highly excited ^{44}Ca nucleus.

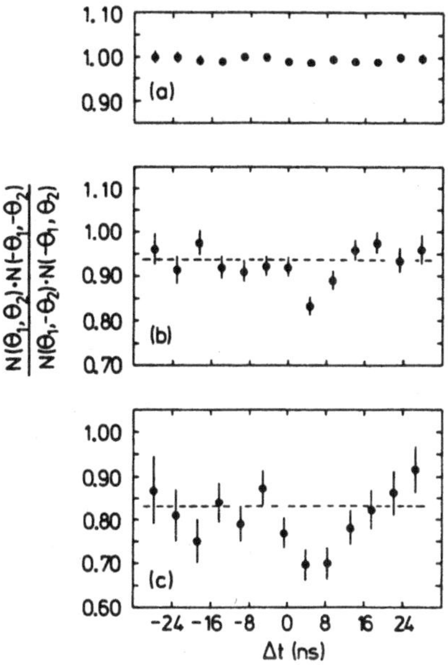

<u>Fig.6.:</u> Double-ratio of coincidence yields (Eq.(1)) as a function of the time-of-flight difference $\Delta t = t(\theta_2)-t(\theta_1)$ für 2-10 MeV neutrons from $^{18}O+^{26}Mg$ at θ_1,θ_2 (see Fig.2), E_i = 88 MeV (panel (a) accidentals from successive pulses of the beam, panel (b) true coincidences) and at θ_1',θ_2', E_i = 65 MeV (panel (c) true coincidences). The average angular separation of coincident neutrons in the close geometry is the same ($\Delta\theta^{close}$ = 20° c.m.) in the two experiments. The dashed lines represent the weighted averages outside of [-5 ns, + 15 ns]. The dip at $\Delta t \approx 5$ ns, corresponding to $\Delta E_{cm} \approx 0$, gives evidence for the quantum-statistical interference effect.

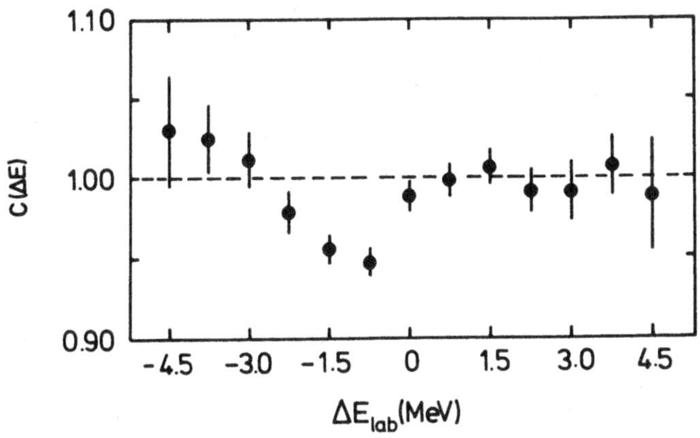

<u>Fig.7.:</u> Correlation function $C(\Delta E \equiv E(\theta_2)-E(\theta_1)) = C_0 \times \sigma^{close}(\Delta E)/\sigma^{far}(\Delta E)$ averaged over both experiments and normalized to unity outside of a 3-MeV window around ΔE = -0.75 MeV (corresponding to $\Delta E_{cm} \approx 0$), for neutrons with E = 3-8 MeV.

For comparison, an enhancement from the final-state attractive interaction would have a half width of 2 $\hbar/a_{nn} \approx 25$ MeV/c in all directions of the momentum difference, as pointed out in Sect.2. This type of enhancement was observed in a nn experiment studying 7.5 GeV/c proton-nucleus collisions[10]. For the present average neutron energy, a transversal momentum width of this size corresponds to an average $\Delta\theta$ = 20° (c.m.) which is equal to our $\Delta\theta^{close}$. A longitudinal momentum width of equal size corresponds to an average ΔE_{lab} = 2.7 MeV (ΔE_{cm} = 2 MeV). This width (FWHM = 2 ΔE_{lab}) is much larger than that of the dip in Fig.7 and even a 5% fraction of the large enhancement[5] (see Fig.4), calculated for simultaneous emission and the present source dimension, would have been detected in this range. This observation of insignificant final-state interaction justifies the use of the elementary purely quantum-statistical formula Eq.(4) in the analysis above. Concerning the use of ΔE or of Δp (longitudinal) = $\Delta E/v_n$ as variables of the correlation function, the former is more adequate in the purely quantum-statistical case, allowing straightforward application of Eq.(4) in the analysis, while the latter would require folding of Eq.(4) with the velocity (v_n) distribution and the response of the detector. Evidently, in the case of dominant final-state interaction the situation is reversed.

5. Implications of the Deduced Decay Width

Decay widths Γ at low excitation energies (E_x < 30 MeV for medium-heavy nuclei) have been determined by means of Ericson fluctuations (which are conceptually related to the present effect[9,11]) and have been found to depend systematically on E_x and atomic number[12]. Extrapolation to the present case yields a range of Γ up to about 300 keV for the primary $\langle E_x \rangle$ = 65 MeV. Weighting with statistical model branchings[13] and assuming a vanishingly small interference effect for nonconsecutively emitted neutrons, we find a total energy average $\langle\Gamma\rangle$ = 80 keV. Since the semi-empirical systematics is valid only within a factor of two[12], the agreement with our experimental $\langle\Gamma\rangle$ = 60 keV is satisfactory. Our experiment thus confirms the systematics of compound nuclear decay widths in a region not accessible by means of Ericson fluctuations.

Assuming validity of the statistical-model branchings, according to which the complete E_X-range is approximately equally weighted by demanding nn coincidences, and assuming that the semi-empirical systematics[12] of Γ roughly holds at $E_x < 30$ MeV, we can deduce from our experimental average Γ a more specific average for E_x = 30–70 MeV and A = 39–44 which amounts to Γ = 110 keV, corresponding to a lifetime $\tau = \hbar/\Gamma = 6 \cdot 10^{-21}$s.

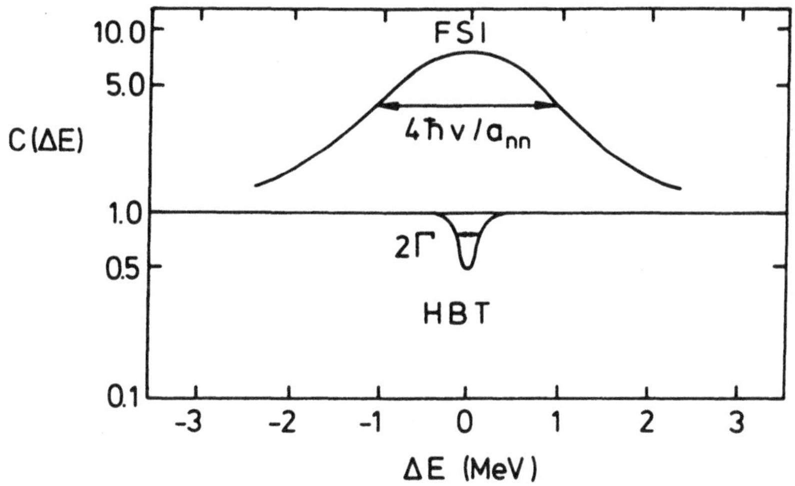

Fig.8: Schematic comparison of the two extreme cases of the nn correlation function at vanishing angular separation. For short lifetimes of the source, such that $\tau \cdot v_n \ll a_{nn}$ (=singlet scattering length = 16 fm), an enhancement results from the attractive final-state interaction (FSI). For large lifetimes, such that $\tau \cdot v_n \gg a_{nn}$, contributions from the FSI become negligibly small as compared to the Hanbury-Brown Twiss (HBT) effect, i.e. the purely quantum-statistical anticorrelation which is due to the dominant spin triplet configuration.

A gross estimate of the extension of the neutron wave packets and hence of the average separation of successive neutrons is given by the length $d = v_n \tau$. For a neutron velocity v_n that corresponds our average E_{cm} = 3.1 MeV, the value of τ deduced above corresponds to d =

150 fm. This length is much larger than the singlet scattering length a_{nn}, which is in accordance with the observed insignificance of the nn final-state interaction. The different characteristics of correlation functions dominated by the final-state interactions and by quantum statistics alone are sketched in Fig.8. The dip observed in our experiment establishes the purely quantum-statistical interference effect for fermions.

References

a) present address: IABG, D-8012 Ottobrunn
b) present address: Deutsches Patentamt, D-8000 München

1) Reports on the progress of this project are found in W.Lippich et al., Annual Report, Beschleunigerlabor München, 1983, p. 19; W.Lippich, Diplomarbeit (München 1984); D.Otten, Diplomarbeit (München 1989); W.Dünnweber et al., Proc.Int.Conf.Nucl.Phys. (Sao Paulo 1989), p.275 and Proc.Int. Workshop on Gross Properties of Nuclei and Nuclear Excitations XVIII (Hirschegg 1990), p.242.

2) W.Dünnweber et al., Phys.Rev.Lett. (in press)

3) D.Evers et al., Nucl.Instr.Meth. 124 (1975) 23

4) S.E.Koonin, Phys.Lett. 70B (1977) 43

5) H.Sato and K.Yazaki, Proc. Japan-US Joint Seminar (Hakone 1980) edited by K.Nakai and A.S.Goldhaber, p.340

6) R.Lednicki and V.L.Lyuboshitz, Yad.Fiz. 35 (1982) 1316 [Sov.J.Nucl.Phys. 35 (1982) 770]

7) C.K.Gelbke and D.H.Boal, Progr.Part.Nucl.Phys. 19 (1987) 33

8) G.I.Kopylov and M.I.Podgoretski, Yad.Fiz. 15 (1972) 392 [Sov.J.Nucl.Phys. 15 (1972) 219]

9) S.E.Koonin, W.Bauer and A.Schäfer, Phys.Rev.Lett. 62 (1989) 1247

10) Y.D.Bayukov et al., Phys.Lett. 189B (1987) 291

11) N.W.Tanner and D.M.Brink, Nature 201 (1964) 806

12) D.Shapira et al., Phys.Rev. C10 (1974) 1063

13) codes GROGI and CASCADE

This work was supported in part by the BMFT

Two-Neutron Correlations in 30 A MeV Heavy-Ion Interactions

B. Norén, B. Jakobsson, A. Kristiansson, A. Oskarsson, M. Westenius. Dept. of Physics, University of Lund, Sölvegatan 14, S-223 62 Lund, Sweden.

and

M. Cronqvist, S. Mattson, M. Rydehell, Ö. Skeppstedt. Dept. of Physics, Chalmers University of Technology, S-412 96 Göteborg, Sweden.

J. C. Gondrand, B. Khelfaoui, S. Kox, F. Merchez, C. Perrin, D. Rebreyend. Institut des Sciénces Nucléaires de Grenoble, Université de Grenoble, 53 ave. des Martyrs, F-38 026 Grenoble Cedex, France.

L. Westerberg. The Svedberg Laboratory, University of Uppsala, Box 533, S-75 121 Uppsala, Sweden.

S. Pratt. Dept. of Physics, University of Wisconsin, Madison, U.S.A.

ABSTRACT.

Preliminary data on two-neutron correlations at small relative momentum in ^{20}Ne+C and ^{20}Ne+Co reactions at 30A MeV show the behaviour expected from an attractive nuclear interaction + quantum statistical effects. The correlation function for n-n differs from the one obtained for p-p data in the sense that a strong positive peak close to zero relative momentum is observed because of the absence of a repulsive Coulomb interaction. Two-proton and two-neutron correlations are consistent with each other, both predicting similarly large apparent emission sources.

1. Introduction.

The original idea, developped in astrophysics, to extract source size information from two-photon correlation measurements[1] has later been utilized in nuclear physics by introducing two-pion[2] and two-proton[3] interactions. The present information about the space-time evolution of the particle emitting sources as obtained from proton correlations during heavy ion reactions is however still ambigous. At high energies, most results indicate that source sizes can be well extracted. Such data[4,5] seem to establish the existence of geometrically determined emission sources, the sizes of which thus depend on the masses of the nuclei and the impact parameter.

The situation is more complicated at medium energies, where the time evolution of the sources and thus the time difference between the two emitted protons ($\Delta\tau$) cannot be neglected[6,7]. Under the assumption that $\Delta\tau$=0, one obtains however in general a surprisingly consistent picture of similarly large emission sources, independent of projectile-target system, impact parameter trigger or phase-space region for the emitted particles.

It has often been stressed that final state interactions dominate the structure of the measured correlation function. The strong effects of mutual (and mean-field) Coulomb interaction on protons in p-p (and p-n) correlations are additional complications. Correlations for n-n, on the other hand, are difficult to measure with compact interferometers mainly due to cross-talk between the detectors (particles scattered from one detector into another).

We report here on an attempt to measure n-n correlations with a setup of four liquid scintillators placed at the same distance from the target (3.5 m) but with various relative distances (0.2-1.2 m).

2. The Experiment.

The complete experimental setup consisted of four hexagonal (~15 cm in diameter and 15 cm thick) Bicron BC-501 liquid scintillators[8] combined with the EMRIC CsI + MWPC interferometer[9] for charged particles. Two of the neutron detectors were placed behind two holes created diagonally to each other at the center of EMRIC whereas the other two detectors were placed at the upper and lower edges of EMRIC. The performances of the EMRIC modules and the neutron detectors are discussed elsewhere[8,9]. Both proton and neutron triggers were accepted. The standard pulse-shape discrimination technique was used to identify neutrons and the RF signal from the accelerator (~1 ns wide) was utilized as a start signal for the neutron time-of-flight (energy) measurement. The detection efficiency was calculated with the Cecil et al. Monte Carlo code[10].

A 30A MeV ^{20}Ne beam from the SARA coupled cyclotron facility was bombarding ~1 mg/cm^2 targets of C and Co. All detectors were placed in the angular interval 24°-34°. The energy threshold was 1.5 MeV for neutrons and 8 MeV for protons. In the n-n correlation case the closest detectors allow relative momenta $q=|\mathbf{p}_1-\mathbf{p}_2|/2$ of $q\geq 1.125$ MeV/c and the most distant pairs q~80 MeV/c. Since no position determination within the neutron detectors is possible we randomize the position of each neutron homogeneously within the front area. Another alternative is to assign all positions at the center of each detector but a comparison between these two methods shows an almost negligible difference.

The correlation function R(q) is given by

$$R(q)+1= C \cdot \frac{\dot{N}_c(q)}{N_{nc}(q)} \tag{1}$$

where N_c stands for correlated and N_{nc} for non-correlated events. The non-correlated events are constructed by taking the first particle from the actual correlated event and the second particle from a previous pair. C is a normalization constant which gives R(q)=0 for high q-values (q>60 MeV/c).

The energy dependent efficiency factors have been introduced in an identical way for correlated and non-correlated events. A complete simulation of all false cross-talk correlation events is in progress but has not yet been introduced.

Particles scattered from interactions in the first detector into a second one may create false events with a non-trivial R(q) structure. All charged particles will however be absorbed in the 2 mm thick stainless steel housing, surrounding the detectors and γ-rays will be rejected in the pulse-shape discrimination tests. The q region of false events due to neutrons scattered into the second detector can be derived for the elastic C(n,n') and H(n,n') channels. From the momentum of the first neutron one can calculate the possible momentum range for the second neutron and thereby determine if the latter is a candidate for cross-talk. No significant difference between R(q) has been found in these momentum regions compared to those with larger or smallermomenta (possible from inelastic scattering).

Another test is to compare R(q) for different detector combinations. In Fig. 1 we show such a comparison. In this figure, close detectors

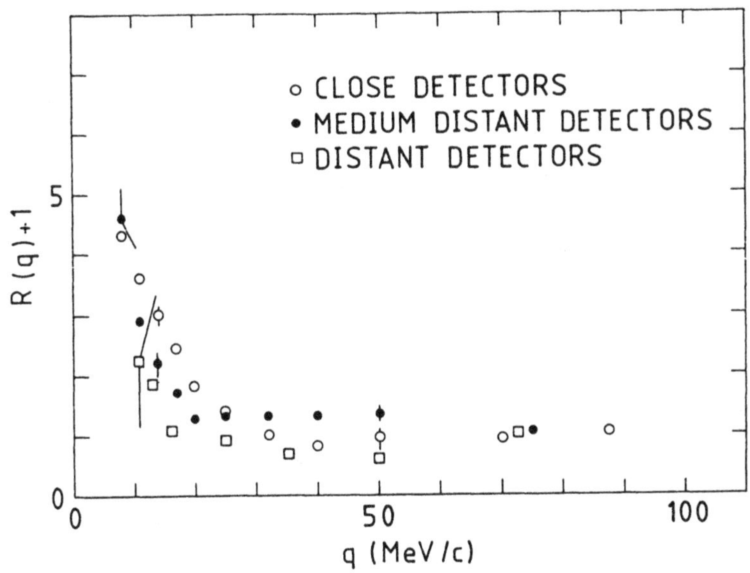

Fig. 1. Correlation functions for different detector combinations. The classification of combinations is explained in the text.

means the two at the center of EMRIC, distant detectors stands for the two at the edges of EMRIC and medium distant detectors involves all other possible detector combinations. There are obviously differences in the peak/background ratio between close and distant detector combinations. This difference is partly due to a natural preference of small q for the close detector combinations which also has to be taken into consideration. Since the momentum region for accepted cross-talk is much broader for close detector combinations we believe that the dominant contribution in the region of small q-values comes from true n-n correlations.

3. The Interference Calculation.

Particles which are randomly emitted with $\Delta\tau=0$ will interfere with each other due to final state interactions or exchange effects. Calculations of the correlation function in this case as well as in a scenario with a long time evolution of the emitting source (evaporation process) have been performed by Pratt and Tsang[11]. The correlation function is in the latter case

$$1 + R(\mathbf{k},\mathbf{q})=\int d^3r\phi^*(\mathbf{q},\mathbf{r})\phi(\mathbf{q},\mathbf{r})F_q(\mathbf{r})$$

where

$$F_q(\mathbf{r})=\frac{\int d^3Rf(\mathbf{k},\mathbf{R},\mathbf{q}=0,\mathbf{r},t\to\infty)}{\int d^3r_1d^3r_2f(\mathbf{p}_1,\mathbf{r}_1,t\to\infty)f(\mathbf{p}_2,\mathbf{r}_2,t\to\infty)} \qquad (2)$$

Here ϕ is the relative wave function, f the Wigner function and $\mathbf{k}$ is the total momentum $\mathbf{p}_1+\mathbf{p}_2$. This approach differs from that of Koonin[3] only in the way that the time integration now may be incorporated.

In the $\Delta\tau=0$ calculations $\mathbf{r}_i$ and $\mathbf{p}_i$ are generated from Gaussian probability distributions. The solid curves in Fig. 1 and 2 shows the result of such calculations.

4. Experimental Results and Conclusions.

Fig. 2 shows R(q) for all neutron detector combinations for ^{20}Ne+C (upper) and ^{20}Ne+Co (lower) collisions. Also the proton-proton correlations, measured with the EMRIC modules at the same two data taking periods are shown. The calculated solid curves represent the $\Delta\tau=0$ calculations as presented above with sources of apparent size of 3, 5 and 7 fm. The error bars contains statistical errors only and we stress that in the lowest q region ($q\lesssim15$ MeV/c) might be seriously affected by cross-talk for which no correction is yet introduced. However we expect the q-region most strongly populated by cross-talk to appear at higher q-values. It also remains to analyze a part of the statistics for Ne+Co collisions. From the discussion in section 2 we believe however that the general structure of the n-n correlation

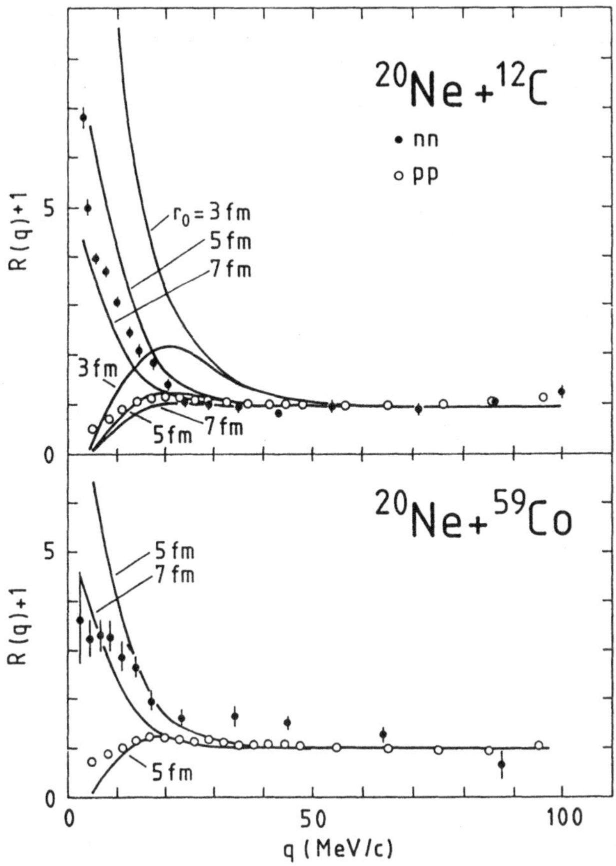

Fig. 2. Correlations for ^{20}Ne+C and ^{20}Ne+Co at 30 A MeV (θ_{mean}=29°). The dots represent the experimental data and the curves show correlation functions calculated for different r_0.

function is correct. The consistency between the p-p and n-n correlation functions is obvious in both reactions. The apparent source size is large, 5-6 fm, and it is the same for a light and medium heavy target.

A preliminary attempt to calculate the space-time evolution of the emission from a pure evaporation scenario gives reasonable but not perfect argreement with the data. Possibly this could be improved with a refined description of the pre-equilibrium nucleon emission.

In conclusion we have observed consistent results between p-p and n-n correlations, confirming the large apparent source sizes independent of the interacting systems which has been reported for p-p correlations earlier[7,12].

References.

1. R. Hanbury Brown and R.Q. Twiss, Phil.Mag. 45, 663(1954).
2. G. Goldhaber et al., Phys.Rev. 120, 300(1960).
3. S.E. Koonin, Phys.Lett. 53, 544(1977).
4. P. Dupieux et al., Phys.Lett. B200, 17(1988).
5. H.Å. Gustafsson et al., Phys.Rev.Lett. 53, 544(1984).
6. D. Ardouin et al., Nucl.Phys. A495, 57c (1989)
7. D. Rebreyend et al., ISN de Grenoble Preprint ISN90.07(1990).
8. M. Rydehell, Thesis, Dept. of Physics, Göteborg 1990.
9. F. Merchez et al., Nucl.Inst. and Methods A275, 133(1989).
10. R.A. Cecil, B.D. Anderson and R. Madey, Nucl.Instr. and Methods 161, 439(1979).
11. S. Pratt and M.B. Tsang, Univ. of Tennessee Preprint (1988).
12. D.A. Cebra et al., Phys.Lett. B227, 336(1986).

TWO-PROTON CORRELATIONS FROM THE $^{58}Ni+^{58}Ni$ REACTION AT E_{inc}= 876.5 MeV

M. KOROLIJA(1,3), D. SHAPIRA(2), J. GOMEZ DEL CAMPO(2), H.J. KIM(2), K. TEH(2,3,4), J.Y. SHEA(3,4), J.P. WIELECZKO(2,3,5), and N. CINDRO(1)

(1) Ruđer Bošković Institute, Zagreb, Croatia, Yugoslavia, 41001

(2) Oak Ridge National Laboratory, Oak Ridge, Tennessee 37831

(3) Joint Institute for Heavy Ion Research, Oak Ridge, Tennessee 37831

(4) Vanderbilt University, Nashville, Tennessee 37235

(5) CEN, Saclay, Gif-sur-Yvette Cedex, France 91191

In several recent papers, light-particle correlations have been used to obtain information on the size, lifetime, and temperature of the reaction zone in an energetic heavy-ion collision [1-4]. The procedure is analogous to the well-known Hanbury-Brown-Twiss interferometer [5] where photon-photon correlations with small relative momentum Δp ($\Delta p = |(\vec{p}_2-\vec{p}_1)/2|$) were used to determine the size of stars. Owing to their nature, the interaction between the two correlated photons can be neglected. On the other hand, two nucleons correlated at small relative momentum will interact in the final state [6]. In the model of Ref. 6 it is the mechanism of final state interaction that yields information on the effective source size ρ, obtained as $\rho = [r_o^2+(v_{rel}\cdot\tau)^2]^{1/2}$, where r_o and τ are the emitter radius and lifetime, respectively. v_{rel} is the velocity of the center of mass of the emitted pair of particles relative to the source. Obviously, the interpretation of the correlation function from the above model is unambiguous only in two limiting cases, i.e., for negligible particle emission times ($\rho \rightarrow r_o$) and for sufficiently large emission times ($\rho \rightarrow v_{rel} \cdot \tau$). In an intermediate situation, exclusive light-particle correlations are needed to obtain v_{rel} to follow the spatial and temporal evolution of the collision system. As the excitation energy of the collision system increases, the statistical production and subsequent decay of particle-unstable resonances should also be included as a source of correlated light particles [7-9].

This contribution gives results obtained from exclusive two-proton correlations from the $^{58}Ni+^{58}Ni$ reaction. The data were taken at the Holifield Heavy Ion Research Facility of the Oak Ridge National Laboratory. The HILI detection system was used [10]. A 1 mg/cm^2 ^{58}Ni target was bombarded with 876.5 MeV ^{58}Ni projectiles. The charge (Z), energy (E_f), and direction ($\theta_f;\phi_f$) of the heavy fragment were obtained from a position-sensitive ionization chamber. Light charged particles were measured using an array of 96 ΔE-E phoswich scintillator detectors. The energy calibration for protons was accomplished by measuring protons from the elastic scattering of the ^{58}Ni beam from a polypropylene target at two different beam energies. The azimuthal and polar angles of each of these detectors were calculated at the central point and then randomized over the detector area. The average angular resolution was $\approx 1^o$. The minimum and maximum polar detection angles were 2° and 22°, respectively.

The correlation function R was calculated from the expression

$$R(\vec{p}_1,\vec{p}_2,\vec{p}_f,E_f)+1 = C*\sigma_{123}(\vec{p}_1,\vec{p}_2,\vec{p}_f,E_f)$$
$$/\sigma'_{123}(\vec{p}_{1,1},\vec{p}_{1,2},\vec{p}_{1f},E_{1f};\vec{p}_{2,1},\vec{p}_{2,2},\vec{p}_{2f},E_{2f}). \quad (1)$$

Here $\sigma_{123}(\vec{p}_1,\vec{p}_2,\vec{p}_f,E_f)$ is the measured yield of triple coincidence events consisting of two protons with momenta $\vec{p}_1$ and $\vec{p}_2$ and a fragment with momentum $\vec{p}_f$ and energy E_f.

$\sigma'_{123}(\vec{p}_{1,1},\vec{p}_{1,2},\vec{p}_{1f},E_{1f};\vec{p}_{2,1},\vec{p}_{2,2},\vec{p}_{2f},E_{2f})$ is the background yield of coincidences generated by two such triple events with components stemming from two different sources. σ'_{123} was constructed following a method outlined in Ref. 11. This method selects background events of the same class as those counted in σ_{123}, allowing for a correct normalization of the correlation function R at both the small and large values of Δp, hence an unbiased estimate of the correlation effects.

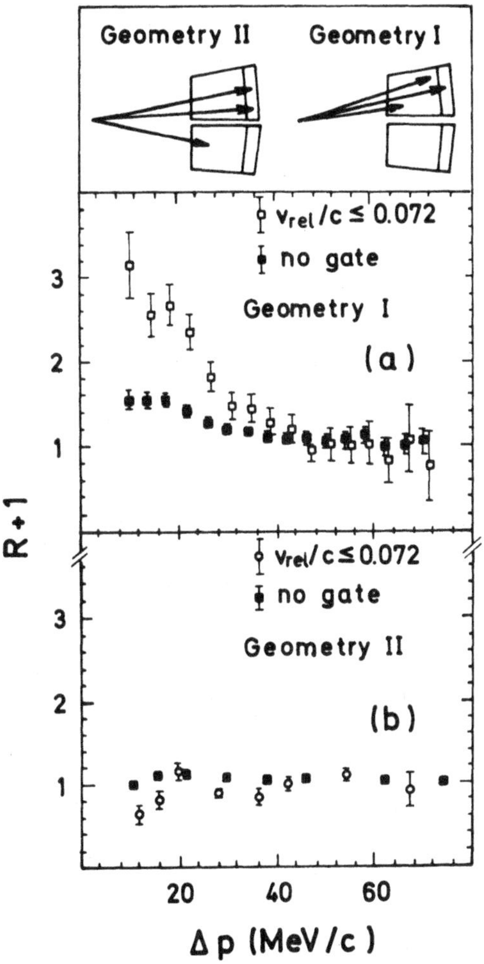

Fig. 1. Proton correlation functions for the $^{58}Ni+$ $+^{58}Ni$ reaction at 876.5 MeV incident energy ((a) and (b)); two different experimental arrangements are shown on top. The proton correlation functions with no gate on the data (open symbols) and with a gate on v_{rel} (full symbols) are shown for geometry I (a) and geometry II (b). Errors reflect statistical uncertainties only.

Figs. 1(a)-(b) show two-proton (p-p) correlation functions. Fig. 1(a) corresponds to the case when all the three particles are detected above the beam axis (geometry I, see top of figure). Fig. 1(b) corresponds to the case when the two light particles and the heavy fragment are detected on the opposite sides with respect to the beam axis (geometry II). The p-p correlation functions include fragments with charges in the range $21 \leq Z \leq 26$. The data are with (open symbols) and without a gate (full symbols) on the relative velocity $|\vec{v}_{rel}| = |\vec{v}_f - \frac{1}{2}(\vec{v}_{p1}+\vec{v}_{p2})|$, where $\vec{v}_{p1}$, $\vec{v}_{p2}$, and $\vec{v}_f$ are the velocities of the two protons and the heavy fragment, respectively. To determine $\vec{v}_f$ from $\vec{p}_f$ at each Z, we have taken for the fragment masses the most probable mass obtained from a distribution calculated with the statistical evaporation code LILITA [12].

Figs. 2(a)-(b) show the v_{rel} distributions for both geometries (I and II). The shaded areas represent the data used in the analysis with a gate on $v_{rel} \leq 0.072$ c. The combined figures 1 and 2 reveal the main feature of the data, i.e., the difference in the p-p correlation functions obtained, respectively, from the two

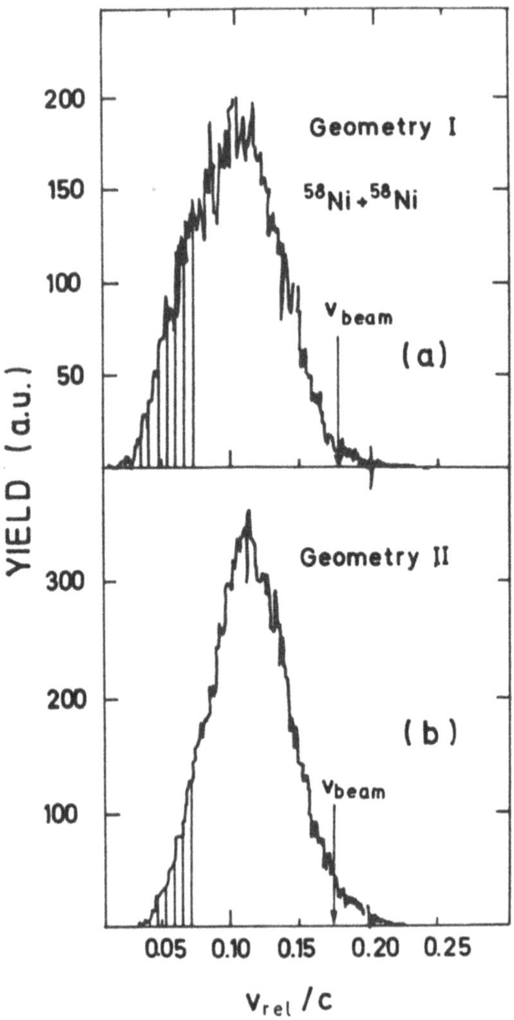

for geometry I (a) and geometry II (b). The arrows show the beam velocity (v_{beam}).

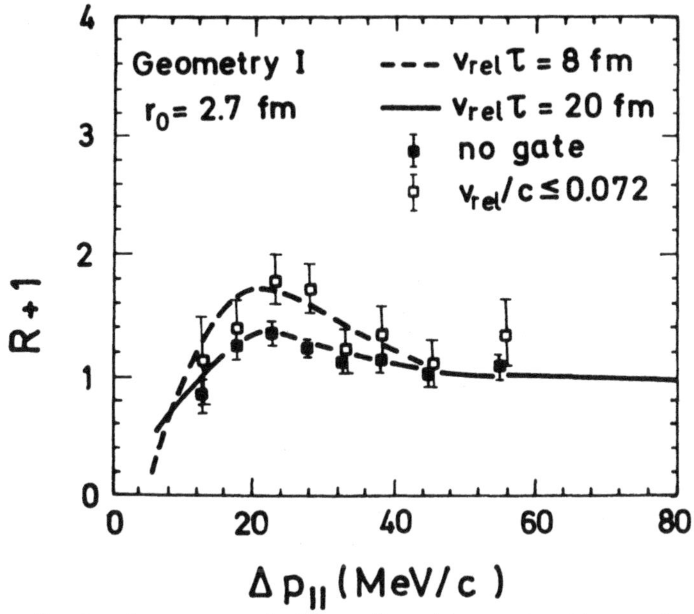

Fig. 3. Proton correlation functions vs the value of $\Delta\vec{p}$ projected onto the v_{rel} axis ($\Delta p_{||}$) for $^{58}Ni+^{58}Ni$ at geometry I. The full and dashed lines correspond to calculations of the proton correlation function from the model of Ref. 6 for $\Delta\vec{p}$ parallel to $\vec{v}_{rel}$. Symbols are as in Fig. 1(a).

geometries using an identical gate on v_{rel} (open symbols in Figs. 1(a) and (b)). This difference tells us that the main contribution to the p-p correlation function comes from such events where the relative angle between the two protons and the fragment is small. Obviously, such a condition cannot be fulfilled for the geometry-II setup (note $\theta_{min} \leq 2^{o}$ for all detected particles). Energy and momentum conservation yield the result that such a situation corresponds to small values of v_{rel}.

Fig. 3 shows the extracted p-p correlation functions compared with the calculations from the model of Ref. 6, for $\Delta\vec{p}$ projected in the $\vec{v}_{rel}$ direction ($\Delta p_{||}$). This model interprets the enhancement in the p-p correlation function as a result of the two-particle final state interaction. The comparison includes only the data obtained in the geometry-I setup: those taken without the gate on v_{rel} (full symbols) and those with the gate on v_{rel} (open symbols). Fitting the two sets of data allows the determination of the two parameters of the model, namely, r_o and $v_{rel} \cdot \tau$. The best fit values obtained are $r_o = 2.7$ fm (Gaussian density), $v_{rel} \cdot \tau = 20$ fm for the ungated data and $r_o = 2.7$ fm, $v_{rel} \cdot \tau = 8$ fm for the gated data. Taking $\langle v_{rel} \rangle = 0.05$ c for the gated data (Fig. 2(a)), we obtain $\tau = 5.3 \times 10^{-22}$ s; taking $\langle v_{rel} \rangle = 0.12$ c for the "no gate" condition, we obtain $\tau = 5.5 \times 10^{-22}$ s. The source size of $r_o = 2.7$ fm corresponds to a 4.3 fm radius of a uniform density distribution. This value is quite compatible with what could be expected as a source size since projectilelike fragments represent the bulk of fragments detected in coincidence with the two protons.

In summary, the enhancement in the p-p correlation function from the $^{58}Ni + ^{58}Ni$ reaction at $E_{inc} = 876.5$ MeV is interpreted as a result of the interaction of the emitted protons in the final state. The p-p correlation function has been found to decrease very rapidly with the increase of the relative angle between the fragment and the two-proton c.m. velocities. This causes the extracted p-p correlation function to be dependent on the kinematics of the measurements. The deduced average particle emission times are in the range of $5\text{-}6 \times 10^{-22}$ s.

This contribution is part of a study under the US-Yugoslav collaboration project JF 939 DOE/IRB, whose financial support is gratefully acknowledged.

References

1. D. Fox et al., Phys. Rev. C38, 146 (1988), and references therein.
2. T.S. Awes et al., Phys. Rev. Lett. 61, 2665 (1988).
3. D.A. Cebra et al., Phys. Lett. B227, 336 (1989).
4. P.A. DeYoung et al., Phys. Rev. C39, 128 (1989).
5. R. Hanbury-Brown and R.Q. Twiss, Nature (London) 178, 1046 (1956).
6. S.E. Koonin, Phys. Lett. 70B, 43 (1977).
7. M.A. Bernstein et al., Phys. Rev. Lett. 54, 402 (1985).
8. P.D. Bond and R.J. de Meijer, Phys. Rev. Lett. 52, 2301 (1984).
9. W.G. Lynch et al., Phys. Rev. Lett. 52, 2302 (1984).
10. D. Shapira et al., Phys. Div. Prog. Rep., 1990, in press, and to be published.
11. W.A. Zajc et al., Phys. Rev. C29, 2173 (1984).
12. J. Gomez del Campo and R. Stokstad, ORNL Report, No. TM7295, 1981 (unpublished).

CORRELATIONS IN EQUILIBRIUM AND NON-EQUILIBRIUM SYSTEMS

J. CUGNON
Université de Liège
Physique Nucléaire Théorique
Institut de Physique au Sart Tilman
Bâtiment B.5
B-4000 LIEGE 1 (Belgium)

1. INTRODUCTION

If correlations can be defined as a non factorizability of the two-body distribution function (we will use this terminology here covering classical distribution functions as well as Wigner functions) into a product of two one-body distribution function, there are basically four sources of correlations, namely :

(1) conservation laws
(2) dynamics
(3) quantum statistics
(4) "spurious" correlations.

The first three sources are rather familiar. We will illustrate the fourth case by a few examples in section 4. Mathematically, correlations are present whenever the following function

$$C\left(\vec{r}_1\vec{p}_1,\vec{r}_2\vec{p}_2\right)=\frac{f_2\left(\vec{r}_1\vec{p}_1,\vec{r}_2\vec{p}_2\right)}{f_1\left(\vec{r}_1\vec{p}_1\right)f_1\left(\vec{r}_2\vec{p}_2\right)}-1 \tag{1.1}$$

is non vanishing. Usually, one turns to simpler functions, obtained by integrating over some of the arguments of the function C in (1.1). For instance, one often considers

$$C\left(\vec{r}_1,\vec{r}_2\right)=\frac{\int d^3p_1\, d^3p_2\, f_2\left(\vec{r}_1\vec{p}_1,\vec{r}_2\vec{p}_2\right)}{\int d^3p_1\, f_1\left(\vec{r}_1,\vec{p}_1\right)\int d^3p_2\, f_1\left(\vec{r}_2,\vec{p}_2\right)}-1 \tag{1.2}$$

or

$$C\left(\vec{p}_1, \vec{p}_2\right) = \frac{\int d^3r_1\, d^3r_2\, f_2\left(\vec{r}_1\vec{p}_1, \vec{r}_2\vec{p}_2\right)}{\int d^3r_1\, f_1\left(\vec{r}_1, \vec{p}_1\right) \int d^3r_2\, f_1\left(\vec{r}_2, \vec{p}_2\right)} - 1 \,. \tag{1.3}$$

One may also integrate on more variables, but one has then to consider always these variables symmetrically on the indices 1 and 2. Otherwise, the function C can be different from zero even if f_2 factorizes.

In section 2, we describe the effect of dynamical correlations in Boltzmann systems, both in equilibrium and non-equilibrium situation. In section 3, we do the same thing for bosons and fermion systems and make a review of our present knowledge of dynamical correlations in nuclear matter. In both cases, we critically examine the interferometric methods which are currently used.

2. CORRELATIONS FOR BOLTZMANN PARTICLES

2.1. Systems at Equilibrium

(a) Infinite systems

In this case, the translational and rotational invariances impose that

$$\int d^3p_1 d^3p_2\, f_2\left(\vec{r}_1\vec{p}_1, \vec{r}_2\vec{p}_2\right) = n^2\left(1 + v(r)\right) \tag{2.1a}$$

and

$$C\left(\vec{r}_1, \vec{r}_2\right) = v(r) \tag{2.1b}$$

where n is the (uniform) equilibrium density and $\vec{r} = \vec{r}_1 - \vec{r}_2$. The function v(r) is intimately connected to dynamics. It vanishes for independent particles. For a normal fluid, it vanishes at some finite distance r , which usually allows to define a correlation length. The pair correlation function v(r) is related to density fluctuations through the relation

$$\frac{\overline{\rho^2} - \bar{\rho}^2}{\rho} = 1 + n \int d^3r\, v(r) \,. \tag{2.2}$$

Density fluctuations are in turn related to the isothermal compressibility :

$$kT\left(\frac{\partial n}{\partial p}\right)_T = 1 + n \int d^3r\, v(r) \;, \tag{2.3}$$

where p is the pressure and T the temperature.

The calculation of the pair correlation function is quite complicated and require fancy methods [1]. These are obtained by summing selected terms in the cluster expansion of the joint distribution function $n_{12} = 1 + \nu$. In the Percus-Yevick approach, one ends with the following integral equation

$$n_{12}(r)\, e^{\beta V(r)} = 1 - n \int d^3s \left(e^{\beta V(s)} - 1\right) n_{12}(\vec{s}) \left[n_{12}\left(|\vec{r} - \vec{s}|\right) - 1\right] , \tag{2.4}$$

where $V(r)$ is the two-body interaction potential and $\beta = 1/kT$. An alternative approach is provided by the hypernetted-chains method, which yields a similar, but somewhat different, integral equation

$$\ln n_{12}(r) = - \beta V(r) + n \int d^3s \left[n_{12}(s) - 1 - \ln n_{12}(s) - \beta V(s)\right]\left[n_{12}\left(|\vec{r} - \vec{s}|\right) - 1\right] . \tag{2.5}$$

In first order, both methods give the same well-known results :

$$n_{12}(r) = \exp\left(-\beta V(r)\right) . \tag{2.6}$$

In second order, for a hard sphere gas, one obtains

$$\begin{aligned} n_{12}(r) &= 0 \quad , \quad r < d \\ &= 1 + \frac{4}{3}\pi d^3 n \left(1 - \frac{3}{4}\frac{r}{d} + \frac{1}{16}\left(\frac{r}{d}\right)^3\right) , \quad d < r < 2d \\ &= 1 \quad , \quad r > 2d , \end{aligned} \tag{2.7}$$

d being the diameter of the spheres. The non-trivial value for $d < r < 2d$ is due to the shielding due to a third partner.

(b) <u>Finite systems</u>

In this case, the equivalent of n_{12} , i.e. the pair distribution function, is influenced by the geometry even in the absence of correlations. One has then

$$n_{12}(r) = \int d^3r_1 \int d^3p_1 \int d^3p_2\, f_2\left(\vec{r}_1 \vec{p}_1 , \vec{r}_1 + \vec{r} , \vec{p}_2\right) = \int d^3r_1\, \rho\left(\vec{r}_1\right) \rho\left(\vec{r}_1 + \vec{r}\right) . \tag{2.8}$$

For a spherical system of radius R , one obtains

$$\begin{aligned} n_{12}(\vec{r}) = h(r) &= 3N\,(N-1)\left[1 - \frac{3}{4}\left(\frac{r}{R}\right) + \frac{1}{16}\left(\frac{r}{R}\right)^3\right] , \quad r < 2R \\ &= 0 \quad , \quad r > 2R \end{aligned} \tag{2.9}$$

and

$$h(r) = \frac{N(N-1)}{\left(\sqrt{2\pi}\, R\right)^3} \exp\left(-\frac{r^2}{2R^2}\right) \tag{2.10}$$

for a gaussian shape.

2.2. Non Equilibrium Systems

(a) Generalities

The evolution of the correlations can be handled classically by the BBGKY formalism and quantum mechanically by the density matrices formalism. Using Wigner transforms, the two formalisms have the same structure. The relevant equation can write (in the weak coupling limit)

$$\left(\partial_t - L_1^0 - L_2^0\right) f_2\left(x_1, x_2\right) = L_{12} f_2\left(x_1, x_2\right) + \int dx_3 \left(L_{13} + L_{23}\right) f_3\left(x_1, x_2, x_3\right) , \tag{2.11a}$$

where L_i^0 is the free Liouvillian for particle i (x_i stands for the set of coordinates $\vec{r}_i , \vec{p}_i$) and where

$$L_{ij} = \vec{p}_i . \vec{\nabla}_i V_{ij} - \vec{p}_j . \vec{\nabla}_j V_{ij} \tag{2.11b}$$

classically. It has a more complex structure in the quantum case (see ref. [1]) for detail).

One is often interested in the asymptotic ($t \rightarrow \infty$, i.e. observable) behaviour of the correlations and more precisely of the correlations between the momenta of the particles. One can see from Eq. (2.11a) that the correlations between two particles are eventually due to their mutual interaction, but also to their interaction with the rest of the system. Eq. (2.11a) is of few practical use. The problem is usually approached by assuming a separation of the interactions : first the interaction involves all the system up to a late time, after which the two particles are assumed to interact independently of the system. This picture is usually denoted by the name final state interaction (FSI).

(b) Classification of the FSI

In fig. 1, we schematically depict the possible FSI between say particles 2 and 3 for instance. In this picture, "1" can be viewed as an arbitrary number of particles and not necessarily as a single one. Case (i) corresponds to no FSI. Case (iia) is the usual FSI. Case (iib) and (iic) are similar, but are less important for the pair (2,3) of particles. Finally (iiia) corresponds to a more complex FSI. It may correspond to just a small perturbation of the FSI between 2 and 3, if the interaction between 1 and 3 is weak. Case (iiia) has some interest when the interaction between 1 and 3 is strong.

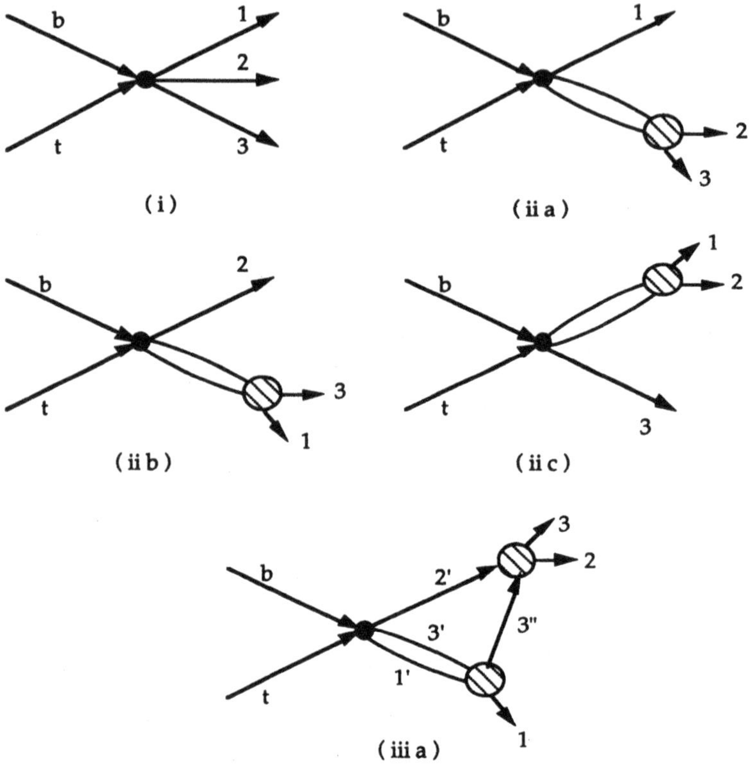

Fig. 1. Classification of three-body final-state interactions processes.

The final state interaction (case (iia)) can be used to study the spatial extension of the system and is then usually treated with semi-classical arguments. However, we think that the scattering theory approach is quite instructive. It is reviewed in the next paragraph.

(c) Collision theory formalism

We closely follow here ref. [2]. Let us assume that the hamiltonian can be divided as

$$H = H_0 + V + v \tag{2.12}$$

where H_0 is the part which can describe the incoming and outgoing asymptotic state of fig. 1(i), V is the interaction responsible for the transition to these states and where v is the interaction between particles 2 and 3, but which acts only in the final states illustrated by fig. 1(i). It can then be shown that the transition amplitude for process (ii)a writes

$$T_{fi} = \left\langle \phi_f^{(-)} \middle| V \middle| \psi_i^{0(+)} \right\rangle + \left\langle \psi_f^{(-)} - \phi_f^{(-)} \middle| v \middle| \psi_i^{0(+)} \right\rangle , \tag{2.13}$$

where

$$\psi_i^{0(+)} = \left(1 + \left(E + i\varepsilon - H_0 - V\right)^{-1} V\right)\left|\chi_i\right\rangle , \quad \phi_f^{(+)} = \left(1 + \left(E + i\varepsilon - H_0 - v\right)^{-1} v\right)\left|\chi_f\right\rangle ,$$

$$\psi_f^{(+)} = \left(1 + \left(E + i\varepsilon - H_0 - V - v\right)^{-1} (V + v)\right)\left|\chi_f\right\rangle , \tag{2.14}$$

and where χ_i and χ_f are unperturbed initial and final states, eigenstates of H_0 with eigenvalue E. The (+) and (-) signs and the infinitesimal ε refer to the usual boundary conditions. The second term in (2.13) is generally negligible since v cannot efficiently create the final states of the primary interaction. The first term can be specified using plausibility arguments. First, the interaction V is limited to a finite volume of radius R. Second, if the coordinates of the system ξ are divided between the relative distance r (between particles 2 and 3) and the remaining coordinates ξ' , the following factorization holds

$$\left\langle \xi \middle| V \psi_i^{0(+)} \right\rangle \approx h_f(r)\, \psi(\xi') .$$

The transition matrix element (2.13) can then be written as (C is a constant)

$$T_{f_i} \approx C \int d^3r\, g_q(r)\, h_f(r) , \tag{2.15a}$$

where $g_q(r)$ is the scattering wavefunction due to the interaction v with the proper boundary condition. The important thing is the fact that the q-dependence is solely contained in the integral (2.15). One, of course, so neglects conservation law effects, which is valid if many particles are emitted. Eq. (2.15) yields

$$\left| T_{f_i} \right|^2 \propto \int d^3r \int d^3r'\, g_q(r) g_q(r') \rho_2(r , r') , \tag{2.15b}$$

where $\rho_2(r, r')$ is the density matrix corresponding to the state $V \mid \psi_i^{(0)} >$ i.e., $\propto h_f(r)h_f(r')$. One usually approximates by taking the diagonal value of ρ_2 only or by assimilating $h_f(r)$ to pair distribution function as estimated in Eq. (2.9) or (2.10). One then gets

$$\frac{d\sigma}{dq} \propto q^2 \left| \int d^3r \, g_q(r) h(r) \right|^2 . \tag{2.16}$$

Two cases are of interest :

(1) At small q , an attractive (repulsive) interaction gives an amplification (depletion) of the wavefunction $g_q(r)$ for $r < a$, the range of the interaction. However, at very low q , this effect is not important, since the wavefunction have no nodes for $r \lesssim 2R$. In that case, the wavefunction is well approximated by

$$g_q(r) \approx \frac{\sin \delta}{qr} \left[1 + \alpha r + \frac{1}{2} q^2 r \left(r_0 - r \right) \right] , \tag{2.17}$$

where one has made use of effective range theory. Using also the effective range expansion of the phase shift

$$k \cot \delta = \alpha + \frac{1}{2} r_0 q + ..., \tag{2.18}$$

one obtains in first order for

$$C(q) = C\left(\vec{p}_1 , \vec{p}_1 + \vec{q}\right) = \frac{\left(\frac{d\sigma}{dq}\right)}{\left(\frac{d\sigma}{dq}\right)_0} - 1 , \tag{2.19}$$

where $\left(\frac{d\sigma}{dq}\right)_0$ is the cross-section without FSI, the following expression

$$C(q) = \frac{1}{\left(\alpha^2 + q^2\right) R^2} . \tag{2.20}$$

In conclusion, the q-dependence is simply related to the scattering length α^{-1} . Note that this is only true if the third term in (2.17) is neglected, which is only valid if $\alpha R << 1$. In other words, if the scattering length is large, the correlation is not sensitive to the other typical lengths as r_0 and R. Note however that if q is not so small, Eq. (2.20) is no longer valid and the correlation will be sensitive to the sign of α and therefore to the attractive or repulsive nature of the interaction.

Small q correlations dominated by the behaviour expressed by Eq. (2.20) has been observed in n-n [3] and in n-p correlations [4]. For n-n correlations, (2.20) should be corrected for quantum statistics, see section 3, but this does destroy the enhancement, but just reduces it. For p-p, the Coulomb interaction makes this enhancement since then $C(q) \to 0$ for $q \to 0$.

(2) At large q , the interaction v may generate resonance states. In this case, using the simplest approximation, one gets

$$\frac{d\sigma}{dq} \propto \frac{\pi(2\ell + 1)}{q^2} \frac{\Gamma \Gamma_i}{(E - E_0)^2 + \frac{\Gamma^2}{4}} , \tag{2.21}$$

where Γ_i is proportional to the probability of forming the resonant state by the primary interaction. The physical link between Eq. (2.15) and (2.19) is due to the fact that the scattering wavefunction is considerably enhanced in the inner region with a Breit-Wigner coefficient. Eq. (2.21) however neglects the structure of h(r) , i.e. finite size effects. The effects have been studied in ref. [5], where the authors used the structure of the wavefunction and the statistical modification of the density of states. They show that both approaches lead to the same conclusion which is not surprising, since the density of states is determined by the number of nodes of the wavefunction. They arrive at

$$C(q) = \frac{3}{2} \frac{1}{R^3 q^2} (2\ell + 1) \frac{d\delta_\ell(q)}{dq} . \tag{2.22}$$

This however neglects the fact the resonant state wavefunction is formed in a finite volume. The influence of this fact has been worked out at long time ago [6] and leads to the following modification

$$C(q) = \frac{3}{2} \frac{1}{R^3 q^2} (2\ell + 1) \left(\frac{d\delta_\ell(q)}{dq} + \frac{d\varphi_\ell(q)}{dq} \right) \tag{2.23}$$

where

$$\varphi_\ell(q) = \tan^{-1} \frac{j_\ell\,(qR)}{n_\ell\,(qR)} \tag{2.24}$$

In most practical cases, this additional term is however small.

It should be noticed that the range a of the interaction does not enter Eq. (2.22), but is rather replaced by a length $\Lambda = d\delta/dq$, which at the resonances reaches $\Lambda_{max} = q^{-1}\ E^*/\Gamma$, where E^* is the energy of one of the particles in the c.m. system. For narrow resonances, Λ_{max} may reach several fm's and be as large as R. The physical meaning of Λ is the range within which the incoming and outgoing waves interfere constructively (even if this interference builds up within a only). In other words if particles 2 and 3 are within Λ at the right energy (E_0), they will interfere strongly.

This leads naturally to the question to know whether the resonance is formed at once or whether particles 2 and 3 are first formed and interact later. This question is irrelevant if Λ is larger or comparable to R. As stressed in refs. [6, 7], this is also reflected by the fact that taking interaction into account appears as modifying the density of states of the compound (2+3) system.

The case of broad resonances (like the Δ's) appears as the opposite case. Then $\Lambda \sim$ 1-2 fm and is much smaller than the size of the system. This is why the effect of correlated pairs is small compared to the background in the analysis of ref. [8]. Here, the question of whether the Δ is formed or not is pertinent. There is however not so clear a separation between primordial and final state interaction in the reactions involving Δ's.

(d) Proximity scattering

The relevant graph (iiia) of fig. 1 has been evaluated by Aitchison and Kacser [9], who arrive at the following expression for low q (for particles 2 and 3)

$$T(q) = \frac{C}{\alpha - iq}\left\{K + \frac{i\gamma}{mp}\,\ell n\left[\frac{(q-q_s)+\left(\frac{m}{M}\right)(p+p_s)}{(q-q_s)-\left(\frac{m}{M}\right)(p-p_s)}\right]\right\}, \tag{2.25}$$

where C , K are constants, α is the inverse scattering length, q_s and p_s are quantities fixed by the kinematics. When the second term is neglected, the Watson formula (2.11a) is recovered. Apparently, the application of formula (2.25) reduces the correlation at low q [10].

3. CORRELATIONS FOR FERMION (BOSON) PARTICLES

3.1. At Equilibrium

(a) Infinite homogeneous systems

Just like for Boltzmann particles, one can write Eq. (2.1). But for quantum statistics, ν is different from zero, even for independent particles, at least for parallel spin directions. In nuclear matter, summing over spin states, one obtains

$$n_{12} = 1 - \frac{9}{2k_F^6}\left(\frac{\sin k_F r}{r^3} - \frac{k_F \cos k_F r}{r^2}\right)^2 . \tag{3.1}$$

This behaviour is illustrated in fig. 2. It should be stressed that correlation

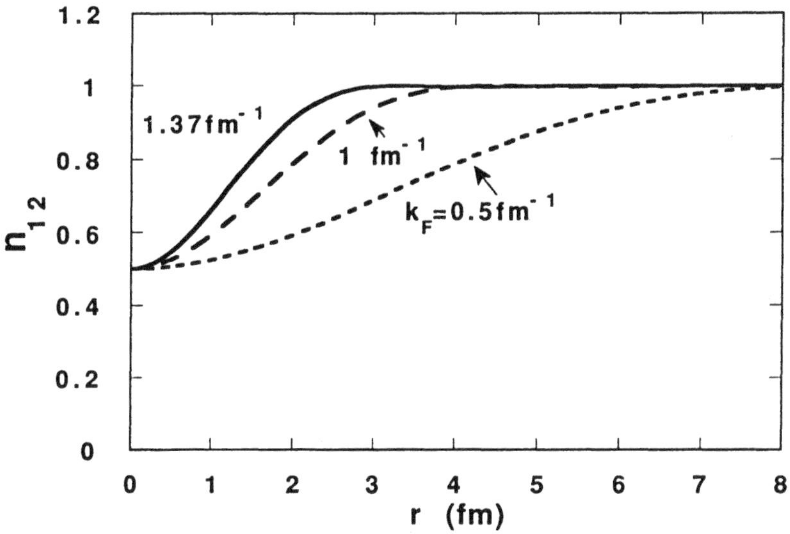

Fig. 2. Correlated pair distribution function in a Fermi gas of nucleons (Eq. (3.1)).

effects increase when the density <u>decreases</u>. This is however in keeping with the fact that the particles are restricted to smaller and smaller momentum and thus to larger and larger de Broglie wave lengths.

For interacting fermions, the function $n_{12}(r)$ is given by

$$n_{12}(r) = \sum_{k<k_F} \sum_{j<k_F} \left|\psi_{kj}(\vec{r})\right|^2 , \tag{3.2}$$

where ψ_{kj} is the correlated wavefunction for particles $\vec{k}$ and $\vec{j}$. Calculations [11] of n_{12} in the Brueckner-Hartree-Fock approximation for the Paris potential are shown in fig. 3. The remarkable fact is that the attractive interaction reduces considerably the correlation length.

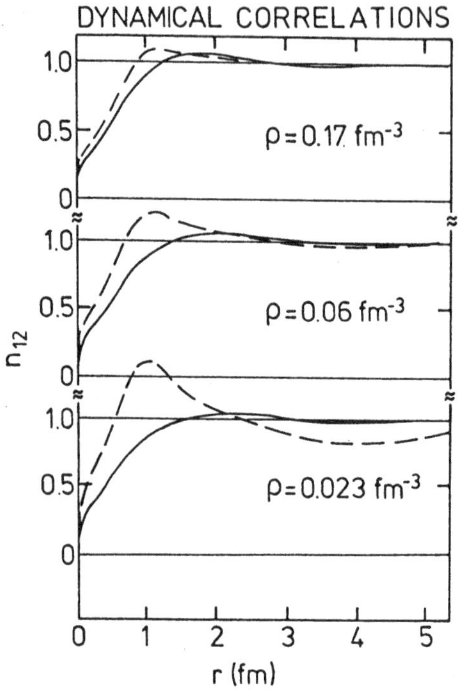

Fig. 3. Correlated pair distribution function in nuclear matter as calculated in ref. [11], for T=0 (full lines) and T=10 MeV (dashed lines).

Interactions are also responsible for modifying the momentum distribution, giving rise to a depletion of the Fermi sphere and a consequent population above the Fermi level. A calculation [12] of the momentum distribution $n(k)$ at the same level of approximation as $n_{12}(r)$ is given in fig. 4. It should be noted that the depletion is quite large (~ 20 %). The recently measured occupation probability in finite nuclei supports this strong effect [13]. However, it is not clear that finite size effects are not cooperating to this depletion. The relation between $n(k)$ and n_{12} (in second order) is given by $(k < k_F)$

$$n(k) = \sum_j \frac{\delta}{\delta n(j)} \int d^3r \left(n_{12}(\vec{r}) - 1\right) . \tag{3.3}$$

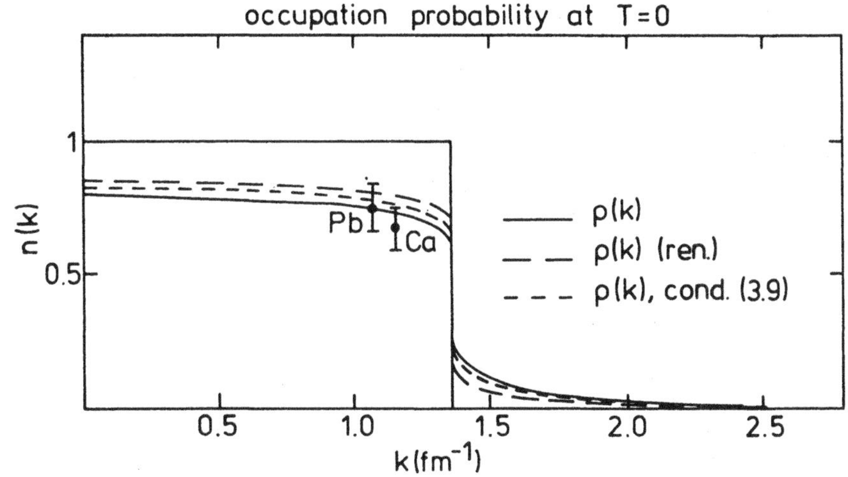

Fig. 4. Occupation probability in cold nuclear matter, as calculated in ref. [12] (full curves). Slashed curves correspond to less accurate approximations. The experimental data of ref. [13] for Pb and Ca nuclei are schematically given.

The correlations are also responsible for corrections to the nucleon mean field beyond the (Brueckner)Hartree-Fock approximation (see ref. [11] for detail). For this correction V_{CO} , one has the following approximate relation

$$V_{CO}(k) \approx - \kappa\, V_{PO}(k) \quad , \quad \kappa = \int_{k < k_F} n(k) \ , \tag{3.4}$$

where V_{PO} is the lowest order Brueckner-Hartree-Fock contribution.

(b) <u>Finite systems</u>

The correlations at small $\vec{p}$ are connected to the spatial extension of the system, as the correlations at small $\vec{r}$ are connected to the momentum extension of the system. Indeed, for independent particles, the two-body Wigner function reads

$$f_2\left(\vec{k}_1\vec{p}_1, \vec{k}_2\vec{p}_2\right) = f_1\left(\vec{k}_1\vec{p}_1\right) f_1\left(\vec{k}_2\vec{p}_2\right)$$

$$\pm f_1\left[\frac{1}{2}\left(\vec{k}_1 + \vec{k}_2\right) - \frac{1}{\hbar}\left(\vec{p}_1 - \vec{p}_2\right), \frac{1}{2}\left(\vec{p}_1 + \vec{p}_2\right) - \frac{\hbar}{4}\left(\vec{k}_1 - \vec{k}_2\right)\right]$$

$$f_1\left[\frac{1}{2}\left(\vec{k}_1 + \vec{k}_2\right) + \frac{1}{\hbar}\left(\vec{p}_1 - \vec{p}_2\right), \frac{1}{2}\left(\vec{p}_1 + \vec{p}_2\right) + \frac{\hbar}{4}\left(\vec{k}_1 - \vec{k}_2\right)\right] . \tag{3.5}$$

Actually, these are rather Fourrier transforms (on the spatial variables) of the Wigner function. One then has, assuming $f_1\left(\vec{r}, \vec{p}\right) = \psi(r)\varphi\left(\vec{p}\right)$

$$\int d^3k_1 \int d^3k_2\, f_2\left(\vec{k}_1\vec{p}_1, \vec{k}_2\vec{p}_2\right) = \int d^3r_1 \int d^3r_2\, f_2\left(\vec{r}_1\vec{p}_1, \vec{r}_2\vec{p}_2\right) =$$

$$\varphi_1\left(\vec{p}_1\right)\varphi_2\left(\vec{p}_2\right) \pm \int d^3K\, \psi\left(\vec{K} - \frac{1}{\hbar}\left(\vec{p}_1 - \vec{p}_2\right)\right) \psi\left(\vec{K} + \frac{1}{\hbar}\vec{p}_1 - \vec{p}_2\right)$$

$$\int d^3k\, \varphi\left(\frac{\vec{p}_1 + \vec{p}_2}{2} - \frac{\hbar}{4}\vec{k}\right) \varphi\left(\frac{\vec{p}_1 + \vec{p}_2}{2} - \frac{\hbar}{4}\vec{k}\right) . \tag{3.6}$$

Assuming ψ and φ to be gaussian functions (with a radius R for ψ), one obtains

$$\int d^3k_1\, d^3k_2\, f_2\left(\vec{k}_1\vec{p}_1, \vec{k}_2\vec{p}_2\right) = \varphi\left(\vec{p}_1\right)\varphi\left(\vec{p}_2\right) \pm \left[\varphi\left(\frac{\vec{p}_1 + \vec{p}_2}{2}\right)\right]^2 e^{-\frac{2\left(\vec{p}_1 - \vec{p}_2\right)^2}{R^2\hbar^2}}$$

$$\approx \varphi\left(\vec{p}_1\right)\varphi\left(\vec{p}_2\right)\left[1 \pm e^{-\frac{2\left(\vec{p}_1 - \vec{p}_2\right)^2}{R^2\hbar^2}}\right] . \tag{3.7}$$

This is at the basis of the Hanbury-Brown/Twiss interferometry [14,19].

3.2. Non Equilibrium Situations

The dynamical evolution of the correlations can be handled in principle with the help of Eq. (2.11) (or similar ones obtained with Green functions), with however a more complicated definition of L'_{ij} (see ref. [1] for detail). However, these equations are not tractable. We just discuss two physical aspects of the dynamical evolution below.

(a) Decay rate of correlations

Instead of single-time correlation function, one can consider two times correlations. The simplest quantity in this perspective is

$$S(\vec{r}, \tau) = \int d^3p_1\, d^3p_2\, f_2\left(\vec{r}_2 + \vec{r}, \vec{p}_1, t_2 + \tau, \vec{r}_2, \vec{p}_2, t_2\right), \tag{3.8}$$

where we disregard the $\vec{r}_2$ and t_2 dependence (correct for a large system). $S(\vec{r}, \tau)$ can then be considered as

$$S(\vec{r}, \tau) = \overline{\rho\left(\vec{r}_2 + \vec{r}, t_2 + \tau\right)\rho\left(\vec{r}_2, t_2\right)}, \tag{3.9}$$

i.e. the correlation between densities a some point in space-time and the density in a near-by point. The bar indicates a quantum average over the state of the system. This quantity also enters in the description of the evolution of density fluctuations [20]. In a classical (boltzmannian) fluid, it can be shown that the time evolution of the quantity $S(\vec{r}, \tau)$ (for long distance r or rather, for a small wave vector) behaves like

$$S(\vec{r}, \tau) \propto \left(\frac{d^2 - 2}{(2\eta')^{d/2}} + \frac{1}{\Gamma^{d/2}}\right)\tau^{-d/2}, \tag{3.10}$$

where η' is the viscosity coefficient, Γ the sound attenuation coefficient and d is the dimensionality of the system. It is natural to find that correlations decay with the viscosity coefficient (i.e. they are destroyed by collisions in heavy ion collisions). The surprising aspect is the power law involving the dimension of the system. Typical values of Γ and η' are of the order of 2-5 $(fm/c)^{-1}$. Relation (3.10) is valid for Boltzmann systems. For Fermi systems, a smaller damping of sound waves is expected. Therefore, a typical survival time could be of the order of 2-3 fm/c. Initial correlations are thus not so much important. This does not mean that correlations are vanishing. They are rebuilt regularly by the collisions themselves.

(b) Remarks on interferometry

For fermions, interferometry is usually based on Koonin's formula [18], which writes

$$C(\vec{q}) = N^{-1}\int d^3r\, d^3p_1\, d^3r_1 \left|\psi(\vec{r}, \vec{q})\right|^2 \varphi\left(\vec{r}_1, \vec{p}_1\right)\varphi\left(\vec{r}_1 + \vec{r}, \vec{p}_1 + \vec{q}\right) - 1 \tag{3.11}$$

where N is the uncorrelated integral. Once again, this formula assumes a decoupling between FSI and primordial interaction. As we indicate in section 2, the basic aspect is the fact that the distribution of the relative distance r is a function of the radial extension of the system, even for independent particles. Formula (3.11) neglects Fermi statistics. At very low density, as it is often referred to, and for two proton interferometry, say $\rho \approx \rho_0/2$ ($k_F \approx 0.9$ fm^{-1}), the Fermi correlations will extend up a few fm's. This would have introduced a sizeable error in relation (3.11). Fortunately, as we showed, dynamics reduces the correlation up to say $r \sim 2$ fm. The wavefunction in (3.11) is more extended and the net effect is rather small.

4. SPURIOUS CORRELATIONS

We discuss a few examples of situations where correlations (which should be redefined in any case) are observed even if there is a priori no knowledge of the two-body distribution function.

(a) Discretization of hydrodynamics

In hydrodynamical calculations, it is customary at the end of the calculation to shift to a (quasi)-particle description which mocks up the density and velocity profile $\left(\rho(r), \vec{v}(\vec{r})\right)$ in order to make more specific predictions. Generally, this is done by dividing the space in cells and filling the cells with particles at random, but in accordance with $\rho(r)$ and $\vec{v}(\vec{r})$. Let us recall that these quantities are only the zeroth and first moment in $\vec{v}$ of the $f_1(\vec{r}, \vec{p}, t)$. The procedure amounts to consider a N-body distribution function of the following type $\left(x = (\vec{r}, \vec{p})\right)$

$$F_N(x_1, \ldots, x_N) = \sum_{\text{cells } \mu} \sum_{i=1}^{N(\mu)} \Theta_\mu(r_i)\, e^{-\left(\vec{p}_i - m\vec{v}_\mu\right)^2 / 2\tau(\mu)} \tag{4.1}$$

where $N(\mu)$ is the number of particles in cell μ ($N(\mu) = \rho(\mu) \times \Delta v(\mu)$), Δv being the volume of the cell and where $\Theta_\mu(r)$ is a function which is roughly equal to one inside cell μ and vanishes outside. Of course, F_N is consistent with the following f_1 distribution function

$$f_1(x_1) = \sum_\mu N(\mu)\, \Delta\left(r_1 - "\mu"\right) e^{-\left(\vec{p}_1 - m\vec{v}_\mu\right)^2 / 2\tau_i(\mu)}, \tag{4.2}$$

where the Δ function means that r_1 should be inside cell μ, and with the $\rho(\vec{r})$ and $\vec{v}(\vec{r})$ profile. The two-body distribution function issued from (4.1) reads

$$f_2(x_1,x_2) \approx \sum_{\mu_1}\sum_{\mu_2} N(\mu_1)\, e^{-\left(\vec{p}_1 - m\vec{v}_{\mu_1}\right)^2 / 2\tau(\mu_1)} N(\mu_2)\, e^{-\left(\vec{p}_2 - m\vec{v}_{\mu_2}\right)^2 / 2\tau(\mu_2)} , \tag{4.3}$$

which is obtained by integrating over $x_3, \ldots, x_N$. Obviously f_2 factorizes into two f_1 factors and therefore no correlation should be observed from this procedure, provided f_1 is constructed as f_2 ! The only correlations can come from conservation laws. For instance, conservation of current improves $\int \rho(\vec{r})\,\vec{v}(\vec{r})\,d^3r$ to be a constant (in the c.m. system). Therefore $\vec{v}(r_1)$ and $\vec{v}(r_2)$ are not completely independent and this would indicate a relationship upon $\vec{v}_{\mu_1}$ and $\vec{v}_{\mu_2}$ above. In the extreme jet picture one would have $\vec{v}_{\mu_1} = -\vec{v}_{\mu_2}$ and factorization is badly broken down. The correlation mentioned in the hydrodynamical calculation of ref. [21] is a good example of these "spurious" correlations.

(b) Mean field correlations

The same considerations hold when one looks at two particles, imposing some constraint on the observables like on the relative angle between the directions of these outgoing particles. Then the relevant quantity is

$$R(p_1\Theta_1, p_2\Theta_2)_{\Delta\varphi} = \int d^3r_1\, d^3r_2 \int d\varphi_1\, f_2\left(\vec{r}_1 p_1\Theta_1, \varphi_1, \vec{r}_2 p_2\Theta_2\varphi_1 + \Delta\varphi\right) . \tag{4.4}$$

If f_2 factorizes (which one has to assume in theories like Landau-Vlassov, predicting f_1 only), one has

$$R(p_1\Theta_1, p_2\Theta_2)_{\Delta\varphi} = \int d^3r_1\, d^3r_2 \int d\varphi_1\, f_1\left(\vec{r}_1, p_1, \Theta_1, \varphi_1\right) f_1\left(\vec{r}_2, p_2, \Theta_2, \varphi_1 + \Delta\varphi\right) . \tag{4.5}$$

Now, even if there is a decoupling between $\vec{r}$ and $\vec{p}$ coordinates $f_1 = \psi(\vec{r})\,\varphi(\vec{p})$, the quantity (4.5) does not factorize into a function of p_1 times a function of $\vec{p}_2$

$$R(p_1\Theta_1, p_2\Theta_2)_{\Delta\varphi} = \int d\varphi_1\, \varphi(\vec{p}_1)\, \varphi(\vec{p}_2) \neq F(p_1\Theta_1)\, F(p_2\Theta_2) . \tag{4.6}$$

Therefore the quantity R bears some information about the function $\varphi(\vec{p})$. Sometimes [22], property (4.6) is called improperly mean field correlations and this was partly justified by the fact that in these particular cases, the main structures of f_1 came from the mean field dynamics.

(c) Reaction plane correlations

This refers to a situation similar to (a), which imposes that there exists a vector $\vec{n}$ which is such that

$$\int \vec{n}.\vec{p}\, f_1(\vec{r},\vec{p})\, d^3p = 0 \,. \tag{4.7}$$

The latter correlation is the largest for quasi-free scattering, as it is well known from two proton correlation experiment at large momentum [23]. See also refs. [24,25] for other examples.

(d) Impact parameter correlations

Quite often, experiments imply a summation over a certain range of impact parameter. This implies that the relevant quantity is

$$\sum_{b\in\Delta b} f_2(r_1p_1, r_2p_2, b) \neq \sum_{b\in\Delta b} f_1(r_1p_1, b) \sum_{b\in\Delta b} f_1(r_2p_2, b)\,, \tag{4.8}$$

even if f_2 factorizes in f_1f_1 for any b. The latter correlation will reflect the b-dependence of the one-body distribution function.

5. CONCLUSION

We have reviewed the main aspects of the two-body correlations for both equilibrium and non-equilibrium situations. We have clarified the physical meaning of low-energy and resonant interferences. We also presented detailed calculations of correlations and related quantities for both equilibrium nuclear matter.

Acknowledgement : We are very thankful to the correlated pair B. Remaud-D. Ardouin for interesting though uncorrelated discussions.

REFERENCES

1. R. Balescu, Equilibrium and Non Equilibrium Statistical Mechanics, J. Wiley, New York, 1975.
2. K.M. Watson, Phys.Rev. 88, 1163 (1952).
3. R. Lednicky, this conference.
4. P. De Young, this conference.

5. B.K. Jennings, D.H. Boal and J.C. Shillcock, Phys.Rev. C33, 1303 (1986).
6. G.C. Philipps, T.A. Griffy and L.C. Biedenharn, Nucl.Phys. 21, 327 (1960).
7. J. Pochodzalla et al., Phys.Rev. C35, 1695 (1987).
8. D. L'Hôte, this conference.
9. I.J.R. Aitchison and C. Kacser, Phys.Rev. 142, 1104 (1966).
10. T.C. Sharma, J.M. Cameron and G.C. Neilson, Nucl.Phys. A202, 59 (1973).
11. A. Lejeune, P. Grangé, M. Martzolff and J. Cugnon, Nucl.Phys. A453, 189 (1986).
12. P. Grangé, J. Cugnon and A. Lejeune, Nucl.Phys. A473, 365 (1987).
13. D. Witt Huberts et al., to be published.
14. R. Hanburg-Brown and R.Q. Twiss, Phil.Mag. 45, 663 (1954) ; Nature 178, 1046 (1956).
15. G. Goldhaber, S. Goldhaber, W. Lea and A. Pais, Phys.Rev. 120, 300 (1960).
16. G. Cocconi, Phys.Lett. 49B, 459 (1974).
17. G.I. Kopylov and M.J. Podgoretsky, Yad.Fiz. 18, 656 (1973).
18. S.E. Koonin, Phys.Lett. 70B, 1 (1977).
19. M. Gyulassy, S.K. Kauffmann and L.W. Wilson, Phys.Rev. C20, 2267 (1979).
20. L. Landau and E. Lifshitz, Statistical Physics, Mir, 1967.
21. H. Stöcker et al., Nucl.Phys. A387, 205c (1982).
22. Z. Basrak, this conference.
23. S. Nagamiya et al., Phys.Lett. 81B, 147 (1979).
24. J.P. Alard, this conference.
25. S. Kox, this conference.

SHORT RANGE CORRELATIONS IN FERMIONIC MOLECULAR DYNAMICS

H. Feldmeier

GSI, Gesellschaft für Schwerionenforschung mbH
Postfach 110552, D-6100 Darmstadt, Germany

1 ABSTRACT

The inclusion of short range correlations due to the strongly repulsive core in the nucleon-nucleon interaction into the trial wave function of the Fermionic Molecular Dynamics (FMD) model is discussed. The FMD model was recently proposed to solve approximately the many-body problem of interacting identical fermions with spin $\frac{1}{2}$. The interacting system is represented by an antisymmetrized many-body wave-function. It consists of single particle states which are localized in phase space and a product of Jastrow correlation functions which suppress the wave function for short distances between pairs of particles. The equations of motion for the parameters characterizing the many-body state (e.g. positions, momenta and spin of the particles, or range of the two-body correlation) are derived from the time-dependent quantum variational principle. The effects of the Pauli principle as well as those of the repulsive core will thus be present in the trajectories of the wave packets for the particles.

2 THE FERMIONIC MOLECULAR DYNAMICS MODEL

Recently I proposed to describe the collision of two nuclei by Fermionic Molecular Dynamics (FMD) [1,2] in which nucleons represented by wave packets interact via two-body forces. The single-particle wave packets, from which an antisymmetrized many-body wave-function is formed, are parametrized as gaussians. In this contribution I want to sketch how short range correlations can be incorporated in order to obtain a more realistic treatment of the strongly repulsive core in the interaction between nucleons.

In FMD it should be possible to overcome the conceptual difficulties of the classical molecular dynamics models [3-8] which arise when they are applied to nucleons. One major problem with these models is that they are appropriate for distinguishable classical particles but we know that nucleons in nuclei and in nuclear collisions are close enough in phase space that the Pauli principle is important. The most common approach to improve on that is to mock the influence of the Pauli correlations by an additional momentum-dependent potential which becomes strongly repulsive if the fermions get too close in phase space [9-11]. The other main problem is the treatment of the repulsive core. If density packets [12, 13] or test particles are used to obtain a smooth mean field one needs to add a collision term in which the momenta of pairs of nulceons are randomly changed from time to time to account for the hard collisions [12-15].

In FMD the Pauli exclusion principle and the hard core can be explicitly taken into account by an appropriate ansatz for the parametrized many-body state. This state is characterized by a set of time dependent parameters $Q(t) = \{ Q_\nu(t),\ \nu = 1, 2, \cdots \}$ which are regarded as the generalized coordinates of the system. For identical fermions which have short range correlations the ansatz for the parameterized many body wave function $|\, Q(t)\,\rangle$ is an antisymmetrized state formed out of the single-particle states $|\, q_k(t)\,\rangle$ given in Eq. (2.3) multiplied with Jastrow correlation functions [16, 17]. In coordinate representation it is given by

$$\{\langle\, \vec{x}_1\,| \otimes \cdots \otimes \langle\, \vec{x}_N\,|\}\,|\,\hat{Q}\,\rangle =$$

$$\frac{1}{N!} \sum_{all\ P} sgn(P)\ \langle \vec{x}_1 | q_{P(1)} \rangle \otimes \langle \vec{x}_2 | q_{P(2)} \rangle \otimes \cdots \otimes \langle \vec{x}_N | q_{P(N)} \rangle \prod_{k<l}^{N} f(c_{k,l}; |\vec{x}_k - \vec{x}_l|), \quad (2.1)$$

which then is normalized

$$|\, Q\,\rangle = \frac{1}{\langle \hat{Q} | \hat{Q} \rangle^{\frac{1}{2}}}\,|\,\hat{Q}\,\rangle\,. \quad (2.2)$$

For each nucleon a single-particle wave packet $|\, q_k(t)\,\rangle$ is introduced which is localized in phase space [18] and contains a two component spinor

$$\begin{aligned} \langle\, \vec{x}\,|\, q_k(t)\,\rangle &= \langle\, \vec{x}\,|\, \vec{r}_k(t), \vec{p}_k(t), a_k(t), \chi_k(t), \phi_k(t)\,\rangle \\ &= \exp\left\{ -\frac{(\vec{x} - \vec{r}_k(t))^2}{2a_k(t)} + i\vec{p}_k(t)\vec{x} \right\} \begin{pmatrix} \cos \frac{\chi_k(t)}{2} \\ \sin \frac{\chi_k(t)}{2} e^{i\phi_k(t)} \end{pmatrix}. \end{aligned} \quad (2.3)$$

The time dependent parameters $q_k(t) = \{\vec{r}_k(t), \vec{p}_k(t), a_k(t), \chi_k(t), \phi_k(t)\}$ define the wave packet number k uniquely. The parameter $\vec{r}_k(t)$ denotes the mean position and $\vec{p}_k(t)$ the mean momentum while the complex width-parameter $a_k(t)$ describes the spreading and its real part has to be positive. The spin of the fermion is represented by a two-component spinor where the real polar angles $\chi_k(t)$ and $\phi_k(t)$ denote the direction of the spin. The two-body correlation functions $f(c_{k,l}; |\vec{x}_k - \vec{x}_l|)$ are parametrized by the correlation parameters $c_{k,l}$ which in general may be regarded as independent dynamical variables for each nucleon-nucleon pair $\{k, l\}$. The task is to find a functional form of $f(c; |\vec{x}|)$ which is suited for the two-body interaction which one wants to use. In the literature one finds for example [19]

$$f(c; x) = \exp\left\{ -\frac{1}{2} \left(\frac{c}{x} \right) e^{c/x} \right\}. \quad (2.4)$$

Due to the restriction of the shape of the wave function (or the Hilbert space) one probably has to use an effective interaction instead of a realistic potential which has been deduced from a phase shift analysis. If one needs an effective interaction anyhow one can try to find a form which can be tackled analytically. This is highly desirable in a molecular dynamics calculation as all two-body matrix elements have to be calculated each time step. Since the wave packets are of gaussian shape one could think of an interaction which has a gaussian radial dependence. In that case one could try to use for the correlation function a form like

$$f(c;x) = 1 - \exp\left\{-\frac{1}{2}\left(\frac{x}{c}\right)^2\right\}. \tag{2.5}$$

This form has the advantage that the kinetic energy is given analytically and if the repulsive core is expressed by a strongly repulsive gaussian also the interaction matrix elements can be expressed in analytic form. To find a suited interaction together with an appropriate correlation function one has to investigate with the trial wave function defined in Eq. (2.1) the nuclear matter case and the two-body scattering problem, including the deuteron.

In the Fermionic Molecular Dynamics model the equations of motion are given by the time-dependent variational principle [20, 21, 22]

$$\delta \int_{t_1}^{t_2} dt \, \langle\, Q(t) \,|\; i\frac{d}{dt} - \underset{\sim}{H} \;|\, Q(t)\,\rangle \;=\; 0 \,, \tag{2.6}$$

where the variation is done with respect to all time dependent parameters $Q_\nu(t)$ which specify the antisymmetrized many-body state $|\,Q(t)\,\rangle$. $\underset{\sim}{H}$ denotes the total hamiltonian of the system. The variation leads to Euler-Lagrange equations which are the equations of motion for the parameters $Q_\nu(t)$. In order to study the general structure of the Euler-Lagrange equations let us define a "Lagrange function" by

$$\begin{aligned} \mathcal{L}\left(Q(t),\dot{Q}(t)\right) &:= \langle\, Q(t)\,|\; i\frac{d}{dt} - \underset{\sim}{H} \;|\,Q(t)\,\rangle \\ &\equiv \mathcal{L}_0\left(Q(t),\dot{Q}(t)\right) - \mathcal{H}\left(Q(t)\right) \end{aligned} \tag{2.7}$$

in which $\dot{Q}(t) = \{\,\dot{Q}_\nu \equiv dQ_\nu/dt,\ \nu = 1,2,\cdots\}$ is the set of generalized velocities. The equations of motion resulting from the variational principle (2.6) are given by

$$\frac{d}{dt}\frac{\partial\mathcal{L}}{\partial\dot{Q}_\nu} = \frac{\partial\mathcal{L}}{\partial Q_\nu}\,, \tag{2.8}$$

where Q_ν includes the parameters $q_k = \{\vec{r}_k, \vec{p}_k, a_k, \chi_k, \phi_k\}$ for all the single particle states and also all correlation parameters $c_{k,l}$ which in a general dynamical treatment have to be complex.

$\mathcal{L}\left(Q(t),\dot{Q}(t)\right)$ is not the usual Lagrange function known from classical mechanics. It depends on both, spatial coordinates and momenta as well as on parameters which have no classical analogue. Furthermore, it contains the generalized velocities $\dot{Q}_\nu(t) \equiv dQ_\nu(t)/dt$ only linearly. Because of this homogeneity one can write the time derivative part as

$$\mathcal{L}_0\left(Q(t),\dot{Q}(t)\right) \equiv \langle\, Q(t) \mid i\frac{d}{dt} \mid Q(t) \,\rangle = \sum_\nu \frac{\partial \mathcal{L}_0}{\partial \dot{Q}_\nu(t)}\, \dot{Q}_\nu(t)\,. \tag{2.9}$$

The hamilton function is defined as the expectation value of the hamiltonian $\underset{\sim}{H}$.

$$\mathcal{H}(Q(t)) := \langle\, Q(t) \mid \underset{\sim}{H} \mid Q(t) \,\rangle\,. \tag{2.10}$$

Please note that the Hamiltonian may contain spin-dependent parts. Since $\mathcal{H}$ and $\partial\mathcal{L}_0/\partial\dot{Q}_\nu$ do not depend on the generalized velocities $\dot{Q}_\nu$ the equations of motion (2.8) can be written in the following general form [21, 22]

$$\sum_\nu \mathcal{A}_{\mu\nu}\, \dot{Q}_\nu = \frac{\partial \mathcal{H}}{\partial Q_\mu}\,, \tag{2.11}$$

where $\mathcal{A}_{\mu\nu}$ is a skew-symmetric matrix which depends on $\{Q_\nu(t)\}$ but not on the velocities.

$$\mathcal{A}_{\mu\nu} := \frac{\partial^2 \mathcal{L}_0}{\partial Q_\mu \partial \dot{Q}_\nu} - \frac{\partial^2 \mathcal{L}_0}{\partial Q_\nu \partial \dot{Q}_\mu}\,. \tag{2.12}$$

In principle the model is herewith uniquely defined. What remains is to work out the equations of motion for $Q_\nu(t)$, to solve them and to interpret the results.

The classical Hamilton's equations of motion can be written in form of Eqs. (2.11) by setting $Q = \{\vec{r}_1, \vec{p}_1, \cdots, \vec{r}_N, \vec{p}_N\}$, where $\vec{r}_k$ and $\vec{p}_k$ denote conjugate variables, and $\mathcal{A}$ has a tridiagonal block structure ($\mathcal{A}_{11} = \mathcal{A}_{22} = 0$, $\mathcal{A}_{12} = -\mathcal{A}_{21} = -1$).

$$\begin{pmatrix} 0 & -1 \\ 1 & 0 \end{pmatrix} \begin{pmatrix} \dot{\vec{r}}_k \\ \dot{\vec{p}}_k \end{pmatrix} = \begin{pmatrix} \partial\mathcal{H}/\partial\vec{r}_k \\ \partial\mathcal{H}/\partial\vec{p}_k \end{pmatrix}. \tag{2.13}$$

The classical limit is the case in which, first, the particles are heavy enough so that their wave packets do not spread appreciably, second, the distances between the particles are larger than the extension of the packets, third, the potential does not vary much on the scale of a packet. It is not difficult to see that in this limit the FMD model goes over into the classical molecular dynamics model where the motion of each packet is given by the classical Eqs. (2.13). All three conditions are not fulfilled in nonrelativistic nuclear physics [23]. Therefore, nonquantum nonrelativistic (i.e. classical) molecular dynamics cannot be used for nuclear collisions!

3 APPROXIMATIONS

The expectation values $\mathcal{L}_0$ (Eq. (2.9)) and $\mathcal{H}$ (Eq. (2.10)) cannot be expressed in simple terms because the many-body state $|\, Q(t)\, \rangle$ is not a single Slater-determinant. A promising approximation is the cluster development [17] in which one keeps only up to two-body terms:

$$\mathcal{H} \approx \sum_{k=1}^{N} \mathcal{H}^{(1)}(k) + \sum_{k<l}^{N} \mathcal{H}^{(2)}(k,l)\,, \tag{3.1}$$

where $\mathcal{H}^{(1)}(k)$ is the single-particle kinetic energy which also contains the zero point motion

$$\mathcal{H}^{(1)}(k) := \frac{1}{2m} \frac{\langle\, q_k \,|\, \underset{\sim}{\vec{k}}^2 \,|\, q_k \,\rangle}{\langle\, q_k \,|\, q_k \,\rangle} = \frac{\vec{p}_k^{\,2}}{2m} + \frac{3}{2m}\left(\frac{1}{a_k + a_k^*}\right) \tag{3.2}$$

and $\mathcal{H}^{(2)}(k,l)$ is the interaction energy which contains Pauli and hard-core correlations

$$\mathcal{H}^{(2)}(k,l) := \frac{\langle\, k,l \,|\, \underset{\sim}{H} \,|\, k,l \,\rangle}{\langle\, k,l \,|\, k,l \,\rangle} - \mathcal{H}^{(1)}(k) - \mathcal{H}^{(1)}(l)\,. \tag{3.3}$$

The two-body cluster wave-function $|\, k,l\, \rangle$ is given by

$$\langle\, \vec{x}_1, \vec{x}_2 \,|\, k,l \,\rangle := \left(\langle\, \vec{x}_1 \,|\, q_k \,\rangle \otimes \langle\, \vec{x}_2 \,|\, q_l \,\rangle - \langle\, \vec{x}_1 \,|\, q_l \,\rangle \otimes \langle\, \vec{x}_2 \,|\, q_k \,\rangle \right) f(c_{k,l}; |\vec{x}_1 - \vec{x}_2|)\,. \tag{3.4}$$

Similar relations hold for $\mathcal{L}_0$.

Three-body terms are probably already prohibitive in molecular dynamics because the computational effort goes like N^3. One should keep in mind that not only $\partial\mathcal{L}_0/\partial\dot{Q}_\nu$ and $\mathcal{H}$ but also their partial derivatives with respect to all Q_ν (see Eqs. (2.11), (2.12)) have to be evaluated for many time-steps.

Another approximation which might proof useful is to assume the correlation parameters $c_{k,l}$ to be real. In that case the variational principle (2.6) requires that $\partial\mathcal{H}/\partial c_{k,l} = 0$ which means that the energy has to be minimized with respect to $c_{k,l}$ for a given set of dynamical variables $\{q_1(t), \cdots, q_N(t)\}$. In the cluster expansion (3.1) this implies that the correlation parameters $c_{k,l}$ become functions of the parameters q_k and q_l and are not dynamical variables anymore:

$$\mathcal{H}^{(2)}(k,l) \equiv \mathcal{H}^{(2)}(q_k, q_l, c_{k,l}) = \mathcal{H}^{(2)}(q_k, q_l, c_{k,l}(q_k, q_l))\,. \tag{3.5}$$

4 THE GROUND STATE OF NUCLEI AND INITIAL CONDITIONS

In the Fermionic Molecular Dynamics model, as usual in quantum mechanics, the ground state is the state with the lowest energy among all states which are contained within the parametrization. Thus, the ground state parameters have to be determined from

$$\frac{\partial}{\partial Q_\nu}\langle\, Q \,|\, \underset{\sim}{H} \,|\, Q \,\rangle = 0\,. \tag{4.1}$$

This ground state will differ from a Hartree Fock state as it contains short range correlations by means of the Jastrow functions. Looking at the equations of motion (2.11) we see that the ground state is completely stationary which means $dQ_\nu/dt = 0$ for all ν because all derivates of $\mathcal{H} \equiv \langle\, Q \,|\, \underset{\sim}{H} \,|\, Q \,\rangle$ are zero. In particular $\dot{\vec{r}}_k = 0$, $\dot{\vec{p}}_k = 0$. Nevertheless the Fermi motion is present. It is the zero point motion in each of the densely packed gaussian wave packets.

Here, the FMD model differs conceptually from all molecular dynamics models for heavy-ion collisions in which one usually gives the distinguishable classical particles a random motion to mock the Fermi motion. One then has to fight against particle evaporation from a state which is considered to be the ground state. Even if one succeeds with some tricks the model is still only for Boltzmann particles obeying Boltzmann statistics and not Fermi-Dirac statistics. Only if the chemical potential is largely negative the two ensembles become equal. For intermediate energy heavy ion collisions this is never the case. The FMD model obeys the Fermi-Dirac statistics because it is defined with antisymmetrized states.

The initial state for heavy ion collisions consists of two Galilei-boosted ground-state nuclei. They are completely stable and do not evaporate particles because FMD is translational and Galilei invariant. Due to the gaussian shapes of the wave packets the ground state will have some granularity. An improved state would be the average over all Euler angles of the FMD state (similar as for deformed Nilsson states). But here we want to use the granularity to set up an ensemble of initial conditions by rotating the ground state into different orientations. The final result is then an average over all runs with the different initial conditions for the Euler angles.

After the collision has taken place excited fragments which are left over will evaporate nucleons till their energy is too low for further emission of particles. If the interaction used in FMD describes the essential features of nuclear ground states the model should provide reliable information on the final fragment distribution. In FMD the fragments are represented by true quantum states with true Fermi motion. This is at variance with classical molecular dynamics where one cannot follow the fragments in time because the evaporation continues until the fragment has become a crystal-like object without any zero point motion.

5 REFERENCES

[1] H. Feldmeier, Proceedings of the NASI conference on 'The Nuclear Equation of State', Peniscola, Spain (1989), eds. W. Greiner and H. Stöcker, (NATO ASI Series B, Physics; Vol. 216A), Plenum Press, New York (1989) p. 375

[2] H. Feldmeier, Nucl. Phys. **A515** (1990) 147

[3] L. Wilets, A.D. MacKellar and G.A. Rinker, Jr., Proc. IV Int. Workshop on Gross Properties of Nuclei and Nuclear Excitations (Hirschegg, Austria, 1976), ISSN 0720-8715, p. 111

[4] A.R. Bodmer and C.N. Panos, Phys. Rev. **C15** (1977) 1342, and A.R. Bodmer, C.N. Panos and A.D. MacKellar, Phys. Rev. **C22** (1980) 1025

[5] J.J. Molitoris, J.B. Hoffer, H. Kruse and H. Stöcker, Phys. Rev. Lett. **53** (1984) 899

[6] T.J. Schlagel and V.R. Phandharipande, Phys. Rev. **C36** (1987) 162

[7] E. Betak, preprint DUBNA E-86-701

[8] R. Schmidt, B. Kämpfer, H. Feldmeier and O. Knospe, Phys. Lett. **B229** (1989) 197

[9] L. Wilets, E.M. Henley, M. Kraft and A.D. MacKellar, Nucl. Phys. **A282** (1977) 341, and L. Wilets, Y. Yariv and R. Chestnut, Nucl. Phys. **A301** (1978) 359

[10] C. Dorso, S. Duarte and J. Randrup, Phys. Lett. **B188** (1987) 287, and C. Dorso and J. Randrup, Phys. Lett. **B215** (1988) 611, and Phys. Lett. **B232** (1989) 29

[11] D.H. Boal and J.N. Glosli, Phys. Rev. **C38** (1988) 1870

[12] J. Aichelin, A. Rosenhauer, G. Peilert, H. Stöcker and W. Greiner, Phys. Rev. Lett. **58** (1987) 1926, and Phys. Rev. **C37** (1988) 2451

[13] G. Peilert, H. Stöcker, W. Greiner, A. Rosenhauer, A. Bohnet and J. Aichelin, Phys. Rev. **C39** (1989) 1402

[14] G.E. Beauvais, D.H. Boal and J.C.K. Wong, Phys. Rev. **C35** (1987) 545

[15] D.H. Boal and J.N. Glosli, Phys. Rev. **C38** (1988) 2621

[16] R. Jastrow, Phys. Rev. **98** (1955) 1479

[17] J.W. Clark, Progress in Particle and Nuclear Physics, Vol. **2**, ed. D. Wilkinson, Pergamon Press (1979), p. 89

[18] F. Arickx, J. Broeckhove, E. Kesteloot, L. Lathouwers and P. van Leuven, Chem. Phys. Lett. **128** (1986) 310

[19] E. Krotschek and K. Takahashi, Phys. Lett. **61B** (1976) 393

[20] A.K. Kerman and S.E. Koonin, Annals of Phys. **100** (1976) 332

[21] E. Caurier, B.Grammaticos and T. Sami, Phys. Lett. **B109** (1982) 150, and S. Drozdz, J. Okolowicz and M. Ploszajczak, Phys. Lett. **B109** (1982) 145

[22] M. Saraceno, P. Kramer and F. Fernandez, Nucl. Phys. **A405** (1983) 88

[23] Jutta Kunz, Klassisches Modell mit Pionenfreiheitsgrad zur Beschreibung hochenergetischer Schwerionenstösse, PhD thesis (1981), Giessen D26

DYNAMICS OF FLUCTUATIONS AND OF THEIR CORRELATION IN FERMI SYSTEMS

J. RANDRUP
Nuclear Science Division, Lawrence Berkeley Laboratory
University of California, Berkeley, California 94720

B. REMAUD and M. FARINE
Laboratoire de Physique Nucléaire, IN2P3-CNRS et Université de Nantes
2 rue de la Houssinière, F - 44072 Nantes 03, France

Abstract :

With a view towards nuclear dynamics at intermediate energies, we develop a transport treatment for systems whose reduced one-body phase-density exhibits a markovian time evolution. A moment expansion of the corresponding Fokker-Planck equation leads to closed equations for the temporal evolution of the ensemble average of the phase-space occupancy f(r,p) and its correlation function. The general properties of the associated relaxation processes are illustrated ; negative correlations for particles with small relative momenta and positive correlations for those with large relative momenta, are discribed for colliding nuclear systems.

1 - Introduction

In view of studying nuclei far from their equilibrium created by heavy-ion collisions, we aim at developing a theory which goes beyond the description of the average phase space trajectories of nuclear systems. The latter approximation is currently widely used under the labels Boltzmann-Uehling-Uhlenbeck (BUU) or Vlasov-Uehling-Uhlenbeck or Landau-Vlasov (see Ref. [1]).

However, in typical nuclear collisions at intermediate energies, which is the energy regime where the two-body collisions play a significant role, there are generally many different final channels available and so the actual fluctuations can be substantial, particularly when fluctuations in final fragment multiplicity occur. It is therefore of great practical interest to further develop the treatment so as to allow for dynamical fluctuations to develop. [2].

Our object of study is the one-body density matrix and we seek to develop a formally well-founded, yet practical, treatment of the fluctuations of this quantity induced by the underlying two-body collisions. The main aim is to show how the fluctuations can be incorporated in a one-body treatment within the framework of transport theory. Our treatment is conceptually similar to what is known in classical fluid mechanics as the Boltzmann-Langevin theory of hydrodynamic fluctuations, i.e. those fluctuations that appear in collective variables, even in steady states, due to the stochastic behavior of the constituents at the molecular level [3]. Bixon and Zwanzig took a step towards extending the theory of hydrodynamical fluctuations to non-equilibrium situations by deriving the Boltzmann-Langevin equation [4], where the stochastic behavior of the fluid is produced by a fluctuating force incorporated into the standard Boltzmann equation. The characteristics of this force can be explicitely related to the collision kernel. Bixon and Zwanzig furthermore showed that the Boltzmann-Langevin equation describes the fluctuations correctly near equilibrium, where it reproduces the standard hydrodynamical results for the fluctuations in pressure and heat current (for a review, see [5]). Specific studies for non-equilibrium systems, such as a gas subjected to a constant heat flux [6], have demonstrated the validity of the hydrodynamic-fluctuation theory in the presence of strong non-equilibrium constraints, as least when the gas is dilute.

The application of this approach to nuclear dynamics was pioneered by Ayik [7] who derived the statistical properties of the residual interaction by the general projection technique of quantum statistics. Recently, Ayik and Gregoire [8] have applied this general scheme to the particular situation where the residual interaction is approximated by the Uehling-Uhlenbeck collision term. In that work, explicit expressions were derived for the dynamical evolution of the fluctuations in the one-body density matrix. The present work can be seen as a further development along this line, in which we adapt standard techniques from transport theory for the derivation of closed equations for the first and second moments of the one-body phase-space distribution. This approach is motivated by the observation that although the full treatment of the fluctuations of the one-body density is too complex to allow practical calculations, even on the most powerful present computers, most problems of actual interest can be addressed without such degree of detail. While the equation derived for the first moment corresponds to the BUU treatment, the equations for the second moments are novel and give access to the variances and covariances of one-body observables within a one-body treatment.

2 - Basic formalism

We consider an ensemble of identical many-body systems, labelled by $n = 1,\ldots,N$, that we describe with their one-body phase-space densities $f^{(n)}(r,p,t)$. We assume that the systems are prepared in the same manner so that their one-body densities are all equal at $t = 0$; however the systems generally differ at the many-body level. Due to the stochastic nature of the residual two-body interaction, the individual many-body systems evolve differently and it is convenient to write the phase-space density of an individual system n on the form :

$$f^{(n)}(\mathbf{r},\mathbf{p};t) = f(\mathbf{r},\mathbf{p};t) + \delta f^{(n)}(\mathbf{r},\mathbf{p};t) \ . \tag{1}$$

Here f(r,p ; t) is the average of the one-particle phase-space densities for all the individual systems in the ensemble,

$$f(\mathbf{r},\mathbf{p};t) \ \equiv \ \prec f^{(n)}(\mathbf{r},\mathbf{p};t) \succ \ = \ \frac{1}{N}\sum_{n=1}^{N} f^{(n)}(\mathbf{r},\mathbf{p};t) \ . \tag{2}$$

In order to characterize the fluctuations of $f^{(n)}$ away from the ensemble average f it is convenient to introduce the following correlation function for the one-particle phase-space occupancy.

$$\begin{aligned} \sigma(\mathbf{r},\mathbf{p};\mathbf{r}',\mathbf{p}') \ \equiv \ \sigma_{ff'} \ &= \ \prec f^{(n)}(\mathbf{r},\mathbf{p})f^{(n)}(\mathbf{r}',\mathbf{p}') \succ - f(\mathbf{r},\mathbf{p})f(\mathbf{r}',\mathbf{p}') \\ &= \ \prec \delta f^{(n)}(\mathbf{r},\mathbf{p}) \ \delta f^{(n)}(\mathbf{r}',\mathbf{p}') \succ \end{aligned} \tag{3}$$

If we consider a one-body observable A(r, p), it is elementary to show that the average of its expectation values on $f^{(n)}$ is equal to its expectation value on f ; but the covariance between two one-body observables A and B reads :

$$\sigma_{\mathcal{AB}} = \ \int \frac{d\mathbf{r}d\mathbf{p}}{h^3} \int \frac{d\mathbf{r}'d\mathbf{p}'}{h^3} \ \mathcal{A}(\mathbf{r},\mathbf{p}) \ \mathcal{B}(\mathbf{r}',\mathbf{p}') \ \sigma(\mathbf{r},\mathbf{p};\mathbf{r}',\mathbf{p}') \ . \tag{4}$$

The aim of the theory is to derive closed equations for the temporal evolution of the ensemble average of phase-space one-body densities f(r,p) and its correlation function. We treat the one-body density as a dynamical variable which undergoes a stochastic behavior, induced by the two-body collisions. We shall not enter here the details of the derivation which one can find in ref [9]. Let us mention that with a Markovian assumption , a Fokker-Planck equation is derived for the phase-space occupancies, the drift and diffusion coefficients of which are expressed in terms of the two-body collisions rates.

2.1 - Occupancies at a single point

Since the two-body interaction is effective only between particles that have different momenta the evolution of the occupancy of a single phase-space cell, including its variance $\sigma_{ff} = \sigma^2_f$ (which is a measure of the fluctuation of the occupancy at that point), can be considered separately from the evolution of the occupancy covariances $\sigma_{ff'}$, which express the correlation between the fluctuations in occupancy at two different points of phase space.

The temporal evolution of f(r,p) can be regarded as a random walk and then the equations of motion for the mean occupancy and the associated variance take the following form [see ref. 9] :

$$\begin{aligned} \frac{d}{dt}f &= \bar{f}\underline{W}^+ - f\underline{W}^- , \\ \frac{d}{dt}\sigma_f^2 &= 2\underline{D} + 2(\frac{\partial V}{\partial f})\sigma_f^2 = \bar{f}\underline{W}^+ + f\underline{W}^- - 2(\underline{W}^+ + \underline{W}^-)\sigma_f^2 , \end{aligned} \tag{5}$$

where $\bar{f} = 1 - f$ as usual and the W's are respectively the loss and gain coefficients :

$$W^-(\mathbf{r}_a, \mathbf{p}_a) = \tag{6}$$
$$\frac{1}{2}\int\frac{d\mathbf{p}_1}{h^3}\int\frac{d\mathbf{p}_2}{h^3}\int\frac{d\mathbf{p}'_1}{h^3}\int\frac{d\mathbf{p}'_2}{h^3}\delta(\mathbf{p}_1 - \mathbf{p}_a)f(\mathbf{r}_a, \mathbf{p}_2)\bar{f}(\mathbf{r}_a, \mathbf{p}'_1)\bar{f}(\mathbf{r}_a, \mathbf{p}'_2)\ w$$

$$W^+(\mathbf{r}_a, \mathbf{p}_a) = \tag{7}$$
$$\frac{1}{2}\int\frac{d\mathbf{p}_1}{h^3}\int\frac{d\mathbf{p}_2}{h^3}\int\frac{d\mathbf{p}'_1}{h^3}\int\frac{d\mathbf{p}'_2}{h^3}f(\mathbf{r}_a, \mathbf{p}_1)f(\mathbf{r}_a, \mathbf{p}_2)\delta(\mathbf{p}'_1 - \mathbf{p}_a)\bar{f}(\mathbf{r}_a, \mathbf{p}'_2)\ w$$

$w = w(12-1'2')$ is the rate at which nucleons initially situated at the phase space points (r_1, p_1) and (r_2, p_2) scatter into phase points (r'_1, p'_1) and (r'_2, p'_2). The factor of one half occurs in order to compensate for the double counting of the final two-particles states. In eq. 5, one readily recognizes the Uehling-Uhlenbeck collision term with an abbreviated but fairly obvious notation. When the system has reached equilibrium, the ensemble moments of the occupancy remain constant in time, i.e. $f = 0$ and $\sigma_{ff} = 0$. The first condition yields the relation $\langle fW^- \rangle = \langle \bar{f} W^+ \rangle$, which amounts to $fW^- \approx \bar{f}W^+$, as would be expected on the basis of detailed balance.

The equilibrium condition for the variance yields $\sigma^2_f = (\bar{f}W^+ + fW^-)/2(W^+ + W^-)$ for the instantaneous equilibrium variance σ^2_f. When the mean value has reached equilibrium, this relation amounts to $\sigma^2_f = f\bar{f}$. This is recognized as the standard result for the variance of the Fermi-Dirac occupancy at equilibrium. In order to examine the approach to equilibrium for the variance of the occupancy, it is instructive to consider its deviation from the instantaneous equilibrium value, $\Delta = \sigma^2 - f\bar{f}$. An elementary calculation readily yields :

$$\dot{\Delta} = \dot{\sigma}^2 - \dot{f}\bar{f} - f\dot{\bar{f}} = -2(\underline{W}^- + \underline{W}^+)\Delta \ . \tag{8}$$

This result demonstrates that the variance of the single-particle occupancy relaxes at twice the rate characterizing the relaxation of the mean occupancy, as is ordinarily the case in transport problems.

The relaxation time t_f can be estimated by recalling that $W^-(r, p)$ is the rate at which a nucleon with the specified momentum p suffers collisions when moving in the neighborhood of r, ie $W^- \approx v/\lambda$, where v is the speed of the nucleon and λ is its mean free path (which also depends on the velocities of the surrounding nucleons). For temperatures of a few MeV, and for particles near the Fermi surface, the mean free path is typically 6-8 fm. Using $\bar{f} \simeq 1/2$, we then find $t_f \simeq 0.4.10^{-22}s$ for the relaxation of the single-particle occupancy. This time is about an order of magnitude shorter than the duration typical of a nuclear collision at intermediate energy and so we would expect the variances in the single-particle occupancies to track closely their adiabatic values (i.e. their instantaneous equilibrium values $\sigma^2{}_f = f\bar{f}$)

2.2 - Covariance between occupancies at different points

The definition of the ensemble covariance is given in (3). The corresponding diffusion coefficient can be calculated by considering the early evolution of the product of the induced changes in occupation at the two space-points considered. Since each of these points can be either an initial state or a final state, there are four distinct types of scattering process contributing and, the equations of motion for the occupancy covariances are then given by :

$$\begin{aligned}
&\dot{\sigma}(\mathbf{r}_a,\mathbf{p}_a;\mathbf{r}_b,\mathbf{p}_b) \qquad (9)\\
&= (f_a f_b W_{ab}^{--} + \bar{f}_a \bar{f}_b W_{ab}^{++} - f_a \bar{f}_b W_{ab}^{-+} - \bar{f}_a f_b W_{ab}^{+-})\, \delta(\mathbf{r}_{ab})\\
&- (W_a^+ + W_a^- + W_b^+ + W_b^-)\, \sigma(\mathbf{r}_a,\mathbf{p}_a;\mathbf{r}_b,\mathbf{p}_b)\\
&- \int \frac{d\mathbf{p}_c}{h^3} \Big[\Big(\bar{f}_b(W_{bc}^{+-} + W_{bc}^{++}) - f_b(W_{bc}^{-+} + W_{bc}^{--})\Big)\, \sigma(\mathbf{r}_a,\mathbf{p}_a;\mathbf{r}_b,\mathbf{p}_c)\\
&\qquad + \Big(\bar{f}_a(W_{ac}^{+-} + W_{ac}^{++}) - f_a(W_{ac}^{-+} + W_{ac}^{--})\Big)\, \sigma(\mathbf{r}_a,\mathbf{p}_c;\mathbf{r}_b,\mathbf{p}_b)\Big] .
\end{aligned}$$

For convenience, we have here used the condensed notation $f_a = f(r_a,p_a)$ and $f_b = f(r_b,p_b)$. Furthermore, we have introduced the following four quantities :

$$W_{ab}^{--}(\mathbf{r}_a,\mathbf{p}_a;\mathbf{r}_b,\mathbf{p}_b) \equiv \frac{1}{2}\int \frac{d\mathbf{p}_1'}{h^3} \int \frac{d\mathbf{p}_2'}{h^3}\, \bar{f}(\mathbf{r}_a,\mathbf{p}_1')\bar{f}(\mathbf{r}_b,\mathbf{p}_2')\; w(\mathbf{p}_a\mathbf{p}_b \leftrightarrow \mathbf{p}_1'\mathbf{p}_2'), \qquad (10)$$

$$W_{ab}^{++}(\mathbf{r}_a,\mathbf{p}_a;\mathbf{r}_b,\mathbf{p}_b) \equiv \frac{1}{2}\int \frac{d\mathbf{p}_1}{h^3} \int \frac{d\mathbf{p}_2}{h^3}\, f(\mathbf{r}_a,\mathbf{p}_1)f(\mathbf{r}_b,\mathbf{p}_2)\; w(\mathbf{p}_1\mathbf{p}_2 \leftrightarrow \mathbf{p}_a\mathbf{p}_b), \qquad (11)$$

$$W_{ab}^{-+}(\mathbf{r}_a,\mathbf{p}_a;\mathbf{r}_b,\mathbf{p}_b) \equiv$$
$$\int \frac{d\mathbf{p}_2}{h^3} \int \frac{d\mathbf{p}_2'}{h^3}\, f(\mathbf{r}_a,\mathbf{p}_2)\bar{f}(\mathbf{r}_b,\mathbf{p}_2')\; w(\mathbf{p}_a\mathbf{p}_2 \leftrightarrow \mathbf{p}_b\mathbf{p}_2'), \tag{12}$$

$$W_{ab}^{+-}(\mathbf{r}_a,\mathbf{p}_a;\mathbf{r}_b,\mathbf{p}_b) \equiv$$
$$\int \frac{d\mathbf{p}_2}{h^3} \int \frac{d\mathbf{p}_2'}{h^3}\, f(\mathbf{r}_a,\mathbf{p}_2)\bar{f}(\mathbf{r}_b,\mathbf{p}_2')\; w(\mathbf{p}_b\mathbf{p}_2 \leftrightarrow \mathbf{p}_a\mathbf{p}_2'). \tag{13}$$

It should be note that $W_{ab}^{-+} = W_{ba}^{+-}$. Furthermore, these quantities are related to those given in (24) by the following sum rules,

$$\begin{aligned} W_a^- = W^-(\mathbf{r},\mathbf{p}_a) &= \int \frac{d\mathbf{p}_b}{h^3}\, f(\mathbf{r},\mathbf{p}_b)\; W_{ab}^{--} = \int \frac{d\mathbf{p}_b}{h^3}\, \bar{f}(\mathbf{r},\mathbf{p}_b)\; W_{ab}^{-+}\,, \\ W_a^+ = W^+(\mathbf{r},\mathbf{p}_a) &= \int \frac{d\mathbf{p}_b}{h^3}\, \bar{f}(\mathbf{r},\mathbf{p}_b)\; W_{ab}^{++} = \int \frac{d\mathbf{p}_b}{h^3}\, f(\mathbf{r},\mathbf{p}_b)\; W_{ab}^{+-}\,. \end{aligned} \tag{14}$$

(15)

The above equation [10] for the covariance $\sigma(r_a,p_a,r_b,p_b,t)$ is fairly involved. However, its main features are easy to understand. The first term on the right-hand side acts as the source term for the correlations arising from the scattering of two particles situated near the same point in space (hence the δ-function in the separation r_{ab}). The strength of the source term is simply the total rate at which particles at the two specified phase-space points are engaged with each other, each point acting either as an initial state or as a final state (hence the four different terms). The second line is the diagonal degradation term acting to destroy the correlations in the course of time, as a consequence of the collisions the two particles suffer subsequent to the particular collision that generated the correlation being propagated. The strength of the degradation is recognized as the total rate of scattering into or out of either one of the phase-space cells a and b. The off-diagonal terms represent the effect of the scattering of one of the particles (a or b) with a third particle c with which the other particle (b or a, respectively) is correlated. This contribution is obviously the most tedious one to calculate ; the extent to which it may be approximated or ignored requires further study.

The propagation of particle induces a change in the phase-space occupancies, which is taken into account by a Vlasov term. If there are no forces associated with the mean-field, and if we ignore the off-diagonal degradation terms, the equation of motion takes on a relatively tractable form,

$$\dot{\sigma}(\mathbf{r}_a,\mathbf{p}_a;\mathbf{r}_b,\mathbf{p}_b;t) = W_{ab}\,\delta(\mathbf{r}_{ab}) - (W_{a+b} + \mathbf{v}_a\cdot\frac{\partial}{\partial\mathbf{r}_a} + \mathbf{v}_b\cdot\frac{\partial}{\partial\mathbf{r}_b})\,\sigma\ , \tag{16}$$

where $W_{ab} = f_a\ f_b\overline{W_{ab}^{+}}\ \ldots$ and $W_{a\ +\ b} = W_a^{++}\ldots$ for brevity. This equation is amenable to solution by elementary means. By performing a Fourier analysis, it can readily be shown that the stationary solution is given by

$$\tilde{\sigma}(\mathbf{r}_a,\mathbf{p}_a;\mathbf{r}_b,\mathbf{p}_b) = \Theta(\mathbf{r}_{ab}\cdot\hat{\mathbf{v}}_{ab})\ \delta(\mathbf{r}_{ab}^{\perp})\ \frac{W_{ab}}{v_{ab}}\ \exp(-\frac{W_{a+b}}{v_{ab}}r_{ab})\ , \tag{17}$$

where $\vec{r}_{ab}^{\perp} = \vec{r}_{ab} - (\vec{r}_{ab}\cdot\hat{v}_{ab})\,\hat{v}_{ab}$ is the two-dimensional projection of the relative position $\vec{r}_{ab}$ onto the plane perpendicular to the direction of relative motion $\hat{v}_{ab} = \vec{v}_{ab}/|\vec{v}_{ab}|$. This result can be understood as follows. The truncation function θ together with the constraint function δ act to ensure that correlations between the two particles can only exist provided they have scattered (so that their relative motion is directed away from their center of mass). If they are indeed moving back-to-back, their correlation falls of exponentially as a function of their separation $|\vec{r}_{ab}|$

with a correlation length $\lambda_{ab}=v_{ab}/W_{a+b}$, which is the average separation the two particles will have achieved by the time either one of them suffers a next scattering. This quantity is related to the two mean free paths by $\lambda_{ab}^{-1} = \lambda_a^{-1} + \lambda_b^{-1}$. The maximum corelation thus occurs right after the particles have scattered and is given by W_{ab}/v_{ab}, which is quite reasonable : the correlation grows in proportion to the engagement rate W_{ab} and is inversely proportional to the relative speed v_{ab}.

3-Illustrative results

The computational effort required for solving the transport equations in their full generality is prohibitively large, and for our present purposes we restrict our considerations to systems with translational invariance. This symmetry allows to solve the problem in the velocity space only. While this simplification reduces the computational complexity considerably, many features of interest in real nuclear processes can still be elucidated.

In order to imitate the momentum-space distribution relevant to a nuclear collision, we shall employ an initial occupancy corresponding to two dislocated Fermi spheres. Let the two spheres be centered at $p = (0,0,\pm p_0/2)$. The occupancy $f(p\ ;\ t = 0)$ is then unity if the separation between p and either one of the two centers is less than the fermi momemtum p_F , otherwise is vanishes.

The dependence of the variance σ_f^2 on the momentum p is illustrated on the RHS in Fig.1, in the parallel and perpendicular directions. Since the system has been prepared so that the local phase space is either fully occupied or totally empty, there is no fluctuation in the initial occupancy function and so σ_f^2 vanishes at $t = 0$.

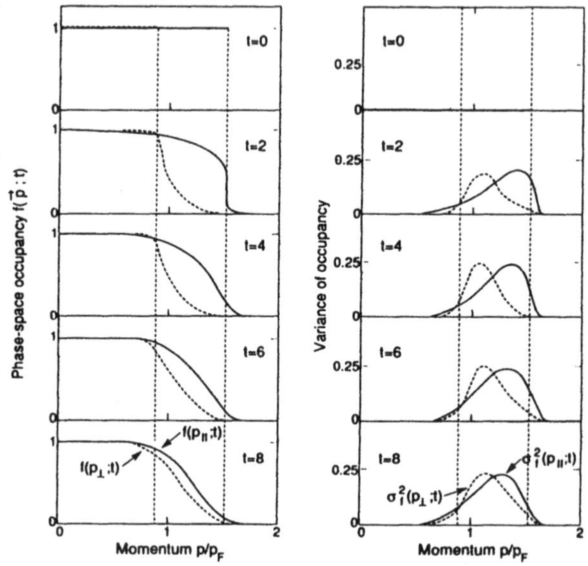

Figure 1 :

For a center separation of $p_0 = p_F$ (corresponding to a bombarding energy of $E_{beam} \propto 37$ MeV per nucleon), the LHS displays the phase-space occupancy $f(p\ ;\ t)$ along the directions parallel and perpendicular to the beam, $p = (0, 0, p_{\parallel})$ and $p = (p_{\perp}, 0, 0)$, respectively, for a number of successive times $t = 0, 2, 4, 6, 8 \cdot 10^{-23}$s. The time evolution of the variance of the phase-space occupancy, $\sigma_f^2(p\ ;t)$, is shown on the RHS.

A detailed understanding of the evolution of the occupancy illustrated above can be achieved by considering the corresponding transport coefficients. The time dependence of the drift coefficient $v(p\ ;\ t)$ and the diffusion coefficient $D(p\ ;\ t)$ is shown in Fig.2 in the parallel and perpendicular directions.

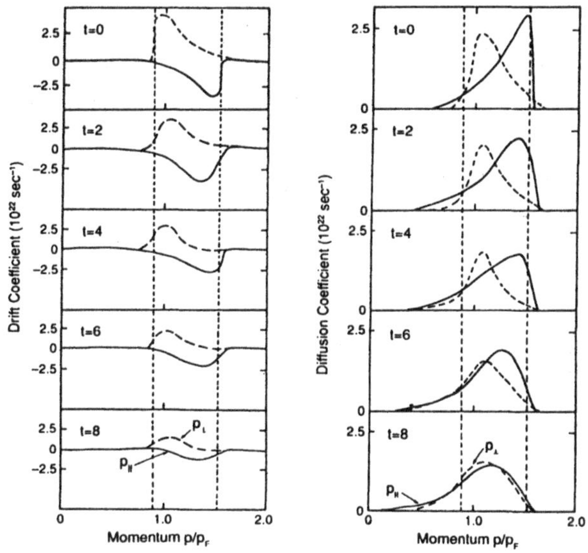

Figure 2 :

For the case shown inf Fig.1, the figure shows the momentum dependences of the drift coefficient for the phase-space occupancy, $V(p\ ;\ t) = \overline{f}W^+ - fW^-$ (LHS) and the diffusion coefficient $D(p\ ;\ t) = \overline{f}W^+ - fW^-$ (RHS).

In order to illustrate the non-local properties of the fluctuations, we consider the covariance $\sigma(p_1,p_2\ ;t)$ as a function of magnitude of the difference between the two momentum arguments, $\Delta p = |p_1 - p_2|$. For a single time at which equilibrium is nearly reached, the LHS of Fig.3 shows the quantity W_{a+b} which acts as a restoring force on the fluctuations. It is shown for specified values of the mean momentum $p = p_a - p_b$, as a function of the magnitude of the difference $\Delta p = |p_a - p_b|$, averaged over the direction p .

It is interesting to note that this quantity does not exhibit much variation, neither as a function of momentum nor in the course of time ; its typical size corresponds to a relaxation time of 1-2 10^{-23}s. This feature may allow analytical parametrizations of W_{a+b} to be made, thus simplifying the treatment considerably.

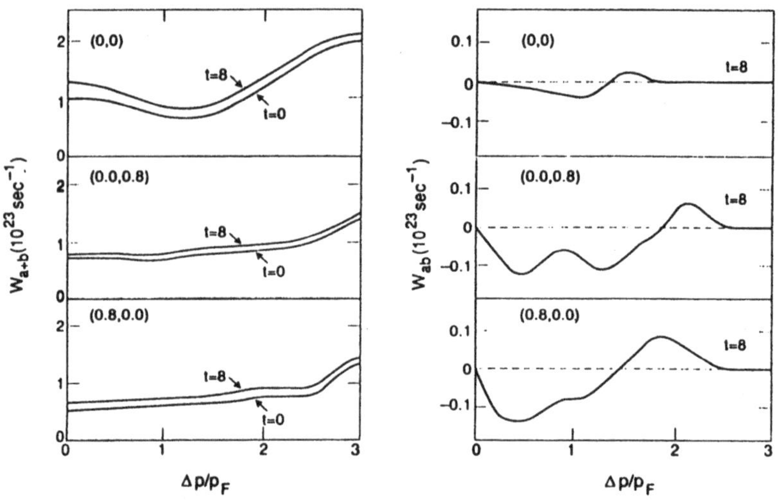

Figure 3 :

LHS : The quantity $W_{a+b}(p, \Delta p)$ degrading the covariance $\sigma(p, \Delta p)$, shown as a function of magnitude of the momentum difference $\Delta p = |p_a - p_b|$, for specified values of the mena momentum $p = p_a + p_b = (p_\perp\ 0, p_{//})$, with the values of $p_\perp p_{//})$ chosen as indicated to lie within the bulk of the occupied region. The quantity is shown at the initial time t = 0 and at the later time t = 8.10^{-23}s when considerable relaxation has occured.
RHS : The corresponding quantity $W_{ab}(p, \Delta p)$ driving the covariance $\sigma(p, \Delta p)$, at t = 8.10^{-23}s.

The behavior of the quantity W_{ab}, which provides the driving force for the fluctuations, is shown on the RHS of Fig. 3. Some structure is evident, namely some anticorrelation for small relative momenta and some positive correlation at large relative momenta. This behavior can be qualitatively understood as follows : In eq. (10) one can split the source term into two parts : The W^{++} and W^{--} terms are responsible for the growth of positive correlations, but are counteracted by the W^{+-} and W^{-+} terms which generate anticorrelations. These two types of term behave differently as functions of $\Delta p = |p_a - p_b|$. In the case of W^{--} and W^{++}, Δp is the relative velocity of the colliding pair of particles ; since the collision rates are directly proportional to the modulus of this quantity, the source of positive correlations is an increasing function of Δp. For the negative source term (see definitions of W^{-+} and W^{+-} in eqs. (13) and (14), Δp is the momentum transferred to a particle during the collision, so that small values of transferred momentum can be obtained in collisions between particles, with large relative velocity which are the most frequent ones. Of course, these elementary considerations must be augmented by the effect of phase-space availability, if quantitative estimates are to be made.

The origin of the positive peak at $\Delta p \simeq 2$ can then be related with the collective depopulation of the outer phase space, for which the positive source terms are large and Pauli blocking less effective. Correspondingly, the negative source term prevails for small Δp.

The resulting covariance $\sigma(p, \Delta p)$ is shown in Fig.4 for momenta situated within the bulk of the occupied region. It exhibits a structure strikingly similar to that of the driving term W_{ab} shown in Fig.3b. This feature is easy to understand : at the time considered the covariance has relaxed to near the adiabatic value given by W_{ab}/W_{a+b} and since W_{a+b} has little structure (Fig. 3a), the resulting covariance tracks W_{ab}. The center frame in Fig.4 includes the covariance at an early time, $t = 2.10^{-23}$s. It is seen that the build-up of the positive correlation at large values of Δp is already well underway, whereas the negative correlation near zero requires a longer time. This latter feature is a result of the fact that the zone considered must first be depopulated so as to allow scattering into it. The displacement of the positive bump reflects the shrinking of the distribution in the parallel direction associated with the establishment of spherical symmetry. Still this feature closely mirrors the development of W_{ab}, so that the correlations remain close to the adiabatic values throughout the evolution.

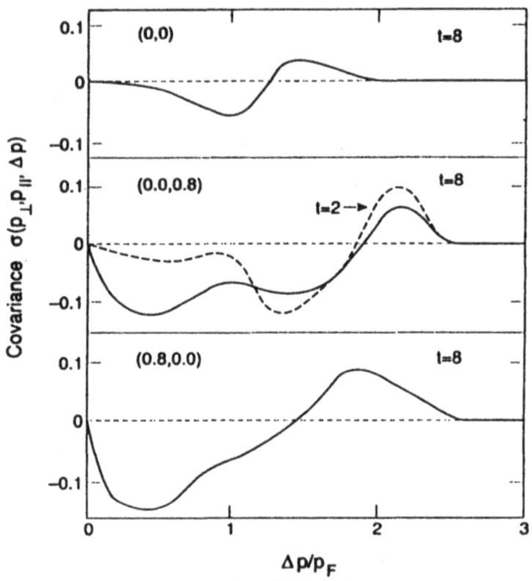

Figure 4 :

The covariance $\sigma(p, \Delta p)$ as a function of the magnitude of the momentum difference Δp, at specified values of the mean momentum $p = (p_\perp, 0, p_{//})$, with the values $(p_\perp, p_{//})$ indicated ; they are chosen as in Fig.3. The solid curves are for the time $t = 8.10^{-23}$s, while the dot-dashed curve in the center frame is for $t = 2$.

For regions at and above the Fermi level, the features noted above also hold, but the amplitude of the correlations is significantly larger. A significant increase of the anti-correlation occurs at $\Delta p = 1.5$. This structure arises from collisions between nucleons that originally populate the region of momentum space situated at large values near the beam direction and scatter into the perpendicular direction where the Pauli blocking is less effective, at least early on.

4 - Conclusions

In the present work, we have developed a formalism for addressing stochastic one-body dynamics within the framework of transport theory. Considering an ensemble of similarly prepared systems, we have derived closed equations of motion for the average reduced one-body phase-space density distribution and the associated correlation function. The equation for the mean distribution corresponds to existing models under the labels BUU or VUU. The formulation of the associated equations of motion for the correlation function is somewhat similar to approaches developed in classical fluid dynamics. Within the context of nuclear dynamics, the incorporation of the seconds moments is an important advance and extends the utility of nuclear one-body models.

Building on earlier work by Ayik [15] towards the incorporation of fluctuations into TDHF theory, Ayik and Gregoire [16] recently made an important advance by considering the semi-classical limit of TDHF. Even though the starting point of that work is identical to that of the present paper, the methods employed are significantly different. In both studies, the fluctuations induced by the two-body collisions is considered as a stochastic process. In the work by Ayik and Gregoire, the phenomenon is treated as a Langevin process, with the entire phase-space distribution considered as a Brownian degree of freedom. The correlation function associated with the one-particle distribution may be then deduced from the properties of the fluctuating force. This approach is known as the Boltzmann-Langevin equation in the theory of classical fluids [12], where it has been applied to the treatment of hydrodynamical fluctuations.

We have developed equations of motion for the variances and covariances of the one-body phase-space density that may be used to elucidate the character of the relaxation processes occuring in nuclear dynamics. Our present studies indicate that the fluctuations relax rather rapidly, within less than 10^{-22}s, towards the equilibrium distribution associated with the instantaneous mean occupancy. This finding may provide the basis for significant simplifications in the analytical treatment of fluid-dynamical fluctuations and of the dynamics of two-body correlations in Fermi systems.

REFERENCES

[1] G.F. Bertsch and S. Das Gupta, Phys. Rep.160 (1988) 190.
J. Aichelin and G.F. Bertsch, Phys. Rev. C 31 (1985) 1730.
H. Kruse, B. Jacak, and S. Stöcker, Phys. Rev. Lett.54 (1985) 289.
C. Gregoire, B. Remaud, F. Sebille, L. Vinet and Y. Raffray, Nucl. Phys. A465 (1987) 317.
[2] W. Bauer, G.F. Bertsch, and S. Das. Gupta, Phys. Rev. Lett. 58 (1987) 863.
[3] L.D. Landau and E.M. Lifshitz, Fluid Mechanics, Pergamon Press, Oxford, 1959.
[4] M. Bixon and R. Zwanzig, Phys. Rev. 187 (1969) 267.
[5] M.H. Ernst and E.G.D. Cohen, J. Stat. Phys. 25 (1981) 153.
[6] M. Malek Monsour, A.L. Garcia, G.C. Lie and E. Clementi, Phys. Rev. Lett. 58 (1987) 874.
[7] S. Ayik, Z. Phys. A 298 (1980) 83
[8] S. Ayik and C. Gregoire, Phys. Lett. 212 B (1988) 269
[9] J. Randrup and B. Remaud, Nucl. Phys.

Multiplicity Distributions and Clan Structures at High Energy Collisions

A. Giovannini

FACTORIAL MOMENTS AND CORRELATIONS IN MESON–NUCLEUS INTERACTIONS AT 250 GeV/c

EHS–NA22 Collaboration

Antwerp/Brussels – Berlin(Zeuthen) – Helsinki – Krakow – Moscow
Nijmegen – Rio de Janeiro – Serpukhov – Tbilisi – Warsaw – Yerevan

M. CHARLET
Department of Physics, Katholieke Universiteit Nijmegen,
Toernooiveld 1, NL-6525 Nijmegen, The Netherlands

ABSTRACT
A sample of 8000 interactions of 250 GeV/c π^+ and K^+ mesons on Al and Au nuclei is used to search for intermittency effects by calculating the normalized factorial moments of order two to four. No significant effect is observed. Bose-Einstein correlations in these interactions are presented, both in terms of the Kopylov-Podgoretskii and the Weiner parametrizations.

1 Introduction

The possibility that particle production in multihadron final states may exhibit so-called "intermittent" behaviour was first put forward by Białas and Peschanski[1]. They suggested to study the normalized factorial moments F_i (i=2, 3, ...) of the multiplicity distributions, and in particular the dependence of these moments on δy, a rapidity interval of decreasing size. Intermittency is observed when the factorial moments exhibit a power law behaviour

$$< F_i > \sim \left(\frac{Y}{\delta y}\right)^{f_i} \qquad \text{with} \qquad f_i > 0 \tag{1}$$

Intermittent behaviour of particle production is now claimed in all studied types of interactions, such as e^+e^-, $h-h$, $h-A$ and $A-A$ interactions[2]. Moreover, it is remarked that the effect seems to decrease with increasing complexity of the collision partners and with increasing complexity of the final state.

The results presented here, are obtained with the european hybrid spectrometer (EHS), irradiated by a beam of π^+ and K^+ mesons of 250 GeV/c momentum. An Al and a Au foil are inserted in the rapid cycling bubble chamber (RCBC), which serves as track detector with 4π angular acceptance for all interactions in the foils. The experimental set-up, the minimum bias trigger and the selection criteria for the nuclear interactions, are described in more detail in[3,4]. The selection criteria aim at isolating a set of well measured and reconstructed inelastic interactions in either the Al or the Au target, thereby eliminating both quasi-elastic and coherent interactions. The events are selected on the basis of the following criteria: the incident particle is well measured in the bubble chamber and its track matches with hits in the upstream wire chamber; the reconstructed vertex position is within one of the nuclear targets; the outgoing tracks are satisfactorily measured with an error on the momentum $\Delta p/p < 0.25$; the accepted loss of tracks, due to measurement or reconstruction problems, is at most one for charge multiplicity up to 10, and at most 20% for higher multiplicities.

Electrons and positrons are identified up to $p_{\text{lab}} \simeq 200$ MeV/c and protons up to $p_{\text{lab}} \simeq 1.2$ GeV/c, by visual inspection of their ionization in the H_2 of RCBC. These particles are not used in the subsequent analysis. Among the accepted charged tracks remains a small admixture of unidentified electrons and positrons from γ-conversions in the foils. This admixture is estimated to be smaller than 2% in the Al and smaller than 7% in the Au sample. Since no statistically significant difference is observed for π^+ and K^+ induced interactions, we use the combined sample of π^+ and K^+ collisions with 4200 interactions on Al and 3700 on Au. The combined samples are denoted as M^+Al and M^+Au.

2 Factorial moments

The scaled factorial moment of order i is defined as

$$< F_i >= \frac{1}{< \overline{n}_m >^i} \left\langle \frac{1}{M} \sum_{m=1}^{M} n_m(n_m - 1) \ldots (n_m - i + 1) \right\rangle \tag{2}$$

with

$$< \overline{n}_m >= \left\langle \frac{1}{M} \sum_{m=1}^{M} n_m \right\rangle \tag{3}$$

whereby the considered rapidity interval Y is subdivided in M equal subintervals, each of size $\delta y = Y/M$. The multiplicity is n_m in bin m ($m = 1 \ldots M$) and the averages $<>$ are taken over all events in the sample. The definition Eq. 2 is often called "horizontal averaging". If non-statistical, self similar fluctuations of many different sizes exist, the δy dependence of the factorial moments is expected to exhibit the power law behaviour Eq. 1. in the limit $\delta y \to 0$. Such behavior shows up as a linear rise of $\ln < F_i >$ versus $(-\ln \delta y)$. Fig. 1 shows the factorial moments of order 2, 3 and 4 for the M^+Al and M^+Au events. A rapidity interval Y of 3 units ($-2 < y < 1$) in the meson–nucleon c.m.s. is used, where the density of particles is almost constant.

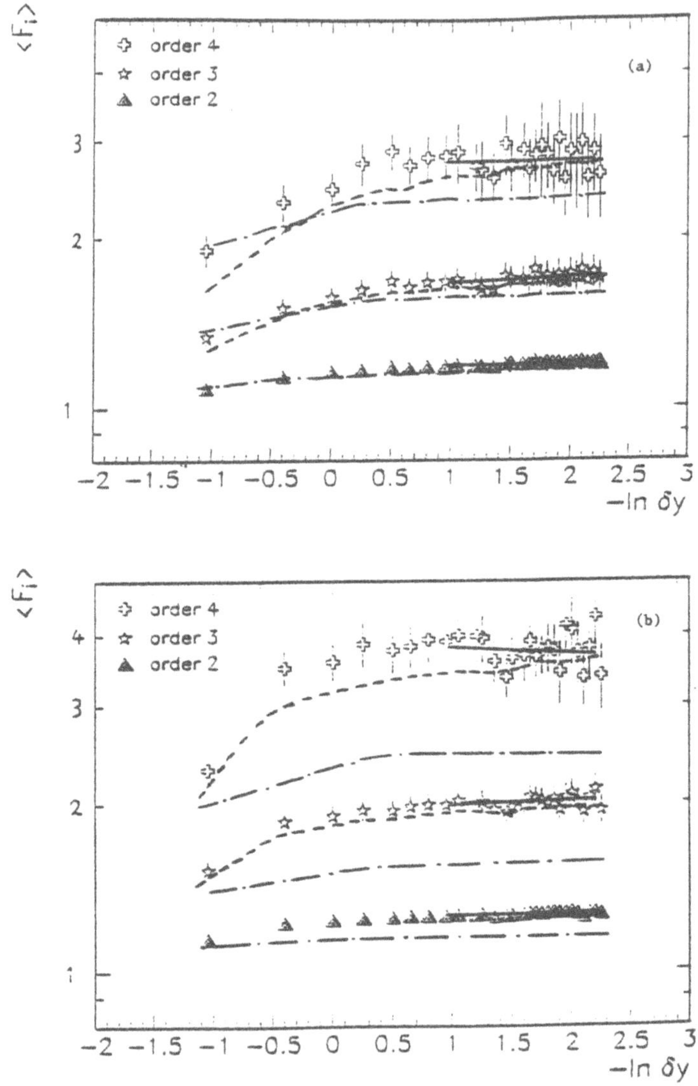

Fig. 1. Moments F_i (i=2, 3 and 4) for **a** M^+Al and **b** M^+Au. Fits to Eq. (1) are shown as straight lines, predictions of the Fritiof model as dash-dotted lines and calculated values of F_3 and F_4 as dashed lines.

From Fig. 1 it is clear that the slope of $\ln F_i$ versus $(-\ln \delta y)$ varies continously with the latter quantity, until $-\ln \delta y \approx 0.7$. We therefore fit the slopes in the interval $0.9 < -\ln \delta y < 2.2$ ($0.4 > \delta y > 0.1$). The results of the fits are collected in Table 1 and given as full lines on Fig. 1. The fitted slopes are small in all cases and compatible with zero within one standard deviation. For comparison, we also show in Table 1 the fitted slopes for 20000 Monte Carlo events per channel, generated according to the FRITIOF model[5] (version 3.0) and subjected to the same cuts as the experimental data. The slopes f_i found in M^+Al and M^+Au collisions are considerably smaller than the ones in M^+p interactions from the same experimental set-up[6].

Table 1. Slopes f_i of the fits to Eq. 2 of the experimental factorial moments F_i and of FRITIOF Monte Carlo events, in the interval $-2 < y < 1$, with the fits performed in the range $0.4 \geq \delta y > 0.1$.

	M^+Al		M^+Au	
	Data	Fritiof	Data	Fritiof
f_2	0.0132±0.0138	-0.0004±0.0067	0.0092±0.0141	0.0047±0.0052
f_3	0.0301±0.0325	0.0114±0.0157	0.0186±0.0313	0.0085±0.0108
f_4	0.0088±0.0825	0.0101±0.0401	-0.0247±0.0648	0.0095±0.084

We are thus led to the conclusion that intermittency effects, if at all present, are weak in positive meson induced interactions on Al and Au nuclei.

Recently, several authors have pointed to the relationship between the factorial moments of order i, and the parametrization of the multiplicity distribution (MD) in terms of a negative binomial distribution (NBD). If the MD is described by the NBD with parameters $\overline{n}$ and $1/k$, then the factorial moments can be expressed in the form

$$F_i = (1 + \frac{1}{k})(1 + \frac{2}{k}) \ldots (1 + \frac{(i-1)}{k}) \tag{4}$$

which implies that the higher order moments F_3, F_4,... can be expressed in terms of the moment F_2, with $\frac{1}{k} = F_2 - 1$. Fig. 1 also shows the factorial moments F_3 and F_4, calculated from F_2 according to Eq. 4. The calculated values follow the trend of the experimental data but fall below the latter ones. The deviation is amplified with increasing order i of the moments.

3 Bose–Einstein correlations

Results were published[7] on Bose–Einstein correlations in K^+p and π^+p interactions at 250 GeV/c, derived from the same experiment with the same set-up as used here. One considers the ratio of the distribution of a suitable variable for like-sign pions, to this distribution for a reference sample. The reference sample used is the

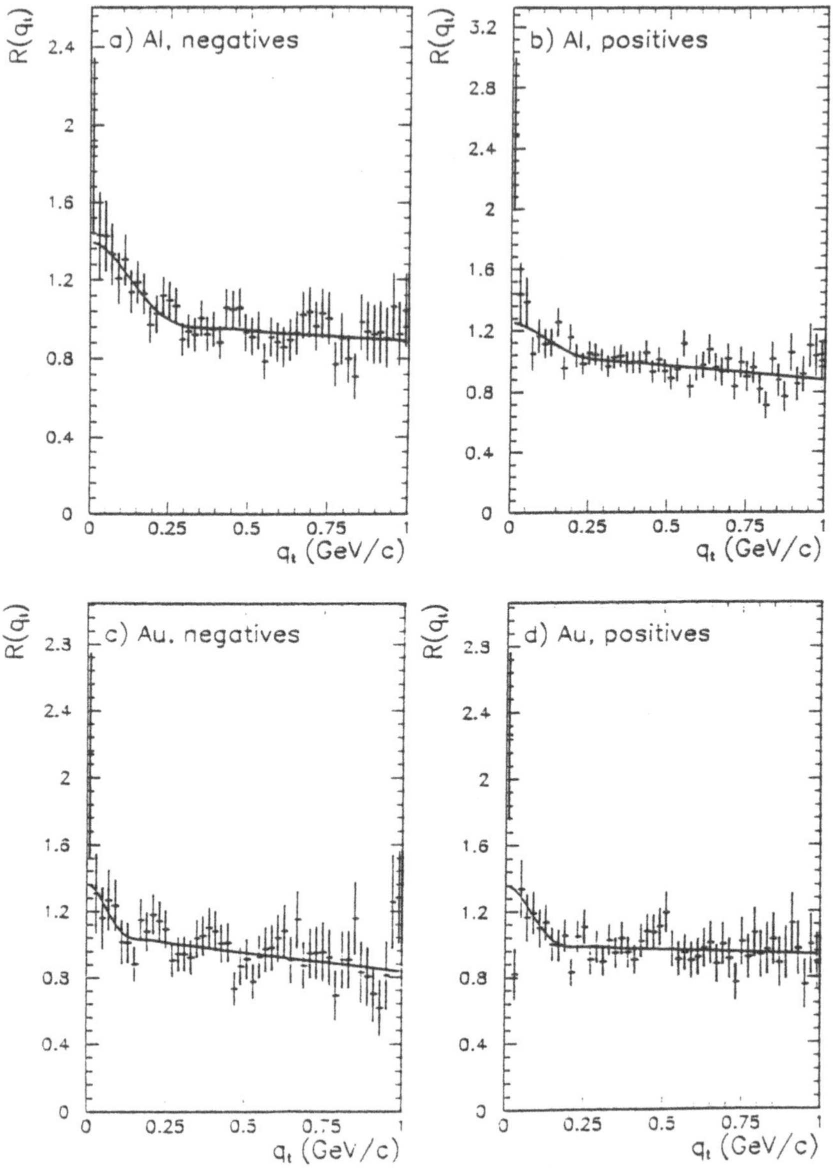

Fig. 2. $R(q_t)$ for $--$ and $++$ combinations, for Al and Au, with fits to Eq. (5) superposed. Events with $6 \leq n_{ch} < 20$ are used.

"mixed" sample, constructed by combining pions from one event with pions of the same charge, randomly chosen in other events with approximately the same charge multiplicity.

The first parametrization which we used is the Kopylov-Podgoretskii one[8]:

$$R(q_T) = \gamma \left[1 + 4\lambda J_1^2(\beta q_T)/(\beta q_T)^2\right] (1 + \delta q_T) \tag{5}$$

at small values of $q_0 = |E_1 - E_2|$. The fit parameters are the normalization γ, the "correlation strength" λ, the spatial dimension of the pion source β (with $r = 0.197\beta$ fm) and δ, introduced to account for the slow variation of the background.

Fig. 2 shows the ratio $R(q_T)$, obtained with the mixed reference sample, for positive and negative like-pion combinations and both targets for the sample with charge multiplicity $6 \leq n_{ch} < 20$. In all four plots of Fig. 2, a significant enhancement is observed at small q_T values. Fits according to Eq. 5 for $q_0 \leq 0.2$ GeV/c are superimposed. The fit parameters are collected in Table 2. Taking into account that a radius $r = 1.4 \pm 0.1$ fm was found in M^+p collisions[7], we observe an increase of r with A.

Table 2. Fit results for the Kopylov-Podgoretskii parametrization (Eq. 5) to the ratio $R(q_t)$ for like-pion distributions for events with $6 \leq n_{ch} < 20$. The results for M^+p interactions are reproduced from ref. 7.

Sample	Charge	λ	β (GeV/c)$^{-1}$	r (fm)	γ
M^+p	$--$	0.30 ± 0.03	7.0 ± 0.5	1.4 ± 0.1	0.91 ± 0.03
M^+Al	$--$	0.41 ± 0.11	12 ± 2	2.3 ± 0.3	0.99 ± 0.05
$6 \leq n_{ch} < 20$	$++$	0.19 ± 0.09	13 ± 4	2.5 ± 0.7	1.05 ± 0.04
M^+Au	$--$	0.28 ± 0.17	24 ± 6	4.8 ± 1.1	1.07 ± 0.04
$6 \leq n_{ch} < 20$	$++$	0.37 ± 0.13	18 ± 4	3.5 ± 0.8	1.00 ± 0.22

Weiner[9] advocated a different approach to Bose–Einstein correlations , based on quantum statistical considerations and assuming the particle fields to be a superposition of a chaotic and a coherent part. The correlation function versus $y_{\text{diff}} = |y_1 - y_2|$ for negative pion pairs in the interval $-1.8 < y < 1$ with a mixed reference sample, is given in Fig. 3 for M^+Al and M^+Au collisions. Fits to the form

$$R(y_{\text{diff}}) = \gamma \left(1 + 2p(1-p)\exp[-(y_{\text{diff}})^2 r^2] + p^2 \exp[-2(y_{\text{diff}})^2 r^2]\right) \tag{6}$$

yeild correlation lengths r is 0.92±0.36 (0.97±0.21) in M^+Al (M^+Au) interactions, to be compared to 1.6 in pp and $p\bar{p}$ interactions[10]. The chaoticity parameter p is only 0.06±0.02, implying that the last term in (6) essentially does not contribute to R.

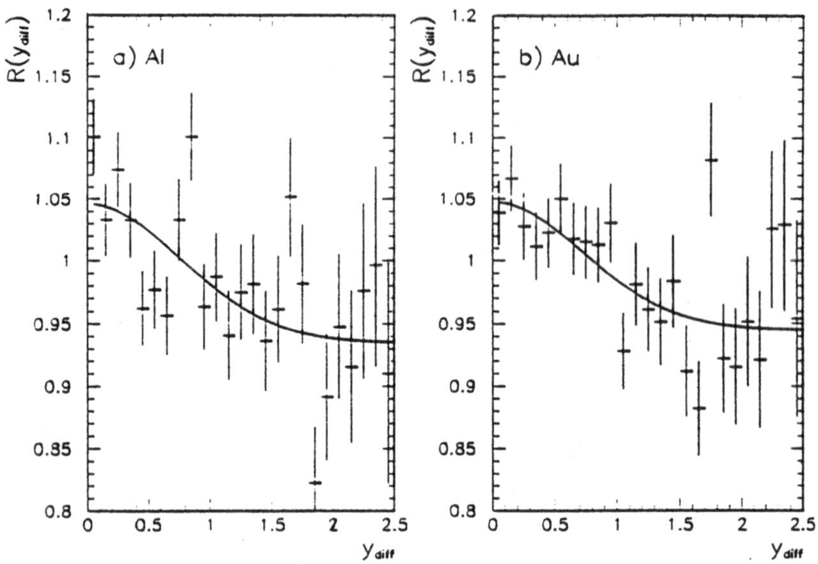

Fig. 3. The ratio $R(y_{diff})$ for $--$ combinations in the *Al* and *Au* samples, with fits to Eq. (6) superposed.

4 References

1. A. Białas. R. Peschanski: Nucl. Phys. **B273** (1986) 703; ibid. **B308** (1988) 857 and Phys. Lett. **B207** (1988) 59

2. W. Kittel, R Peschanski: *Review on intermittency in particle production*, Nijmegen preprint HEN-325 and Saclay preprint SphT/89-143

3.M. Adamus et al.: Z. Phys. C - Particles and Fields, **C32** (1986) 475

4. I.V. Ajinenko et al.: Z. Phys. C - Particles and Fields, **42** (1989) 377

5. B. Andersson, G. Gustafson, B. Nilsson-Almqvist: Lund preprint LU TP 86-3

6. I.V. Ajinenko et al.: Phys. Lett. **B235** (1990) 373

7. M. Adamus et al.: Z. Phys. C - Part. and Fields, **37** (1988) 347

8. G.I. Kopylov, M.I. Podgoretskii: Sov. J. Nucl. Phys. **15** (1972) 219; **18** (1974) 336: G.I. Kopylov: Phys. Lett. **B50** (1974) 472

9. R.M. Weiner: Phys. Lett. **B232** (1989) 278 and references therein

10. K. Kulka, B. Lörstad: Z. Phys. C - Particles and Fields **45** (1990) 581

III. Unlike Particle Correlations

Temperature Measurements at Intermediate Energy Heavy-Ion Collisions

J. Pochodzalla

TEMPERATURE MEASUREMENTS ON A THERMALIZED SOURCE BY PARTICLE CORRELATIONS

H. Dabrowski*[1], D. Ardouin[1], Ph. Eudes[1], D. Goujdami[1], F. Guilbault[1], C. Lebrun[1], B. Remaud[1], F. Sébille[1].

P. Lautridou[2], R. Boisgard[2], J. Québert[2].

A. Péghaire[3].

1 - Laboratoire de Physique Nucléaire, IN2P3-CNRS et Université de Nantes,
2 rue de la Houssinière - F-44072 Nantes Cedex 03 - France.

2 - Centre d'Etudes Nucléaires de Bordeaux - Gradignan. IN2P3-CNRS et
Université de Bordeaux I, Le Haut-Vigneau - F-33170 Gradignan - France.

3 - Laboratoire GANIL, IN2P3-CNRS BP 5027 - F-14021 Caen Cedex - France.

Abstract :

From backward angle correlation measurements for the reaction ^{40}Ar + ^{Nat}Ag at 44 MeV/u, an emission temperature of the target-like fragment is extracted by means of the relative population of excited states of emitted light clusters. The temperature $T = 2.2^{+0.3}_{-0.2}$ MeV and $T = 2.9 \pm 0.7$ MeV values are obtained respectively for the ^{4}He and ^{6}Li cases. Our finding is consistent with the emission of a thermalized system after excitation energy relaxation.

* On leave from I.P.N. KRAKOW (Poland)

The possibility to attain limits of complete equilibration of the degrees of freedom has raised crucial questions concerning the statistical and dynamical properties of hot reaction zone formed in collisions between heavy-nuclei at intermediate energy beams. These properties can be investigated via measurements of light cluster or nucleons emission. Difficulties arise at high excitation energy due to the decreasing of decay time by nucleon emission and competiting time scales with other relaxation processes[1] : like preequilibrium particles emission and multifragmentation processes. Thus, in this energy domain, the comparison of particle emission with statistical concepts requires an exact knowledge of the emitting source and its relation to the excitation energy partition during the collision process. For example, such problems have appeared when looking at intermediate mass fragments or nucleon emission [2,3] : reduction of excitation energies and emission barriers respectively were found necessary in the frame of a statistical description. For such a description, the knowledge of the temperature or the density of states and its relation to the thermal energy part of the deposited excitation energy is very important in order to understand the collision processes.

In the present work, we measured the emission temperature from the relative population of excited states of emitted light clusters. The method has been recently proposed in papers [4,5]. The fact that almost all the temperature measurements, thus deduced, give the T values in the vicinity 5 MeV for different incident energies, particle species and associated multiplicities [5 - 9] has led to the conclusions [9,10,1] of possible dynamical limitations to the stored thermal energy. The above relative population method has however been used in inclusive measurements which leave unknown the exact origin of the emitting source. As a matter of fact, in a normal kinematics reaction, it is highly probable that forward-angle measurements are viewing sources of several mixed origins like compound nucleus, projectile and target fragments in addition to pre-equilibrium emission. Thus, such measurements can be altered by contributions from several energy partitions and non complete thermal equilibrium stages.

Experimentally, it appears possible to clarify the above source mixing problem by using an asymmetric system : either by measuring forward-angle emission using reverse kinematics or backward-angle emission in normal kinematics, it is possible to diminish the influence of pre-equilibrium emission as well as presumably thermal emission arising from the light partner [11,12,13]. We have measured the emission temperature of ^{4}He nuclei from the relative population of the unbound-state at 20.1 MeV and the ground-state as observed in backward angle measurements using a beam of ^{40}Ar (E/A = 44 MeV) on a silver target. This allows a comparison with similar measurements[5] at forward angles and a temperature determination not significantly affected by side-feeding processes[14].

Our experiment was performed at the GANIL facility. The 20.1 MeV state decay[15] was observed by measuring coincident proton and triton detected by a multi-detector ensemble of 24 closed CsI scintillators set at $\theta = 90^\circ - 150^\circ$ polar angle. The energy threshold was 5 MeV for protons. A complete description of the experimental set-up is given elsewhere[16]. Fig.1 shows the inclusive ^{4}He single spectra obtained at three different average angles. The observed energy and angular dependences were fitted by a moving source parametrization superposing two maxwellian distributions with velocity v and slope parameter T values : (v = 0.018c, T = 3.6 MeV) and (v = 0.07c, T = 8.9 MeV) respectively. The latter component was introduced in order to better reproduce the high-energy tail of the spectra (Fig.1). Those sets of parameters have been used to determine the detection efficiency (see below). However, for our whole backward measurements, the lower-velocity component, presumably related to a target-like fragment, is largely dominant : thus, an almost satisfactory fit is obtained for a unique source with v = 0.034c and T = 5.2 MeV. The comparison with the compound system velocity (v_{CM} = 0.08c) reflects the domain of impact parameter which we are sensitive to, through the inherent geometry averaging of an inclusive measurement.

We present our coincidence data (Fig.2) in terms of the correlation function R(q) defined by the ratio of coincidence yields $Y_{12}(\vec{p}_1,\vec{p}_2)$ over singles yields $Y_1(\vec{p}_1)$ and $Y_2(\vec{p}_2)$: $1 + R(q) = C[Y_{12}\ (\vec{p}_1,\vec{p}_2)/Y_1(\vec{p}_1)\ Y_2(\vec{p}_2)]$ (1) where $\vec{p}_1$, $\vec{p}_2$ are the laboratory momenta of the proton and triton ; q is the momentum of relative motion and C is a normalization constant which value does not affect our temperature determination. The correlation function is obtained by summing over all energies and angles corresponding to a given momentum q. Similarly to previous forward-angle similar measurements [17], we observe in the q = 12 MeV/c - 22 Mev/c interval (hatched region on the q axis of the fig 2), a peak corresponding to the 20.1 MeV excitation energy for ^4He. At variance with the result obtained in ref 17, the 21.1 MeV and 22.1 MeV excited states are not clearly visible due firstly to a weak population and secondly to some fluctuations observed in the evolution of the correlation-function at higher relative momenta. The coincidence yield $Y_{4\ He}$, corresponding to the decay of the 20.1 MeV unstable state, was extracted assuming that the p - t coincidence yield is given by $Y_{pt} = Y_{4\ He} + Y_b$ where Y_b denotes the "background-yield". This background is indicated by the solid line (2) in Fig.2. Two extreme boundaries (lines 1 and 3) are also shown which will be used to estimate the temperature uncertainty resulting from this empirical choice. Assuming a thermal distribution from two emission sources with the above quoted characteristics and following exactly the method of ref.17, the primary relative population of the two considered states have been deduced from the calculated decay-yield and the efficiency function using a Monte-Carlo simulation of the detector. From the integration of the experimental yields over the excitation energy range corresponding to hatched q region (Fig.2), the relative population of the ^4He excited (E* = 20.1 MeV) and ground state corresponds to an emission temperature of $T = 2.2 ^{+0.3}_{-0.2}$ MeV. The uncertainty quoted here results from the uncertainties of the background (lines 1 and 3 in Fig.2). It must be emphasized, as already shown in ref.17, that the extracted emission temperature, based on relative populations of states, is not very sensitive to the source characteristics, provided relative efficiencies are well determined. These uncertainties are of negligible importance against the above background uncertainties. In order to demonstrate the sensitivity of the correlation-function on the T-value, fig. 2 also shows (dashed lines) the relative yields of the 20.1, 21.1 and 22.1 MeV excited states computed from some temperature assumptions using the intermediate background line (2) and gaussian distributions with widths representative of the effective observed shape for the most visible structure. From this comparison, we conclude that the $T = 2.2^{+0.3}_{-0.2}$ MeV value is compatible with the weak relative population of the higher states. As an example, one sees that a T value as high as 2.9 MeV (dashed-dotted line) is clearly inconsistent with our relative population data.

Fig.3 shows our measured α-d correlation-function for the same reaction. It exhibits two peaks which correspond to the ^{6}Li state at 2.186 MeV ($J^{\pi} = 3^{+}$, $\Gamma = 24$ keV, $\Gamma_{\alpha}/\Gamma_{tot} = 1.00$) and the state at 4.31 MeV($J^{\pi} = 2^{+}$, $\Gamma = 1.3$ MeV, $\Gamma_{\alpha}/\Gamma_{tot} = 0.97$) moderatly pertubed by an higher state at 5.65 MeV. Though a less accurate temperature determination, due to the smaller energy separation of these two states and subsequent population perturbation by sequential decay [14,17], it is instructive to make a comparative determination of the emission temperature with previous forward-angle measurements[5)]. Using the background shown in Fig.3 (dashed-dotted line) and normalizing our calculations in order to reproduce the experimental yield over the excitation energy range corresponding to q = 30 - 55 MeV/c, it was possible to explore the possible temperature variations compatible with the observed yield for the 4.31 MeV state (Fig.3 full and dashed lines). A $T = 2.9 \pm 0.7$ MeV value is in agreement with our data.

To summarize our results, an emission temperature of $T = 2.2\ ^{+0.3}_{-0.2}$ MeV for ^4He and of 2.9 ± 0.7 MeV for ^6Li excited fragments was obtained in our backward angle measurements. This result differs significantly from the values T = 4.5 - 5 MeV usually observed [5,6,17] in similar measurements at forward angles. From the above discussion of the chosen geometry of detection and the characteristics of the singles spectra, we postulate that this reflects the observation of an almost unique source (target fragment). In order to address the question wether a statistical equilibration has been achieved, the examination of the slope parameter of the energy distributions has long been used. From the above quoted values, it is observed that this parameter is higher than our temperature determination, similarly to forward angle data. This difference has received interpretations[18,19] involving the dynamics of the collision. Schematically , the slope parameter reflects the average over the de-excitation cascade and thus the dynamics of the collision process while the "chemical temperature" determined from the relative population is established at the break-up time of the system when fragments are formed. Values as low as 3 MeV have been predicted[19] offering little sensitivity to the incident energy. Another reason of this difference could be the influence of lifetime effects in the ratio measurement which could modify the strength of the resonances according to the excitation energy of the source[16]. It has been established [20,21,1] that at the beginning of the collision, the two nuclei share the excitation energy equally : part of this energy is stored as collective excitation and the damping of this mode through mainly pre-equilibrium emission is a necessary condition to reach thermal equilibration of the system. The two resulting fragments can thus be produced with different temperatures.

A reliable dynamical calculation is thus necessary to investigate quantitatively such processes. We have used the Landau-Vlasov theory which has been shown [22,23,9] to give a good description of the excitation energy evolution and mass-velocity correlations in heavy-ion collisions. Our calculations have been made when pre-equilibrium emission has disappeared (t = 200 fm/c) and for a range of impact parameter (b = 4 - 6 fm) representative of our inclusive data. The calculated excitation energies after separation are 2.1 MeV/A and 0.9 MeV/A for projectile and target like-fragments respectively. Under the assumption of thermal energies, this would be equivalent to temperatures respectively of 4 MeV and 2.7 MeV in good agreement with temperatures measured at backward angles and forward angles respectively. (Meanwhile, an excitation energy of 3.2 MeV/A is calculated for b = 1fm where a compound system is formed). In the frame of these calculations, reduced values of excitation energy are attained when the separation of fragments is achieved because of competition with a copious nucleon emission in a comparable time-scale (10^{-22}s). Due to this large transfer of available energy into intrinsic degrees of freedom before separation, one can also predict a weak sensitivity of the measured temperatures to the incident energy as observed experimentally[5-9]. As already evoked, the slope parameter reflects an average over the whole emission process or equivalently the excitation energy dependence of the emission probability. Its exact interpretation and comparison with temperature determinations require emission calculations for specific light-cluster species. Particularly, this implies a correct description of dynamical density fluctuations which could play an important role at the surface of the hot nucleus for giving birth to light clusters. These effects have not received a theoretical approach yet.

In final conclusions, for the first time we have measured the temperature emission of highly excited light clusters, emitted mostly from target like source, produced in collisions of ^{40}Ar on ^{Nat}Ag, at beam energy 44MeV/u. The temperature deduced from the population ratio of excited states are : $2.2\ ^{+0.3}_{-0.2}$ MeV for ^{4}He, and T = 2.9 ± 0.7 MeV for ^{6}Li emission respectively. Using a dynamical theory, we propose a low temperature interpretation in terms of a strong relaxation of the excitation energy prior to fragment separation. This measurement provides new important information on the energy partition in heavy-ion collisions and set some limits on the ability of the statistical description of hot nuclei prior to fragment separation or emission.

* Experiment performed at Grand Accélérateur National d'Ions Lourds (GANIL).

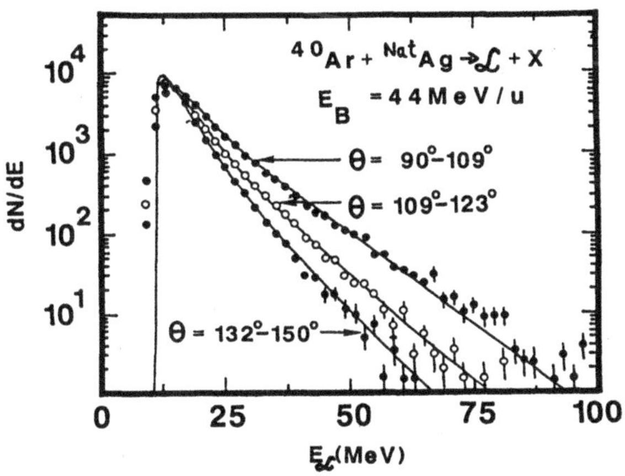

Fig. 1 : Energy distributions of α particles measured in the Ar + Ag reaction for 3 different laboratory angles. The lines indicate moving source parametrizations employing the superposition of two sources (see text).

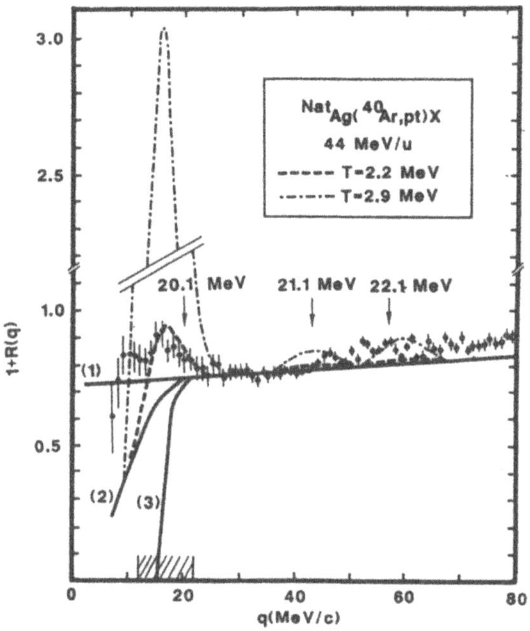

Fig. 2 : p-t correlation function. Excited states in ^{4}He are indicated with excitation energy as measured in ref. 15 . The solid lines correspond to the chosen background correlation-function (see text). The dashed line corresponds to the relative population of the 20.1, 21.1, 22.1 MeV ^{4}He excited states at temperature T = 2.2 MeV ; dash dotted line corresponds to T = 2.9 MeV.

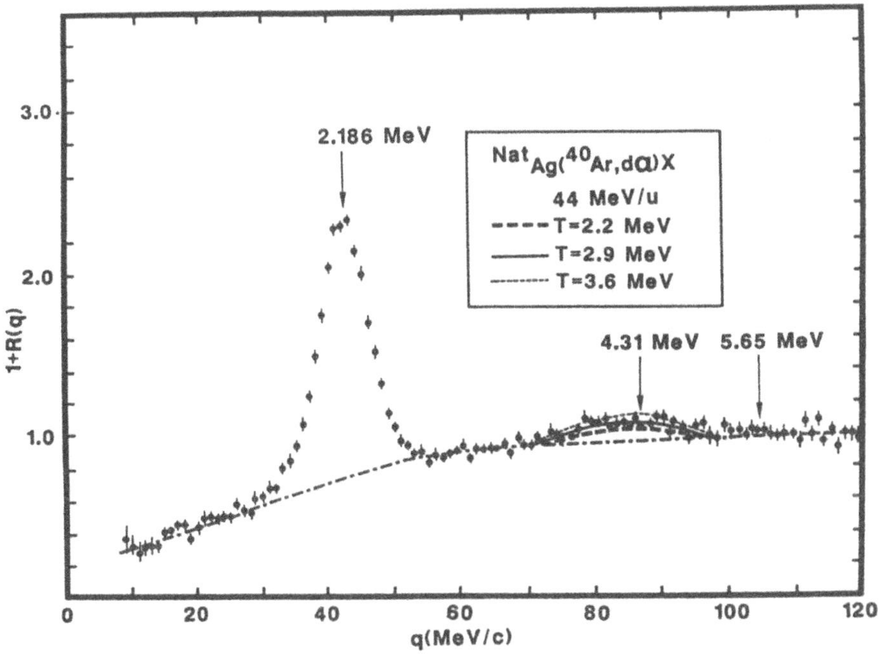

Fig. 3 : d- α correlation function. Excited states of ^{6}Li are indicated. The solid and dashed lines represent the calculated yields for three different temperatures solid T = 2.9 MeV, thin dashed T = 3.6 MeV and heavy dashed line correspond to T = 2.2 MeV. The dashed-dotted line represent the background correlation-function.

REFERENCES

1) E. Suraud, C. Grégoire, B. Tamain, Progress of Nucl. Part. Science, 23, (1989) 357.

2) J. Pochodzalla et al, Phys. Rev. C 40, (1989) 2918.

3) R. Lacey et al, Phys. Lett. 191 B, (1987) 253. M.F. Rivet et al, Phys. Rev. C 25 (1982) 2430.

4) D.J. Morissey et al, Phys. Lett. 148 B (1984) 423.

5) J. Pochodzalla et al, Phys. Rev. Lett. 55, (1985) 177.

6) Z. Chen et al, Nucl. Phys. A 473, (1987) 564.

7) D. Cebra et al, Phys. Lett. 227 B, (1989) 336.

8) F. Saint-Laurent et al, Phys. Lett. 159 B, (1985) 244.

9) E. Suraud et al, Phys. Lett 229 B, (1989) 359.

10) W. Trautmann et al, to be published.

11) H. Morgenstern et al, Phys. Rev. Lett. 52, (1984) 1104.

12) M. Gonin et al, Phys. Rev. C 38, (1988) 135.

13) D. Ardouin et al, Nucl. Phys. A (in press).

14) Z. Chen and C.K. Gelbke, Phys. Rev. 38, (1988) 2630.

15) S. Fairman, W.E. Meyerhof, Nucl. Phys. A206, (1973) 1 I.Y. Barit, V.A. Sergeev, Yad. Fiz. 13 (1971) 1230 [Sov. J. Nucl. Phys. 13 (1971) 708].

16) P. Lautridou et al, submitted for publication.
J. Québert et al, in Proceedings of the XXVII Winter Meeting on Nuclear Physics (Bormio, Italy, 1989) ed. by. I. Iori (University Of Milano) Supp. N. 69 (1989) p. 220.

17) J. Pochodzalla et al, Phys. Lett. 161 B, (1985) 275, Phys. Rev. C 35, (1987) 1695.

18) H.W. Barz, H. Schulz and G. Bertsch, Phys. Lett. 217 B, (1989) 397.

19) D. Boal, J. Glosli and C. Wicentowich, Phys. Rev. 62 (1989) 737.

20) R. Dayras et al, Phys. Rev. Lett. 62, (1989) 1017.

21) J.L. Wile et al, Phys. Rev. C 40, (1989) 1700.

22) B. Remaud et al, Nucl. Phys. A 447, (1985) 555.

23) M.F. Rivet et al, Phys. Lett. 215 B (1988) 55.

Temperature Measurements with Gamma-rays and Unlike-particle Correlations

D. J. Morrissey
National Superconducting Cyclotron Lab
and Department of Chemistry
Michigan State University
East Lansing, MI, USA 48824

Abstract

Measurements of correlated "unlike" particles have been performed to extract the temperature of the emitting system. This technique is based on the idea that the populations of excited states in small composite fragments are temperature dependent and decay into correlated particles. The first and now most complete measurements detected gamma-rays correlated with small nuclei. This work has been extended to measurements of neutrons and charged particles correlated with similar small nuclei by our group and others. A brief discussion of the technique is presented along with an overview of recent results.

1. INTRODUCTION

Traditional measurements of the temperature of a hot nucleus have relied on determining the "kinetic temperature" from the exponential slope of kinetic energy spectra of neutrons, protons and alpha particles. It also has been know for some time that excited (large) nuclei also emit larger more complicated nuclei, although in diminishing numbers.[1] The suggestion was made that because the populations of any excited states of the more complicated nuclei should also depend on the temperature of the emitting nucleus they should provide an independent tool for determining that temperature.[2] Such measurements are akin to those used to determine stellar temperatures by observing the relative populations of atomic levels. The initial studies of nuclear reactions were with bound states of light nuclei, e.g., ^{6}Li and ^{7}Li, observed through their gamma-ray decay.[2,3] Subsequent measurements have been made of higher-lying bound states that emit gamma rays,[4,5] and unbound states that emit charged particles[8,9] and neutrons.[11-15]

The excited state population method for nuclear temperature measurement relies on the Boltzmann factor for a system with energy levels in equilibrium at a finite temperature. The ratio, R, of the initial populations of two states of a nucleus in thermal equilibrium is written as:

$$R = \frac{2j_{upper} + 1}{2j_{lower} + 1} \; e^{-\;\Delta E/kT} \tag{1}$$

where j_{upper} and j_{lower} are the angular momenta of the higher-lying and lower-lying states, respectively, ΔE is the energy gap between the two states and kT is the nuclear temperature. However, this ratio is not always easily measured because the lifetimes of excited states are short and feeding may be important. For example, the fraction of the total population f_n in a given bound excited state can be written:

$$f_n = \frac{(2j_n + 1)e^{-\;\Delta E_n/kT}}{\Sigma_i(2j_i + 1)e^{-\;\Delta E_i/kT}} \tag{2}$$

where the sum over i in the denominator, usually called the partition function, includes all the states of the system. Such a fraction might be determined experimentally from the number of γ-rays "correlated" with a residual cold nucleus. Unfortunately, the observed fraction of a specific nucleus in a given excited state is not generally equal to the fraction of those nuclei *initially* in that state due to γ-ray branching and feeding from other (bound and unbound) levels. Methods developed to address this difficulty primarily rely on measuring a broad range of levels in similar nuclei and extracting an overall (and hopefully consistent) temperature.

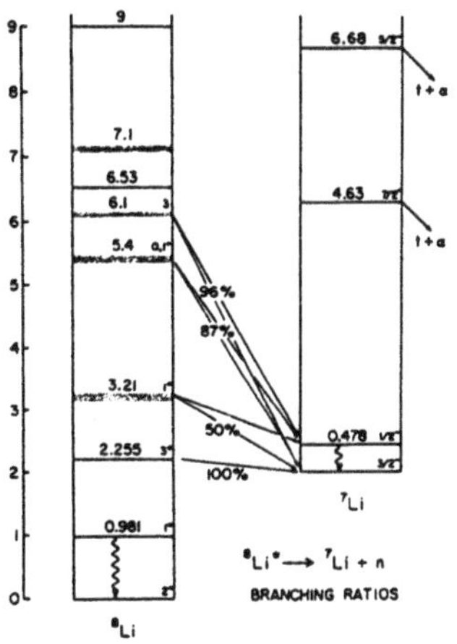

Figure 1: Schematic diagram of the bound and unbound levels in ^{8}Li and ^{7}Li. The known values of the neutron branching to the ground state of ^{7}Li are indicated in percents.

Figure 1 shows the bound and unbound levels in ^{7}Li and ^{8}Li nuclei. Notice that these nuclei have a set of interconnected levels that decay by the emission of gamma rays to the ground states, of neutrons primarily to the ground state in the neighbor and by charged particles.[16] Thus, when the temperature is very low we can expect population of the bound states and their concomitent decay. However, when the temperature kT becomes comparable to 2-3 Mev then neutron emitting states in ^{8}Li will be populated and will feed the ground state of ^{7}Li. Understanding and unraveling these connections requires that the relative primary yields of ^{7}Li and ^{8}Li be known. This is a complication that is not present at lower temperatures. Thus, at higher temperatures it is advisable and in fact necessary to measure as large a number of excited and ground state populations as possible. In this way, the complicated linkages can, in principle, be unraveled. However, up to the present, experimenters have relied on model calculations to fit the pattern observed at high bombarding energies.

In the present report, I will present some of the results obtained by our group using a variety of techniques to explore the use of correlated particles for temperature measurements. First, studies of the basic assumptions of the technique made by measuring the populations of γ-ray emitting states from compound nuclear reactions are presented. Then the results from what might be considered more traditional correlation studies based on neutron charged-particle and two charged-particle detection are presented. Overall, the technique is well founded and has added an important dimension to studies of heavy-ion induced reactions.

2. EXPERIMENTAL TECHNIQUES

2.1 Gamma-ray Charged Particle Measurements

The first suggestion as to how to extract a nuclear temperature from the relative populations of excited states was to measure the bound states in very light nuclei.[2)] The bound states decay by γ-ray emission, of course, and in the very light nuclei are not feed by γ-ray cascades. In a sense these studies, in which γ-rays are detected in coincidence with residual nuclei, represent the limit in terms of the "unlikeness" of the correlated particles. However, because these studies have been rather well developed and represent the only cases in which the technique has been *tested* by application to known systems[3)] I will briefly present some of our results. The measurements are very traditional in that the charged particles can be detected in a silicon telescope, but discrete Z and A resolution is necessary. The gamma rays can be detected in individual detectors or in an array.[4)]

The calibration reactions produced compound nuclei with (previously) well established excitation energies and angular momenta. The population of the single and three bound states in ^{7}Be and ^{10}B, respectively, were measured for the reaction of ^{14}N with ^{12}C at a series of bombarding energies.[17)] The populations were found to rise with bombarding energy (i.e., compound nucleus excitation energy or nuclear temperature) up to $E/A \approx$ 10 MeV. At higher bombarding energies the relative populations remained constant or decreased slightly. This is most likely due to feeding of these states by sequential decay from particle unbound levels in neighboring nuclei. One expects that these levels, which generally lie at energies greater than a few MeV, will become populated at higher temperatures. In this reaction system angular momentum significantly lowers the amount of thermal energy due to the small moment of inertia.

In a second calibration reaction the populations of bound states in a wide range of final nuclei produced in the reaction of ^{40}Ar with ^{12}C ($E/A = 8$ MeV) were measured.[3)] The temperatures extracted from the relative populations are shown in figure 2. All of the values are in good agreement with the predicted temperature of the equilibrated compound nucleus.

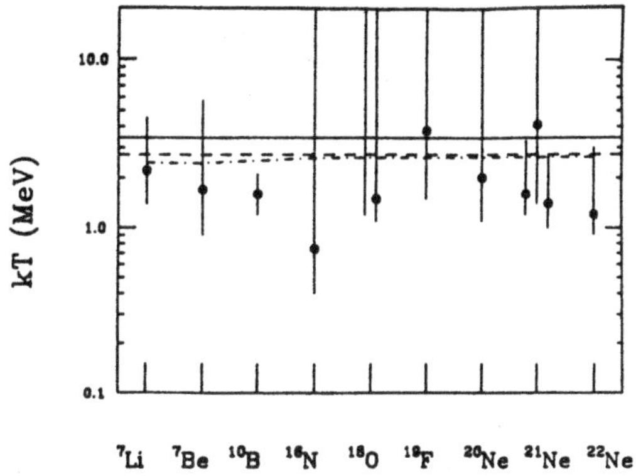

Figure 2: The population temperatures extracted from the γ-ray fractions of complex fragments from the decay of excited compound nuclei produced by the reaction of ^{40}Ar with ^{12}C at E/A = 8 MeV.[3)] The solid line lies along the calculated temperature of the nonrotating compound nucleus, 3.44 MeV, the dashed line along that of the rotating spherical compound nucleus, 2.73 MeV, and the dot-dash line along that estimated for the rotating scission configurations.

2.2 Neutron-Charged Particle Systems

The population of unbound levels that decay by neutron emission can be measured by analyzing the relative velocity distributions of neutrons and charged particles. The neutrons have to be detected in close geometry to the charged particles due to kinematic focussing by the primary fragment velocity. In fact, the colinear geometry has been extensively used for neutrons emitted with low relative kinetic energies.[14,12)] This generally leads to observation of both kinematic solutions and provides a double check on the size of the population. The neutrons have to positively identified and a very accurate measurement of their velocity is required for the analysis. Of course, discrete Z and A resolution for the charged particles is necessary along with accurate energy and relative angle measurements. It is also important to determine the background in an unambiguous way.

The emission of ^7Li from the reaction of ^{14}N with silver (E/A = 35 MeV) has been extensively studied. Notice in figure 1 that this nucleus has only one bound state but is also fed by neutron emission from several levels in ^8Li. There is also a neutron unbound level in ^7Li at 7.456 MeV, so this nucleus is a good candidate for a neutron charge-particle correlation study. We have reported the results of such a study using a set of silicon charged-particle telescopes and discrete liquid-scintillation neutron detectors.[11)] Some indication of the typical results are shown in figure 3 in which the importance of the background measurement are clear. We determined the background by measuring the total coincident neutron distribution as a function of relative angle. These results are interesting in their own right and shed light on the overall reaction, however, they are crucial to extracting the population of the excited states because they number of thermal (uncorrelated) neutrons usually exceeds the correlated neutrons by an order of magnitude.

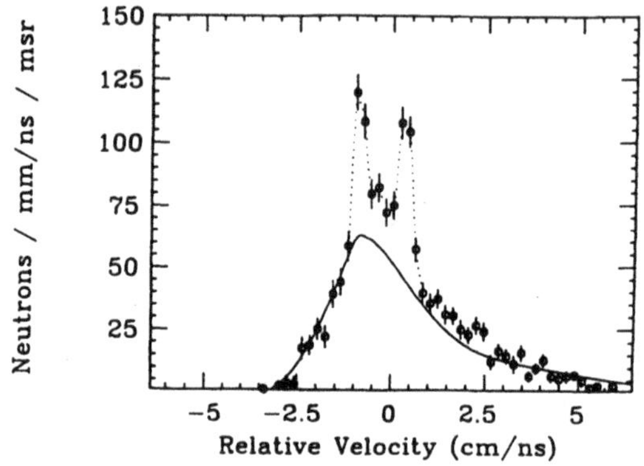

Figure 3: The yield of neutrons at 50^0 in coincidence with ^7Li fragments as a function of the relative velocity (neutron velocity minus ^7Li velocity). The peaks correspond to forward and backward emission from the 2.255 MeV state in ^8Li. The solid line shows the measured background due to emission of thermal neutrons by the reaction partners while the dots are to guide the eye.

For the case of ^7Li, measurements have been made of the bound state populations, an unbound state population, and feeding by neutron decay of an unbound level in another nucleus. All of these measurements are consistent with a single temperature of about 3 MeV when the effects of the total predicted feeding is considered. This temperature is also consistent with charged-particle correlation measurements if one limits the comparison to similar data (i.e., the 50^0 data with $55 \leq E_1 + E_2 \leq 100$ MeV in ref. 7). Thus, the importance of measuring a broad range of states that are linked by various emission channels should be clear.

2.3 Two Charged Particle Systems

The decay of unbound levels in light nuclei, such as the high lying states in ^7Li, into two charged particles has been extensively studied by another group at MSU. Any such studies generally require very accurate measurement of two charge particles in very close geometry. The colinear geometry exploited in the neutron work is not possible and large hodoscopes have been developed (see, e.g. refs. 6 - 9 and a contribution to these proceedings). We have used a slightly different technique in which a multihit tracking chamber was used in conjunction with a close packed array of plastic scintillators.[10] Such measurements of two charged-particle coincidences usually allow observation of a broad range of nuclei similar to the gamma ray measurements.

Taking a global view of the data in the literature, the measurements of the populations of nuclear levels seem to rise until the temperature is roughly 4 to 5 MeV. After that point, the populations of the levels do not depend on the bombarding energy nor the projectile/target combination. We have investigated this (saturation) effect in a study of charged-particle correlations from the reaction of 500 MeV protons with silver (A=107, 109) and ^{9}Be.[10] The results for the temperature derived from the level populations are indicated in Table I.

Table 1: Temperatures extracted from the level populations observed at 65^0 from reactions induced by 500 MeV protons.

Target	Pair	Temperature (MeV)
Ag	p-t	5.6 ± 1
Ag	d-t	5.6 ± 1
Ag	d-α	5.6 ± 1
Ag	t-α	5.6 ± 2
^{9}Be	d-α	< 4

The results indicate that once again the limiting temperature seems to have been attained. Unfortunately, the amount of beam time for the beryllium target was limited and the number of coincidences was similarly small. Only the limit for the temperature indicated in the table was obtained. The related data on source size showed no significant difference between the two targets.[10]

3. CONCLUSION

The measurement of the populations of excited states of light nuclei through the observation of correlated particles has become an important technique for determining the nuclei temperature. The kinds of correlated particles range from gamma rays, to neutrons, to charged particles. The technique has been "calibrated" in that the populations of bound states emitted from compound nuclei agree with the predictions. The populations of unbound states have been measured in higher energy reactions and lead to temperatures on the order of 4 to 5 MeV. This seems to be a general result supported by measurements of different kinds of states in many different nuclei. Whether this represents a limiting temperature of nuclear matter or the freeze-out temperature of expanding nuclear matter remains to be seen.

4. ACKNOWLEDGEMENTS

Three students should be recognized as being the people mainly responsible for this data, C. Bloch for the work with neutrons, J.H. Lee for the recent γ-ray work, and D.A. Cebra for the charged-particle coincidence work. It is also my pleasure to acknowledge that all of this work was done in collaboration with W. Benenson and his group at Michigan State University. In addition, the work with neutron detection was done in collaboration with A. Galonsky's group and that detecting charged particles was done in collaboration with the groups headed by G.D. Westfall at Michigan State, R.S. Tickle at the University of Michigan and R. Korteling at Simon Fraser University. This review was supported by the National Science Foundation under Grant PHY-89-13815.

REFERENCES

1) L.G. Moretto, Nucl. Phys. **A247** (1975) 211.

2) D.J. Morrissey, et al., Phys. Lett. **B 148** (1984) 423.

3) J.H. Lee, et al., Phys. Rev. C **41** (1990) 1562, 2406, and references therein.

4) H.M. Xu, et al., Phys. Lett. **B 182** (1986) 155, and Phys. Rev. C **40** (1989) 186.

5) J. Gomez del Campo, et al., Phys. Rev. Lett. **61** (1988) 290.

6) J. Pochodzalla, et al., Phys. Rev. Lett. **55** (1985) 177.

7) C.B. Chitwood, et al., Phys. Lett., **172B** (1986) 27.

8) J. Pochodzalla, et al., Phys. Rev. C **35** (1987) 1695.

9) Z.P. Chen, et al., Phys. Rev. C **36** (1987) 2297, see also Z.P. Chen and C.K. Gelbke, Phys. Rev. C **38** (1988) 2630, and references therein.

10) D.A. Cebra, et al., Phys. Lett. **B 227** (1989) 336.

11) C. Bloch, et al., Phys. Rev. C **36** (1987) 203.

12) A. Kiss, et al., Phys. Lett. **184B** (1987) 149.

13) F. Deak, et al., Nucl. Phys. **A464** (1987) 133.

14) F. Deak, et al., Nucl. Instrum. Methods **A413** (1985) 1.

15) F. Deak, et al., Phys. Rev. C **39** (1989) 733.

16) F. Ajzenberg-Selove, Nucl. Phys. **A490** (1988) 1, and references therein.

17) D.J. Morrissey, et al., Phys. Rev. C **34** (1986) 761.

18) C. Bloch, et al., Phys. Rev. C **37** (1988) 2469.

ON THE TEMPERATURE CONCEPT FOR INTERMEDIATE-VELOCITY EMISSION IN HEAVY-ION REACTIONS

H. FUCHS and H. HOMEYER
Hahn-Meitner-Institut Berlin GmbH
Bereich Kern- und Strahlenphysik
Glienicker Str. 100, D 1000 Berlin 39, Germany

ABSTRACT

Problems connected with temperature parameters determined from intermediate-velocity particles emitted in medium-energy heavy-ion collisisons are rediscussed. Arguments are given that these parameters probably have little to do with real thermal motion, and a consistent interpretation avoiding the temperature concept is outlined. Its ingredients are direct and sequential projectile breakup, Fermi motion and nucleon-nucleon single scattering.

A fascinating aspect of nucleus-nucleus collisions is the possibility to produce hot nuclear matter either by complete fusion or by incomplete fusion, the latter mechanism extending to such interesting phenomena as the fire ball or locally equilibrated hot nuclear matter neighbouring cold matter ("hot spot"). Though the very existence of these phenomena has not been strictly proven, many attempts have been made to obtain an experimental measure of nuclear temperatures and to relate the results to the fire ball or hot spot or similar pictures.

There are essentially two methods which are used to measure nuclear temperatures. The first one exploits the spectra of emitted particles which are fitted with Maxwell-Boltzmann distributions. For evaporation from compound nuclei one finds thereby indeed temperatures T consistent with the known excitation energies E^* applying the relation $E^*=aT^2$, with $a=(A/10)\pm10\%$. In addition to this emission, at higher bombarding energies ($E/A>10$ MeV) the fits yield a second emitting source with a velocity about half that of the projectile. It can be thought of as the fire ball and emits mainly nucleons. Fig. 1 shows as an example the slope parameters or temperatures of these sources extracted[1] from the proton spectra at various bombarding energies. There is a pronounced increase of this parameter with energy, attaining about 10 MeV at $E/A=35$ MeV and 20 MeV at $E/A=94$ MeV.

The second method is based on the yield ratio of excited states (see also ref. 1), preferentially the widely spaced unbound states appearing as resonances in the correlations between pairs of light particles (p,d,t,α). After some corrections for sequential feeding the yield ratio is, in principle, compared to the Boltzmann factor exp ($-\Delta E^*/T$) to give the temperature T. Up to now this method was mainly applied to emission angles and energies where the intermediate source (the suspected fire ball) is dominant. The obtained results (see ref. 1 and refs. quoted therein) confront us with two major problems. (i) At each energy the second method yields a much lower temperature parameter than the slope parameter of the first method. (ii) The obtained value is always between 4 and 5 MeV and thus does not depend on the bombarding energy, though the excitation energy per nucleon of the intermediate source (fire ball) should increase in proportion to the bombarding energy.

To remedy these inconsistencies, several special mechanisms have been proposed. The energies in the spectra of the emitted particles are assumed (2.) to contain an important contribution due to collective non-thermal (explosive?) motion. The temperature parameter from the resonance yield ratio is taken as the real temperature of the fire ball, but only after its expansion, when freezing-out into its final fragments (1,3,4.). The quantum-statistical model, which considers the thermal and chemical equilibrium in the fire ball nucleon gas is able to describe the data, but the decisive parameter, viz. the freeze-out density, turned out to be, for instance, only 0.04 of normal nuclear density[3]. The source radius of about 7 fm (upper limit) determined at the same time[3] then implies a number of at most 8 nucleons only in the source[5]. It is hard to conceive the statistical emission of light-particle pairs like α-α, τ-τ, α-d, α-p, ... from such a restricted nucleon reservoir. Thus the inconcistency is only pushed to a different place. Moreover, the adapted freeze-out density turns out to depend on the bombarding energy, which again is hard to understand. Finally, the continuous part of the relative-energy spectra[1] strictly contradicts the assumption of collective motion.

Facing these difficulties it seems worthwile to consider alternative reaction pictures which do not imply a thermal interpretation of the observed phenomena. In this sence we shall discuss in the following the two principal phenomena which lead, when ascribed to an intermediate source, to contradictory apparent temepratures. As a starting point, we abandon the idea of a common origin ("source") of the different ejectiles observed at intermediate velocities.

Let us begin with the **nucleon energy spectra**. Within the concept of

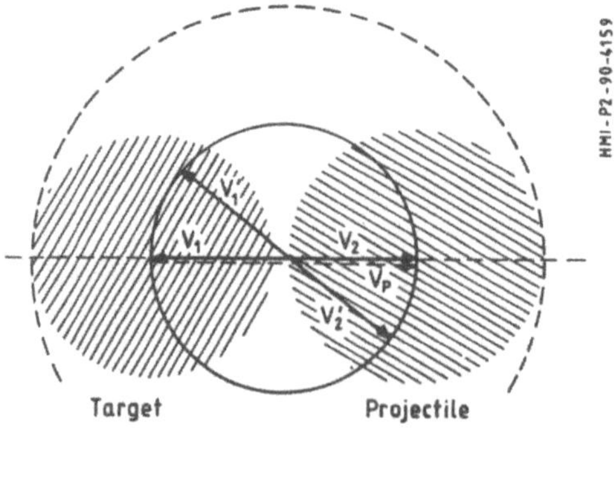

Fig. 1 *Velocity space representation of quasifree nucleon-nucleon collisions. Without Fermi motion, the post-collision velocities distribute over the spherical surface indicated by the solid circle. Due to the Fermi motion (hatched areas for target/projectile Fermi sphere) they fill the sphere up to the dashed circle.*

the exciton model, Blann et al.[6] have identified the intermediate-velocity source found in the nucleon spectra with the first collision between a projectile nucleon and a target nucleon, because their c.m.s. (or "source" system) trivially has a velocity half that of the projectile, on the average. (As Blann noted, there should follow other, slower sources in the process of equilibration.) Let us consider the first collision as a quasifree nucleon-nucleon scattering to obtain a rough estimate of emission energies. If there were no Fermi motion of the two nucleons, their c.m. velocity would be (non-relativistic) half the projectile velocity v_0, and the velocity of either of the two nucleons in their c.m.s. has the same absolute value, $v = v_0/2$, of course. After collision, the direction is arbitrary, but the modulus stays unchanged, the velocities filling the spherical surface indicated in **fig. 1** depicting the relations in velocity space. The corresponding kinetic energy of the nucleon in the c.m.s. then is one forth the projectile energy per nucleon, $E_0/4A$. (Of course, E_0 is the projectile energy minus the Coulomb barrier V_C.). Considering now the Fermi motion of the nucleons prior to the collision, its effect is to smear out the energy distribution before and after scattering. The nucleon energy in the c.m.s., $E_0/4A$, now is an average, and one has to add the average kinetic energy representing the Fermi motion, $1.5\sigma_0^2/m$, to obtain $<E> = 1.5\sigma_0^2/m + E_0/4A$. (This is essentially the additive relation between the mean squares of uncorrelated linear momenta, and we have assumed a Gaussian distribution of the nucleon momenta to represent the Fermi motion, with the dispersion σ_0^2 giving a slope parameter σ_0^2/m for the corresponding kinetic energies. With the generally adopted[7] value σ_0 of around 90 MeV/c we have $1.5\sigma_0^2/m = 12$ MeV as the contribution of the Fermi motion. The mean kinetic energies so estimated should be compared to the slope parameters of the intermediate source fits (1). These are made with (volume emission) Maxwell distributions, for which the mean kinetic energy $<E>$ is 3/2 times the slope parameter T, so with our estimate of $<E>$ we would have a slope parameter

$$T = 8\ \text{MeV} + E_0/6A. \tag{1}$$

This is inserted as a line into **fig. 2** showing the collection of experimental slope parameters for protons. In the intermediate-energy range (E/A = 20 to 200 MeV) it is quite close to the data, indeed, and demonstrates why one should find such high values. The remaining small discrepancies are caused by the crude assumptions entering into the estimate (1): The scattering nucleon will come preferentially from the nuclear surface, thus representing, in the average, lower momenta of the Fermi motion and requiring a some-

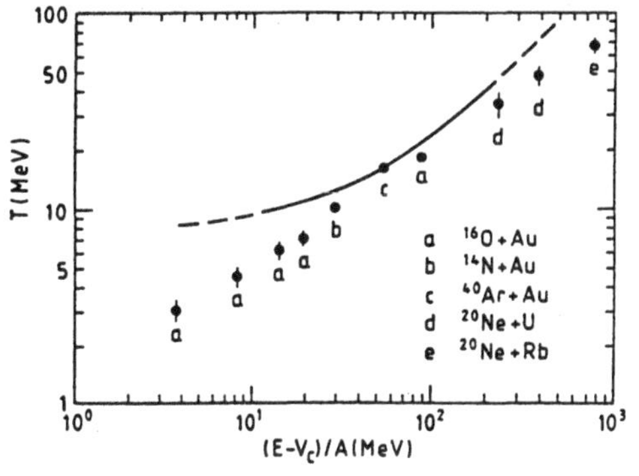

Fig. 2 : Experimental slope parameters of the intermediate-velocity source in proton-spectra, collected by ref. 1 for various heavy-ion systems as function of the bombarding energy E_0 over the Coulomb barrier V_C. The curve represents the estimate (1) obtained for quasifree nucleon-nucleon scattering.

what smaller σ_0. Secondly the phase space to be entered by the collisions is partly blocked, in particular at the low bombarding energies (E/A > 20 MeV), leading to a sensible reduction of momentum spreading. At high energies (E/A > 200 MeV) subnucleonic degrees of freedom may come into play.

In sum, the important point is that the slope parameter T of the intermediate velocity component does not necessarily reflect a temperature and hence a thermalized source but rather, by virtue of the first nucleon-nucleon collision, the projectile energy per nucleon. Actually it has been already pointed out by Sebille et al.[8] on the basis of Landau Vlasow calculations that there may be Fermi jet type preequilibrium particles with the appearance of thermal emission from a source of high temperature without anything like an equilibration being implied by their model. Even the promoters of the hot intermediate source noted[9] the possibility that a dominant part of the corresponding nucleon emission may be, in fact, caused by a single nucleon-nucleon collision. Later they abandoned this possibility because their calculations[10] of double-differential cross sections for quasifree nucleon-nucleon scattering, though satisfactory for measured spectra from $^{16}O + Au$, partly disagreed in the case of the $^{16}O + Al$ system. However, the discrepancies concern essentially those parts of the spectra which, in the $v_0/2$ system, correspond to velocities above $v_0/2 + v_F$ (with v_F as Fermi velocity) and thus are very sensitive[11] to the approximations made on these distributions and to other approximations made (PWBA, no Pauli blocking, etc.). For the bulk of the intermediate-velocity emission the calculations are rather compatible with the predominance of single nucleon-nucleon scattering.

Let us have also a look onto the multiplicity of the nucleons in the intermediate-velocity component. Reliable information on this multiplicity in specified reaction channels is scarce. For 30 MeV/nucleon ^{20}Ne on ^{165}Ho Hilscher et al.[12] deduced an intermediate-source neutron multiplicity of 2 for central collisions, and probably less for more peripheral ones. This is a representative result. Protons generally are somewhat less frequent, so there are at most 4 nucleons in the intermediate source, corresponding to 2 (quasifree) nucleon-nucleon scattering acts. This number does not appear unreasonable.

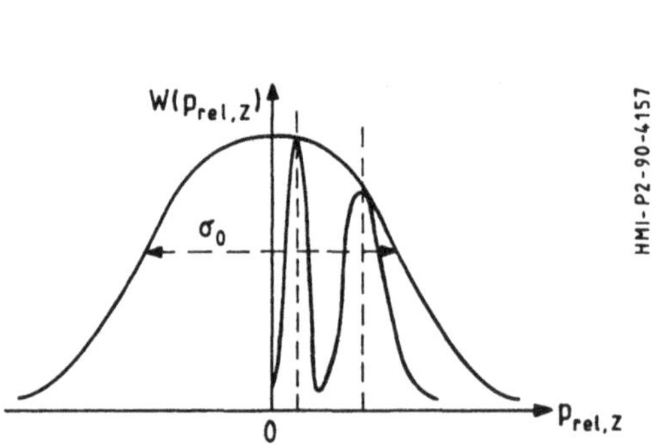

Fig. 3: The distribution of relative momentum between two clusters in the projectile due to the Fermi motion modulates the intensity of resonances present in the two-cluster system.

What about the **composite particles** observed with intermediate-source characteristics? With exception of the α-particles they are generally by an order of magnitude less frequent than nucleons (see for instance refs. 3, 4, 13). As far as **intermediate-mass fragments** are concerned, they display in the experimental correlations with heavy residues the characteristics of final light partners of strongly damped collisions[13,14]. The α-**particles**, in turn, are the main product of the sequential decay of excited projectile-like fragments after such collisions, but also a direct dynamic (non-thermal) projectile dissociation[15,16] contributes to the intermediate velocity component[17].

Now let us turn to the temperature parameters extracted from the **relative population of unbound fragment states** seen in light-particle correlations. Whatever projectile bombarding energy is used, they come out with about a constant value, 4 to 5 MeV. This intriguing feature is reminiscent of an other observation made in heavy-ion reactions: the momentum spreading of quasi-elastic projectile-like fragments, described by the reduced momentum width $\sigma_0=90$ MeV/c, comes out independent of projectile and bombarding energy[7,18]. (Most of the apparent energy dependence below 50 MeV/ nucleon disappears when transforming back to the projectile system on top of the barrier[18]). Also the isotope distributions stay remarkably constant[19], apart from a certain variation of the neutron excess attributed [20] to a changing neutron pick-up probability. These observations could not be reconciled with the idea of a thermal decay of excited projectile-like fragments[21], since this would imply an excitation independent of bombarding energy. Meanwhile it is accepted[7] that the observed distributions reflect essentially the probability distributions of the corresponding quantities within the projectile ground state. For the momentum dispersion this is provided by the statistical combination of the nucleon momenta implied by their Fermi motion and characterized by a spreading width $\sigma_0=90$ MeV/c. The A, Z distribution is determined by the probability to find a given number of neutrons/protons in certain subvolumes of the projectile.

When discussing the intermediate-velocity sources we have already emphasized that - at least for composite particles - fragments resulting from the projectile may play an important role [14,15]. This should also be valid for fragments in particle-unbound states appearing as resonances in two-fragment correlations and does not only concern the sequential population of resonant states [22], but - in contrast to the reception [23] of such arguments - also their direct production out of the projectile in its ground state. Taking ^{16}O projectiles as an example, direct transfer reactions like (^{16}O, ^{15}O), (^{16}O, ^{14}N), or (^{16}O,^{12}C) are established in which the ejectiles are not found exclusively in their ground states but also in excited states including particle-emitting states, and massive-transfer reactions like (^{16}O, ^{4}He) are established, too. So there should also be reactions like (^{16}O, ^{5}He), (^{16}O, $^{6}Li^*$) or (^{16}O, ^{8}Be), all giving rise to two correlated light fragments. More strictly, if one develops the ^{16}O ground state into two-cluster wave functions like ^{12}C x ^{4}He or ^{10}B x ^{6}Li (these components are sampled by the direct production of the respective fragments) it is not sufficient to include the configurations where the clusters are in their ground states. Principally one finds components with excited clusters also in the projectile ground state, prior to any excitation of the projectile.

If an excited fragment like, for instance, $^{6}Li^*$ consists in turn of two smaller clusters (d + α), their asymptotic relative-energy distribution with the well-known resonances samples the statistical distribution of relative energy (E_{rel}) as present between α and d clusters in the projectile ground state because of the Fermi motion. (This is shown schematically in fig. 3 and may be considered as an analog to final states in a stripping reaction like (d, p) where the intensity of the corresponding peaks is modulated by the neutron momentum distribution presented by the deuteron due to the Fermi motion). Approximating again the (three-dimensional) momentum distribution of each cluster by a Gaussian,

$$\exp(-p^2/2\sigma^2) = \exp(-Em_0/\sigma_0^2),$$

the two-fragment distribution is the product,

$$\exp[-(E_d + E_\alpha)m_0/\sigma_0^2] = \exp[-(E_{rel} + E_{cms})m_0/\sigma_0^2],$$

where E_{cms} is the kinetic energy of the center of mass of the two clusters. Thus the relative energy distribution has the same slope parameter σ_0^2/m_0 as the nucleons. It determines the peak ratio of resonances and looks like a Boltzmann distribution with a temperature of 8MeV, but is entirely due to the Fermi motion. We note that in a case where the correlation measurements were extended up to ejectile velocities close to that of the projectile, where quasi-elastic fragmentation as described prevails, the temperature parameters extracted[24] approach 8 or 9 MeV, indeed (see **fig. 4**).

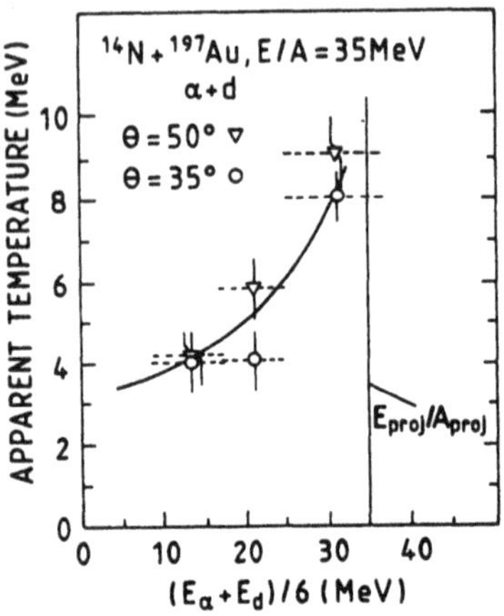

Fig. 4: The slope parameter T of the relative population of resonances approaches the Fermi motion value of ~8 MeV in the quasi-elastic fragmentation regime (data from ref. 24)

As far as the intermediate angles - the domain of most correlation measurements - are concerned, the tails of the quasi-elastic projectile fragmentation described are less negligible than usually supposed. They are, for instance, discernible[25] in the $(E_1 + E_2)$ spectra even at 35° for 35 MeV/nucleon ^{14}N on ^{197}Au. But clearly there are very important contributions at intermediate velocities. Rather astonishingly deep-inelastic collisions are found to persist at bombarding energies in the Fermi energy domain [10,26,27]. Many of the relevant features are reproduced by a realistic dynamical model[14] which treats the projectile as a classical compound of α-particles and follows the trajectories of all α-particles subject to mutual α-α interaction and to α-particle target interaction, the latter including a conventional friction force. The model[15] could qualitatively reproduce the energy damping of projectile fragments observed at large scattering angles. During the whole process of damping the projectile continuously is able to fragmentate in a manner similar to that in quasi-elastic reactions. The model [15] includes this dynamical fragmentation degree of freedom by virtue of the friction force acting to a very different degree on the projectile constituents. Thus it was also able to describe quantitatively the intermediate-velocity contribution to α-α correlations in a specific case[17]. As the model contains also an equivalent to Fermi motion, it is interesting to consider the influence of the motion on the relative-energy distribution of two-cluster fragments, as shown in **fig. 5b** for ^{8}Be ejectiles. The slope reflects essentially the "Fermi" motion of the two clusters in the projectile and grows only very weakly with increasing damping (decreasing ejectile energy), even though the excitation energy of the mother fragment ^{20}Ne increases notably (**fig. 5a**). These statements are only qualitative, but also in reality the excitation energy of the projectile-like fragment, rising due to the damping, will add a certain contribution to the slope parameter on top of the 8 MeV due to the Fermi motion. But there is an inverse effect caused by the sequential decay which becomes more important with rising excitation energy and populates preferentially the lower excited states/resonances. Even more important, the equilization of the neutron excess during the deep-inelastic collision leads to rather neutron-rich projectile-like fragments which perform a series of neutron evaporations and thus enrich the ground and low excited states/resonances. Yield ratio calculations [1,28] taking sequential feeding into account have shown that the lower excited states may well acquire relative abundances rising with increasing source excitation. We therefore believe it to be a reasonable hypothesis that the lower slope parameter found in the intermediate-velocity regime, T=4 to 5 MeV, as compared to the projectile fragmentation, is caused by the sequential decay which is more important after damping.

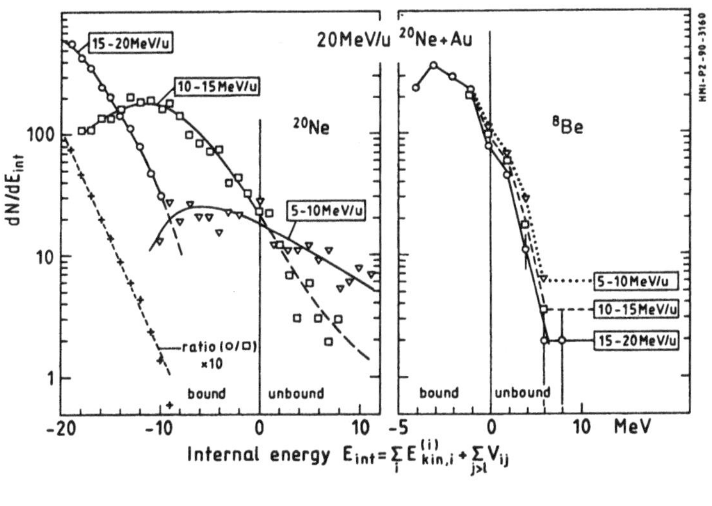

Fig. 5: *Distribution of internal energy (kinetic plus potential energy) of ^{20}Ne and ^{8}Be ejectiles calculated with the dynamical model[19] for different damping of the relative motion, indicated by intervals of ejectile energy. (Due to the classical nature of the model, the states are continuously distributed, and the ^{8}Be g.s. is bound).*

To **summarize**, we have outlined a reaction scenario which exploits a number of features established in nuclear reactions and which qualitatively explains the prominent pecularities found in socalled temperature measurements.

The scenario does not resort to anything like thermalization, but rather takes care of Fermi motion and sequential decay as determining, to a varying degree, the relative energy distributions in correlations of two light fragments. The (inclusive) nucleon emission cross sections, on the other hand, reflect rather the kinematics of the first projectile-nucleon target-nucleon collision. Of course, some of the features mentioned need to be substantiated by quantitative calculations, while the appeal of the outlined scenario is that it circumvents the contradictions present in the intermediate-source picture already on the level of qualitative considerations.

We are indebted to Dr. K. Möhring for valuable discussions.

References

1) Z. Chen, C.K. Gelbke, W.G. Gong, Y.D. Kim, W.G. Lynch, M.R. Maier, J. Pochodzalla, M.B. Tsang, F. Saint-Laurent, D. Ardouin, H. Delagrange, H. Doubre, J. Kasagi, A. Kyanowski, A. Péghaire, J. Péter, E. Rosato, G. Bizard, F. Lefèbvres, B. Tamain, J. Québert, Y.P. Viyogi, Phys. Rev. C36, 2297 (1987)

2) C.B. Chitwood, D.J. Fields, C.K. Gelbke, D.R. Klepsch, W.G. Lynch, M.B. Tsang, T.C. Awes, R.L. Ferguson, F.E. Obenshain, F. Plasil, R.L. Robinson, G.R. Young, Phys. Rev. C34, 858 (1986)

3) J. Pochodzalla, C.K. Gelbke, W.G. Lynch, M.R. Maier, D. Ardouin, H. Delagrange, H. Doubre, C. Grégoire, A. Kyanowski, W. Mittig, G. Bizard, F. Le-fèbvres, B. Tamain, J. Québert, Y.P. Viyogi, W.A. Friedman, D.H. Boal, Phys. Rev. C35, 1695 (1987)

4) D. Fox, D.A. Cebra, J. Karn, C. Parks, A. Pradhan, A. van der Molen, J. van der Plicht, G.D. Westfall, W.K. Wilson, R.S. Tickle, Phys. Rev. C 38, 146 (1988)

5) H.W. Barz, J.P. Bondorf, J.A. Lopez, H. Schulz, Phys. Rev. C37, 2910 (1988)

6) M. Blann, B. Remington, Phys. Rev. C 37, 2231 (1988)

7) C. Grégoire, B. Tamain, Ann. Phys. Fr. 11, 323 (1986)

8) F. Sebille, B. Remaud, Nucl. Phys. A420, 141 (1984)

9) T.C. Awes, G. Poggi, C.K. Gelbke, B.B. Back, B.G. Glagola, H. Breuer, V.E. Viola, Phys. Rev. C 24, 89 (1981)

10) T.C. Awes, S. Saini, G. Poggi, C.K. Gelbke, D. Cha, R. Legrain, G.D. Westfall, Phys. Rev. C 25, 2361 (1982)

11) J. Hüfner, M.C. Nemes, Phys. Rev. C 23, 2538 (1981)

12) D. Hilscher, H. Rossner, A. Gamp, U. Jahnke, B. Cheynis, B. Chambon, D. Drain, C. Pastor, A. Giorni, C. Morand, A. Dauchy, C. Stassi and G. Petit, Phys. Rev. C 36, 208 (1987)
13) D.J. Fields, W.G. Lynch, T.K. Nayak, M.B. Tsang, C.B. Chitwood, C.K. Gelbke, R. Morse, J. Wilczynski, T.C. Awes, R.L. Ferguson, F. Plasil, F.E. Obenshain, Phys. Rev. C 34, 536 (1986)
14) B. Borderie, M. Montoya, M.F. Rivet, D. Jouan, C. Cabot, H. Fuchs, D. Gardes, H. Gauvin, D. Jacquet, F. Monnet, Phys. Lett. B205, 26 (1988)
15) K. Möhring, T. Srokowski, D.H.E. Gross, H. Homeyer, Phys. Lett. B. 203, 210 (1988)
16) W. Terlau, M. Bürgel, A. Budzanowski, H. Fuchs, H. Homeyer, G. Röschert, J. Uckert, R. Vogel, Z. Phys. A 330, 303 (1988)
17) J. Uckert, A. Budzanowski, M. Bürgel, H. Fuchs, H. Homeyer, W. Terlau, Phys. Lett. B 206, 190 (1988)
18) W.A. Friedman, Phys. Rev. C 27, 569 (1983)
19) M. Buenerd, C.K. Gelbke, B.G. Harvey, D.L. Hendrie, J. Mahoney, A. Menchaca-Rocha, C. Olmer, D.K. Scott, Phys. Rev. Lett. 37, 1191 (1976)
20) H. Homeyer, M. Bürgel, Ch. Egelhaaf, H. Fuchs, G. Thoma, Z. Phys. A 319, 143 (1984)
21) C.K. Gelbke, D.K. Scott, M. Bini, D.L. Hendrie, J.L. Laville, J. Mahoney, M.C. Mermaz, C. Olmer, Phys. Lett. 70 B, 415 (1977)
22) P.D. Bond, R.J. de Meijer, Phys. Rev. Lett. 52, 2301 (1984)
23) Z. Chen, C.K. Gelbke, J. Pochodzalla, C.B. Chitwood, D.J. Fields, W.G. Lynch, M.B. Tsang, Nucl. Phys. B 186, 280 (1987)
24) C.B. Chitwood, C.K. Gelbke, J. Pochodzalla, Z. Chen, D.J. Fields, W.G. Lynch, R. Morse, M.B. Tsang, D.H. Boal, J.C. Shillock, Phys. Lett.
B 172, 27 (1986)
25) Z. Chen, C.K. Gelbke, J. Pochodzalla, C.B. Chitwood, D.J. Fields, W.G. Gong, W.G. Lynch, M.B. Tsang, Nucl. Phys. A473, 564 (1987)
26) Ch. Egelhaaf, H. Fuchs, A. Gamp, H. Homeyer, Phys. Rev. C 24, 759 (1981)
27) G. Imme, G. Raciti, Y. Cassagnou, R. Dayras, R. Legrain, E.C. Pollaco, A.D. Panagiotou, N. Papadakis, L. Rodriguez, F. Saint-Laurent, M.G. Saint-Laurent, N. Saunier, N. Vodinas, in "Third International Conference on Nucleus-Nucleus Collisions", St. Malo (France) 1988 Contributed Papers ed. by Estève et al., GANIL, Caen (France), p. 107 (1988)
28) D.J. Fields, C.K. Gelbke, W.G. Lynch, J. Pochodzalla, Phys. Lett. B 187, 257 (1987)

II. Correlations in Nuclear Matter at Equilibrium and Out-of-Equilibrium — Experimental and Theoretical Approaches (continued)

SHORT RANGE CORRELATIONS IN NUCLEAR MATTER AT EQUILIBRIUM *

A. Polls and J. Boronat
Departament d'Estructura i Constituents de la Matèria, Universitat de Barcelona, E-08028 Barcelona, Spain

A. Ramos
TRIUMF, 4004 Wesbrook Mall, Vancouver,B.C., Canada V6T 2A3

W. H. Dickhoff
Department of Physics, Washington University, St. Louis, Missouri 63130, USA

Abstract: The effects of short-range correlations on the single-particle spectral functions, the occupation probabilities and the pair distribution functions of nuclear matter are studied. Calculations have been performed in the framework of Green functions and variational theories, whose main characteristics are briefly discussed. Semirealistic interactions derived from the Reid's soft-core interaction have been used. The momentum distribution is calculated in several ways showing in all cases a depletion of about 13% around normal nuclear matter densities. The effects of the statistical and short-range correlations on the different pair distribution functions are compared.

---

* This research was supported by CAYCIT Grant No. PB85-0072-C02-00 (Spain) and by the Condensed Matter theory Program of the division of Materials Research of the U.S. National Science Foundation under Grant No. DMR-8519077 (at Washington University) which also provided computer time for the calculations which were partly performed at the Pittsburgh Supercomputer Center. Additional support was provided in part by NATO under Grant No. RG 85/0684.

1. Introduction

Since many years, the subject of correlations has been one of the main topics in nuclear physics. Even though this meeting is basically devoted to correlations between particles emitted in high energy heavy-ion collisions, which are mainly affected by final state interactions, it is interesting to report on the nuclear structure effects of nucleon-nucleon (N-N) correlations in equilibrium nuclear matter (NM) because of its relevance in all nuclear phenomena. The discussion is developped in the framework of microscopic many-body theories, namely Green functions [1] and correlated basis functions (CBF).[2]

The starting point is to consider a non-relativistic Hamiltonian

$$H = \sum_i \frac{-\hbar^2}{2\,m}\nabla_i^2 + \sum_{i<j} V(ij) + \sum_{i<j<k} V(ijk), \tag{1}$$

in which all the mesonic and other effects are absorbed in the two and three-nucleon potentials.[3–5] At present, neither $V(ij)$ nor $V(ijk)$ are completly known. Realistic models of $V(ij)$ have a long range part given by the one pion exchange potential and a shorter range part determined by fitting experimental phase shifts. The short-range behaviour of any realistic interaction is strongly repulsive and gives origin to the so called short-range correlations (SRC), which are the most important correlations for the bulk properties of nuclear systems and will constitute the central subject of this talk.

The SRC are to a large extent the main responsible of the breakdown of the mean field picture of nuclei, which has been corroborated by the recent analysis of the exclusive (e,e'p) electron scattering data [6,7] ,i.e. of the presence of highly excited 2p1h (two particle-one hole) configurations producing a decrease in the shell-model occupation probabilities of the states below the Fermi level and a fragmentation of the single particle (sp) strength. The "depletion" and "fragmentation" of the sp states can in turn be related to the "quenching" and "spreading" of the quasielastic peak of the longitudinal response which is measured in the inclusive (e,e') experiments [8,9]. These two facts provide an unambiguous signature of correlation effects in nuclei. Relevant theoretical quantities to analyze these experiments are therefore the single particle spectral functions and the occupation probabilities which have been recently investigated[10–12] . Other quantities naturally affected by the N-N correlations are the pair distribution functions, which give the probability to find two particles at a given distance. The strong repulsion of the interaction at short distances precludes two particles of being to close to each other, in consequence the distribution functions present a "hole" at short distances. [13]

The main purpose of this contribution is to discuss, in a rather qualitative way, the effects of the SRC on the quantities mentioned above. The techniques employed to study the many-body system are based either in perturbation or variational methods and all of them must deal with the hard-core of the N-N interaction. In this paper, results for nuclear matter concerning both perturbative and variational methods are presented. In the next section we briefly sketch the formalism needed for the analysis and discuss the influence of SRC on the spectral functions and momentum distributions, which in this case have been calculated in the framework of Green functions formalism [10]. Sect. 3 contains results for the one-body density matrix and the different pair distribution functions calculated in the framework of variational theory.[14–15]

The calculations have been performed for a two-body model interaction called v_2 which is defined as the central part of the $^3S_1 - ^3D_1$ channel of the Reid soft core potential. This interaction has many features in common with a Lennard-Jones type interaction. It is also useful to introduce the potential $v_2^{l=0}$ which is the v_2 interaction allowed to operate only in S-waves. The potential $v_2^{l=0}$ leads to a sp energy spectrum similar in shape to those obtained for realistic nuclear interactions although it is shifted to positive energies.

2. Short range correlations and spectral functions

The hole spectral function $S_h(k,\omega)$ is the probability of removing a particle with momentum **k** from the target system of A particles leaving the resulting (A-1) system with an energy $E^{A-1} = E_0^A - \omega$. Analogously, the particle spectral function $S_p(k,\omega)$ is the probability of adding a particle with momentum **k** and leaving the resulting (A+1) system with an energy $E^{A+1} = \omega + E_0^A$. The Lehmann representation , which expresses the sp propagator in terms of the spectral functions, gives

$$S_h(k,\omega) = \sum_n |< \psi_n^{A-1} \mid a_k \mid \psi_0^A >|^2 \, \delta(\omega - (E_0^A - E_n^{A-1})), \tag{2}$$

$$S_p(k,\omega) = \sum_m |< \psi_m^{A+1} \mid a_k^\dagger \mid \psi_0^A >|^2 \, \delta(\omega - (E_m^{A+1} - E_0^A)), \tag{3}$$

where $\mid \psi_0^A >$ is the ground state of the target, and $\mid \psi_n^{A-1} >$ and $\mid \psi_m^{A+1} >$ are excited states of the A-1 and A+1 systems with energies E_n^{A-1} and E_m^{A+1} respectively.

In an infinite system of independent particles moving in an average field, each nucleon of momenta k has a well defined energy $\epsilon(k)$ and, therefore, the corresponding spectral function reduces to a delta function $\delta(\omega - \epsilon(k))$. In the presence of correlations this strength is distributed over several energies as a result of the interaction which couples the sp motion to more complicated states. The shape of the spectral function is then broadened and its strength redistributed with respect to the δ-function behaviour of the free system.

In a correlated many-body system the sp propagator is determined by the interaction with the other particles by means of the self-energy as it is implied by the Dyson equation. Since the sp self-energy, describing the modified properties of a particle in the medium (dressed particle), is determined from the two body effective interaction between dressed particles, and this requires in turn, the knowledge of the propagator, we are dealing with a coupled problem which must be solved self-consistently. This procedure leads to the so called Self-Consistent Green's Function (SCGF) theory.[10] Due to the presence of the strong repulsion in the N-N interaction the minimum meaningful approximation to the effective interaction is the ladder approximation which in the SCGF formalism will include particle-particle as well as hole-hole propagation. In principle, a complete self-consistent calculation implies the use of fully dressed sp propagators in the solution of the ladder equation for the effective interaction. At present, self-consistency has been required only for the quasi-particle energy, i.e., at the level of the real on-shell part of the self-energy

$$\epsilon(k) = \frac{k^2}{2\,m} + Re\Sigma(k, \epsilon(k)). \tag{4}$$

When this averaged self-consistency is achieved, the complete energy dependence of the self-energy can be studied.

Using the Lehmann representation and the Dyson equation one obtains the following equations

$$S_h(k,\omega) = \frac{1}{\pi}\frac{Im\Sigma(k,\omega)}{(\omega - \frac{k^2}{2m} - Re\;\Sigma(k,\omega))^2 + (Im\;\Sigma(k,\omega))^2} \qquad \omega < \epsilon_F \tag{5}$$

$$S_p(k,\omega) = -\frac{1}{\pi}\frac{Im\;\Sigma(k,\omega)}{(\omega - \frac{k^2}{2m} - Re\;\Sigma(k,\omega))^2 + (Im\;\Sigma(k,\omega))^2} \qquad \omega > \epsilon_F \tag{6}$$

which show that the spectral functions are completly determined by the full energy dependence of the self-energy. The inclusion of hh propagation in the ladder equation allows the coupling of sp degrees of freedom to 2h1p (two hole-one particle) and more complicated states and therefore it is a necessary condition to calculate $S_h(k,\omega)$ in this approach. From the hole and particle spectral functions one can also calculate the occupation probability

$$n(k) = \int_{-\infty}^{\epsilon_F} d\omega\; S_h(k,\omega) \tag{7}$$

and the depletion

$$1 - n(k) = \int_{\epsilon_F}^{\infty} d\omega \; S_p(k,\omega), \tag{8}$$

respectively. The depletion of the Fermi sea implies that for a given $k < k_F$ a fraction of strength has been shifted to energies above ϵ_F. Since the number of particles must be conserved, i.e. the integral over n(k) is constant, the $S_h(k,\omega)$ should show strength for momenta $k > k_F$, in particular at energies far below ϵ_F.

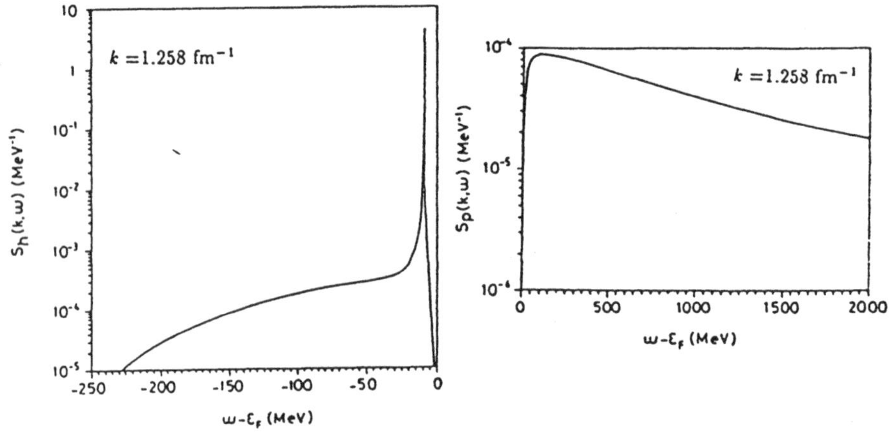

Fig.1. Hole and particle spectral functions at $k_F = 1.4\ fm^{-1}$ for $k = 1.258 fm^{-1}$.

Results for the spectral functions are given in Fig.1 for $k = 1.258 fm^{-1}$ at a density corresponding to $k_F = 1.4\ fm^{-1}$. Due to the correlations, the strength is distributed along a large energy interval. As this momentum is smaller than k_F, most of the strength is kept by $S_h(k,\omega)$. In fact, $n(k = 1.258 fm^{-1}) = 86\%$ (see Eq.(7)). The remaining strength is moved beyond ϵ_F and belongs to $S_p(k,\omega)$. This shows that it is possible to add a particle in a sp-state which would be fully ocuppied in a mean field picture. $S_p(k,\omega)$ is smeared out over a very large energy domain which is intimately connected with the character of the interaction which couples sp degrees of freedom to very high-lying 2p1h, etc. states. This removal of sp strength to very high energy is observed for all momenta in the same amount emphasizing the short range nature of the effect (see also $S_p(k,\omega)$ in Fig. 2). The $S_h(k,\omega)$ shows, for $k < k_F$, a sharp peak (quasi-particle peak) located at the quasi-particle energy that concentrates most of the hole-strength (83% for

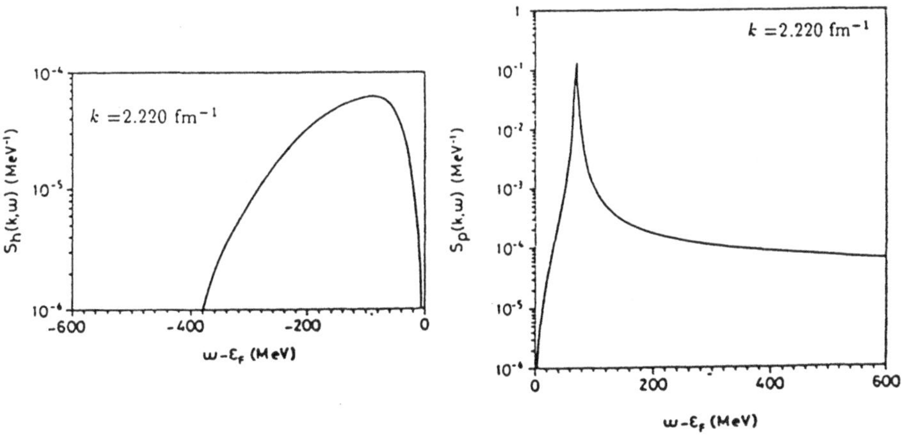

Fig.2. Hole and particle spectral functions at $k_F = 1.4 fm^{-1}$ for $k = 2.22 fm^{-1}$.

$k = 1.258 fm^{-1}$). When k approaches k_F the quasi-particle peak becomes sharper the closer the momenta is to k_F. In the limit $k = k_F$, the peak is just a δ-function whose strength $(Z(k_F))$ defines the discontinuity of the momentum distribution at k_F.

Fig. 2 shows the spectral functions for $k = 2.22 fm^{-1}$. As k becomes larger than k_F, the quasi-particle peak moves to energies larger than ϵ_F, i.e. belongs to $S_p(k,\omega)$, and contributes to the depletion (see Eq. (8)). In this case, $1 - n(k = 2.22 fm^{-1}) = 98.9\%$, and the strength below the quasi-particle peak is 85.7%. A smooth distribution is observed for $S_h(k,\omega)$ indicating the possible ocurrence of knock-out processes from states above the Fermi sea, which would be empty in an uncorrelated system. The integration of $S_h(k,\omega)$ gives $n(2.22\ fm^{-1}) = 1.1\%$ The independent calculation of $n(k)$ (Eq.(7)) and $1-n(k)$ (Eq. (8)), constitutes a check to the accuracy of the calculation. The hole spectral function has also recently been calculated using CBF theory for the Urbana V_{14} interaction.[11,12] A comparison of the spectral functions calculated with the two completly different methods shows many similarities. The main difference is the somewhat stronger broadening for momenta below k_F in the CBF calculation since a force was used which is also realistic at medium and long-range. For momenta above k_F the hole spectral functions are practically identical. This shows that the SRC are the main responsible for the high momentum components in the wave function.

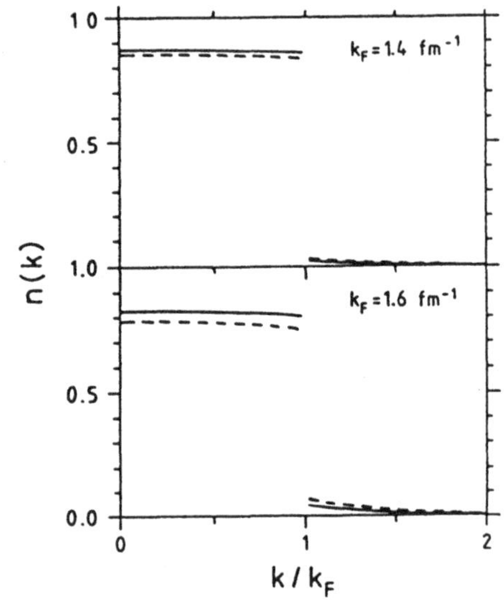

Fig.3. Momentum distribution at $k_F = 1.4 fm^{-1}$ and $k_F = 1.6 fm^{-1}$ as calculated from Eq. 7 (full line). The dashed line refers to a Goldstone expansion which includes pp as well as hh propagating terms in the ladder equation.

The momentum distributions produced by the $v_2^{l=0}$ interaction are shown in Fig.3 for densities corresponding to $k_F = 1.4 fm^{-1}$ and $1.6 fm^{-1}$. The continuous line corresponds to the calculation using the hole spectral function (Eq. (7)) and the dashed line corresponds to a Goldstone expansion of the expectation value of the occupation number operator in terms of pp and hh ladder diagrams.[16]

Due to the correlations there is an appreciable amount of particles promoted outside of the Fermi sea. As a reasonable measure of those correlations one can consider the 13% (18%) depletion for $k_F = 1.4\ fm^{-1}(1.6 fm^{-1})$ at $k = 0$. The SRC are the main responsible for the tail of n(k). The depletion of the states below k_F increases when the density increases. This is in agreement with the intuitive notion that the particles will experience their mutual repulsion at shorter distance more frequently at higher density which results in an increase of high momentum components in the ground state. As a consequence of this result a decrease of the discontinuity at k_F is obtained with increasing density ($z(k_F = 1.4 fm^{-1}) = 0.83, z(k_F = 1.6 fm^{-1}) = 0.75$). The smaller value for $z(k_F)$ obtained with realistic interactions can be ascribed to the tensor correlations whose influence is specially important in redistributing sp strength around k_F. Preliminary results for the Reid soft core obtained in the framework of SCGF give a depletion of 18% at $k = 0$.[19] Almost all the existing calculations for realistic interactions give a very similar value for $n(0)$ independently of the interaction or the method used for the calculation [11,12,17–19] and should provide a reasonable comparison with occupation numbers obatined in finite nuclei.[7]

3. Short range correlations and the pair-distribution function

Another alternative to study the many-body problem is based on the variational method. In this case, instead of building the correlations perturbatively, one starts with a trial wave function:

$$\psi(NM) = (S \prod_{i<j} F_{ij})\, \phi_{FG}, \tag{7'}$$

where ϕ_{FG} is the free Fermi gas (FG) wave function and F_{ij} is a correlation operator,which incorporates to the wave function the correlations induced by the potential. S is a symmetrizer. The operatorial and radial dependence of the correlation operator is linked to the structure of the interaction. For a realistic potential F_{ij} contains central, tensor and other correlations. For the model potential v_2, it is enough to consider a Jastrow type wave function,

$$\psi(NM) = \prod_{i<j} f(r_{ij})\phi_{FG}. \tag{8'}$$

The next steps are to choose a proper $f(r_{ij})$ and to learn how to calculate with this wave function. The value of $f(r_{ij})$ is almost zero inside the hard-core potential, so as to avoid two particles being too close and approaches unity for large r, so as to ensure the cluster decomposition property.[20] The expectation value $< H >$ of the Hamiltonian with respect to $\psi(NM)$ is to be calculated and minimized with respect to the choice of $f(r)$. Ideally, this involves the solution of the Euler equation $\delta < H > /\delta f(r) = 0$, which determines the optimal Jastrow correlations. A common alternative, less demanding, is to adopt a simple parametrized form for $f(r)$. In this case we have taken a parametrized correlation function which has previously been used in the literature and whose paremeters have been chosen so as to minimize $< H >$ at each density.[21] Once a wave function has been specified one has to face the calculation of the desired expectation values. The Fermi hypernetted-chain (FHNC) procedure allows one to calculate a given expectation value (e.g. n(k), the pair distribution function, etc.) by solving a set of coupled,

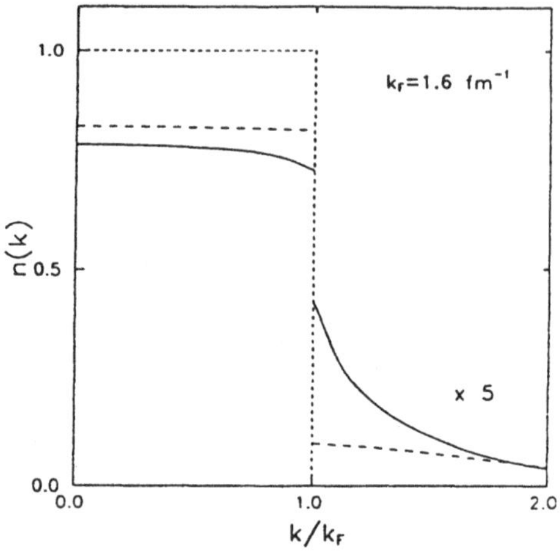

Fig. 4. Momentum distribution produced by V_2 at $k_F = 1.6 fm^{-1}$ calculated in FHNC/0 approximation (dashed line). The continuous line refers to a Goldstone expansion which includes propagation of particles and holes.

nonlinear integral equations which resums cluster expansions in terms of $f^2 - 1$ or $f - 1$.

Fig.4 shows the momentum distribution produced by v_2 at a density corresponding to $k_F = 1.6 fm^{-1}$, calculated with the variational wave function (dashed line) in FHNC/0 approximation [15] and with the Goldstone diagram expansion[16] (continuous line). The variational calculation gives $n(0) = 0.825$ and $z(k_F) = 0.80$, the momentum distribution is almost flat below k_F. The correlated wave function is too simple to describe realistically the properties of particles close to the Fermi surface and corrections based on perturbation theory using CBF, should be implemented in order to obtain the correct analytic behaviour at k_F.[22] The kinetic energy obtained by integrating the momentum distribution is 73.5 Mev, the difference with the kinetic energy (31.8 Mev) of a FG at the same density can be ascribed to the correlations.

As it has been said in the introduction, another quantity strongly affected by SRC is the pair distribution function which is the correlation function between the position of two particles and can be defined as the following expectation value:

$$g(r_{12}) = \rho^{-2} \sum_{i<j} < \psi_0 \mid \delta(\mathbf{r}_1 - \mathbf{r}_i)\, \delta(\mathbf{r}_2 - \mathbf{r}_j) \mid \psi_0 >, \qquad (9)$$

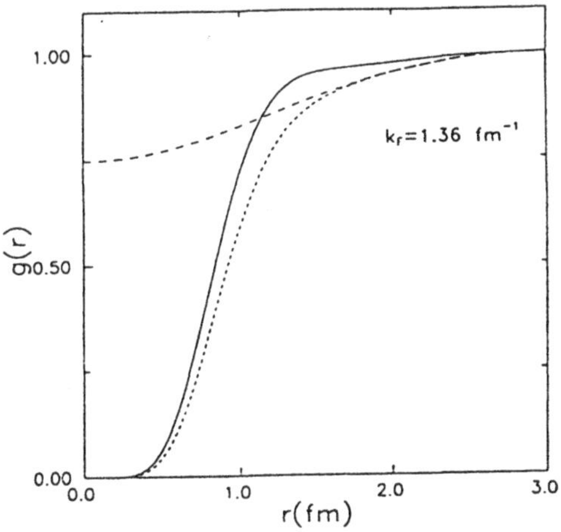

Fig.5. Pair distribution function produced by V_2 at $k_F = 1.36 fm^{-1}$ calculated in different approximations . See the text for further explanations

where $| \psi_0 >$ stands for the ground state of the system and the translationally invariance of the NM has already been taken into account and therefore the pair distribution function depends only on the distance between the two particles. The normalization has been chosen such that $g(r) \rightarrow 1$ when $r \rightarrow \infty$. In the variational approach, $| \psi_0 >$ is approximated by the trial wave function of Eq.(7). Introducing the appropiate spin-isospin projection operators , one obtains the different spin-isospin pair distribution functions $g^{\tau_1 \tau_2}_{\sigma_1 \sigma_2}(r)$. Afterwards, g(r) is expressed as the average in the spin-isospin space of the different $g's$,

$$g(r) = \frac{1}{\nu^2} \sum_{\sigma's,\tau's} g^{\tau_1 \tau_2}_{\sigma_1 \sigma_2}(r), \tag{10}$$

where ν is the level degeneracy (i.e., $\nu = 4$ for nuclear matter).

Fig.5 shows $g(r_{12})$ produced by v_2 at a density corresponding to $k_F = 1.36$-fm^{-1} when the Jastrow variational wave function (Eq. (8)) is used. The long-dashed line corresponds to $g(r)$ for a FG and therefore only contains the statistical correlations which already produce a "hole" to $g(r)$ at the origin. The short-dashed line stands for a calculation of $g(r)$ to second order of the cluster expansion of $g(r)$,

$$g^{(2)}(r) = f^2(r)\, g_{FG}(r), \tag{11}$$

where $g_{FG}(r)$ is the distribution function of a Fermi gas. The continuous line has beeen obtained in the FHNC/0 approximation, i.e. neglecting the so called elementary diagrams. The difference between the continuos and the short-dashed line measures the correlations effects of the remaining particles which are not included in $g_{FG}(r)$. The effect of the SRC is to produce a deeper hole in $g(r)$ at the origin. The many-particle correlation give some structure in the region where the potential is atractive. When the density increases this structure transforms in a well defined peak.

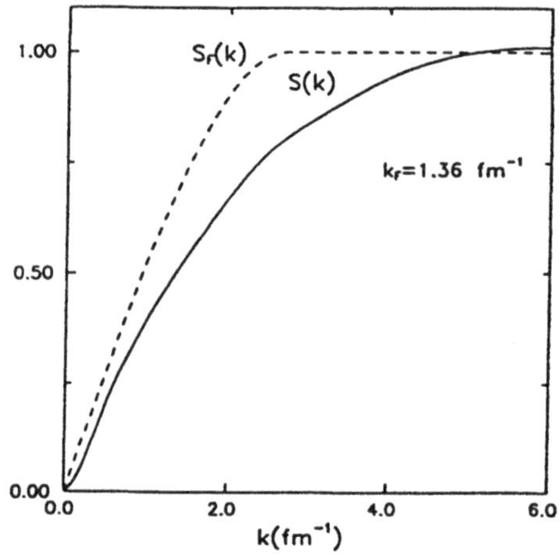

Fig. 6. Structure functions produced by v_2 at $k_F = 1.36 fm^{-1}$. The continuos line refers to the FHNC/0 approximation. The dashed line corresponds to to s(k) for FG.

One can also look at the distribution functions in the momentum space by studying the corresponding structure functions:

$$S(k) = 1 + \rho \int d^3\mathbf{r}\; e^{i\mathbf{k}\cdot\mathbf{r}}(g(r) - 1). \tag{12}$$

In the case of the proton-proton distribution function, it is known that it can be extracted from the longitudinal structure function $S_L(k)$ obtained by integrating respect to the energy loss the longitudinal response $S_L(k, \omega)$ which is measured in inclusive (e,e') experiments.[13] The structure functions are shown in Fig. 6, also for $k_F = 1.36 fm^{-1}$. The FG approximation (dashed line) is equal to 1 for $k > 2\ k_F$, the correlated $S(k)$ (continuous line) shows structure at large momenta, i.e. the "hole" of $g(r)$ is reflected in smooth oscillations of $S(k)$ at large values of k.

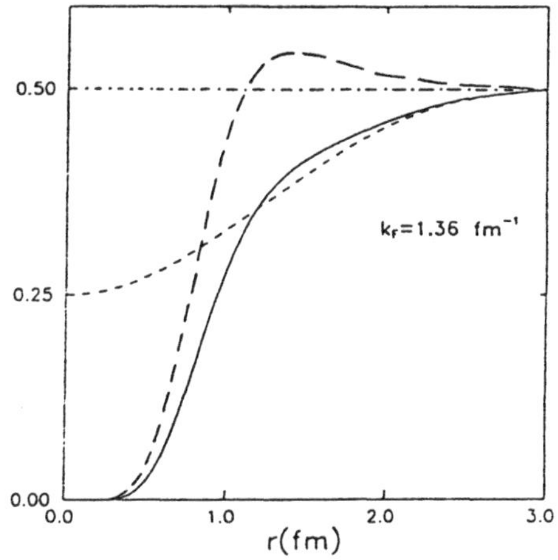

Fig. 7. Different pair distribution functions produced by v_2 at $k_F = 1.36 fm^{-1}$. The continuous, short-dashed,long-dashed and dashed-point lines correspond to $g^{\uparrow\uparrow}(r)$, $g_{FG}^{\uparrow\uparrow}(r)$, $g^{\uparrow\downarrow}(r)$, $g_{FG}^{\uparrow\downarrow}(r)$ respectively.

The distribution functions for particles with parallel or antiparallel spin are presented in Fig. 7. The degeneracy factor $1/\nu^2$ has been absorbed in the definition of $g^{\uparrow\uparrow}(r)$ and $g^{\uparrow\downarrow}(r)$, in such a way that $g(r) = g^{\uparrow\uparrow}(r)+g^{\uparrow\downarrow}(r)$. The influence of the SRC is larger in $g^{\uparrow\downarrow}(r)$. In fact in a FG there is no correlation between particles of opposite spin, and with the prescription defined above, its value is constant and equal to 0.5. On the contrary, the statistical correlations are present in $g_F^{\uparrow\uparrow}(r)$ and the effect of SRC is smaller.

Summarizing, some illustrative applications of two relevant microscopic many-body techniques have been briefly discussed. Although the problem of nuclear matter is not completly solved, there has been an enormous progress in the last years and sophisticated microscopic many-body techniques are available. Now, they should be used in more complex situations, such as finite nuclei or also situations far from equilibrium. Finally non-nucleonic degrees of freedom and relativistic effects should be incorporated to the theory whenever they are necessary.

References

1) A.L. Fetter and J.D. Walecka, *Quantum Theory of Many-Particle Systems*, (Mc Graw-Hill, 1971).
2) E. Feenberg, *Theory of Quantum Fluids*,(Academic Press, New York,1969).
3) R.V. Reid, *Ann. Phys.* **50**,411(1968).
4) R.B. Wiringa, R.A. Smith and T.L. Ainsworth,*Phys. Rev.* **C29**,1207(1984).
5) R. Schiavilla, V.R. Pandharipande and R.B. Wiringa,*Nucl. Phys.* **A449**,-219(1986).
6) E.N.M. Quint et al., *Phys. Rev. Lett.* **58**,1088(1987).
7) P.K.A. de Witt Huberts, *J. Phys.*G**16**,507(1990).
8) C.C. Blatchley et al., *Phys. Rev.* **C34**,1243(1986).
9) A. Polls, A. Ramos, W.H. Dickhoff, in *Topical Workshop on Two Nucleon Emission Reactions*, Proc. EIPC, Vol. 3. O. Benhar and A. Fabrocini ,eds. (Giardini, Pisa, 1990) in press.
10) A. Ramos, A. Polls and W.H. Dickhoff, *Nucl. Phys.* **A503**,1(1989).
11) O. Benhar, A. Fabrocini, S. Fantoni, *Nucl. Phys.* **A505**,267(1989).
12) O. Benhar, A. Fabrocini, S. Fantoni, *Phys. Rev.* **C41**, R24(1990).
13) R. Schiavilla et al., *Nucl. Phys.* **A473**,267(1987).
14) S. Fantoni and S. Rosati, *Nuovo Cimento* **A25**,593(1975).
15) S. Fantoni, *Nuovo Cimento* **A44**,191(1978).
16) A. Ramos, W. Dickhoff and A. Polls, preprint 1990.
17) P. Grange, J. Cugnon and A. Lejeune, *Nucl. Phys.* **A473**,365(1987).
18) M. Baldo at al., *Phys. Rev.* **C41**,1748(1990).
19) B. E. Vonderfecht, A. Ramos, A. Polls and W. H. Dickhoff, work in progress.
20) J. W. Clark, *Prog. Part. Nucl. Phys.* **2**,89(1979); and references therein.
21) O. Benhar et al., *Phys. Lett.* **B64**,395(1976).
22) S. Fantoni and V. R. Pandharipande, *Nucl. Phys.* **A427**,473(1984).

σ AND ρ MESON STRENGTH DISTRIBUTIONS FROM IN MEDIUM CORRECTED Π-Π CORRELATIONS AND THE NUCLEAR EQUATION OF STATE

G. CHANFRAY,
Institut de Physique Nucléaire, 43, Bd du 11 Novembre 1918
F-69622 VILLEURBANNE-CEDEX,

Z. AOUISSAT, P. SCHUCK
Institut des Sciences Nucléaires, IN2P3-CNRS et Université Joseph Fourier,
53 Avenue des Martyrs, F-38026 GRENOBLE-CEDEX

and

W. NÖRENBERG
GSI, Postfach 110552, D-6100 DARMSTADT, 11

INTRODUCTION

The most important part of the intermediate range nucleon-nucleon attraction is attributed to the exchange of non interacting as well as interacting two pions in the $I = J = 0$ channel. The corresponding 2π exchange potential, or part of it, is often simulated by a sigma-meson exchange [1] with mass of about 600 MeV, which is then used as a basic input to nuclear matter calculation. This may give in particular a justification to the popular σ-ω model [2]. However a number of question remains concerning the microscopic nature of such an effective sigma meson. The point we want to emphasize in this paper concerns the possible modification of the sigma in the medium since we think that it is made out of pions whose propagations are strongly altered by the medium (see also [3]). However also the ρ-meson may undergo similar in medium corrections which possibly are related to recent experimental findings [4].

We first will discuss, with a very simple model, the problem of pion propagation and pion-pion interaction in the medium and the occurence of highly collective pion-delta modes (pionic branch, pisobar, quasi pion) in agreement with experimental data [5, 6]. They also should have some very important consequences on the dynamics of relativistic heavy collision [7]. We then address the very important question of the nuclear matter equation of state where collective pionic effects are included to reconstruct an effective sigma exchange whose mass and coupling constant are density dependent.

A SIMPLE MODEL FOR PIONIC MODES IN NUCLEAR MATTER

The propagation of a pion in nuclear matter differs from its free propagation since it is self-interacting with the medium. The pion propagator has the form

$$D_\pi(\mathbf{k},\omega;\rho) = (\omega^2 - k^2 - m^2_\pi - \Sigma\ (\mathbf{k},\omega;\rho))^{-1} \qquad (1)$$

In the high energy domain ($\omega > m_\pi$) it is sufficient to consider, as a first reasonable approximation, contributions to the self-energy Σ ($\mathbf{k}$; $\omega;\rho$) arising from the p-wave coupling of the pion to virtual-delta-hole states. In addition it is very important to take into account short range correlations usually characterized by the parameter $g'_{\Delta\Delta} \cong$ 0.5 and leading to a screening of the delta polarization

$$\Sigma = k^2\ \Pi^o(k,\ \omega;\rho)/(1- g'_{\Delta\Delta}\ \Pi^o(k,\ \omega;\rho))$$

where Π^o is given by :

$$\Pi^o = \frac{4}{9}\left(\frac{f^*_{\pi N\Delta}}{m_\pi}\right)^2 (v_{\pi N\Delta}(k))^2\rho\ \left[\frac{1}{\omega-\varepsilon_{\Delta k}+i\eta} - \frac{1}{\omega+\varepsilon_{\Delta k}}\right] \qquad (2)$$

In the above formulas $f^*_{\pi N\Delta}$ =2.0 and $v_{\pi N\Delta}$ (k) are respectively the $\pi N\Delta$ coupling constant and the form factor. We have assumed that the nucleons are static (i.e. $\varepsilon_{\Delta k} = \sqrt{M_\Delta{}^2 + k^2} - M_N$) and have neglected delta width and coupling to 2p-2h states. These approximations should not modify very much the total strength carried by pion-states but just reduce the spreading of the strength distribution. It is then straightforward to show that the pion propagator writes :

$$D_\pi(\mathbf{k},\omega;\rho) = \frac{Z_1(\mathbf{k};\rho)}{\omega^2-(\Omega_1(\mathbf{k},\rho))^2} + \frac{Z_2(\mathbf{k},\rho)}{\omega^2-(\Omega_2(\mathbf{k},\rho))^2} \qquad (3)$$

where Ω_1,Ω_2 ($\Omega_1 < \Omega_2$) and Z_1, Z_2 ($Z_1 + Z_2 = 1$) are very simple analytic functions of **k** and ρ. We see that due to its coupling to deltas the pion is no longer an eigenstate of the system but becomes a mixture of two collective pion-delta states with strength Z_1 (Z_2) for the lower (upper) branch $\Omega_1(\Omega_2)$ (see Fig. 1). At low momentum the lower branch is purely pionic whereas the upper branch is mainly made out of delta and the two branches exchange their structure at a certain momentum transfer of order 2-3 m_π c. The lower branch (often called pionic branch) is space-like and is in principle accessible by looking at the longitudinal spin-isospin response function on the delta regions. In spite of its simplicity this model gives the position and the strength of the pionic branch very close to the results of much more detailed calculations. On the other hand, the recent (^{3}He,T) experiment [5] seems to give a clear indication in favor of the existence of these pionic modes [6].

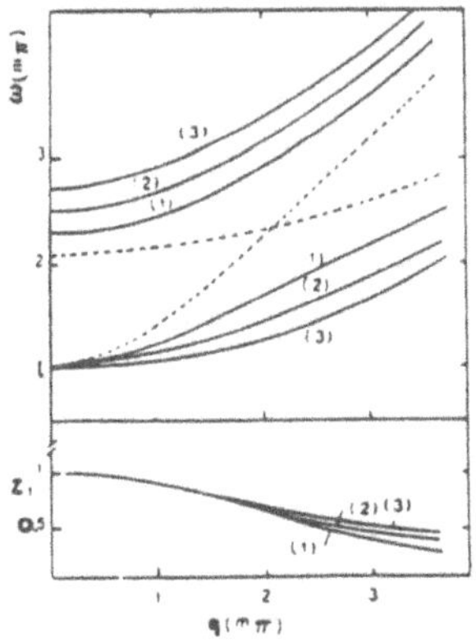

Fig. 1

Ω_1, Ω_2, Z_1 at density ρ = 0..5 ρ_o(1), $\rho = \rho_o$(2), ρ = 1.5 ρ_o(3)

IN MEDIUM Π-Π PROPAGATION

We now turn to the very interesting problem of the possible modification of the pion-pion interaction in the medium [7] . This is very important since we know that correlated pion-exchange gives an important contribution to the intermediate range nucleon-nucleon interaction. The point is that in the process of generating the in medium pion-pion T-matrix, intermediate two-pion states have to be replaced by two-quasi pion states whose propagators are given by equation (3).

It follows that the free two-pion propagator

$$G^o(k,E) = \frac{1}{\omega_k} \frac{1}{E^2 - 4\omega_k{}^2 + i\,\eta} \tag{4}$$

where E is the total energy in the CM system and **k** the momentum of a pion has to be replaced by its in medium analogue :

$$G(\mathbf{k},E) = \frac{(Z_1(k))^2}{\Omega_1(k)} \frac{1}{E^2-4(\Omega_1(k))^2+i\eta}) + \frac{(Z_2(k))^2}{\Omega_2(k)} \frac{1}{E^2-4(\Omega_2(k))^2+i\,\eta}$$

$$+ Z_1(k)\ Z_2(k) \frac{\Omega_1(k)+\Omega_2(k)}{\Omega_1(k)\ \Omega_2(k)} \frac{1}{E^2-(\Omega_1+\Omega_2)^2+i\eta} \tag{5}$$

Due to the fact that $\Omega_1(\mathbf{k})$ is considerably decreased with respect to $\omega_{\mathbf{k}} = (k^2 + m_\pi{}^2)^{1/2}$ we expect that an accumulation of strength will occur at low energy $E > 2m_\pi$. We illustrate this feature by calculating the strength function Im $T_{\pi\pi}$ using two different models fitting π-π phase shifts up to $E = 7\ m_\pi$. The model I (ref. 8) consists of an energy dependent separable potential

$$V(q,q'.E) = g\ v(q)\ v(q')/(E-M) \tag{6}$$

with $g = 0.755\ m_\pi{}^{2-1/2}$, M = 897 MeV and $v(q) = (1 + q^2/q^2{}_c)^{-1}$,$q_c$=2.7 m_π. The model II /9/ treats the π-π channel with a double separable potential and explicit ly includes coupling to $K\bar{K}$ channel which is supposed to be responsible for the existence of the $S^*(993)$ resonance. The results of the calculations are shown in Fig. 2.

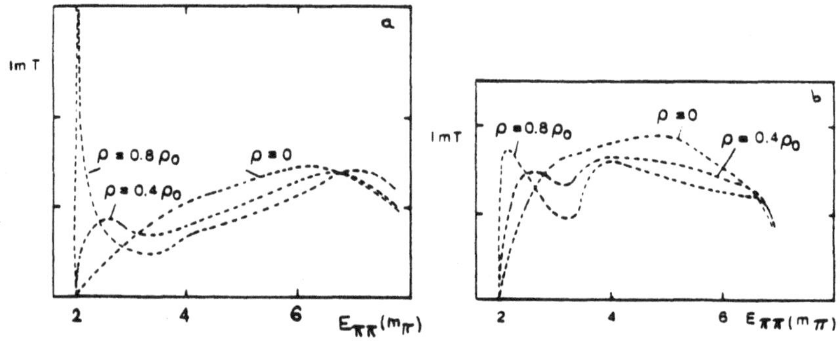

Fig. 2
Strength function for model I (2a) and model II (2b) for free pion and in medium pions at density $0.4\rho_o$ and $0.8\ \rho_o$.

We see that in model I a sharp resonance appears at low energy. Medium effects are even able to produce a bound state at E = 1.97 m_π for normal nuclear matter density. In model II the reshaping of the strength function is still present though attenuated. Clearly the result may depend on the off-shell extrapolation of the π-π interaction and these uncertainties should be removed with a more microscopic model. It is, however, manifest that medium effects should influence the correlated part of the two-pion exchange potential which is usually simulated by a simple σ exchange with mass and coupling constant without any density dependence. We also mention that unexpected peaks show up in (π,2π) data [10] and resist theoretical explanation even though all standard medium effects are correctly taken into account. A possible interpretation is the decay in flight of the above resonance which might be produced in the medium when the incident pion scatters on a virtual pion in the nucleus. This would constitute another experimental signature of collective pisobar states in nuclei.

It is straightforward to incorporate the in medium changes also into the ρ-channel. We used for this the relativistic 2π propagator (5) and readjusted the parameters of the separable force in such a way as to reproduce the ℓ = 1, I = 1 phase shifts (M = 826.7 Mev; $g = 0.505\ m_\pi{}^{-1/2}$; q_0 = 0.428 fm). In Fig. 3 we show the corresponding strength distribution for the free case (ρ = 0), for the case of nuclear matter at equilibrium ($\rho = 1.0\ \rho_o$), and at a compressed state ($\rho = 1.5\ \rho_o$). We see that a similar scenario as in the σ-channel occurs though attenuated : strength is pulled out of the ρ-resonance and shifted to threshold as the matter density increases; at the same time the original ρ-meson is weakened and pushed to higher energies [11]. We tentatively conclude that the build up of strength close to the 2m_π threshold corresponds to what has been seen in the experiment by Roche et al. [4]. On the other hand the experimental error bars are too large to extract from the data a possible upward shift of the ρ. More precise experiments are clearly in need.

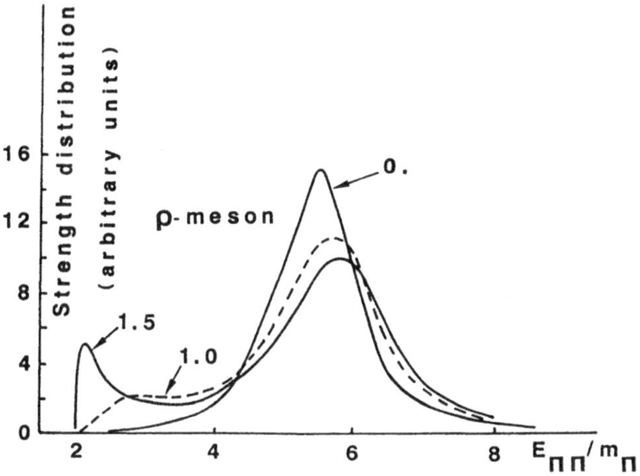

Fig. 3
In medium corrected strength distributions of the ρ-meson at three different densities ρ/ρo(= 0., 1., 1.5)

NUCLEAR MATTER EQUATION OF STATE

As already mentioned the origin of the intermediate range nucleon-nucleon attraction is essentially the uncorrelated and correlated two pion exchange processes in the I = J = 0 channel. A standard procedure is to simulate these contributions by a simple σ exchange potential and reinject it in some G-matrix calculations. Deltas in the intermediate state are usually included since we know that they are extremely important. But we also know that pions are strongly affected by the medium in particular through their mixing with delta states. For those reasons we believe that a realistic description of the binding mechanism in nuclear matter has to incorporate all these highly collective effects. In other words this means that we have to dress the pion in the medium and reconstruct at each density the popular effective sigma meson whose coupling constant and mass should consequently explicitely depend on the density. In order to illustrate these ideas we consider a model Hamiltonian which consists of T_N (kinetic energy the nucleon) T_π (kinetic energy ot the pions) and $H_{\pi NN}$ (standard p-wave πNN and πNΔ Hamiltonians). To be complete we also include a ρ-exchange potential. Short range correlations are incorporated by adding a quartic residual interaction in the NN spin-isospin channel

$$V_{g'}(1,2) = (v_{\pi NN}(q))^2(g'_{LNN}\,\sigma_1.q\,\sigma_2.q + g'_{TNN}\,\sigma_1 \times q\,\sigma_2 \times q)\,\tau_1.\tau_2 \quad (7)$$

and similar expressions for the NΔ and ΔΔ interactions. The various g' are generated from a Reid-soft core potential [12]. We choose a strong rho coupling C_ρ = 2.2 in the NN channel and weaker coupling C_ρ = 1.2 for NΔ and ΔΔ channels. The model for g' is certainly oversimplified but has the advantage to give numbers which are very close to standard values used by a number of authors to account for the phenomenology of quasielastic and delta responses as well as position and strength of the Gamow-Teller resonance. For instance

$q = 0$ $\qquad g'_{LNN} = g'_{TNN} = 0.62 \quad g'_{LN\Delta} = g'_{TN\Delta} = g'_{L\Delta\Delta} = g'_{T\Delta\Delta} = 0.41$

$q = 350$ MeV/c $\quad g'_{LNN} = g'_{TNN} = 0.7; \quad g'_{LN\Delta} = g'_{L\Delta\Delta} = 0.5; \quad g'_{TN\Delta} = g'_{T\Delta\Delta} = 0.45$

Using Green's function techniques it is possible to calculate the expectation value of the various pieces of the Hamiltonian at the desired level of approximation (Mean-Field and RPA in practice). We obtain for the pion channel the standard correlation energy per nucleon

$$\frac{E^{corr}{}_L}{A} = \frac{1}{\rho} \int \frac{d^3k}{(2\pi)^3} \int \frac{id\omega}{2\pi} e^{i\omega\eta^+} \int \frac{d\lambda}{\lambda} \frac{[\lambda^2 V_L(k,\omega)\Pi^o(k,\omega)]^2}{1-\lambda^2 V_L(k,\omega)\Pi^o(k,\omega)} \tag{8}$$

and a similar expression for the transverse channel. Here V_L is the full interaction in the longitudinal spin-isospin channel, i.e. $V_L = g' + k^2/(k^2 - m^2 - \omega^2)$and Π^o which includes form factor is the sum of the first order nucleon-hole and delta-hole propagators. The actual calculation is performed after a Wick rotation. This allows us to relax everywhere the static approximation (for form factors; g', ρ exchange...) without any further complication. The next step is the inclusion of pion-pion interaction. This problem can be reformulated in terms of the exchange of a pion pair coherent state [3]. We thus have to solve a gap equation

$$\Delta_{\mathbf{k}} = -\int \frac{d^3q}{(2\pi)^3} v_{\mathbf{k},\mathbf{q}} \; K_{\mathbf{q}} \tag{9}$$

where $v_{\mathbf{k},\mathbf{q}}$ is the π-π potential extracted with the correct normalization from model I. $K_{\mathbf{q}}$ is the anormal (or pairing) pion density which can be calculated from Green's function techniques

$$K_{\mathbf{q}} = \langle 0 | b_{a\mathbf{k}} \, b_{a-\mathbf{k}} | 0 \rangle \qquad a = 1, 2, 3 \tag{10}$$

$$= \int \frac{id\omega}{2\pi} e^{i\omega\eta^+} \frac{\Delta_q - \Sigma(q,\omega)/2\omega_k}{\omega^2 - \omega_q{}^2 + \Delta_q{}^2 - (1+\Delta_q/\omega_q)\Sigma(q,\omega)} \tag{11}$$

Here $\Sigma(\mathbf{q},\omega)$ is the irreducible pion-self energy which is $\Sigma = q^2\Pi^o/(1-g'\Pi^o)$ when $g'_{NN} = g'_{N'\Delta} = g'_{\Delta\Delta} = g'$. The exact expression for different g' is a little more complicated and will not be given here for simplicity. We also omit in eq. (11) the role of the so-called Hartree energy. Its effect is actually taken into account in the calculation and always gives a small correction. The exact solution of eqs. (10-11) is rather difficult to obtain since the gap modifies the pion propagator itself. A first reasonable approximation is to neglect the influence of the gap in the pion propagator, i.e. $\Delta_k{}^2 << \Sigma(k,\omega)$, $\Delta k << \omega_k$. All the integrals are dominated by the momentum region 400 MeV/c for which the gap is smaller than 20 MeV at normal density. This number is much smaller than $\omega_k \cong$ 400 MeV/c and $\sqrt{\Sigma} \cong k \cong$ 400 MeV/c. It would be, however, interesting to evaluate more precisely these highly non linear effects at high density. Even with this approximation inclusion of π-π interaction in the total energy is still tricky to calculate. To have a first estimate (in particular the behavior with density) we treat its effect in perturbation theory. We thus find the following contribution to the binding energy (see also Fig. 4)

$$<H_{\pi\pi}> = -\frac{1}{2} G_{o\pi\pi} \frac{\rho}{\rho_o} \frac{(\delta n_\pi)^2}{(1-\Gamma)^2} \qquad (12)$$

where $G_{o\pi\pi}$ = 38 MeV is an energy scale which depends on the π-π potential parameters. δn_π is a dimensionless number

$$\delta n_\pi = -\frac{3}{\rho} \int \frac{d^3q}{(2\pi)^3} \int \frac{i\,d\omega}{2\pi}\, e^{i\omega\eta^+} \frac{v(q)}{2\omega_q} \frac{\Sigma(q,\omega)}{\omega^2-\omega_q{}^2-\Sigma(q,\omega)} \qquad (13)$$

the factor (1 - Γ) comes from the rescattering of the π-π system in the medium.

$$\Gamma = \frac{G_{o\pi\pi}}{\rho_o} \int \frac{d^3q}{(2\pi)^3} \int \frac{i d\omega}{2\pi}\, e^{i\omega\eta^+} \frac{v^2(q)}{\omega^2-\omega_q{}^2-\Sigma(q,\omega)} \qquad (14)$$

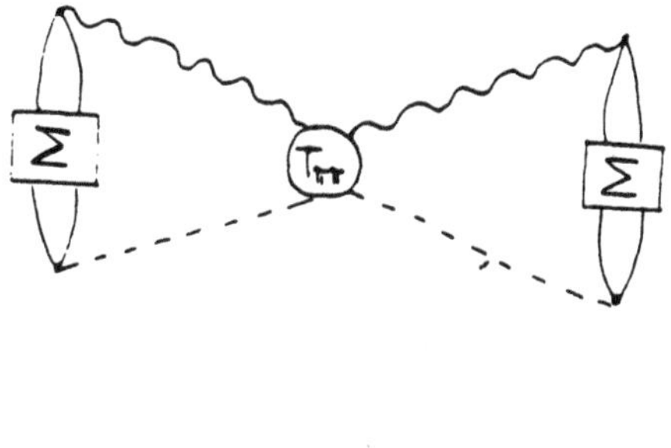

Fig. 4
Basic contribution to the correlated pion exchange

In the above form, the model will produce pion condensation at about two times the normal density [13]. Its proximity destroys all kinds of saturation mechanism. As it is well known these precursor effects are extremely sensitive to the precise value of g'. Moreover it has been shown that those effects which are at work to produce pion condensation also should produce an increase of g' when density increases [14]. Detailed G-matrix calculations seem to indicate that it is really the case when the so-called induced interaction is taken into account [14]. g' is a rather flat function below ρ_0 and starts to increase at ρ_0 (or slightly below). We thus believe that full microscopic understanding of saturation really needs a detailed knowledge of short range correlations. To illustrate the extreme sensitivity to the density dependence we modify g' (q) to g'(q,ρ) according to (see ref. [14])

$$g'_L(q;\rho) = g'_L(q) + \left[1+\exp\left(\frac{0.8\,\rho_0-\rho}{0.1\rho_0}\right)\right]^{-1} \cdot \;(1-g'_L(q))\;\frac{\rho-0.8\,\rho_0}{\rho_D-\rho_0} \qquad (15)$$

where ρ_D is a slope parameter. For $\rho_D \cong 4\,\rho_0$ this increases g'_{LNN} (q = 400 MeV/c) by only ten per cent at $\rho = 2\,\rho_0$ but the effect on saturation properties is very pronounced.

To have the total binding energy we add this correlation energy to the mean field expression for the kinetic energy. Short range repulsion is described by omega exchange potential with $g_\omega = 6.5$ (i.e. $E_\omega = 47.2\ \rho/\rho_0$ MeV). We show the results on Fig. 5.

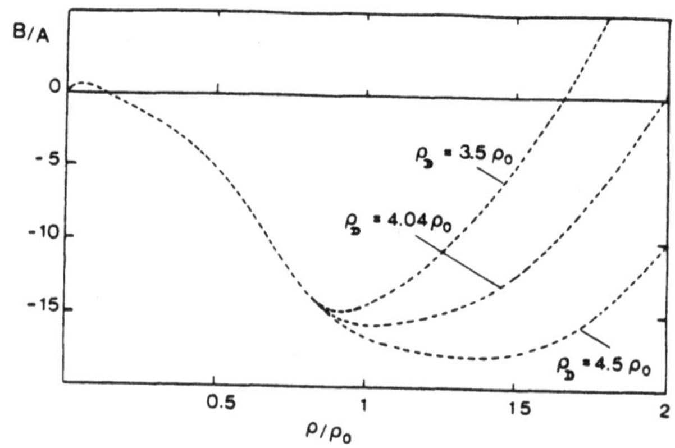

Fig. 5

Equation of state corresponding to our theory. Different density dependences of the g' parameter have been assumed expressed by the slope parameter ρ_D(see eq.(15))

Evidently there are some uncertainties in the calculation. Our many-body treatment suffers from the approximation in the sense that full self-consistency is not yet reached. In addition we have included in the expression of Π^{0} some relativistic corrections in the pion channel affecting its high momentum high-energy behaviour which sould be improved. This means that the absolute values of various pieces entering the calculations might be modified in a more sophisticated treatment. Nevertheless we believe that the saturation mechanism linked to the behaviour of the pion in the medium together with the subsequent density dependence of g' is a key point for a real understanding of the nuclear matter equation of state.

CONCLUSION AND DISCUSSIONS

In this work we treated the in medium correction to pion-pion correlations and its influence on the nuclear equation of state. We showed that medium corrections to the π-π scattering introduces a strong density dependance of the σ(S') meson which is often introduced to mock up π-π correlations. The effect of the medium consists in a strong build up of strength close to the $2m_\pi$ threshold which may even lead to a π-π bound state in matter. The same scenario, though attenuated due to the less attractive force, holds for the channel of the ρ-meson. The strength found at low energies (Fig. 3) may eventually serve to explain the enhanced cross section of dilepton production at the $2m_\pi$ threshold found in recent experiments [4]. This density dependence of the σ-meson together with a density dependence of the g' parameter are sufficient to create a saturation mechanism when the groundstate energy is calculated. The way we calculate the groundstate energy is essentially based on RPA in ring approximation supplemented by a BCS like treatment of the π-π interaction which is very easily incorporated into the RPA formalism. We here have therefore been concerned with ph, Δ-h and π-π correlations whereas the short range correlations have been treated in an ad hoc manner in adding to the groundstate energy the expectation value of a density independent repulsive force (ω -meson). This procedure is of course unrealistic and in principle short range (Brückner type) pp correlations and long range ph, Δ-h and π-π correlations should be treated on the same footing (this should include a correct treatment of the overcounting problem of the second order contribution to the groundstate energy). The purpose of this work was essentially pedagogical in order to show that the correlations we have been treating here and which in most cases are neglected in other approaches to the nuclear equation of state can indeed give important keys to the nuclear saturation mechanism.

Acknowledgements. We like to thank P. Guichon for the insight on the two level model treated in the first paragraph.

REFERENCES

1. R. Machleidt, K. Holinde and Ch. Elster, Phys. Rep. 149 (1987) 1
2. B.D. Serot and J.D. Walecka, Adv. Nucl. Phys. Vol. 16 (1986)
3. P. Schuck, G. Chanfray and W. Nörenberg, Nucl. Phys. A495 (1989) 355c

4 G. Roche et al. Phys. Rev. Lett. 61 (1988) 1069
C. Naudet et al., Phys. Rev. Lett. 62 (1989) 2652
G. Roche et al., Phys. Lett. B226 (1989) 228
5. C. Ellegaard et al., Phys. Rev. Lett. 50 (1983) 1745
D. Contardo et al., Phys. Lett. 168 B (1986) 331
6. J. Delorme and PAM Guichon, Proceedings of the 10th Session d'Etudes d'Aussois, Preprint Lyon, LYCEN 8902, (1989)
7. P. Schuck, W. Nörenberg and G. Chanfray, Z. Phys. A330 (1988) 119
8. J.A. Johnstone and T.S.H. Lee, Phys. Rev. C34 (1986) 243
9. J.W. Durso, A.D. Jackson and B.J. Verwest, Nucl. Phys. A345 (1980) 471
10 N. Orion et al., Nucl. Phys. A492 (1989) 509
11. This work is part of "Etat lié de deux pions dans la matière nucléaire " par Z. Aouissat. Compte-rendu de stage DEA, ISN, Université de Grenoble, Juin 1990
12. E. Oset, H. Toki, W. Weise, Phys. Rep. 33C (1982) 281
13. B. Friman and E.M. Nyman, Nucl. Phys. A302 (1978) 365
14. H. Müther, Prog. Part. Nucl. Phys. 14 (1985) 123
W.H. Dickhoff and H. Müther, Nucl. Phys. A473 (1987) 394

TWO-PARTICLE CORRELATIONS IN NUCLEAR MATTER

G. RÖPKE

Sektion Physik, Universität Rostock, Universitätsplatz 3
Rostock, GDR - 2500

M. SCHMIDT

Institut für Meereskunde der AdW der DDR, Seestraße 15
Rostock, GDR - 2530

TH. ALM

Sektion Physik, Universität Rostock, Universitätsplatz 3
Rostock, GDR - 2500

ABSTRACT

A quantum statistical approach to hot nuclear matter is given on the basis of a nonrelativistic potential approach. Kinetic equations for the single particle distribution function as well as for correlations (two-particle distribution function) are derived from a general approach to non-equilibrium.

In equilibrium, effective density-dependent binding energies and scattering phase shifts are determined using a separable representation of the Paris n-n potentials. At the Mott density, bound states (deuterons) are dissolved due to the Pauli blocking. The equation of state as well the composition is considered for symmetric as well as for asymmetric nuclear matter. Furthermore, the Gorkov equations are solved in the region of nucleonic suprafluidity.

Despite some attemps to include in-medium effects and bound state formation into kinetic equations, a systematic treatment of cluster formation and correlation in non-equilibrium has still to be worked out.

1. INTRODUCTION

The occurrence of bound states in nuclear matter such as deuterons, ^3H, ^3He, and α particles is of importance to describe the properties in the temperature region T of the order 10 MeV and density values $n < n_0$, $n_0 = 0.17\ fm^{-3}$ [1,2]. This concerns not only the equation of state and the verification of the nuclear matter phase transition [3-8], but also the problem of composition of nuclear matter. Furthermore, correlations between nucleons and/or nuclei are observed from light and complex particle emission in the course of heavy ion collisions at intermediate energies, for references see [9]. The recent paper is aimed to discuss the problem of correlation and bound state formation in hot dense matter.

For this, one has to go beyond the level of the Vlasov-Uehling-Uhlenbeck (VUU) equation or, in the case of equilibrium, the Bethe-Brueckner-Goldstone theory. Both fail to describe the correct low density limit of hot matter where the composition in equilibrium is governed by the law of mass action.

In general, one should use a relativistic description and has to consider the subnucleonic degrees of freedom to describe the nucleon-nucleon interaction. However, we will use a nonrelativistic potential approach which may be appropriate at temperatures up to some 10 MeV and density values below n_0. Although the spectra of light nuclei in heavy ion reactions, or the fragmentation of hot matter near the phase transition region are nonequilibrium phenomena, it seems to be necessary to have a correct description of the equilibrium first.

The present contribution is aimed to give a consistent description of bound state formation and correlation in equilibrium within the frame of a systematic quantum statistical approach. Furthermore, some aspects of nonequilibrium are pointed out, in particular the derivation of kinetic equations including cluster formation.

2. QUANTUM STATISTICAL APPROACH TO DENSE MATTER

We start from a description of hot dense matter where the nucleons (N, P) are considered as elements interacting via a nucleon-nucleon potential V. The Hamiltonian is given by

$$H = \sum_1 E(1)c_1^+c_1 + \frac{1}{2} \sum_{12,1'2'} V(12,1'2')c_1^+c_2^+c_{2'}c_{1'} \ , \qquad (1)$$

where i=1,2,... denotes momentum $\underline{p}_i$, spin σ_i and isospin (species) τ_i of the nucleon, c_i^+, c_i are creation or annihilation operators, respectively. $E(1) = p_1^2/2m_1$ is the kinetic energy. V(12,1'2') can be taken as a so-called realistic nucleon-nucleon potential. We will use a separable representation of the well known PARIS-potential [11] in the form given in [12,13].

We include S-, P-, and D-waves. The parameters of the potential have been chosen to reproduce the half-offshell T-matrix associated with the PARIS-potential. It is a nice feature of the resulting separable potential that it does not generate additional poles or resonances in the T-matrix, which has a simple structure allowing for analytical solutions. Therefore, it is rather suitable for many-body calculations.

The macroscopic nonequilibrium state of nuclear matter is characterized by the mean values $\langle B_n \rangle^t$ of a set of relevant observables B_n. They are evaluated with the statistical operator $\varrho(t)$ obeying the von Neumann equation

$$\frac{\partial}{\partial t} \varrho - \frac{1}{i\hbar} [H, \varrho] = 0 \ . \qquad (2)$$

The problem to formulate the correct initial conditions is solved introducing a relevant statistical operator $\varrho_{rel}(t)$ which maximizes the entropy

$$S[\varrho] = - k_B \ \mathrm{Tr}\{\varrho \ln \varrho\} \qquad (3)$$

under the requirement of given mean values

$$\langle B_n \rangle^t = \mathrm{Tr}\{\varrho_{rel}(t) \ B_n\} \ . \qquad (4)$$

The solution is given by the generalized Gibbs state

$$\varrho_{rel}(t) = \exp\{-\sum_n F_n(t)B_n\} / \mathrm{Tr} \exp\{-\sum_n F_n(t)B_n\} \ , \qquad (5)$$

where the Lagrange multipliers $F_n(t)$ are determined by (4).

The initial condition ϱ_{rel} is incorporated into the von Neumann equation (2) using Abel's theorem [14]. Then the nonequilibrium statistical operator is found by formal integration [14]

$$\varrho(t)=\varrho_{rel}(t)-\lim_{\varepsilon\to 0}\int_{-\infty}^{t} dt' e^{\varepsilon(t-t')}\frac{d}{dt}\left\{U(t,t')\varrho_{rel}(t')U(t',t)\right\} . \quad (6)$$

The time dependence of the relevant observables is given according to

$$\frac{d}{dt}\langle B_n\rangle^t = \mathrm{Tr}\left\{\varrho(t)\frac{i}{\hbar}[H,B_n]\right\}. \quad (7)$$

It is obvious that this approach describes correctly the special case of the thermal limit, where the set of the relevant observables B_n reduces to the integrals of motion $\{H,N_c\}$.

3. SINGLE-PARTICLE APPROACH TO NON-EQUILIBRIUM

A kinetic equation for the single-particle distribution function is derived [15] specifying the set of relevant observables according to

$$n(\underline{p},\underline{r}) = \int \frac{d^3q}{(2\pi)^3}\; e^{i\underline{qr}}\, c^+_{p-q/2}\, c_{p+q/2} \quad ; \quad (8)$$

the mean values

$$\langle n(\underline{p},\underline{r})\rangle^t = f_1^W(\underline{p},\underline{r},t) \quad (9)$$

are the well-known Wigner distributions.

According to (6) the time derivation of the Wigner distribution function consists of two parts

$$\frac{d}{dt} f_1^W(\underline{p},\underline{r},t) = -D^{Vlasov}(\underline{p},\underline{r},t) + I^{Coll}(\underline{p},\underline{r},t) \quad . \quad (10)$$

The Vlasov term describes the reversible part of the evolution characterized by $\varrho_{rel}(t)$. Assuming locality in space and time ($\varrho_{rel}(t')\approx \varrho_{rel}(t)$) the collision term can be written as

$$I^{Coll} = 2\pi \sum_{p_1'p_2p_2'} |t_2(p_1p_2,p_1'p_2')-t_2(p_1p_2,p_2'p_1')|^2\, \delta_{p_1+p_2,p_1'+p_2'} \,*$$
$$*\,\delta(E_{p_1}+E_{p_2}-E_{p_1'}-E_{p_2'})\left[f_{p_1}f_{p_2}(1-f_{p_1'})(1-f_{p_2'})-f_{p_1'}f_{p_2'}(1-f_{p_1})(1-f_{p_2})\right]. \quad (11)$$

Thus, eq. (10) is the well known VUU equation. However, the VUU approach has some short comings. The equilibrium solution is the Fermi distribution function, the entropy (3) reduces to the Boltzmann entropy of an ideal Fermi gas, and we have only conservation of the kinetic energy. Consequently it does not describe the formation of bound states like d, ^3H, ^3He or α. To overcome these shortcomings we have to include correlations of higher order.

4. QUANTAL CORRELATIONS IN NON-EQUILIBRIUM

We extend the set of relevant observables B_n including two particle density operators. In particular, for a homogeneous system, we have

$$\{ B_n \} = \{ c_k^+ c_k ; c_{p-q/2}^+ c_{k+q/2}^+ c_{k-q/2} c_{p+q/2} \} \cdot \tag{12}$$

The relevant statistical operator is now

$$\rho_{rel} = \frac{1}{Z_{rel}} \exp \{ - \sum_k F(k) c_k^+ c_k - \sum_{pkq} F(pkq) c_{p-q/2}^+ c_{k+q/2}^+ c_{k-q/2} c_{p+q/2} \} \cdot \tag{13}$$

The Lagrange parameters F(k), F(pkq) are determined according to eq.(4) from the given mean values (distribution function)

$$f(k) = \langle c_k^+ c_k \rangle , \tag{14}$$

$$f(pkq) = \langle c_{p-q/2}^+ c_{k+q/2}^+ c_{k-q/2} c_{p+q/2} \rangle \cdot \tag{15}$$

Applying a perturbational expansion with respect to F(pkq) the Wick theorem and a Matsubara Green's function technique can be used to evaluate f(k) and f(pkq). In the lowest order with respect to F(pkq) we have for the single particle distribution function

$$f(k) = \int \frac{d\omega}{\pi} f(\omega) \operatorname{Im} G(k,\omega) \approx [e^{F(k)}+1]^{-1} , \tag{16}$$

whereas the two particle distribution function reads in ladder approximation

$$f(p,k,q) = \sum_{nP} \tilde{\psi}^*_{nP}(p-q/2,k+q/2)\left[e^{\tilde{E}_{nP}}-1\right]^{-1}\tilde{\psi}_{nP}(p+q/2,k-q/2) \ , \qquad (17)$$

$\tilde{\psi}_{nP}$, $\tilde{E}_{nP}$ are the solutions of a wave equation

$$(F(p)+F(k)-\tilde{E}_{nP})\,\tilde{\psi}_{nP}(p,k)+2\sum_q F(p,k,q)\,\tilde{\psi}_{nP}(p+q,k-q) = 0 \ . \qquad (18)$$

For the further evaluation, we will treat the special case of local equilibrium. It is obvious that our quantum statistical approach avoids the problems of orthogonality of single particle states and bound states (correlation). The Fermionic character of the constituents of the bound states is reflected by the Pauli blocking term in the two-particle Green's function as shown in the next section.

5. GENERALIZED BETH-UHLENBECK FORMULA

In order to include explicitly two-particle correlations into the equation of state of nulear matter we use a Green's function approach [16] for thermal equilibrium. The equation of state is determined with the spectral function A(1,E)

$$n(\mu_N,\mu_P,T) = \sum_{i=N,P} \int \frac{dE}{2\pi} f_i(E,\mu_i)\, A_i(1,E,\mu_{N,P}) \ .$$

To allow for the formation of bound states it is necessary to include the imaginary part of the self-energy Σ into the spectral function. It then reads

$$A(1,E) = \frac{2\pi\,\delta(E-e(1))}{1-(d/dz\ \mathrm{Re}\ \Sigma(1,z))\big|_{z=e(1)+i\varepsilon}}$$

$$- 2\ \mathrm{Im}\ \Sigma(1,E+i\varepsilon)\,\frac{d}{dE}\,\frac{P}{E-E(1)-\mathrm{Re}\ \Sigma(1,E+i\varepsilon)} \ . \qquad (19)$$

In order to account for two-particle clustering we have to approximate Σ by the following Feynman diagrams

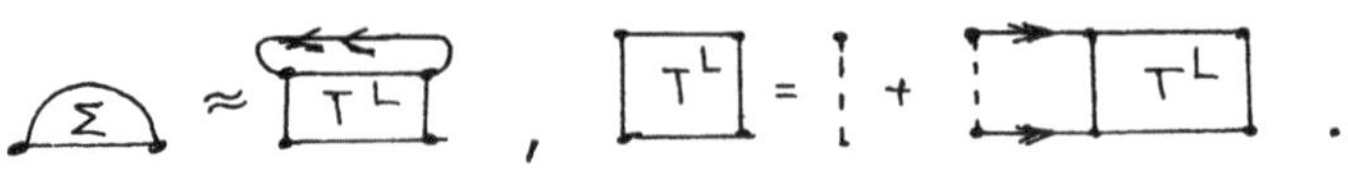

(20)

The T-matrix $T^L(1234,z)$ follows from a Bethe-Goldstone equation

$$T(1234,z) = V(1234) + \sum_{56} V(1256) \frac{1-f(e(5))-f(e(6))}{e(5)-e(6)-z} T(5634,z) \quad (21)$$

and describes a two-particle state under the influence of the surrounding matter, which weakens the interaction via the Pauli operator (1-f-f) and the quasiparticle effects.

As a consequence of the decomposition of the spectral function A(1,E) the total density n splits into a contribution of uncorrelated quasiparticles n_{FREE} which are still present in the Brueckner-Bethe-Goldstone theory and an additional contribution n_{CORR} accounting for correlated pairs, i.e. deuterons and scattering states of nucleons

$$n(\mu_{N,P},T) = n_{FREE}(\mu_{N,P},T) + 2\, n_{CORR}(\mu_{N,P},T) \; ; \quad (22)$$

$$n_{FREE} = \sum_1 f(e(1)) \quad , \quad n_{CORR} = \int dE/2\pi \; g(E)\, F(E) \, . \quad (23)$$

We can cast n_{CORR} in the form of a Beth-Uhlenbeck formula, i.e. it is possible to express the two-particle part of the density in terms of generalized density-dependent binding-energies E_B and scattering phase-shifts δ_a [17]

$$n_{CORR} = 3 \sum_{K>K_{MOTT}} g\,(E_B(K,\mu_{N,P},T))$$

(bound-state part)

$$+\; 2/\pi \sum_{\alpha} c_{\alpha} \int dE\; g(E) \sum_{K} \sin^2\delta_{\alpha}(E,K,\mu_{N,P},t)\; d/dE\; \delta_{\alpha}(E,K,\mu_{N,P},T) \, .$$

(scattering contributions, singlet) + triplet contributions

(24)

The binding energy E_B and the scattering phase shifts δ_{α} follow from the on-shell thermodynamic T-matrix, and in the low density limit they coincide with the usual deuteron binding energy and the nucleon-nucleon phase shifts, respectively. They become density dependent due to the Pauli operator in the Bethe-Goldstone equation (Figs. 1,2).

The classical Beth-Uhlenbeck formula (no density and temperature dependence in E_B and δ_α) predicts a gradual increase of n_{CORR} with the total density n. However, this is clearly in contradiction with the empirical finding that there are no clusters at nuclear matter density and low temperatures.

Calculating n_{CORR} with the generalized Beth-Uhlenbeck formula the results coincide with the classical Beth-Uhlenbeck formula only for low density values. For higher densities a maximum of n_{CORR} is reached and at still higher densities it rapidly decreases with n until it is negligible small at saturation density (Fig. 3). This suppression of correlated pairs at high density values is called Mott effect and it is well known in plasma physics.

The equation of state is modified by correlation effects (Fig. 4), in particular in the low density region.

6. SUPRAFLUIDITY IN NUCLEAR MATTER

Starting with the equation of state of nuclear matter (24) as derived from the ladder approximation for the T-matrix (20) it is obvious that the Bose distribution function in the density contribution of bound states can lead to a Bose condensation of deuterons at $E_B=2\mu$.

Concerning the scattering contribution to (24), the Bose pole at $E=2\mu$ is generally compensated by a zero of the Pauli-operator at this energy. However, a critical temperature T_c may be found where the T-matrix pole defining the bound state's energy reappears at the positive real axis thus producing a divergency in the scattering contribution [17, 16]. This phenomenon is well known as break down of the T-matrix approximation [19].

In Fig. 5 the critical temperature is given as a function of the Fermi energy.

Below this critical temperature the system is in a condensate state that may be well described by the BCS-theory of superconductivity. It predicts the occurence of an energy gap in the single particle spectrum that is due to the formation of Cooper pairs.

For nuclear matter this phenomenon has been discussed by several authors [20,21] especially in connection with the possible realization of a suprafluid phase in astrophysical objects like neutron stars [21,10, 22]. The exact magnitude and position of the gap strongly depend on the interaction. Both the attractive part of the nucleon-nucleon interaction as well as hard-core effects have to be taken into account.

The usual way to derive the selfconsistent gap equation is the introduction of anomal propagators into the Hartree-Fock decoupling of the equation of motion for the one-particle Green's function [24]. This leads to the Gorkov equation and to the well known BCS gap equation

$$\Delta_0(k) = -\frac{1}{2}\int k'^2 dk' \frac{\langle k'|V_0|k\rangle \Delta_0(k')\tanh\left[\beta/2\,\left(E(k')^2 + \Delta_0(k)^2\right)^{1/2}\right]}{\left(E(k')^2 + \Delta_0(k')^2\right)^{1/2}} . \quad (25)$$

This derivation can be generalized to incorporate pairing in other partial waves than $^1S_0$ too [25]. Triplet-pairing was discussed first by Tamagaki [26] using a generalized Bogoljubov transformation at T=0. His work was motivated by the large positive phase shifts at high energies in the $^3P_2$-$^3F_2$ channel indicating the possible formation of Cooper pairs at the corresponding Fermi energies. The $^3P_2$-gap (within the approximation of maximum m_j-coupling [26] is then given by

$$\Delta\, {}^3P_2 = -\frac{1}{2}\int k'^2 dk' \langle k'|V\;|k\rangle\, \Delta\, {}^3P_2(k') \,*$$

$$* \int d\hat{k}' \;\frac{3/8\pi \sin^2\vartheta \tanh\left[\beta/2\,\left(E(k')^2 + 3/8\pi\sin^2\vartheta\,\Delta^3P_2(k')^2\right)^{1/2}\right]}{\left(E(k')^2 + 3/8\pi\sin^2\vartheta\;\Delta^3P_2\;(k')^2\right)^{1/2}} . \quad (26)$$

We used a separable representation of the Graz realistic nucleon-nucleon potential [27] for the determination of the $^1S_0$- and $^3P_2$-pairing gaps (25,26).

In Fig. 6 the $^3P_2$-N-N gap is given in comparison with the $^1S_0$-N-N and $^1S_0$-P-P gaps in asymmetric nuclear matter (neutron star like situation, $\alpha \approx 1$). It is to be seen that this $^3P_2$-N-N pairing should become important at much higher densities than $^1S_0$-N-N pairing and that the $^3P_2$-gap is much smaller. However it should be of some importance in neutron star phenomena due to the high density in the core of a neutron star.

For attempts to apply the BCS-formalism to finite nuclei too using a local density approximation see [29].

7. ASYMMETRIC NUCLEAR MATTER

In asymmetric nuclear matter, the chemical potentials of neutrons μ_N and protons μ_P are different. As a new parameter, the neutron excess coefficient is introduced defined by

$$\alpha = (n_N - n_P) / (n_N + n_P) \quad .$$

It can be shown that the Pauli blocking of deuterons does not depend on α for a nondegenerate system. At ordinary temperatures, the degeneration density is higher than the Mott density, and then we have to consider the α dependence of the scattering phase shifts.

As general tendency, the density dependent phase shifts become smaller with increasing α, see [23]. Thermodynamic properties of asymmetric matter are of importance for astrophysical objects such as neutron stars [30].

8. LADDER-HARTREE-FOCK APPROXIMATION

The extension of the Beth-Uhlenbeck formula to include correlations of higher orders and the formation of nuclei with mass number $A>2$ is a difficult task. The equation of state including higher clusters is obtained considering the scattering states as well as the in-medium effects in Hartree-Fock approximation [31].

It has been found that due to correlations (bound states) the critical temperature T_c is decreasing, see Fig. 7. The abundance of higher nuclei in correspondence to astrophysical elemental abundances has been discussed in [32].

9. MEDIUM DEPENDENT NUCLEON-NUCLEON CROSS SECTION

In-medium effects are of importance too for the nucleon-nucleon cross section. It is derived from the T-matrix that takes into account

Pauli blocking and effective mass effects, see [16]. At high densities substantial deviations from the free nucleon-nucleon cross sections are obtained, see Fig. 8.

From the collision term of the Boltzmann equation a mean free path is obtained and has been discussed including in medium effects (for references see [16]).

10. CLUSTER-VUU EQUATION

The extension of the set of relevant observables including correlations has been considered for the special case of deuteron formation [15].

In the low density limit a Vlassov term has been derived where the mean field is determined not only by the free nucleons but also by deuterons. Similiarly, the collision term contains different reaction channels. It is a task of future work to derive kinetic equations including correlations for higher densities where in medium effects are of importance.

11. SUMMARY

Present state of treatment of correlations in nuclear matter:

equilibrium

	in medium effects	
	neglected	included
single particle approximation	ideal Fermi gas	Brueckner theory
correlations ↗ bound states	law of mass action	ladder-Hartree Fock
correlations ↘ scattering states	Beth-Uhlenbeck (A=2)	generalized Beth-Uhlenbeck (A=2)
non-equilibrium		
single particle approximation	Boltzmann equation	VUU, eff. mass, medium dep. cross s.
correlations ↗ bound states	Cluster-Boltzmann	(Cluster-VUU)
correlations ↘ scattering states	(noneq.2-part. distr.function)	?

REFERENCES

1. H. Stöcker and W. Greiner, Phys.Rep. 137, 277 (1986) and references therein
2. V.D. Toneev, H. Schulz, K.K. Gudima and G. Röpke, Sov.J.Part.Nucl. 17, 485 (1986) (Fiz.Elem.Chastits At.Yadra 17, 1093 (1986)) and references therein
3. M. Brack and P. Quentin, Phys.Lett. 52B, 154 (1974)
4. D.Q. Lamb, J.M. Lattimer, C. Pethick and D.G. Ravenhall, Phys.Rev. Lett. 41, 1623 (1978); Nucl.Phys. A360, 459 (1981)
5. P. Danielewicz, Nucl.Phys. A314, 465 (1979)
6. G. Röpke, L. Münchow, H. Schulz: Phys.Lett. 110B, 21 (1982)
7. B. Friedman and V.R. Pandharipande, Nucl.Phys. A361, 502 (1981)
8. A. Lejeune, P. Grange, M. Martzloff and J. Cugnon, Nucl.Phys. A453, 189 (1986)
9. C.K. Gelbke and D.H. Boal, Prog.Part.Nucl.Phys. 19, 33 (1987)
10. G.E. Brown, Editor, Phys.Rep. 163, 1 (1988)
11. M. Lacombe, B. Loiseau, J.M. Richard, R. Vinh Mau, J. Pires and R. de Tourreil, Phys.Rev. C21, 861 (1980)
12. L. Mathelitsch, W. Plessas and W. Schweiger, Phys.Rev. C26, 65 1982)
13. J. Haidenbauer, W. Plessas, Phys.Rev. C30, 1822 (1984)
14. D.N. Zubarev, Nonequilibrium statistical thermodynamics, Moscow, Nauka, 1987
15. G. Röpke, H. Schulz, Nucl.Phys. A477, 472 (1988)
16. M. Schmidt, G. Röpke, H. Schulz, to appear in Ann.Phys.
17. R. Zimmermann and H. Stolz, phys.stat.sol.(b) 131, 151 (1985)
19. L.P. Kadanoff, G. Baym, Quantum Statistical Mechanics, Benjamin, 1962
20. L. Amundsen and H. Ostgaard, Nucl.Phys. A472, 487 (1985)
21. J.M.C. Chen, J.W. Clark, E. Krotscheck and R.A. Smith, Nucl.Phys. A451, 509 (1986)
22. T. Takatsuka, Prog.Theor.Phys. 71, 1423 (1984)
23. Th. Alm, G. Röpke and M. Schmidt, to be published
24. R.K. Su, S.D. Yang and T.T.S. Kuo, Phys.Rev. C35, 1539 (1988)
25. Th. Alm, G. Röpke and M. Schmidt, to appear in Wiss.Zeitschrift Universität Rostock
26. R. Tamagaki, Prog.Theor.Phys. 44, 905 (1970)
27. J. Haidenbauer, W. Plessas, Phys.Rev. C30, 1822 (1984)
28. Y. Yamaguchi, Phys.Rev. 95, 1628 (1954)
29. H. Kucharek, P. Ring, P. Schuck, R. Bengtsson, M. Girod, Phys.Lett. B216, 249 (1989)
30. R.K. Su, S.D. Yang, G.L.Li, T.T.S. Kuo, Mod.Phys.Lett. 1, 71 (1986)
31. G. Röpke, M. Schmidt, L. Münchow and H. Schulz, Nucl.Phys. A399, 587 (1983)
32. G. Röpke, Phys.Lett B185, 318(1987)

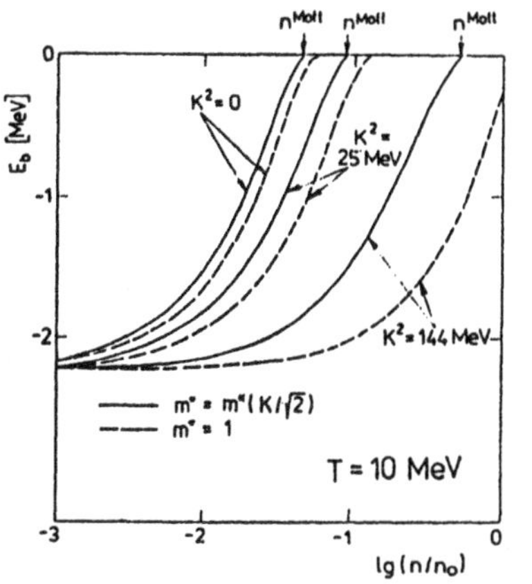

Fig. 1
The medium dependent deuteron binding energy as a function of the total density n for given temperature $T = 10$ MeV and several energies K^2 of the deuteron's center of mass motion. The solid lines show the results when using a momentum dependent effective mass approximation. For comparison the results are shown (dashed lines) for which the effective mass corrections are neglected. The Mott densities $n^{Mott}(K, T)$ at which the deuteron is dissolved are indicated.

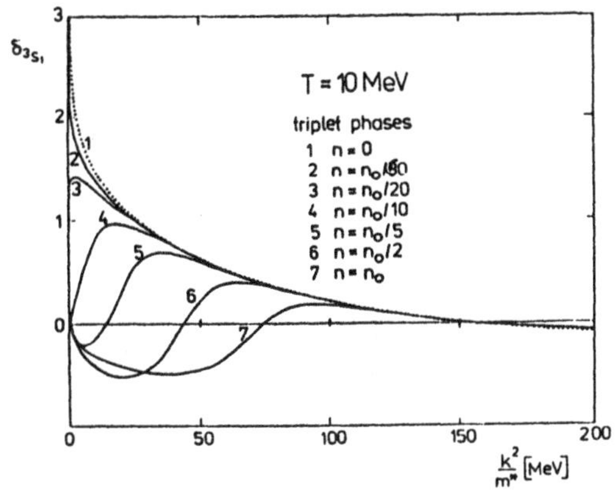

Fig. 2
The same as Fig. 4 but for the triplet phases. Note, however, that the phase jumps by $-\pi$ when the deuteron dissolves for density values $n > n_0/20$.

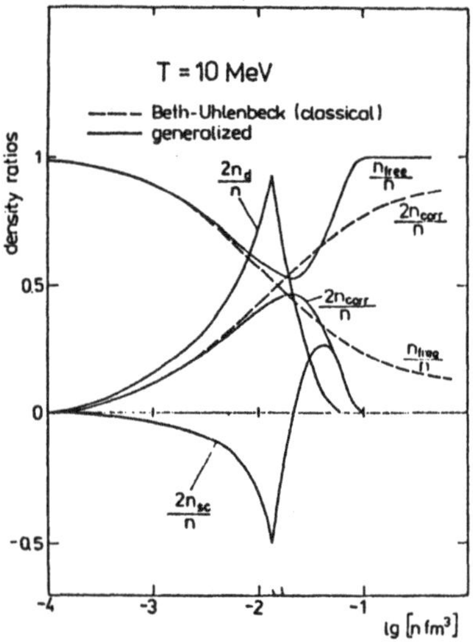

Fig. 3 The composition of nuclear matter as a function of the density n for given temperature $T = 10$ MeV. The solid and dashed lines show the results of the generalized and classical Beth–Uhlenbeck approach, respectively. Note the distinct behavior of n_{free} and n_{corr} predicted by the two approaches in the low and high density limit!

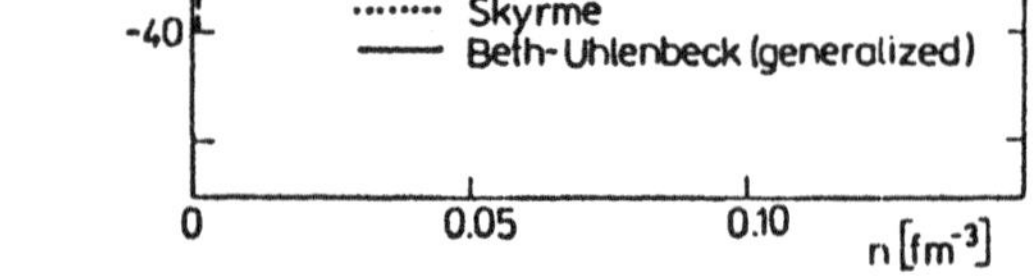

Fig. 4 The chemical potential μ as a function of the density n for given temperature $T = 10$ MeV. The solid curve gives the result of the generalized Beth–Uhlenbeck approach. The dashed curve represents the results of the Brueckner theory for finite temperature. For comparison is shown the prediction obtained by means of a Skyrme type interaction

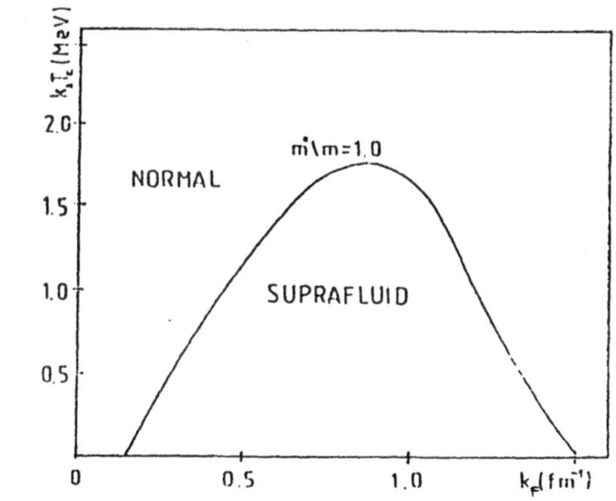

Fig. 5 Critical temperature T_c as function of the Fermi momentum K_F of neutrons for $^1S_0$-pairing

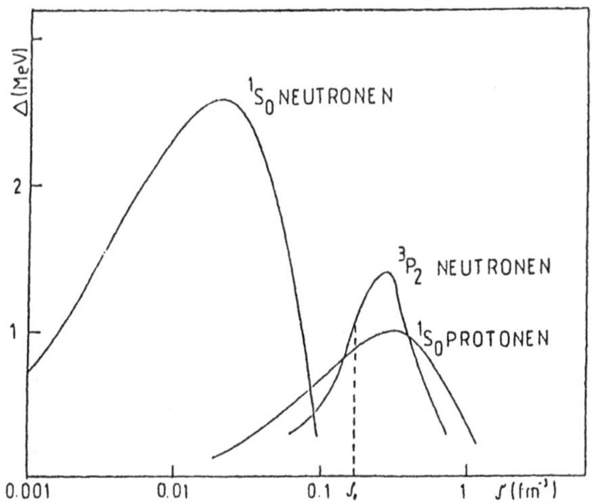

Fig. 6 $^1S_0$-neutron gap, $^1S_0$-proton gap and $^3P_2$-neutron gap as function of the total nucleon density (ρ_0: saturation density)

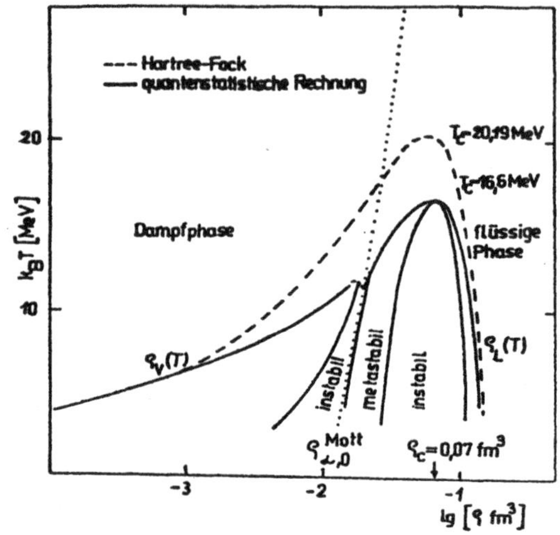

Fig. 7 liquid-vapor phase transition

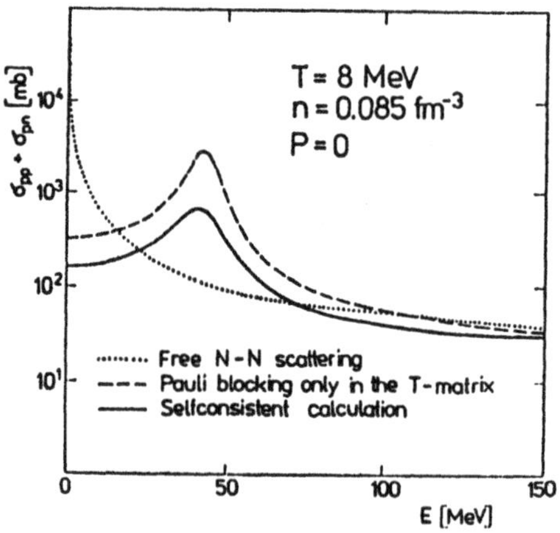

Fig. 8

The nucleon–nucleon cross section as a function of the relative energy. The dotted curve gives the result of the free scattering. The dashed line represents the result of an approximate calculation in which only the Pauli blocking in the *T*-matrix is taken into account. The solid curve shows the result of the selfconsistent calculation within the generalized Beth–Uhlenbeck approach.

"Two-proton correlation functions for equilibrium and non-equilibrium emission"

C.K. Gelbke, W.G. Gong, W. Bauer, N. Carlin*, R.T. de Souza, Y.D. Kim, W.G. Lynch, T. Murakami**, G. Poggi***, M.B. Tsang, and H.M. Xu
National Superconducting Cyclotron Laboratory and Department of Physics and Astronomy, Michigan State University, East Lansing, MI 48824, USA

S. Pratt
Department of Physics, University of Wisconsin, Madison, WI 53706, USA

D.E. Fields, K. Kwiatkowski, R. Planeta, V.E. Viola, Jr. and S.J. Yennello
Indiana University Cyclotron Facility, Indiana University, Bloomington, IN 47405, USA

1. ABSTRACT

Two-proton correlation functions are compared for equilibrium and non-equilibrium emission processes investigated, respectively, in "reverse kinematics" for the reactions $^{129}Xe+^{27}Al$ and $^{129}Xe+^{122}Sn$ at E/A=31 MeV and in "forward kinematics" for the reactions $^{14}N+^{27}Al$ and $^{14}N+^{197}Au$ at E/A=75 MeV. Observed differences in the shapes of the correlation functions are understood in terms of the different time scales for equilibrium and preequilibrium emission.

2. INTRODUCTION

Two protons, emitted at small relative momenta from an excited nuclear system, carry information about the space-time characteristics of the emitting source [1-11]. The shape of the two-proton correlation function reflects the interplay of the short-range attractive nuclear interaction, the Pauli exclusion principle, and the long range repulsive Coulomb interaction between the two emitted protons. The short range nuclear interaction is dominated by the attractive singlet S-wave nuclear interaction which leads to a pronounced maximum in the

* Present address: Instituto de Fisica, Universidade de São Paulo, C. Postal 20516, CEP 01498, São Paulo, Brazil
** Present address: Department of Physics, Kyoto University, Kyoto 606, Japan
*** Permanent address: Dipartimento di Fisica dell' Universita and INFN, Largo Enrico Fermi 2, 50125 Firenze, Italy

two-proton correlation function at relative momentum q≈20 MeV/c, when the average distance upon emission is of the order of 10 fm or less [1]. The long range Coulomb interaction and the Pauli exclusion principle give rise to a minimum at q=0. Some directional information can be provided by antisymmetrization effects [1,7,8].

The average distance between two coincident protons upon emission depends on the spatial dimension, d, and the average lifetime, τ, of the emitting system; it is of the order of $d+\bar{v}\tau$, where $\bar{v}$ is the average velocity of the emitted protons. For the decay of equilibrated compound nuclei with temperatures below 5 MeV, estimated lifetimes are larger than several hundred fm/c [12]. As a consequence, the average distance between emitted particles is much larger than the size of the emitting nucleus and the effects of the Coulomb interaction and the Pauli principle should dominate. On the other hand, preequilibrium light particle emission in intermediate energy heavy ion collisions is calculated to proceed on much shorter time scales [13,14] and average particle separations may reflect the spatial dimension of the emitting system rather than the emission rate. Here, the nuclear interaction should be prominent.

We have performed a comparative study of two-proton correlation functions measured for preequilibrium and equilibrium emission processes. When light projectiles impinge on heavy target nuclei, preequilibrium emission is strongly enhanced at small angles, whereas equilibrium emission dominates at backward angles, see e.g. refs. [5,10,11]. Preequilibrium emission was studied in "forward kinematics" for the reactions $^{14}N+^{27}Al$ and $^{14}N+^{197}Au$ at E/A=75 MeV. Equilibrium emission was studied at a lower energy, E/A=31 MeV, in "reverse kinematics" for the reaction $^{129}Xe+^{27}Al$ and for the nearly symmetric system $^{129}Xe+^{122}Sn$.

3. EXPERIMENTAL DETAILS

The experiments were performed with beams from the K1200 cyclotron from the National Superconducting Cyclotron Laboratory at Michigan State University. For reactions induced by ^{129}Xe, we used ^{27}Al and

^{122}Sn targets with areal densities of 5.6 and 5.3 mg/cm^2, respectively. For reactions induced by ^{14}N, we used ^{27}Al and ^{197}Au targets with areal densities of 15.0 and 15.9 mg/cm^2, respectively. Light particles were detected with two ΔE-E detector arrays consisting of 300-400 μm thick silicon ΔE-detectors and 10 cm long CsI(Tl) or NaI(Tl) E-detectors. An array consisting of 37 Si-CsI(Tl) telescopes [15] was centered at the polar and azimuthal angles of θ=25° and ϕ=0°; each of its detectors had a solid angle of ΔΩ=0.37 msr and a nearest neighbor spacing of $\Delta\theta$=2.6°. Another array consisting of 13 Si-NaI(Tl) telescopes was centered at θ=25° and ϕ=90°; each of its detectors had a solid angle of ΔΩ=0.5 msr and a nearest neighbor spacing of $\Delta\theta$=4.4°. Coincidence and downscaled singles data were taken simultaneously. Energy calibrations are accurate to better than 2%. Typical detector energy resolutions were of the order of 2% and 1% for protons of 40 MeV and 100 MeV, respectively. All data were corrected for random coincidences.

4. BASIC EQUATIONS

The experimental two-particle correlation function, R(q), is defined in terms of the coincidence yield, $Y(\vec{p}_1,\vec{p}_2)$, and the single particle yields, $Y(\vec{p}_1)$ and $Y(\vec{p}_2)$:

$$\sum Y(\vec{p}_1,\vec{p}_2) = C\big(1+R(q)\big)\sum Y(\vec{p}_1)Y(\vec{p}_2) \ . \tag{1}$$

Here, $\vec{p}_1$ and $\vec{p}_2$ are the laboratory momenta of particles 1 and 2, and q is the relative momentum of the particle pair. For each experimental gating condition, the sums on both sides of Eq. 1 are extended over all energy and detector combinations corresponding to the given bins of q. The normalization constant, C, is determined by the requirement that R(q)=0 for large relative momenta.

Our theoretical analysis is based on the expression [7,8]:

$$1+R(\vec{P},\vec{q}) = \int d^3r F_{\vec{P}}(\vec{r})\,|\phi(\vec{q},\vec{r})|^2 \ , \tag{2}$$

where $\phi(\vec{q},\vec{r})$ is the relative two-proton wavefunction and $F_{\vec{P}}(\vec{r})$ is defined by:

$$F_{\vec{P}}(\vec{r}) = \frac{\int d^3R f(\vec{P}/2,\vec{R}+\vec{r}/2,t_>) f(\vec{P}/2,\vec{R}-\vec{r}/2,t_>)}{|\int d^3r f(\vec{P}/2,\vec{r},t_>)|^2} \ . \tag{3}$$

The Wigner function $f(\vec{p},\vec{x},t_>)$ is the phase-space distribution of particles of momentum $\vec{p}$ at position $\vec{x}$ at some time, $t_>$, after the emission process:

$$f(\vec{p},\vec{r},t_>) = \int_{-\infty}^{t_>} dt \ g(\vec{p},\vec{r}-\vec{v}_{\vec{p}}(t_>-t),t) \ . \tag{4}$$

Here, $g(\vec{p},\vec{r},t)$ denotes the probability of emitting a particle with momentum $\vec{p}$ from location $\vec{r}$ and at time t.

The relative importance of antisymmtrization and the nuclear and Coulomb interactions depends on the size of the emitting system. In order to provide a quantitative comparison of these effects, we have calculated two-proton correlation functions for Gaussian sources of negligible lifetime,

$$g(\vec{p},\vec{r},t) = \rho_0 \exp(-r^2/r_0^2)\delta(t-t_0) \tag{5}$$

in which the nuclear interaction and the Pauli principle were turned off successively. These calculations are compared in Fig. 1 for a number of representative radius parameters (r_0=2.5, 5, 10, 20 fm) . The solid curves show the results of the full calculations which include the nuclear and Coulomb interactions and the Pauli exclusion principle. The dotted lines show calculations for which the nuclear interaction has been turned off; these calculations still include the two-proton Coulomb interaction and the antisymmetrization of the relative wave function. The dashed curves represent calculations for which both the nuclear interaction and the antisymmetrization of the relative wave function have been turned off. For radius parameters, $r_0 \lesssim 10$ fm, both the antisymmetrization of the relative wave function and the nuclear interaction have important effects on the detailed shape of the

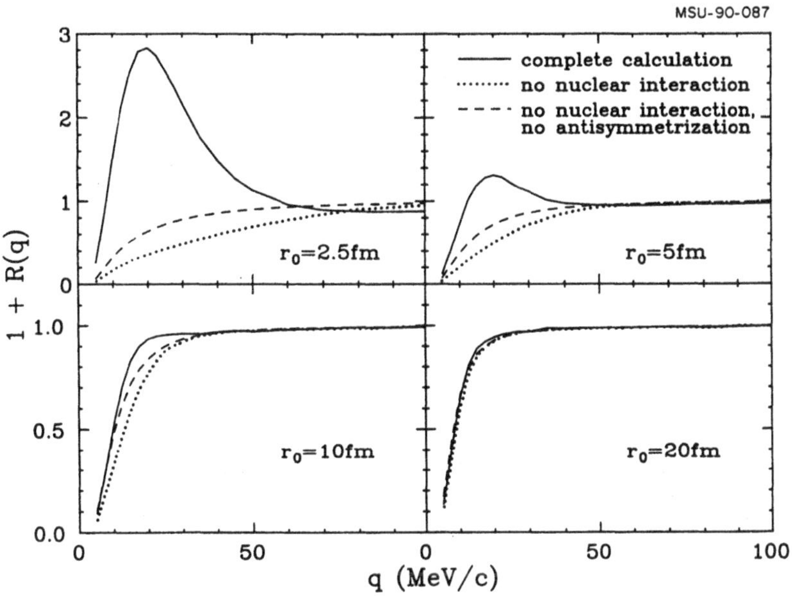

Fig. 1. Two-proton correlation functions calculated for short-lived Gaussian sources of representative radius parameters, r_0=2.5, 5, 10, 20 fm. The solid curves show the result of the full calculations; the dotted curves show calculations for which the nuclear interaction is neglected; the dashed curves shown calculations for which the nuclear interaction and the Pauli principle are neglected.

calculated correlation functions. For much larger source dimensions, $r_0 \gtrsim 20$ fm, the Coulomb interaction dominates and the Pauli principle can be neglected.

5. ANGLE AND ENERGY INTEGRATED CORRELATION FUNCTIONS

Two-proton correlation functions corresponding to sums over all detectors and all proton energies above the applied software threshold of 10 MeV are compared in Fig. 2. The correlation functions measured for the ^{14}N-induced reactions exhibit a minimum at q=0 MeV/c and a pronounced maximum at q≈20 MeV, see top panels of Fig. 2. In contrast, the correlation functions measured for the ^{129}Xe-induced reactions exhibit only a minimum at q≈0 MeV/c, see bottom panels of Fig. 2. For

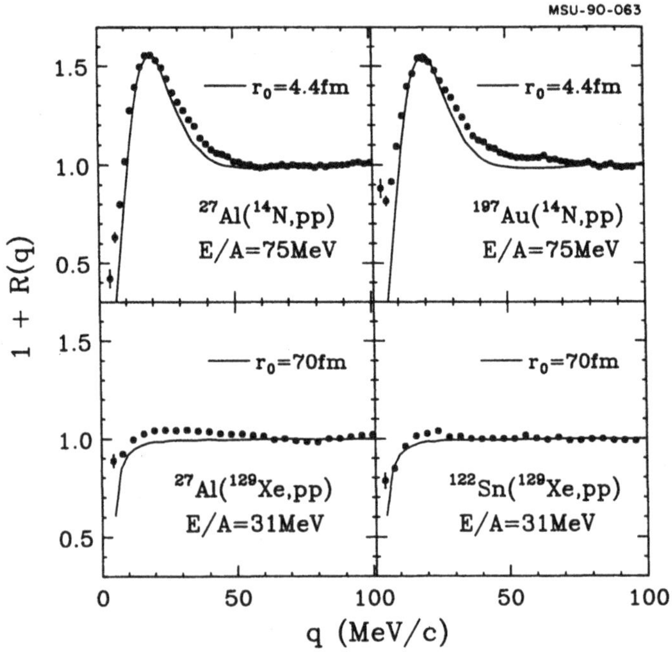

Fig. 2. Comparison of energy integrated two-proton correlation functions measured for the reactions $^{14}N+^{27}Al$ and $^{14}N+^{197}Au$ at E/A=75 MeV (top panels) and the reactions $^{129}Xe+^{27}Al$ and $^{129}Xe+^{122}Sn$ at E/A=31 MeV (bottom panels). The solid curves represent correlation functions predicted for Gaussian sources of negligible lifetime for the indicated radius parameters, r_0.

orientation, the solid lines show theoretical correlation functions predicted for Gaussian sources of negligible lifetime with the radius parameters r_0=4.4 and 70 fm for the reactions induced by ^{14}N and ^{129}Xe, respectively. These source parameters should be compared to the equivalent radius parameters for Al and Au nuclei, $r_0(Al)\approx2.5$ fm and $r_0(Au)\approx4.4$ fm , which are obtained from tabulated [16] rms charge radii, $\langle r^2\rangle^{1/2}$, using the approximate relation $r_0=0.82\langle r^2\rangle^{1/2}$. For the ^{14}N-induced reactions, a source radius of r_0=4.4 fm is not necessarily unreasonable. However, it is astonishing that this radius parameter exhibits no obvious dependence on the size of the target nucleus. A purely geometrical interpretation of the correlation function is, therefore, in doubt. For the large source parameter, $r_0 \approx 70$ fm, used to describe the correlation functions measured for the ^{129}Xe induced reactions, a purely geometrical interpretation is clearly unphysical and the lifetime of the emitting system must play a major role.

6. DEPENDENCE ON TOTAL MOMENTUM

Two-proton correlation functions are known to exhibit strong dependences on the energy of the emitted particles [2-6,8] or, equivalently, on the total momentum, $\vec{P}=\vec{p}_1+\vec{p}_2$, of the coincident particle pair. Figures 3 and 4 show correlation functions for representative ranges of the total momentum, P, for the reactions ^{14}N+^{27}Al and ^{14}N+^{197}Au, respectively. Consistent with previous measurements, the maximum at $q \approx 20$ MeV/c becomes more pronounced for

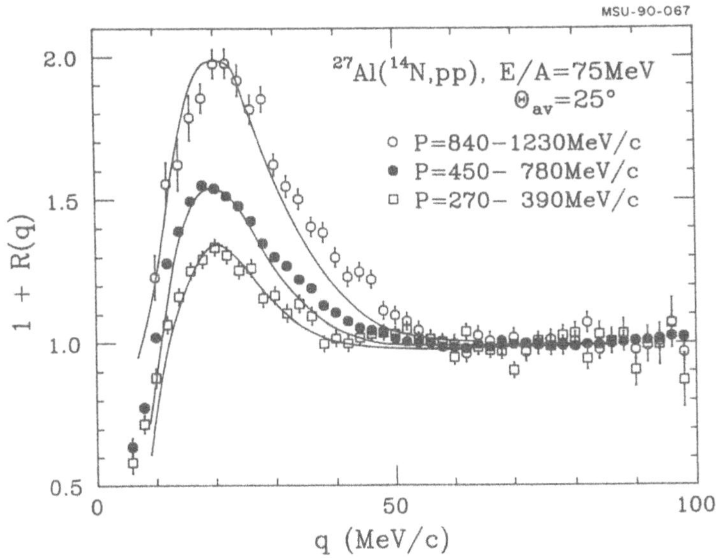

Fig. 3. Two-proton correlation functions measured for the reaction ^{14}N+^{27}Al at E/A=75 MeV. The gates placed on the total momenta, P, of the coincident particle pairs are indicated. The solid curves represent calculations for Gaussian sources of negligible lifetime, assuming a dependence of the radius parameter, r_0, on total momentum, P, as shown by the open points in Fig. 6 and folding in the response of the experimental apparatus.

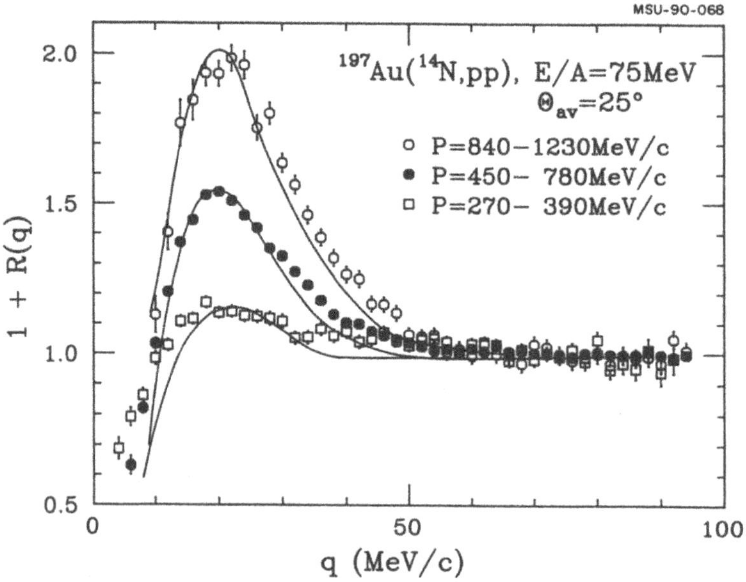

Fig. 4. Two-proton correlation functions measured for the reaction $^{14}N+^{27}Al$ at E/A=75 MeV. The gates placed on the total momenta, P, of the coincident particle pairs are indicated. The solid curves represent calculations for Gaussian sources of negligible lifetime, assuming a dependence of the radius parameter, r_o, on total momentum, P, as shown by the solid points in Fig. 6 and folding in the response of the experimental apparatus.

larger total momenta, i.e. for the emission of more energetic particles. For the lowest momentum gate, P=270-390 MeV/c, the correlation functions are distinctly different for the two targets. For the ^{27}Al target, a clear maximum at q≈20 MeV/c is measured. For the ^{197}Au target, on the other hand, this maximum is barely visible and the shape of the correlation function resembles that measured for evaporative processes.

Correlations measured for the ^{129}Xe-induced reactions exhibit a less dramatic dependence on the total momentum of the detected proton pair. Figure 5 shows two-proton correlation functions for two representative momentum gates, measured for the reactions $^{129}Xe+^{27}Al$ (upper panel) and $^{129}Xe+^{122}Sn$ (lower panel). The momentum gates represented by the solid and open points correspond to protons emitted

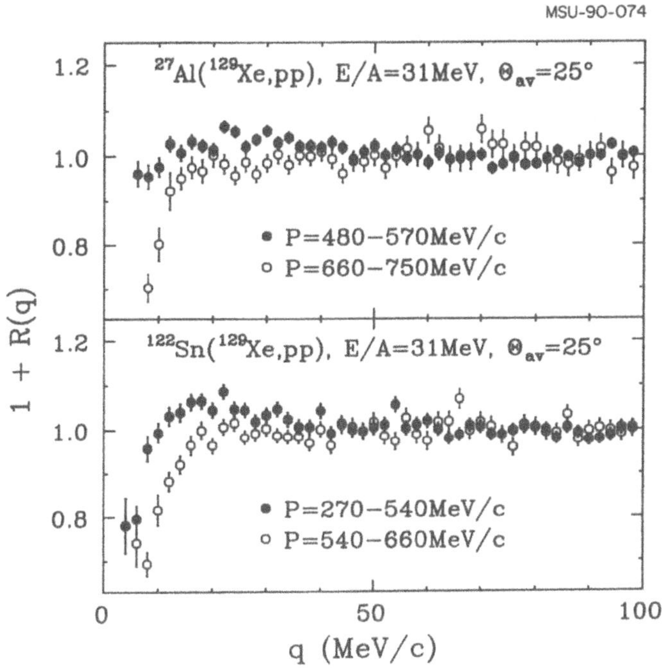

Fig. 5. Two-proton correlation functions measured for the reactions ^{129}Xe+^{27}Al and ^{129}Xe+^{122}Sn at E/A=31 MeV. The gates on the total momenta, P, of the coincident proton pairs are indicated; solid and open points represent center-of-mass energies below and above the compound nucleus Coulomb barriers.

with kinetic energies below and above the compound nucleus Coulomb barriers, respectively. As can be expected from qualitative time scale arguments, sub-barrier emission results in a reduction of the minimum at $q \approx 0$ MeV/c. At sub-barrier energies, the correlation functions are also more sensitive to attenuations and/or distortions from sequential decays of primary fragments emitted in particle unbound states [5,12] and from deflections in the Coulomb field of the heavy reaction residue. Because of these additional complications, we will refrain from a more detailed analysis of two-proton correlation functions for particles emitted with sub-barrier energies.

In order to describe the dependence of the measured two-proton correlation functions on total momentum in terms of a simple variable, we have constructed experimental correlation functions for a number of narrow gates placed on the total momentum P. For each such correlation function we extracted the source parameters, r_0, of Gaussian sources of negligible lifetime, Eq. (5), for which the predicted correlation functions were consistent with the data. The dependence of the extracted parameters, r_0, on the total momentum of the detected particle pair is shown in Fig. 6. The error bars indicate estimated systematic errors. The solid curves in Figs. 3 and 4 show correlation functions calculated for Gaussian sources for which the radius parameter r_0 was assumed to exhibit the dependence on total momentum shown in Fig. 6. In these calculations, appropriate averages over total momentum and the calculated resolution of the hodoscope were taken into account. The overall trends of the data are well described. However, for the $^{14}N+^{197}Au$ reaction, the adopted parametrization fails to reproduce the exact shape of the minimum at q≈0 MeV/c for the low momentum gate, P=270-390 MeV/c. Here, contributions from much larger

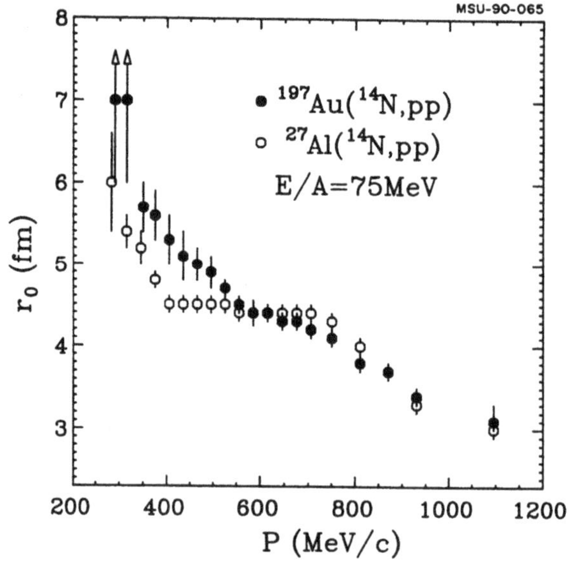

Fig. 6. Radius parameters, r_0, for Gaussian sources of negligible lifetime extracted from two-proton correlation functions gated by different total momenta, P, of the coincident particle pairs for ^{14}N induced reactions on ^{27}Al and ^{197}Au at E/A=75 MeV. The error bars indicate estimated systematic errors.

sources or, alternatively, from long lived evaporative processes would be required to fit the data. We did not attempt to improve the description by assuming a superposition of sources of different dimensions or life-times.

While the average correlation functions measured for the $^{14}N+^{27}Al$ and $^{14}N+^{197}Au$ reaction are very similar (see Fig. 2), significant differences surface when one explores the dependence on the total momentum of the proton pairs. Such more subtle differences, already apparent in the raw data shown in Figs. 3 and 4, are clearly revealed in Fig. 6. For the $^{14}N+^{197}Au$ reaction, the extracted source dimensions exhibit a nearly monotonic increase with decreasing total momentum of the detected particle pair. For the $^{14}N+^{27}Al$ reaction, on the other hand, the extracted source dimensions are rather constant over the range of $P \approx 400$-750 MeV/c. The extracted source dimensions are comparable for the two targets at high total momenta, $P \gtrsim 800$ MeV/c, indicating that very energetic particles are emitted by comparable processes. At low total momenta, $P \lesssim 500$ MeV/c, the extracted source dimensions are considerably larger for the reactions on ^{197}Au than for the reactions on ^{27}Al, most likely indicating larger contributions from slow evaporation processes for the ^{197}Au target.

7. EVAPORATIVE EMISSION

We have used the statistical model of ref. [12] to construct Wigner functions for evaporative emission from equilibrated compound nuclei. In this model, the average particle emission is calculated from the Weisskopf formula and cooling of the compound nucleus is calculated from the average mass and energy emission rates. Sub-barrier emission is not incorporated because of the use of the sharp cut-off approximation for the inverse cross sections. For simplicity, the level density is approximated by that of an ideal Fermi gas.

In Fig. 7, results of these evaporation calculations are compared with two-proton correlation functions measured for the $^{129}Xe+^{27}Al$ and $^{129}Xe+^{122}Sn$ reactions. The curves represent calculations which were folded with the resolution of the experimental apparatus and averaged

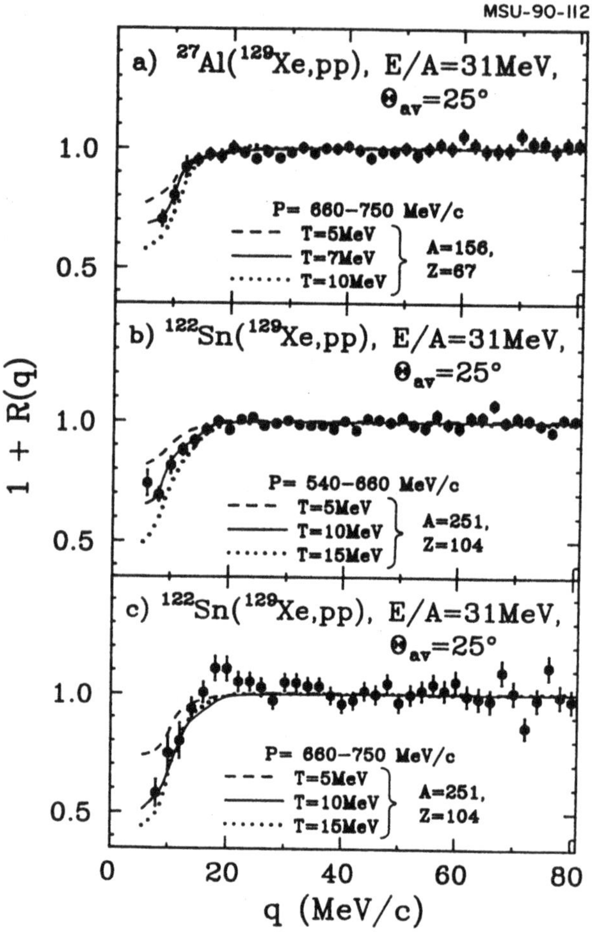

Fig. 7. Two-troton correlation functions measured for the $^{129}Xe+^{27}Al$ (part a) and $^{129}Xe+^{122}Sn$ (parts b,c) reactions at E/A=31 MeV for the indicated gates on the total momenta, P, of the proton pairs. The curves represent calculations for evaporative sources with parameters indicated on the figure.

over the appropriate momentum bins using the experimental proton yields as relative weights. The parameters used in these calculations (initial mass, A, charge, Z, and temperature, T, of the decaying nucleus) are indicated in the figure; normal nuclear density was assumed. In order to illustrate the sensitivity of the calculated emission rates to the initial temperature of the emitting system, we used the compound nucleus values for A and Z, but treated the temperature as a free parameter. For complete fusion of $^{129}Xe+^{27}Al$ and $^{129}Xe+^{122}Sn$, temperatures of 8.2 and 10.3 MeV, respectively, are calculated if one assumes the level density of an ideal Fermi gas of normal nuclear matter density; the more common relation, $T^2=(8\ MeV)E^*/A$, gives values of 5.8 and 7.3 MeV. However, the equilibrated emitting systems should have temperatures which are somewhat lower than those calculated for compound nuclei formed in complete fusion reactions [5,17-19] since some energy is carried away by preequilibrium emission. Good agreement between calculations and data is obtained for initial temperatures of about 7-10 MeV, see Fig. 7.

We have explored the dependence of the two-particle correlation function on the angle, $\Psi = \mathrm{acos}(\vec{P}\cdot\vec{q}/Pq)$, between the relative and total momentum vectors of the two-proton pair to search for clues on the source lifetime and shape [1,7,8,11]. Emission from long-lived systems produces Wigner functions elongated in the longitudinal direction [7,8]. Because of the reduced Pauli anti-correlation in this direction, the longitudinal correlation function ($\Psi\approx0°$ or 180°) of a long-lived source may be enhanced compared to the transverse correlation function ($\Psi\approx90°$), unless the average particle separations become so large that sensitivity to antisymmetrization effects is lost. Figure 8 shows longitudinal and transverse two-proton correlation functions measured for the $^{129}Xe+^{27}Al$ (top panel) and the $^{129}Xe+^{122}Sn$ reactions. The longitudinal correlation functions, shown by solid points, were evaluated for the gate $|\cos\Psi|\geq0.77$ (corresponding to the angular cuts of Ψ=0°-40° or 140°-180°). The transverse correlation functions, shown by open points, were evaluated for the gate $|\cos\Psi|\leq0.5$ (corresponding to the angular cut of Ψ=60°-120°). For improved statistical accuracy, the gates on the total momenta of the proton pairs were made wider than in Fig. 7; the values are indicated in the figure. No statistically significant difference between longitudinal and transverse correlation functions is visible. However, this result does not contradict theoretical expectations for evaporation from long-lived compound nuclei. The solid and dotted curves in Fig. 8 show longitudinal and

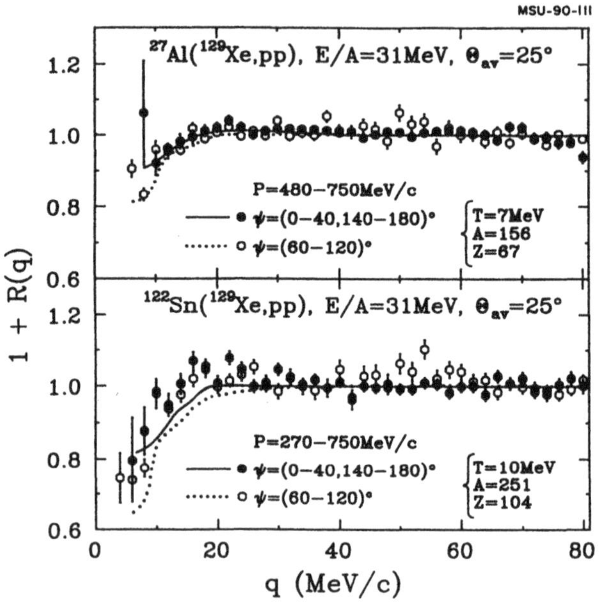

Fig. 8. Longitudinal (Ψ=0°-40° or Ψ=140°-180°) and transverse (Ψ=60°-120°) two-proton correlation functions measured for the reaction ^{129}Xe+^{27}Al (top panel) and ^{129}Xe+^{122}Sn (bottom panel) at E/A=31 MeV. The normalizations were determined from the Ψ-integrated data.

transverse correlation functions calculated for evaporative emission using the parameters indicated in the figure. The calculations were averaged over the appropriate momentum bins and folded with the resolution of the experimental apparatus. The predicted differences between transverse and longitudinal correlation functions are of the order of a few percent and, therefore, below the statistical sensitivity of the present experiment.

8. NONEQUILIBRIUM EMISSION

We have used the semiclassical Boltzmann-Uehling-Uhlenbeck (BUU) equation [20,21] to obtain Wigner functions for non-equilibrium emission processes. The BUU equation describes the temporal evolution of the one-body density under the influence of the nuclear mean field and individual nucleon-nucleon collisions. The Pauli exclusion principle is included semiclassically by suppressing collisions into occupied regions of phase space.

In our calculations, we approximated the mean field potential by a density-dependent Skyrme-type parametrization,

$$U(\rho) = -124(\rho/\rho_0) + 70.5(\rho/\rho_0)^2. \quad (6)$$

This parametrization is conventionally referred to as a "stiff" equation of state with nuclear compressibility, K=380 MeV. The in-medium nucleon-nucleon cross section is approximated by the energy-dependent free nucleon-nucleon cross section, parametrized from experimental data. The calculations were terminated at a time of 160 fm/c after the nuclei came into contact. Nucleons were considered as emitted when the surrounding density fell below $\rho=\rho_0/8$, where ρ_0 denotes the density of normal nuclear matter.

In Fig. 9, preliminary results of such calculations, averaged over

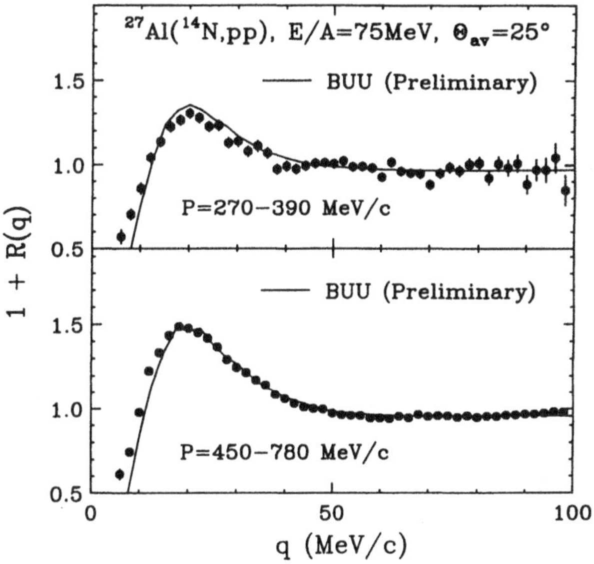

Fig. 9. Comparison of experimental and theoretical two-proton correlation functions for $^{14}N+^{27}Al$ collisions at E/A=75 MeV for the indicated gates on the total momenta, P, of the proton pairs. The calculations are explained in the text.

impact parameter and appropriate momentum bins, are compared to the experimental correlation functions for the $^{14}N+^{27}Al$ reaction. Because of lack of statistical accuracy, we only present calculations for the lower two momentum bins shown in Fig. 3. These preliminary results are encouraging: the calculations reproduce the overall magnitude of the measured correlation function, indicating that the overall time-scale of the collision process is correctly predicted by the BUU equation.

9. ACKNOWLEDGEMENTS

This work is based upon work supported by the National Science Foundation under Grant numbers PHY-86-11210 and PHY-89-13813 and the Department of Energy under grant number DE.FG602-88ER.40404.A000. WGL acknowledges the receipt of U.S. Presidential Young Investigator Award and NC acknowledges partial support by the FAPESP, Brazil.

10. REFERENCES

1. S.E. Koonin, Phys. Lett. 70B,43(1977).
2. W.G. Lynch et al., Phys. Rev. Lett. 51,1850(1983).
3. Z. Chen et al., Phys. Rev. C36,2297(1987).
4. Z. Chen et al., Phys. Lett. B186,280(1987).
5. J. Pochodzalla et al., Phys. Rev. C35,1695(1987).
6. J. Pochodzalla et al., Phys. Lett. B174,36(1986).
7. S. Pratt and M.B. Tsang, Phys. Rev. C36,2390(1987).
8. T.C. Awes et al., Phys. Rev. Lett. 61,2665(1988).
9. P.A. DeYoung et al., Phys. Rev. C39,128(1989).
10. J. Quebert et al., Proc. Sympos. Nucl. Dynamics and Nucl. Disassembly, Dallas, April 1989, ed. J.B. Natowitz, World Scientific, Singapore 1989, p. 337.
11. D. Ardouin et al., University of Nantes, Internal Report LPN-89-02.
12. W.A. Friedman and W.G. Lynch, Phys. Rev. C28,16(1983).
13. J. Aichelin and G. Bertsch, Phys. Rev. C31,1730(1985).
14. W. Cassing, Z. Phys. A329,471(1988).
15. W.G. Gong et al., Nucl. Instr. and Meth. A268,190(1988).
16. B.A. Brown et al., J. Phys. G10,1683(1984).
17. R. Wada et al., Phys. Rev. C39,497(1989).
18. D.X. Jiang et al, Nucl. Phys. A503,560(1989).
19. H.M. Xu et al., Phys. Rev. C40,186(1989).
20. G.F. Bertsch and S. Das Gupta, Phys. Rep. 160,L89(1988), and references given there.
21. A detailed account of the numerical details is given in W. Bauer, Michigan State University report MSUCL-699, submitted to Phys. Rev. C.

ANALYSIS OF PREEQUILIBRIUM TWO-PARTICLE CORRELATIONS AT LARGE RELATIVE MOMENTA

Z. Basrak[1,2]

D. Ardouin[2], P. Schuck[3], A. Peghaire[4], F. Saint-Laurent[4], H. Delagrange[4], H. Doubre[4], C. Gregoire[4], A. Kyanowski[4], W. Mittig[4], J. Peter[4], Y.P. Viyogi[5], J. Quebert[6], C.K. Gelbke[7], W.G. Lynch[7], M. Maier[7], J. Pochodzalla[7], G. Bizard[8], F. Lefebvres[8], B. Tamain[8], B. Remaud[2], F. Sebille[2]

[1]Ruđer Bošković Institute, 41001 Zagreb, P.O.B.1016, Yugoslavia

[2]Laboratoire de Physique Nucleaire, IN2P3/CNRS et Universite de Nantes, 2 rue de la Houssiniere, F-44072 Nantes Cedex 03, France

[3]ISN, 53 avenue des Martyrs, F-38026 Grenoble Cedex, France

[4]GANIL; B.P. 5027, F-14021 Cean Cedex, France

[5]VECC, Calcutta 700064, India and GANIL

[6]CEN Bordeaux, F-33170 Gradignan, France

[7]NSCL, Michigan State University, East-Lansing, MI 48824, USA

[8]LPC Universite de Caen, F-14032 Cedex, France

ABSTRACT

Both the polar and the azimuthal dependence of particle-particle correlation functions at large relative momenta are measured at 60 MeV/A for asymmetric (Ar+Au) and nearly symmetric (Ar+Ti) systems. The experimental data and dynamical simulations performed using the Landau-Vlasov semiclassical equation show persisting strong mean-field effects at 60 MeV/A. The general correlation-function pattern is perturbed via dissipative effects of nucleon-nucleon collisions. Theoretical studies are carried out also at lower energy (25 MeV/A) and compared with existing experimental data.

INTRODUCTION

Two-particle correlations at large relative momenta can yield precious information on the dynamics of heavy-ion collisions. At 800 MeV/n these observables show that even 50% of the collision cross section for light systems is due to pure nucleon-nucleon scattering[1]. In contrast to this, at 25 MeV/n the experimental results were entirely explained by mean-field effects, namely, by collective and transverse motion of nucleons in the reaction plane [2]. It was interesting to investigate the region above the Fermi energy in order to see whether two-body effects show up and whether the nuclear mean field is still of some importance, especially because the disappearance of complete fusion was attributed to the weakening of the nuclear mean field around 45 MeV/n.

In the present work [3] the correlation of light particles at large relative momenta was studied both experimentally (60 MeV/n) and theoretically (25 and 60 MeV/n). A reliable theory should include contributions from both nuclear mean field and residual nucleon-nucleon collisions. It is possible to consider such a situation by using a dynamical theory, like the semiclassical Landau-Vlasov approach with the Ühling-Uhlenbeck collision term.

EXPERIMENT

Systems Ar+Au and Ar+Ti at 60 MeV/n

The 60 MeV/n argon beam supplied by the GANIL facility was used to study correlation functions between light particles for a heavy asymmetric (Ar+Au) and a nearly symmetric (Ar+Ti) system. Particle telescopes discerned Z=1 and 2 isotopes and were mounted onto two half-planes crossing at the beam axis: one of them was movable around the axis, covering the relative azimuthal angle between half-planes from $\Delta\Phi = 0^{\circ}$ to 180° in steps of 45°. In such a way, both the polar (Θ) and the azimuthal ($\Delta\Phi$) dependence of correlated emission of light particles was investigated at large relative angles between pairs of detectors (see inset in Fig. 5 for definition of angles).

The experimental correlation function is defined as the ratio of the coincident cross section $\sigma_{xy}(\Theta_1, \Theta_2, \Delta\Phi)$ between particles x emitted in the (Θ_1, Φ_1) direction and particles y emitted in the $(\Theta_2, \Phi_1 + \Delta\Phi)$ spatial direction and the singles cross sections $\sigma_x(\Theta_i)$, $\sigma_y(\Theta_2)$ of the respective particles:

$$R(\Theta_1, \Theta_2, \Delta\Phi) = \frac{\sigma_{xy}(\Theta_1, \Theta_2, \Delta\Phi)}{\sigma_x(\Theta_1) \cdot \sigma_y(\Theta_2)} \quad . \tag{1}$$

The main features of the present data are as follows:

For both investigated systems, the experimental correlation functions exhibit a nearly isotropic azimuthal dependence $R(\Delta\Phi)$ (see Fig. 1). This feature is largely independent of the nature of particles in coincidence and particular values of polar angles Θ_1 and Θ_2 of the detectors. For complex particles, however, a slight enhancement of opposite-side emission is observed and could be understood as being due to linear-momentum conservation [2].

Similarly, the polar experimental correlation functions $R(\Theta)$, obtained by fixing the trigger detector at Θ_1 and the relative polar angle $\Delta\Phi$, roughly exhibit the same pattern, independently of the choice of the angles $\Theta_1, \Delta\Phi$, or the nature of particles in coincidence (see Fig. 2). The function $R(\Theta)$ displays a maximum around $\Theta = \Theta_2 \approx 70^{\circ}$ for the heavy system (Ar+Au) and 30° for the lighter system (Ar+Ti), respectively.

The analysis was performed only for preequilibrium particles ($E > 20$ MeV for protons). Lower-energy particles ($E < 25$ MeV) gave structureless correlation patterns.

Fig. 1. Light-particle azimuthal correlations $R(\Delta\Phi)$ measured for the Ar+Au (left-hand side) and Ar+Ti (right-hand side) reactions at 60 MeV/n. The lines are just to guide the eye.

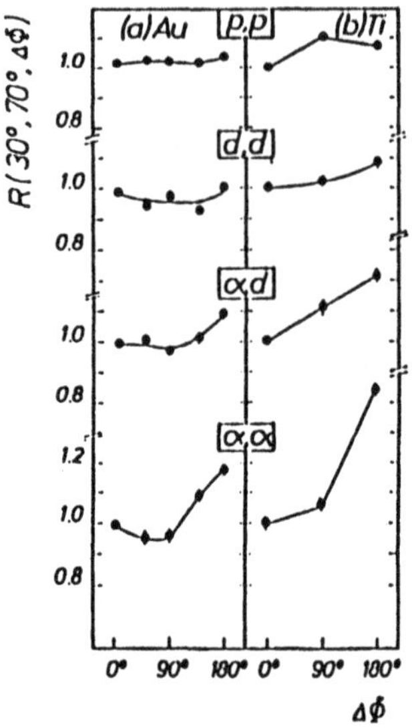

Fig. 2. Light-particle polar correlations $R(\Theta)$ at fixed $\Delta\Phi = 0^{\circ}$ (left-hand side) and 180° (right-hand side) for Ar+Au (left pannel) and Ar+Ti (right pannel) reactions at 60 MeV/n. The curves are merely to guide the eye.

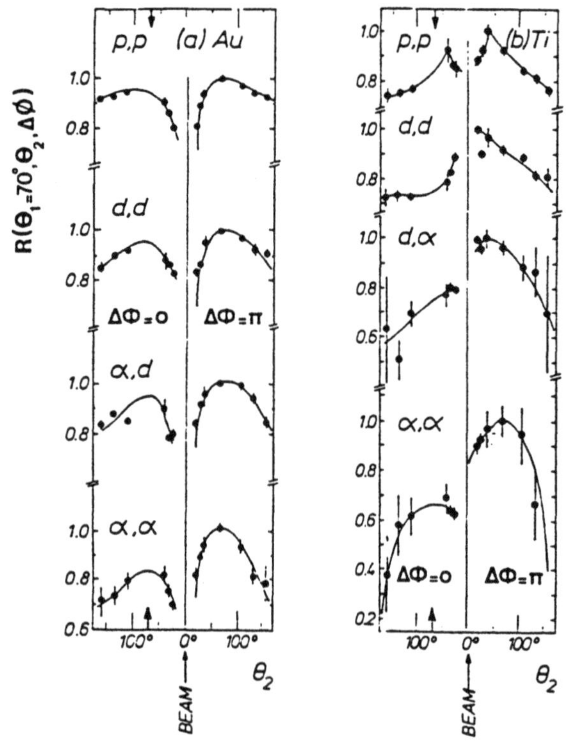

The global independence of the observed correlation functions R(ΔΦ) and R(Θ) on the detection geometry and the nature of particles in coincidence suggests that one is probing an average matter flux distribution rather than measuring genuine correlations. Therefore, one-body effects should give a considerable contribution to the observed correlation functions and the possible correct theoretical approach appears as the Landau-Vlasov equation which, on a semiclassical basis, entirely describes one-body dynamics, introducing besides the nuclear mean field also two-body effects via an explicit collision term. In the present investigation the Nantes/GANIL version [4] of the Landau-Vlasov code was used and each nucleon was represented by many test particles.

System O+Au at 25 MeV/n

At 25 MeV/n the azimuthal correlation function R(ΔΦ) displays a chain-curve behaviour with a very strong suppression of the emission out of the trigger plane [2]. This effect is pronounced for heavier emitted particles. The observation was entirely explained [2] as a superposition of random Maxwellian emission and ordered collective motion of nucleons, originating in either a rotating or a laterally moving emitting source.

THEORETICAL SIMULATIONS

We consider each test particle as carrying part of the nucleonic probability. Therefore, the ensemble-averaged emission in the given direction of observation can be interpreted as emission probability out of the collision zone, i.e. a singles distribution.

In principle, there are two distinct contributions to the correlation yields: a long-range and a short-range one.

The long-range contribution is automatically taken into account by the self-consistent field of the Landau-Vlasov equation since each test particle evolves in the mean field of all the others. On the other hand, the short-range contribution is important only in critical situations, such as pairing or resonance. The short-range contribution can be safely discarded in the present experimental conditions, so that the two-particle emission probability can be considered as the product of singles probabilities:

$$\frac{\sigma_{xy}(\Theta_1,\Theta_2,\Phi_1,\Phi_2)}{\sigma_x(\Theta_1)\,\sigma_y(\Theta_2)} \doteq \frac{\sigma_x(\Theta_1,\Phi_1)}{\sigma_x(\Theta_1)}\;\frac{\sigma_y(\Theta_2,\Phi_2)}{\sigma_y(\Theta_2)} \; . \tag{2}$$

In the actual experiment, the orientation of the reaction plane is undetermined and only the relative azimuthal angle between the detectors $\Delta\Phi = \Phi_2 - \Phi_1$ is known. To compare calculation with experiment, the above expression should be folded over the variable Φ_1 which leads to a correlation function:

$$R(\Theta_1,\Theta_2,\Delta\Phi) = \int d\Phi \;\frac{\sigma_x(\Theta_1,\Phi)}{\sigma_x(\Theta_1)}\;\frac{\sigma_y(\Theta_2,\Phi+\Delta\Phi)}{\sigma_y(\Theta_2)} \; . \tag{3}$$

The subintegral function $P(\Theta,\Phi) = \sigma(\Theta,\Phi)/\sigma(\Theta)$ is the singles-emission probability and can be readily obtained using dynamical Landau-Vlasov simulations.

Fig. 3 shows the singles-emission probability for the Ar+Au interaction, displaying a pronounced bump on the target side for $\Theta \approx 70^{\circ}$ and a smaller one at the projectile side for the same Θ. Careful examination of the large bump shows that it contains a considerable contribution coming from projectile test particles that were captured by the mean field of the target, deviated by it and ejected to the target side (see Fig. 4 for b = 5 fm and several projectile test-particle trajectories). Similarly, the small projectile-side bump is partly due to target test particles captured by the much weaker and more perturbed projectile mean field.

The expression (3) is evaluated by making cuts in the singles emission probability (compare Fig. 3) at given $\Theta = \Theta_1$ or Θ_2 values, and by folding the cut curves with an appropriate $\Delta\Phi$ shift for each impact parameter b and by averaging the obtained values over b. Following such a procedure, the calculated azimuthal and polar correlation functions are shown in Figs. 5 and 6. The model agrees very well with the observed azimuthal isotropy at 60 MeV/n and the strong out-of-plane suppression of emission at 25 MeV/n. For the polar correlation function the model perfectly reproduces the position and the width of the maxima for both the Ar+Au and the Ar+Ti systems. When the contribution from the two-body collision term is switched off, the main pattern of the polar correlation function is still reproduced with the nuclear mean field only (dashed curve in Fig. 6). The two-body collisions merely enlarge the curve without altering the position of the maximum.

CONCLUSIONS

The global features of the experimental azimuthal and polar correlation functions are satisfactorily reproduced by dynamical Landau-Vlasov simulations. The ordered collective motion of nucleons, which was attributed to the "third intermediate velocity source", finds its microscopic explanation: the flow of nucleons is mediated by the nuclear mean field and two-body collisions show up via their dissipative effects. To summarize, the global behaviour of correlation functions is due to the strongly persisting and well-defined nuclear mean field confirming the validity of the mean-field concept at 60 MeV per nucleon.

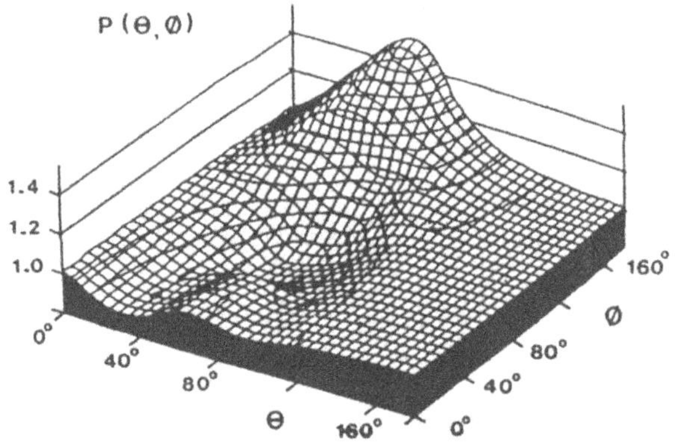

Fig. 3. Calculated prediction of the singles emission probability P(Θ,Φ) for the Ar+Au reaction at 60 MeV/n. The yields are averaged over the impact parameter at t = 240 fm/c. The projectile is impinging on the Φ=0 side of the reaction plane.

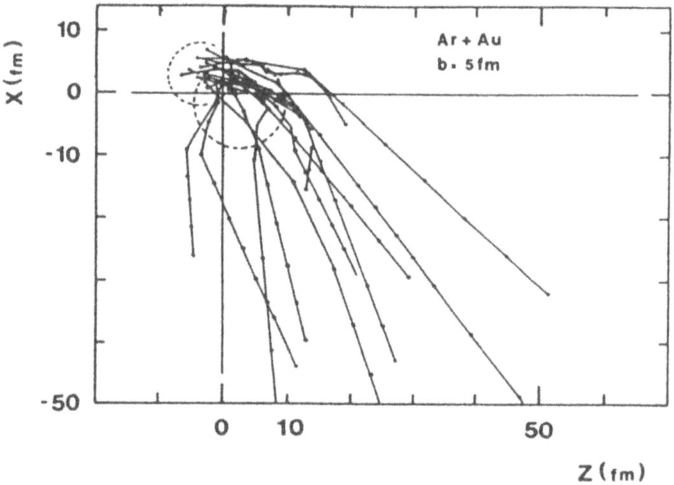

Fig. 4. Calculated trajectories of projectile test particles emerging at $\Theta_{CM} = 50^o \pm 20^o$ and $\Phi = 180^o \pm 20^o$ for the Ar+Au reaction at 60 MeV/n and an impact parameter of 5 fm.

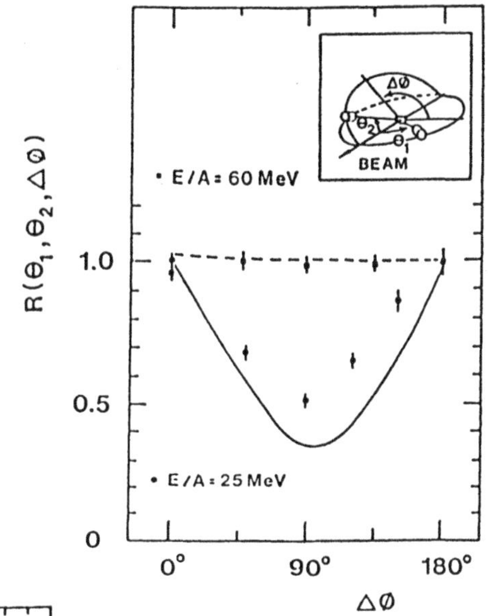

Fig. 5. Comparison of experimental (circles - O+Au at 25 MeV/n, squares - Ar+Au at 60 MeV/n) and calculated (solid line - O+Au, dashed line - Ar+Au) azimuthal correlation functions R(ΔΦ).

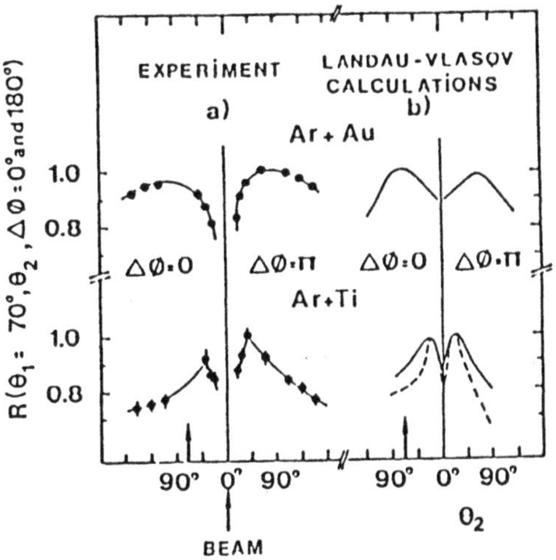

Fig. 6. Comparison of experimental (left-hand side) and Landau-Vlasov predicted (right-hand side) polar correlation functions P(Θ) for the Ar+Au and Ar+Ti reactions at 60 MeV/n. The curves through the data are merely to guide the eye.

REFERENCES

1. S. Nagamiya et al., Nucl. Phys. A335, 517 (1980).
2. M.B. Tsang et al., Phys. Lett. 148B, 265 (1984).
3. D. Ardouin et al., Nucl. Phys. A (in press).
4. C. Gregoire et al., Nucl. Phys. A465, 317 (1987).

ESTIMATION OF THE SIZES OF HOT NUCLEAR SYSTEMS FROM PARTICLE-PARTICLE LARGE ANGLE KINEMATICAL CORRELATIONS*

J.L. LAVILLE+, A. BADALA++, R. BARBERA++, G. BIZARD+, D. DURAND+, G.M. JIN+++, A. PALMERI++, G.S. PAPPALARDO++ and E. ROSATO++++

+ Laboratoire de Physique Corpusculaire - LA34-ISMRA-IN2P3-CNRS-Université de Caen -14032 Caen Cedex (France)
++ Istituto Nazionale di Fisica Nucleare, Corso Italia 57, 95129 Catania (Italy)
+++ GANIL, BP 5027, 14021 CAEN Cedex (France)
++++Dipartimento di Scienze Fisiche e Sezione INFN, Università di Napoli, Mostra d'Oltremare, Pad. 20, 80125 Napoli (Italy)

ABSTRACT

Light fragment emission, when triggered by large transverse momentum protons shows specific kinematical correlations due to recoil effects of the excited emitting source. Such effects have been observed in azimuthal angular distributions of He-particles produced in collisions induced by 94 MeV/u ^{16}O ions on Al, Ni and Au targets. A model calculation assuming a two-stage mechanism (formation and sequential decay of a hot source) gives a good description of these whole data. From this successfull confrontation, it is possible to estimate the size of the emitting system.

1. Introduction

The formation and decay of very excited nuclear sub-systems is of specific interest in the intermediate energy range of a few ten MeV/u. At low energy, the main process is the fusion/fission reaction which leads generally to a cold residue nucleus after de-excitation by evaporation of few particles.

This is not so at higher beam energies where the collision is governed by geometrical effects (i.e. the "fireball" phenomenon[1]). This reactionnal scenario, which corresponds to the participant spectator picture is more complicated at intermediate energies where dissipative effects may wash out the initial collision geometry. Nevertheless, if we assume such a participant/spectator mechanism, the final state of the reaction is at least composed of three nuclear parts, i.e. three sources : a projectile-like one, a target-like one and the hot participant zone which are characterized by different "sizes" (by "size", we mean the number of constituant nucleons), velocities and "temperatures".

One method to measure the size of nuclear interacting systems is to perform particle-particle correlation at low relative momentum using interferometry concepts. This approach leads to the space-time extent of the reaction from a dynamical point of view[2]. An other way consists in using momentum balance effects to identify the emitting source.

---

* Experiment performed at GANIL

Within the frame of the participant-spectator picture implying three components, the first problem is to disentangle the various final state channels in order to isolate the specific contribution of the hot source. Such an aim needs to perform exclusive experiments, i.e. to be at least able to identify the major part of the decay products of the reaction. As we shall see in this paper, it is possible to derive the velocity, "temperature" (slope parameter) and size of the source from the experimental angular and velocity distributions, with the help of a convenient trigger which selects central collisions. In the present case, that trigger was a high transverse momentum proton[3,4].

The use of large area multidetectors at GANIL enabled us to detect light particles emitted in collisions induced by ^{16}O projectiles impinging on Al, Ni and Au targets, for a beam energy of 94 MeV/u, over a very large solid angle.

2. Experimental details and data overview

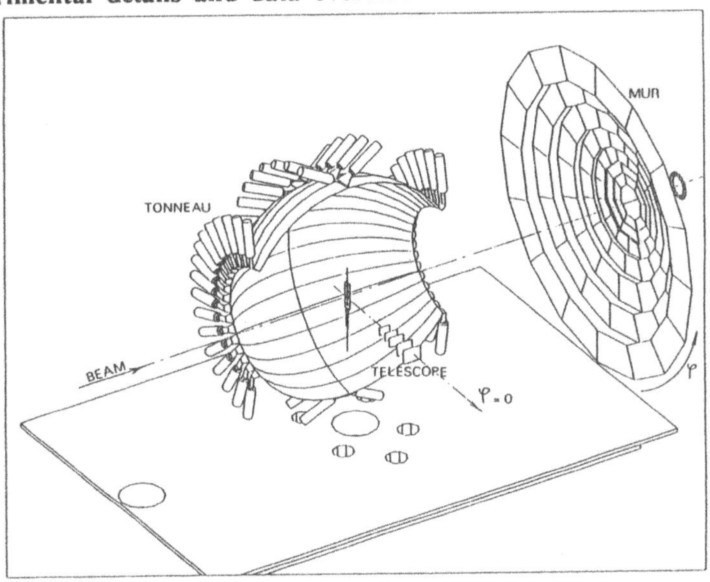

Fig. 1 : Experimental set-up.

An exhaustive description of the experimental lay-out has been done in previous papers[3,4]. We recall here the main features of the apparatus (fig. 1) : light nuclei (from Z=1 to 8) were detected by two large area multidetectors, the "MUR" and "TONNEAU" installed in the vacuum chamber "NAUTILUS" at GANIL. They covered a polar angle θ range between 3° and 150° with full azimuthal acceptance for the "MUR" ($3° \leq \theta \leq 30°$) ; with half azimuthal acceptance ($90° < \phi < 270°$) for the "TONNEAU" ($30° \leq \theta \leq 150°$). Both multidetectors were able to provide θ, ϕ and velocity measurements of the detected particles.

The triggering high energy (~150 MeV) proton was detected by a range telescope set at $\phi = 0°$ for $\theta_p = 70°$, 90° and 120°.

In this paper, we shall derive the source size estimate using only coincident events between a high energy proton and He-particles detected in multi-detectors in the system ^{16}O (^{27}Al, p He)X.

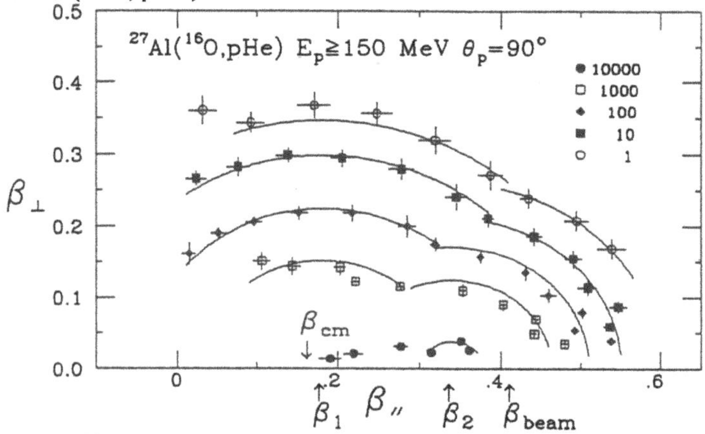

Fig. 2 : Contour plot of the invariant cross section of He velocities for proton-triggered events.

The reaction mechanism is illustrated by fig. 2 presenting an invariant contour plot of He-particle velocities for such coincident events : the emission of He-particles occurs from two sources, the first one with ß~.36 ($ß_{beam}$=.417), the second one with a velocity close to the center-of-mass one. Using a classical moving source procedure, we found for the latter a temperature of 20.±1 MeV, and a velocity ß=0.18±-0.01. This value agrees with the calculated one $ß_F$=0.19 from the fireball model of ref.[1]. This suggests a participant/spectator scenario. However the third (target-like) component is not observed because of the experimental velocity threshold. A double component contribution is also visible in the He-particle velocity spectra (fig. 3). We see that the projectile-like component disappears for polar angles θ_{He} larger than ~20°.

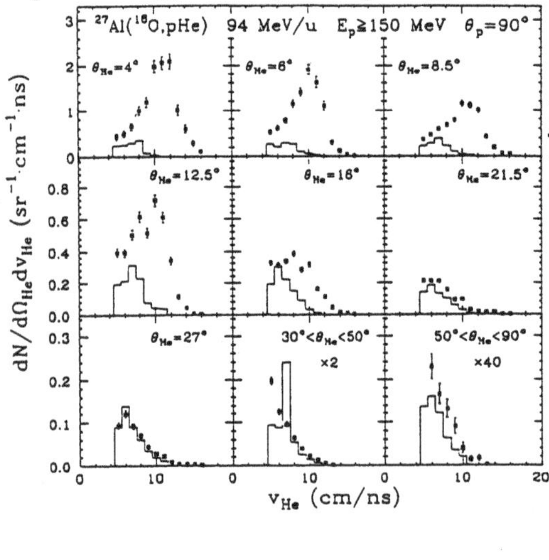

Fig. 3 : He particles velocity spectra for proton-triggered events. The histograms represent the model calculation detailed in ref.[4].

The most interesting characteristics of these data stands in the azimuthal distributions of He-particles displayed in fig. 4-a for θ_{He} ranging from 4° to 90°. We clearly notice that these distributions peak at $\phi \sim 180°$, i.e. in the opposite direction of the triggering proton which carries away at 90° a high transverse momentum of > 500 MeV/c, and the ϕ–asymmetry varies with θ_{He}. This variation is quantified in table I. The quoted asymmetries are defined by :

$$a(\theta_{He}) = \frac{F(\phi=\pi) - F(\phi=0)}{F(\phi=\pi) + F(\phi=0)}$$

where :

$F(\phi) = A(\theta_{He}) + B(\theta_{He})\cos\phi$ is a parametrization of the ϕ-distributions. They are reported for various target and trigger angle combinations ; these $a(\theta_{He})$ values significantly increase between 4° and 30° for every system, but they do not seem to be target dependant. So we may point out that the whole He distributions exhibit the same qualitative features whatever the targets (Al, Ni or Au) or trigger angles (θ_p=70°, 90° or 120°).

The flatness of the homolog ϕ distributions presented in fig. 4-b for inclusive events (i.e. He emission without the proton trigger request) shows that this typical behaviour is not due to an experimental bias.

Fig. 4 : He-Particles azimuthal distributions :
a) for proton-triggered events. The histograms represent the model calculation detailed in ref.[4].
b) for inclusive He emission.

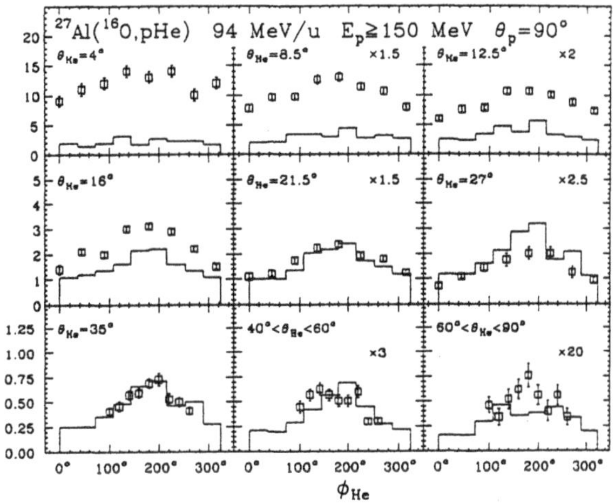

Fig. 4a)

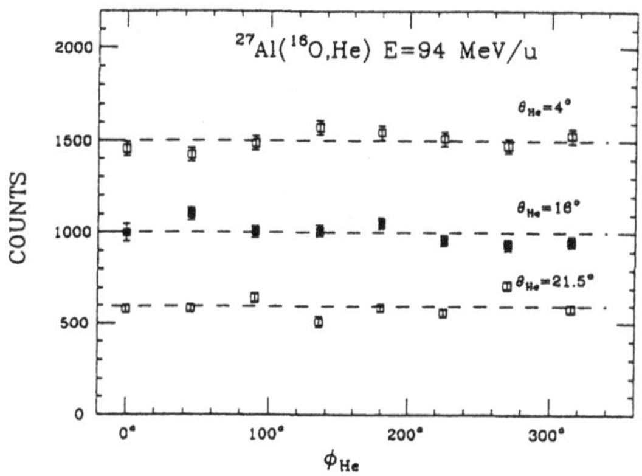

Fig. 4b)

TABLE 1 : Asymmetry parameters $a(\theta_{He})$ of azimuthal He distributions

System (target/θ_p)	Al /70°	Al /90°	Ni /70°	Ni /90°	Au /70°	Au/90°
θ_{He}						
4°	.19±.05	.16±.06	.11±.03	.10±.07	.27±.10	.11±.11
6°	.25±.03	.29±.06	.22±.02	.10±.05	.19±.07	.10±.10
9°	.31±.02	.25±.04	.26±.03	.20±.04	.25±.05	.29±.05
12°5	.38±.03	.26±.04	.29±.03	.26±.03	.28±.03	.16±.05
17°5	.47±.03	.39±.05	.38±.03	.27±.04	.37±.04	.29±.06
23°	.40±.04	.33±.06	.30±.04	.21±.06	.13±.05	.18±.04
29°	.43±.06	.30±.12	.20±.04	.19±.08	.05±.10	.04±.06

3. Source size estimate from the correlations measurements

From the observed ϕ behaviour we may assume the following scenario : both proton and He particle are emitted by a hot participating zone, and the ϕ-angular correlations reflect recoil effects of the emitting source. Following this picture, the mass of the source can be kinematically related to the experimental ϕ-distributions. The histograms in fig. 3 and 4-a arise from a model calculation assuming a two-step reaction process : formation and sequential decay of a highly excited zone. We have used the fireball geometry for the first stage and the sequential Weisskopf theory for the second one. This calculation is detailed in ref[4]. We observe a good agreement with both velocity and ϕ spectra only for large polar angles. We cannot expect that such an approach could reproduce the forward contribution, since it only describes the low velocity source decay. This agreement is obtained without any fitting parameter. The mean number of participating nucleons, i.e. the "size" of the zone, which derives from the model is N_s~30 (13 nucleons from the ^{16}O projectile nucleus, 17 from the Al target one). In order to illustrate the strong dependance of the ϕ–distribution upon the sub-system "mass", fig. 5 displays three calculated ϕ-angle spectra together with the experimental one : the Ns=30=13+17 value appears to be a realistic estimate. From this calculation, we can also derive the mean impact parameter (~2 fm) for the class of collisions selected by the detected trigger energies.

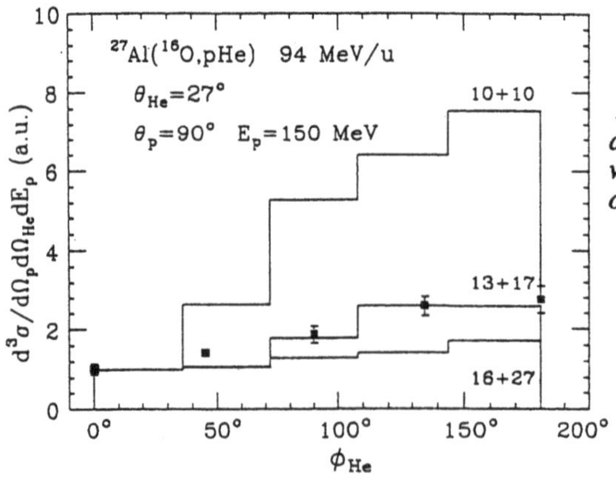

Fig. 5 : A typical experimental ϕ-distribution (points) confronted with the calculation for three sets of nucleon numbers (see text).

4. Discussion

A question has now to be raised : is the proposed mechanism of formation and sequential decay of a hot zone the only possible one to describe the He-p correlations ? An alternative explanation could be a simple quasi-free knock-on process between an α-particle from the projectile and a proton from the target: if so, the observed He emission should reflect the internal Fermi motion of the initial nuclei. Experimental evidence for a similar process have been reported[5] at much higher energy (800 MeV/u) and in the system Ar+Au at 60 MeV/u[6]. Kristiansson and al., who analysed[7] particle coincidence data for the systems ^{12}C + targets at 85 MeV/u suggested a possible direct scattering contribution in the α–α case, but not in the p-p one. A dynamical calculation of such a process appears to be very difficult if one wants to account for a realistic description of internal He and proton Fermi motion, off-shell quasi-elastic cross sections... We only attempt to estimate if such a process could contribute from a purely kinematical point of view. Fig. 6 shows a comparison of the experimental ϕ and velocity spectra with a crude calculation of a free α-p scattering, assuming a gaussian-like Fermi-momentum distribution of the α and proton inside their respective initial nuclei. We used the prescription of ref[8] for a correct treatment of energy-momentum conservation. The outgoing proton is ejected at θ_p=90°, with a kinetic energy of 150 MeV as in the experiment. Since forward ϕ spectra show (see table I and fig. 4a) significant asymmetries which cannot be accounted by sequential decay of the hot source, we supposed that the θ_{He} = 4° cross-section is exhausted by this quasi-elastic contribution. Following this assumption, the calculated distributions at 4° (fig. 6a,b) are normalized to the data.We see that such a knock-on α-p mechanism qualitatively reproduces the velocity and ϕ distributions at θ_{He} = 4° ; but, at θ_{He} = 21°5, it only accounts for the slight high velocity component.

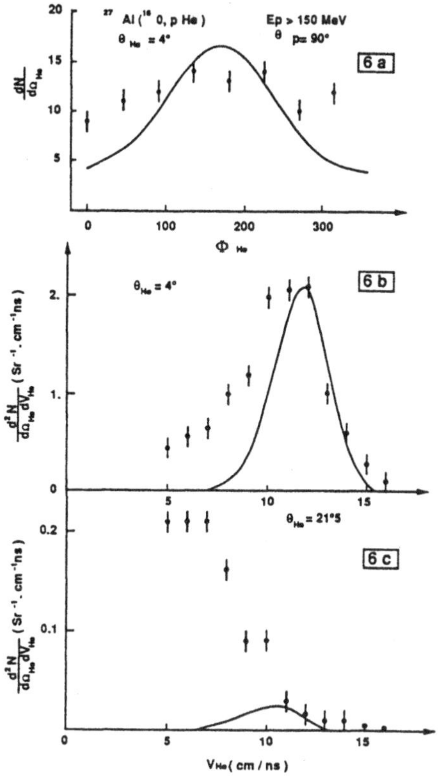

Fig. 6 : Azimuthal (a) and velocity (b,c) distributions confronted with a simple knock-on hypothesis (continuous curves) for θ_{He}=4° (a,b) and θ_{He}=21°5.

5. Conclusion

We have shown that it is possible to derive the number of nucleons of an excited reaction zone from specific angular correlations. These results clearly rest upon the hypotheses of the model calculation ; however the good agreement with the whole data in the kinematical zone of the hot source particle decay justifies such an approach. Finally, the presented analysis can also provide an estimate of the collision impact parameter.

References

1. J. Gosset et al, *PR* C16 (1977) 629
2. D. Ardouin et al, *Z. Phys.* A329 (1988) 505
3. S. Aiello et al, *Europhys. Lett.* 6 (1988) 25
4. D. Durand et al, *Nucl. Phys.* A511 (1990) 442
5. S. Nagamiya et al, *Phys. Lett.* B81 (1979) 147
   J. Tanihata et al, *Phys. Lett.* B100 (1981) 121
6. D. Ardouin, *Proc. of the XXV Int. Wint. Meet. on Nucl. Phys. Bormio* (1987)
7. P. Kristiansson et al., *Phys. Lett.* B155 (1985) 31
8. R. Shyam and J. Knoll, *Nucl. Phys.* A426 (1984) 606

Π – P CORRELATIONS AND Δ MEASUREMENTS IN RELATIVISTIC HEAVY-ION COLLISIONS

D. L'HOTE[a], C. CAVATA[a], J. GOSSET[a], M.-C. LEMAIRE[b], J. POITOU[a], J.-P. ALARD[c], J. AUGERAT[c], R. BABINET[a], N. BASTID[c], F. BROCHARD[d], P. CHARMENSAT[c], M. DEMOULINS[a], P. DUPIEUX[c], L. FRAYSSE[c], P. GORODETSKY[d], B. LUCAS[a], J. MARRONCLE[c], G. MONTAROU[c], M.J. PARIZET[c], C. RACCA[d], A. RAHMANI[c], Y. TERRIEN[a], O. VALETTE[a].

[a] *DPhN/SEPN, CEN Saclay, F-91191 Gif sur Yvette Cedex.*
[b] *LNS, F-91191 Gif sur Yvette Cedex.*
[c] *LPC Clermont-Ferrand, BP 45, F-63170 Aubière.*
[d] *CRN, BP 20 CR, F-67037 Strasbourg Cedex.*

ABSTRACT

Observables related to pion-proton correlations measured in nucleus-nucleus collisions in the energy range of a few hundred MeV per nucleon are presented. A special emphasis is put on delta measurements and on pion "flow" studies using the Diogene detector. In what concerns delta measurements, the technique for background estimate and subtraction is discussed in detail. It is based on the assumption that the nucleon-pion pairs from the background are not correlated, while those emitted by detas are correlated. The pion "flow" main trends are presented. Their determination requires the measurement of a second type of baryon-pion correlations. They come from the fact that the momenta of the particles measured in an event are correlated because each of them is correlated to the impact parameter vector.

1. Introduction

1.1 Relativistic heavy ion experiments and $\pi - p$ correlations

The study of dense and hot nuclear matter properties is a major goal of relativistic ($\sim 0.1GeV < E_{inc.}/A <\sim 2GeV$) heavy ion experiments. The pioneering experiments at the Berkeley Bevalac[1,2], together with the first theoretical investigations using thermal[3], phase-space[4], cascade[5-7], hydrodynamics[8,9]... models showed that only little could be learned on nuclear matter from inclusive data. From a mathematical point of view, this is because most of the measurable information (i.e. the n-body distribution function in p space for each type of particle) is lost when we look at inclusive data (i.e. the one-body distribution function). The latter results from an integration of the former, which suppresses the *correlations*. Note however that even using an ideal detector that measures all what goes out from each collision, we still measure neither the **r** part of the n-body distribution function, nor its time evolution.

In the field of relativistic nucleus-nucleus collisions, the main progress made to overcome the inclusive experiments limitations was the construction of large solid angle multiparticle detectors. Among them, let me quote the Plastic Ball[10] and the Streamer Chamber[11] installed at the Bevalac in Berkeley, Diogene[12] at Saturne in Saclay, and the 4π-Detector[13] now being built at SIS in Darmstadt. Using such devices, many correlations can be reached since the multiplicities vary from 2 to more than 200 ! (note however the limitations due to the statistics). During the last years, the correlations which have been mostly used to obtain significant informations on the dynamics of the reaction are rather trivial: they are due to the fact that the particles in a given event are correlated because the momentum of each of them is *correlated to the impact parameter* **b** of the event. The impact parameter **b** contains two informations: its magnitude b, and the azimuth of the reaction plane ϕ_b. I will tackle the question of reaction plane determination in the second part of this talk. In what concerns the impact parameter magnitude, the main point is that in most experiments, it is estimated using the participant or the spectator nucleon multiplicities[1,2,11,14-15] . Of course, many other correlations can be studied experimentally, and many of them are presented in this workshop.

In this talk, I would like to concentrate mainly on the measurements of two types of correlations between pions and protons which have been investigated using Diogene. The first one comes from the delta decays and the second one from the pion "flow". Note that both can be investigated because of the specificity of the Diogene detector (a cylindrical set of drift chambers surrounding the target and placed in a magnetic field[12]) which can measure with large statistics both positive and negative pions together with protons, deuterons, etc.... In what concerns the π-p correlations due to Δ's, only results from nucleus-nucleus collisions will be presented. A study of Δ production in p+A reactions at Diogene is also in progress[16,17]. Before presenting these results, let us mention two other kinds of $\pi - p$ correlations which contain a very interesting physical information. The first one is the correlation between the pion multiplicity and the participant proton multiplicity. It has been studied at the Streamer Chamber[11,18,19] and at the "Magnetic spectrometer and multiplicity array[20]" at Berkeley; at Dubna[21,22]; and at Diogene for He+A, p+A, and (Ne or Ar)+A collisions[23,16,24]. The most significant result from such studies is the mean pion multiplicity $<N(\pi)>$ dependence as a function of the participant proton multiplicity. If we assume the latter to be proportional to the system volume, this dependence can help to determine whether pions are emitted by the surface or the volume of the system, hence giving an information on their absorption and production mechanism[19,25,26]. The second one is the correlation between the pion momenta and the participant (or spectator) proton multiplicity. As the latter can be used to estimate the impact parameter[14-15], looking at pion spectra with a selection on proton multiplicity is a way to study the effect of the centrality of the collision on these spectra. Such studies have been performed in several experiments at Berkeley[18,20,25,27-31] and at Dubna[21,22]. An interesting result is that the pion spectra, like the proton one, becomes more and more isotropic when the participant proton multiplicity increases. It indicates that the pions, like the protons, are sensitive to the enhancement of stopping and equilibration of the system when the impact parameter decreases.

1.2 The Diogene experiments

The Diogene detector[12] is made of: i) a Pictorial Drift Chamber (PDC) placed in a magnetic field, which measures *in coincidence* light charged particles (π^+, π^-, p, d,...) emitted from 20° to 132° and above energy thresholds of $\sim$ 25 MeV for the pions and $\sim$ 45 MeV for the baryons (participants); ii) a Plastic Wall which measures the particles below 6°. In what follows, I will present only data from the PDC. The spectator physics measured with the wall is reported in ref. 32. The measured reactions are Ne+(NaF, Nb, or Pb) and Ar+(Ca, Nb, or Pb) for bombarding energies ranging from $E_{inc}/A = 200\,MeV$ to $E_{inc}/A = 1000\,MeV$. 100 to 300 Kevents were stored for each $(A_P, A_T, E_{inc.})$ combination. The trigger selected non peripheral reactions. Their centrality was determined from their proton-like multiplicity $M(\tilde{p})$ (a proton-like is a proton free or bound in a composite). We estimated the mean squared reduced impact parameter $\tilde{b}^2 = b^2/(R_1 + R_2)^2$ corresponding to a given $M(\tilde{p})$ interval from the total cross section in this interval[14] ($R_1 = 1.12A_P^{1/3}$ and $R_2 = 1.12A_T^{1/3}$ are the projectile and target radii in fm).

Other correlations than the $\pi - p$ ones have been investigated at Diogene. Among them, let me quote the $p - p$ correlations due to final state interaction and Fermi statistics which are related to the source size[33-34], and $\pi - \pi$ interferometry[35]. In addition, as already mentioned, all the results which require the impact parameter magnitude estimate (through the multiplicity selection), or which are related to the reaction plane reconstruction, need the measurement of correlations. In particular, let me quote the baryon flow observables[36-40].

2. The search for delta resonances through $\pi - p$ correlations

2.1 Introduction

Most relativistic nuclear matter models include not only nucleons and mesons but also deltas as basic ingredients[41-49]. For instance, ter Haar and Malfliet have shown that including the Δ-degree of freedom in their relativistic Dirac-Brueckner calculations changes the saturation point[44,45]. The study of Δ (and π) properties in nuclear matter gave rise to many experimental[52-55] and theoretical[41,46,50,51,55-64] investigations, for which were invoked the "pionic branch" or "pisobar", and the possibility that the width and the mass of the in-medium deltas would be different from the free ones. Using relativistic heavy ion collisions, such effects could be investigated for hot and dense nuclear matter. For that purpose, the first observables that we would like to look at are the delta yields, masses, widths and spectra. However, the purpose can be reached only if our knowledge of the nucleus-nucleus collision process makes us able to understand what is the connection between the in-medium and the measured deltas.

There is a second type of reasons why measuring Δ's should help us to progress in this field: it is a way to constrain models of the A+A collision, to have a better knowledge of the collision process, and to understand pion yields and spectra.

If we look at the nucleus-nucleus collision as a superposition of nucleon-nucleon collisions, we expect the delta resonance to play an important role in the dynamics of the collision because the inelastic part of the total NN cross section consists mainly in pion production which is dominated by delta creation and decay. At 800 MeV for instance, pion production represents roughly half of the total NN cross section. Hence, many pions are created during the collision, which can interact with nucleons once again through delta formation and decay, etc.... It has been even predicted that the amount of in-medium pions would be much larger than what is expected from the free NN production because of the π dispersion relation modification in the nuclear matter[46,50,65]. A correct prediction of the Δ yields, widths, masses and spectra should be a way to check the validity of the models that describe the collision, or to constrain their free parameters. At the present time, if we go beyond the thermal models involving an equilibrated fluid of nucleons, deltas, pions, etc..., we have models in which pions and deltas are treated in a cascade mode. They include pure cascade models[5-7,66] and cascade+mean field models like BUU[44,68], VUU[67] or QMD[70]. Most of them take into account the deltas and the pions through the following binary reactions: $NN \rightleftharpoons N\Delta$, $N\Delta \rightleftharpoons N\Delta$, and $N\pi \rightleftharpoons \Delta$. Such models represent a huge progress when compared to equilibrium calculations, but they still simplify the $\pi - N - \Delta$ dynamics. Just note, following Siemens et al.[71], that the pions should not be treated as classical particles because their wave packets overlap several nucleons simultaneously. The forthcoming models are based on transport equations derived from relativistic and quantal Lagrangians[71-76]. Delta measurements should put constraints on their Δ-field part.

To measure the Δ's in relativistic heavy ion collisions can also help investigate the reaction process because, when compared to nucleons or composites, their different mass, binding energy, size, lifetime, etc... makes them a specific tool. For instance, phenomenological concepts similar to those of coalescence (a neutron and a proton make a deuteron if they are close enough in phase space[77]) could be relevant (a pion and a proton make a delta if they are close enough in **r** space and if their relative momenta correspond to the delta decay kinematics), etc....

Finally, knowing the delta yield could help to understand the pion spectra. The question of the influence of the delta decay kinematics on the pion spectra has been addressed a long time ago. Cugnon et al.[5] made the remark that the steeper slopes for pion when compared to proton spectra for $\theta_{CM} = 90°$ could be due to the fact that pions would come from the decay of deltas in a $(\Delta - p)$ gas at a given temperature. Barz et al.[78] investigated the influence of the thermal motion and of the collective flow on the delta decay peak in pion spectra. More recently, the Berkeley Streamer Chamber group found two components in the $\theta_{CM} = 90°$ spectra of negative pions[25,27,79]. They suggested that the low energy component was due to the delta decay, and the high energy one to a thermal pion gas. An alternative explanation was proposed in which the pion spectra are produced by a sequential freeze-out in an expanding fireball[80]. Let us quote also the prediction that such spectrum shapes would be a signal of abnormal nuclear matter[81]. Clearly, to measure the delta yield and spectra should help to progress on this question of the delta decay contribution to pion spectra.

The goal of Δ measurements would be to determine all the observables one can think of: their production cross section, multiplicity distribution, mass distribution and spectrum (and also the correlations between those observables !). But, the difficulty to measure a wide resonance in a multiparticle final state makes that, with the statistics obtained at Diogene, only a part of the whole program is reachable. Here, the first delta signals measured in non-peripheral heavy ion collisions will be presented, and the limitations and possible future developments will be discussed. The systems concerned will be Ne+NaF at E/A=600 and 800 MeV, and Ar+Ca at E/A=600 MeV.

2.2 The study of (π,p) invariant mass distributions

As Diogene measures in coincidence π^+, π^- and p, it is possible to search for Δ^{++} and Δ^0. Hence, we looked for a bump in the invariant mass (M(π,p)) distributions for (π^+,p) or (π^-,p) pairs in the measured events[24,82]. If the deltas were free ones and without phase space constraints (the Δ masses that can be reached in an elementary reaction are partly determined by the energy-momentum distribution of the particles in the entrance channel), one would expect an average mass $M_\Delta = 1232\,MeV$ and a width $\Gamma_\Delta = 115\,MeV$. Fig. 1 (dark circles) gives an example of the (π^+,p) invariant mass distribution we get. Similar shapes were found for other (A_P , A_T) systems or bombarding energies, and for (π^-,p) pairs. The maximum around 1165 MeV cannot be attributed to the deltas because a similar maximum is obtained for *uncorrelated* pairs (open circles on figure 1). These pairs contain a pion from one event and a proton from *another* event, thereby the momenta of the two particles are not correlated by the delta. One could argue however that the delta decay kinematics can generate a bump even in uncorrelated M(π,p) distributions because the π and the p from the two events can come from the decays of two Δ's. But the continuous line on figure 1 proves that the general shape of the distribution is not mainly due to delta decays. It is a phase-space simulation using the CERN program "FOWL", for the same proton multiplicities as the experimental ones, in which one pion is emitted and the detector acceptance cuts are imposed. *No deltas* are in the simulation.

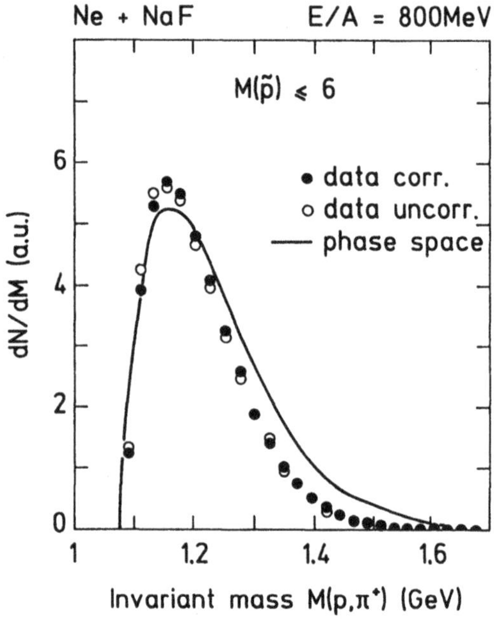

Figure 1 The dark circles are the invariant mass distribution for (π^+,p) pairs measured in Ne+NaF collisions at an incident energy of 800 MeV per nucleon. Only events with a proton-like multiplicity smaller than 6 have been taken to select intermediate impact parameters. The open circles are the same distribution for *uncorrelated* pairs (see text). The line is a phase space simulation with pions and protons but *without deltas*.

A reason why no Δ bump emerges from the experimental spectrum in fig. 1 is that the $M(\pi, p)$ distribution contains a large contribution from "mixed" pairs (i.e. a π or a p from a Δ decay and a p or a π from another Δ, or not originating from a Δ). To subtract this background contribution, and also the contribution of pairs where both π and p do not come from a Δ, we looked for a signal in the distribution of the difference $D = dN/dM_c - \alpha.dN/dM_{nc}$ where dN/dM_c is the invariant mass distribution for correlated pairs (i.e. the π and the p come from the same event) and dN/dM_{nc} is the distribution of uncorrelated pairs (i.e. the π and the p come from two different events which have the same pion and proton multiplicities). The normalization factor α is chosen so that D = 0 in the $M(\pi, p)$ interval $[1078, 1140MeV]$ where the contribution of the delta is expected to be weak. Fig. 2 gives examples of the D distribution. Clear bumps appear, which prove that the protons and the pions are correlated. Similar D distributions were obtained for Ar+Ca and Ne+NaF reactions at 600 MeV per nucleon.

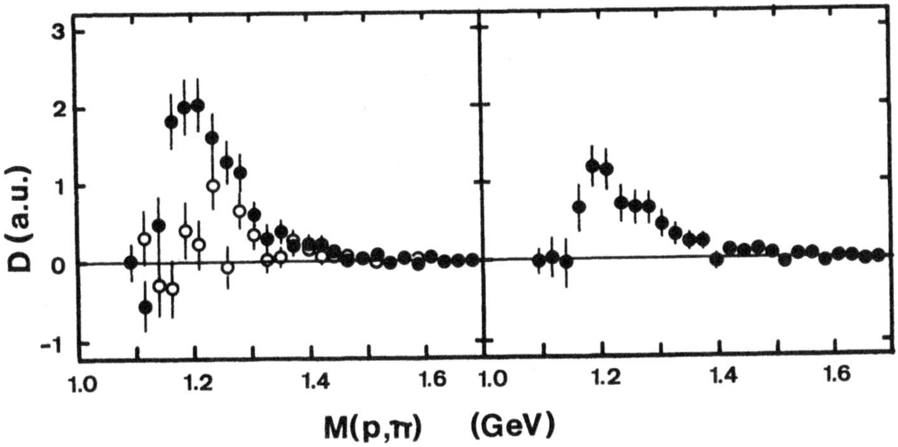

Figure 2 Difference D between the correlated and the uncorrelated (π,p) invariant mass distributions (see text) for (π^+,p) (dark circles) and (π^-,p) (open circles) pairs. The system is Ne+NaF (E/A=800MeV) for low (left side) and high (right side) $\tilde{p}$ multiplicities.

The fact that the π-p correlation is due to delta decays is confirmed by the comparison between (π^+, p) and (π^-, p) D distributions. The latter give a weaker signal than the former as can be seen on fig. 2. This is also true for Ne+NaF and Ar+Ca reactions at 600 MeV per nucleon, and for other proton-like multiplicity selections. Simple isospin considerations assuming charge invariance, charge symmetry, and delta dominance explain this result. Evaluating the appropriate Clebsch-Gordan coefficients, one gets[83]:

$$\begin{array}{llllllll}
\sigma(pp & \rightarrow & \Delta^{++} n) & = & (3/4)\sigma_1 & Br(\Delta^{++} \rightarrow \pi^+ p) & = & Br_{3/2} \\
\sigma(pp & \rightarrow & \Delta^{+} p) & = & (1/4)\sigma_1 & Br(\Delta^{+} \rightarrow \pi^+ n) & = & (1/3)Br_{3/2} \\
\sigma(pn & \rightarrow & \Delta^{+} n) & = & (1/4)\sigma_1 & Br(\Delta^{+} \rightarrow \pi^0 p) & = & (2/3)Br_{3/2} \\
\sigma(pn & \rightarrow & \Delta^{0} p) & = & (1/4)\sigma_1 & Br(\Delta^{0} \rightarrow \pi^- p) & = & (1/3)Br_{3/2} \\
\sigma(nn & \rightarrow & \Delta^{0} n) & = & (1/4)\sigma_1 & Br(\Delta^{0} \rightarrow \pi^0 n) & = & (2/3)Br_{3/2} \\
\sigma(nn & \rightarrow & \Delta^{-} p) & = & (3/4)\sigma_1 & Br(\Delta^{-} \rightarrow \pi^- n) & = & Br_{3/2}
\end{array} \quad (1)$$

where σ_1 is the total $NN \rightarrow \Delta N$ cross section for isospin 1, and $Br_{3/2}$ is the total branching ratio for $\Delta \rightarrow \pi N$. If we assume that the deltas are produced only via NN collisions, we can calculate the ratio of the mean number of Δ^0's to the mean number of Δ^{++}'s in the system of the two colliding nuclei, from the numbers of *pp*, *nn* and *pn* pairs:

$$\frac{\Delta^0}{\Delta^{++}} = \frac{(1/4)N(N-1)/2 + (1/4)ZN}{(3/4)Z(Z-1)/2} \approx \frac{1}{3}\frac{N}{Z}(2+\frac{N}{Z}) \tag{2}$$

for N=Z this ratio is 1. Furthermore, as only one third of the Δ^0's decays in $\pi^- p$, we expect the number of $\pi^- p$ pairs from Δ^0's to be three times smaller than the number of $\pi^+ p$ pairs from Δ^{++}. The assumptions made here are rather rough but they give a convenient estimate.

2.3 How to estimate the delta production cross section ?

To determine the relationship between the spectra presented in Figs. 1 and 2 and the delta yield, we shall first assume that we can measure all the nucleons and pions in the whole phase space. We postulate that the final state of the system we measure is a mixture of free pions, nucleons and deltas. Finally, we first restrict ourselves to events that contain a fixed number of free nucleons ($M_N - M_\Delta$), pions ($M_\pi - M_\Delta$), and deltas (M_Δ), without any correlation between the three types of particles (it is a reasonable assumption for physical events with rather large multiplicities). Hence the measured events are made of free N's and π's (the "background") and of N's and π's coming from delta decays. Let us call $f(\mathbf{p}_N, \mathbf{p}_\pi)$ (it will be written $f_{N\pi}$ for simplicity) the distribution function in $\mathbf{p}$ space of all the (N,π) pairs. We shall determine $f_{N\pi}$. The (N,π) pairs of the background have a distribution function $F(\mathbf{p}_N, \mathbf{p}_\pi)$ ($=F_{N\pi}$), and the (N,π) pairs that come from delta decays (the N and the π come from the *same* Δ) have a distribution function $\Phi(\mathbf{p}_N, \mathbf{p}_\pi)$ ($=\Phi_{N\pi}$). We can write:

$$\Phi(\mathbf{p}_N, \mathbf{p}_\pi) = \int f(\mathbf{p}_\Delta, m_\Delta) f(\mathbf{u})\, \delta^3(\mathbf{p}_N - \mathbf{L}(\mathbf{v}_\Delta/c,\, p(m_\Delta).\mathbf{u},\, m_N)$$

$$.\delta^3(\mathbf{p}_\pi - \mathbf{L}(\mathbf{v}_\Delta/c,\, -p(m_\Delta).\mathbf{u},\, m_\pi)\, d^3\mathbf{p}_\Delta\, dm_\Delta\, d^2\mathbf{u} \tag{3}$$

where $f(\mathbf{p}_\Delta, m_\Delta)$ is the distribution of the Δ momentum and mass, and $f(\mathbf{u})$ is the distribution of the unit vector $\mathbf{u}$ along the momentum of the nucleon from the delta decay in the delta c.m.. $\mathbf{L}(\mathbf{v}/c,\, \mathbf{p}',\, m)$ is the Lorentz transform which changes the momentum $\mathbf{p}'$ of a particle with a mass m, in the reference frame of velocity $\mathbf{v}/c$, into its momentum in the laboratory. $\mathbf{v}_\Delta/c$ is the velocity of the Δ with a momentum $\mathbf{p}_\Delta$ and a mass m_Δ. We call $f(\mathbf{p}_N) = f_N$ and $f(\mathbf{p}_\pi) = f_\pi$ the marginal distribution functions of $f_{N\pi}$:

$$f(\mathbf{p}_N) = \int f(\mathbf{p}_N, \mathbf{p}_\pi)\, d^3\mathbf{p}_\pi \quad \text{and} \quad f(\mathbf{p}_\pi) = \int f(\mathbf{p}_N, \mathbf{p}_\pi)\, d^3\mathbf{p}_N \tag{4}$$

$F(\mathbf{p}_N) = F_N$ and $F(\mathbf{p}_\pi) = F_\pi$ are the marginal distribution functions of $F_{N\pi}$, while $\Phi(\mathbf{p}_N) = \Phi_N$ and $\Phi(\mathbf{p}_\pi) = \Phi_\pi$ are those of $\Phi_{N\pi}$. We assume that all those functions are normalized to 1. The basic idea of our analysis is that the background (N,π) pairs are uncorrelated, while the (N,π) pairs that come from delta decays are correlated, hence:

$$F_{N\pi} = F_N.F_\pi \quad \text{and} \quad \Phi_{N\pi} \neq \Phi_N.\Phi_\pi \tag{5}$$

Experimentally, when we take a (N,π) pair, we have five possibilities, each of them having a probability P_i ($i = 1$ to 5) to occur :

1. N and π from background: $P_1 \propto (M_N - M_\Delta)(M_\pi - M_\Delta)F_N\, F_\pi$
2. N from background and π from Δ: $P_2 \propto (M_N - M_\Delta)M_\Delta F_N\, \Phi_\pi$
3. N from Δ and π from background: $P_3 \propto (M_\pi - M_\Delta)M_\Delta \Phi_N\, F_\pi$
4. N from Δ and π from *another* Δ: $P_4 \propto M_\Delta(M_\Delta - 1)\Phi_N\, \Phi_\pi$
5. N and π from *the same* Δ: $P_5 \propto M_\Delta\, \Phi_{N\pi}$

$f_{N\pi}$ is the sum of those five probabilities multiplied by the appropriate nomalization coefficient. The calculation gives:

$$f_{N\pi} = \frac{1}{M_N\, M_\pi}\{[(M_N - M_\Delta)F_N + M_\Delta\Phi_N]\,[(M_\pi - M_\Delta)F_\pi + M_\Delta\Phi_\pi]$$

$$+M_\Delta(\Phi_{N\pi} - \Phi_N\Phi_\pi)\} \tag{6}$$

This distribution cannot be measured because it is impossible to keep M_Δ fixed in the experimental data. If we let M_Δ vary, and if we assume that it is not correlated to the N and π momenta, the average of relation (6) gives:

$$f_{N\pi} =$$

$$\frac{1}{M_N\, M_\pi}\{[(M_N - <M_\Delta>)F_N + <M_\Delta>\Phi_N]\,[(M_\pi - <M_\Delta>)F_\pi + <M_\Delta>\Phi_\pi]$$

$$+\sigma^2_{M_\Delta}(F_N - \Phi_N)(F_\pi - \Phi_\pi) + <M_\Delta>(\Phi_{N\pi} - \Phi_N\Phi_\pi)\} \tag{7}$$

where $<M_\Delta>$ is the mean Δ multiplicity and $\sigma^2_{M_\Delta}$ is the variance of M_Δ. A similar equation can be obtained if not only M_Δ can vary, but also M_N and/or M_π. In this equation, the mean values M_Δ and $\sigma^2_{M_\Delta}$ are replaced by ratios of mean values (M_Δ divided by M_N and/or M_π). The distribution function $f^{nc}_{N\pi}$ of uncorrelated pairs (the N and the π are taken in two different events) is just the product of the two marginal functions of $f_{N\pi}$. From Eqs. (4) and (7), we get:

$$f_N = (1/M_N)[(M_N - <M_\Delta>)F_N + <M_\Delta>\Phi_N] \tag{8}$$

$$f_\pi = (1/M_\pi)[(M_\pi - <M_\Delta>)F_\pi + <M_\Delta> \Phi_\pi] \tag{9}$$

hence,

$$f_{N\pi}^{nc} = f_N f_\pi = \frac{1}{M_N M_\pi}\{[(M_N - <M_\Delta>)F_N + <M_\Delta> \Phi_N]$$

$$\cdot [(M_\pi - <M_\Delta>)F_\pi + <M_\Delta> \Phi_\pi]\} \tag{10}$$

The method presented in section *2.2* for a Δ signal extraction consists in plotting the invariant mass distribution which corresponds to the function $d_{N\pi} = f_{N\pi} - \alpha f_{N\pi}^{nc}$. As a matter of fact, the correspondence between the distribution functions in $\mathbf{p}_N \otimes \mathbf{p}_\pi$ space and their invariant mass distribution is *linear*. From Eqs. (7) and (10) we get:

$$d_{N\pi} = \frac{1}{M_N M_\pi}\{\sigma_{M_\Delta}^2 (F_N - \Phi_N)(F_\pi - \Phi_\pi) + <M_\Delta> (\Phi_{N\pi} - \Phi_N \Phi_\pi) + (1-\alpha)$$

$$\cdot[(M_N - <M_\Delta>)F_N + <M_\Delta> \Phi_N][(M_\pi - <M_\Delta>)F_\pi + <M_\Delta> \Phi_\pi]\} \tag{11}$$

for $\alpha = 1$ we have

$$d_{N\pi} = \frac{1}{M_N M_\pi}\left[\sigma_{M_\Delta}^2 (F_N - \Phi_N)(F_\pi - \Phi_\pi) + <M_\Delta> (\Phi_{N\pi} - \Phi_N \Phi_\pi)\right] \tag{12}$$

At that point, let us stress that the distribution obtained using uncorrelated pairs (Eq. (10)) is *not* the distribution of the background $F_N F_\pi$ because of the pair combinatorics. The integrals of $d_{N\pi}$ and of the corresponding invariant mass distribution are zero for $\alpha = 1$. But in §*2.2* we have chosen α in order that the integral of the invariant mass distribution is zero only in the interval $[1178, 1140 MeV]$ where the Δ contribution is expected to be very small. We can calculate α using a more restrictive condition which must yield the same value: α must be such that $d_{N\pi}$ is zero in the area of $\mathbf{p}_N \otimes \mathbf{p}_\pi$ space where $\Phi_N = 0$ and $\Phi_\pi = 0$ ($\Rightarrow \Phi_{N\pi} = 0$). This condition implies, using Eq. (11), that in this area:

$$[\sigma_{M_\Delta}^2 + (1-\alpha)(M_N - <M_\Delta>)(M_\pi - <M_\Delta>)]\, F_N F_\pi = 0 \tag{13}$$

If F_N and F_π are different from zero in this area, we can calculate α from $\sigma_{M_\Delta}^2$, M_N, M_π and $< M_\Delta >$. If the shape of the M_Δ distribution is known, we can derive $< M_\Delta >$ from α.

An important point is that if we take α from Eq. (13), or if we take any other value of α, the difference $d_{N\pi}$ (Eq. (11)) will never contain only the contribution of the (N, π) pairs that come from the Δ decays, i.e. a term proportional to

$\Phi_{N\pi}$. As a matter of fact, we see in Eq. (11) that $d_{N\pi}$ contains in addition four other terms proportional to $F_N F_\pi$ (the pure background), and to $F_N \Phi_\pi$, $\Phi_N F_\pi$ and $\Phi_N \Phi_\pi$ (the "mixed" contributions). Taking α from Eq. (13) suppresses the term proportional to $F_N F_\pi$ but not the other ones. Any choice of α can suppress other terms among those four, but not all of them. Hence, the invariant mass distribution corresponding to $d_{N\pi}$, D (see §*2.2*), will also contain a contribution from other (N, π) pairs than the "pure" Δ decay pairs.

Fortunately, this is no longer true if we make following assumptions:

$$F_N = \Phi_N \quad \text{and} \quad F_\pi = \Phi_\pi \tag{14}$$

we can even request a weaker assumption: that only the invariant mass distributions of $f_{N\pi}$ and $f_{N\pi}^{nc}$ behave as if Eq. (14) was true. These assumptions are rather reasonable since they do not concern the correlations, but only the marginal distributions. From Eqs. (7) and (14), we get:

$$f_{N\pi} = \left(1 - \frac{< M_\Delta >}{M_N M_\pi}\right) F_N F_\pi + \frac{< M_\Delta >}{M_N M_\pi} \Phi_{N\pi} \tag{15}$$

$$f_N = F_N \ ; \ f_\pi = F_\pi \quad \Rightarrow \quad f_{N\pi}^{nc} = F_N F_\pi \tag{16}$$

$$d_{N\pi} = f_{N\pi} - \alpha\, f_{N\pi}^{nc} = \left(1 - \alpha - \frac{< M_\Delta >}{M_N M_\pi}\right) F_N F_\pi + \frac{< M_\Delta >}{M_N M_\pi} \Phi_{N\pi} \tag{17}$$

If we define α in order that $d_{N\pi}$ is zero for the (N, π) pairs which do not come from Δ's, that is for $\Phi_{N\pi}$=0, we obtain from Eq. (17):

$$d_{N\pi} = \frac{< M_\Delta >}{M_N M_\pi} \Phi_{N\pi} \tag{18}$$

$$\alpha = 1 - \frac{< M_\Delta >}{M_N M_\pi} \tag{19}$$

and with this choice for α, we have:

$$< M_\Delta > = (1-\alpha) M_N M_\pi \quad ; \quad < M_\Delta > / M_\pi = (1-\alpha) M_N \tag{20}$$

As a conclusion, Eq. (20) gives the delta yield and from Eq. (18) we can obtain the deltas shape (average mass and width) and spectrum, (they are related to $f(\mathbf{p}_\Delta, m_\Delta)$ in Eq. (3)) and polarization (related to $f(\mathbf{u})$ in Eq. (3)). However, in practice, limitations arise due to the fact that the statistics is limited, neutrons and neutral pions are not measured (but the Δ^{++} study is not affected by this limitation), and the detector -like all the detectors- has acceptance cutoffs.

2.4 The delta measurements taking into account the acceptance cutoffs

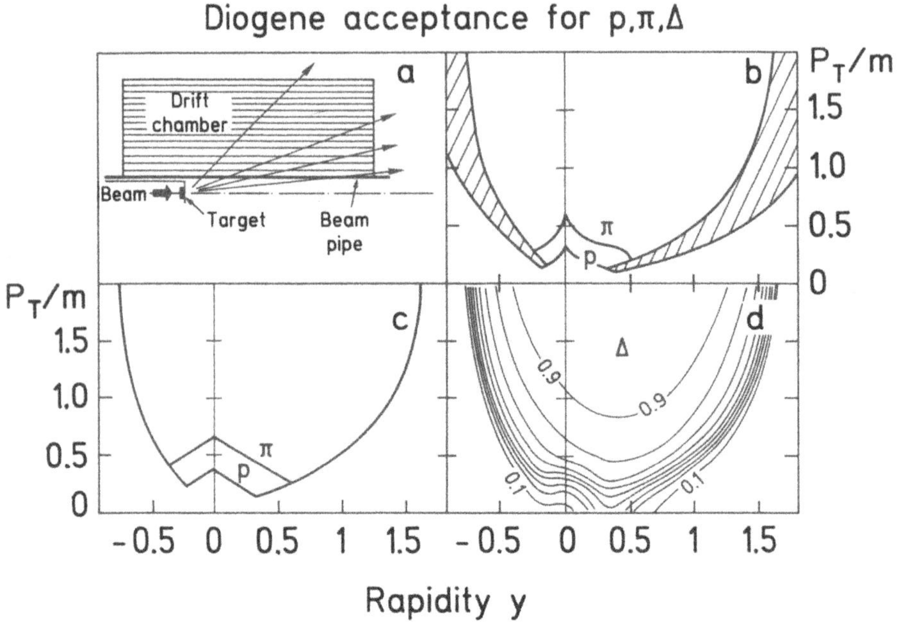

Figure 3 In the upper left corner (a), a schematic view of the detector longitudinal cross section (only the upper part is drawn) shows the origins of its acceptance cuts. The low energy particles are stopped in the target, the target holder or the beam pipe. The arrows indicate outgoing particles emitted at different polar angles θ, and the horizontal lines are the drift chamber anode wires: the particles emitted at small θ's don't hit enough wires to be measured with accuracy. In the upper right corner (b), the resulting boundaries of the acceptance area for π's and p's are shown in the (y $-$ p_T/m) plane. y is the rapidity and p_T/m the transverse momentum divided by the mass. Only particles located above the boundaries are measured. The lines around $y = 0$ correspond to the low energy cuts, while the hatched zones correspond to the forward and backward angular cuts. Their boundaries correspond to 4 and 9 wires hit. In the lower left corner (c), the software clean cuts boundaries used for data analysis are shown. They are simpler than those of (b), and they are more restrictive than all the experimental cuts due to the different cases encountered in the experiments (varying target or beam pipe sizes and materials). Only the particles located above those lines are taken in the data analysis. In the lower right corner (d), contour lines for constant delta measurement efficiency obtained from the π and p acceptances of (c) are given. The lines correspond to 0.1, 0.2, 0.3, ... , 0.9 efficiencies, but only those corresponding to 0.1 and 0.9 are labelled.

Figure 3 gives the detector acceptance for protons, pions and deltas. The two lower parts ((c) and (d)) are the most important for our Δ studies. The abscissa is the p, π, or Δ rapidity ($y = \text{argtanh}(p_{\parallel}/E$, $p_{\parallel}$ being the momentum component along the beam axis, and E the total energy), and the Y-axis is the transverse momentum p_T divided by the mass of the particle (it is the average mass, 1232 MeV for the Δ's). On fig. 3-c, the "software" acceptance cuts are given; their parametrization is:

A p or a π is taken if its lab. polar angle is such that:

$$20° < \theta < 132° \tag{21}$$

and if:

for protons:	or for pions:
$y < 0$ and $p_T/m > 0.36 + 0.72\,y$	$y < 0$ and $p_T/m > 0.66 + 0.77\,y$
$y > 0$ and $p_T/m > 0.36 - 0.80\,y$	$y > 0$ and $p_T/m > 0.66 - 0.63\,y$ (22)

those acceptance boundaries are more restrictive than the true ones, but they are much simpler. Fig. 3-d gives the Δ efficiency contours obtained assuming a Δ mass Breit-Wigner distribution ($M_\Delta = 1232\,MeV$, $\Gamma_\Delta = 115\,MeV$) and an isotropic decay in the Δ center of mass. The efficiency at a given delta rapidity and transverse momentum is calculated according to the principle that a delta is taken if the proton and the pion from its decay are inside the acceptance given by Eqs. (21,22) and shown on fig. 3-c. As a result, a delta which is outside the π and p acceptances can be sometimes measured. This is the case for $p_T/m = 0$. Similarly, a delta which is inside the π and p acceptances can sometimes not be measured.

The smooth efficiency cuts of fig. 3-d must be taken into account for Δ yields, spectra, etc... estimations. In order to estimate the proportion of deltas lost, we made simulations using a thermal source of deltas in the center of mass of the target and projectile system. For $E_{inc}/A = 800MeV$, and for the (reasonable) temperature $T = 80MeV$, one finds that the proportions of π, p, and Δ which are lost are respectively 17%, 34% and 46%. Using these values for correcting the experimental data, we estimated the ratio of Δ^{++} to π^+ yields (in average) from Eq. (20). Note that a part of the π^+'s come from Δ^+. For the system Ne+NaF at 800 MeV per nucleon, we found, as a preliminary result, a value of ~ 0.4. A next step will be to correct for the systematic error due to the fact that we used the mean proton multiplicity for estimating the ratio (using Eq. (20)), while the equation is valid for a fixed proton multiplicity.

3. Pion "flow"

3.1 Introduction

The collective transverse sidewards flow of nuclear matter was first predicted by hydrodynamics models[8,9]. Later, intranuclear cascade[5] and models involving mean field plus two-body dynamics[44,67,68,70] also predicted some flow. It can be described as a collective energy-momentum transfer to the nucleons, parallel to the reaction plane. It has raised much interest because it is less affected by the thermal unordered motion of the nucleons than the usual double differential cross sections. Hence it would be more sensitive to nuclear matter properties. It has been first established by Plastic Ball measurements[85] and also by Streamer Chamber measurements[86]. Experimentally, the flow is often measured by first estimating event by event the reaction plane azimuth. Following Danielewicz and Odyniec[84], it is the azimuth of the vector:

$$\mathbf{Q} = \sum_i (y_i - y_0)\mathbf{p}_{\perp i} \tag{23}$$

The sum runs over the p-likes $\tilde{p}$ of the event, y_i and $\mathbf{p}_{\perp i}$ being the rapidity and the transverse momentum of the ith $\tilde{p}$, and y_0 the mean rapidity of all the $\tilde{p}$'s in the event. The particle which is used to increment the flow observables distributions is not put in the summation in order to avoid the auto-correlations[84]. The second step consists in studying the one-body distribution function of nucleons or composites in a reference frame *bound to the reaction plane*. Its axes are usually labelled as follows: p_z is along the projectile velocity, p_x lies in the reaction plane and p_y is perpendicular to the reaction plane. The experimental observables in this reference frame are the baryon triple differential cross sections, or their moments. The flow is a preferential emission of the forward (resp. backward) emitted baryons towards positive (resp. negative) p_x in the c.m.. It can be determined by looking at the mean p_x of the baryons as a function of their rapidity y (ref. 84): when y increases $< p_x >$ is first negative, then positive; and it is zero for $y = y_{c.m.}$ ($y_{c.m.}$ is the participant nucleons center of mass rapidity). An alternative method consists in looking at the particle azimuthal distributions[39,40,87]. The azimuth of a particle, φ, is the angle between its transverse momentum and the reaction plane. Experimentally, it is estimated using φ', the angle between the transverse momentum and $\mathbf{Q}$. In φ (or φ') baryon distributions with a rapidity selection, the flow shows up as follows: for backward rapidities ($y < y_{cm}$) there is a minimum at $\varphi = 0°$ (and a maximum at $\varphi = 180°$), while for forward rapidities, there is a maximum at $\varphi = 0°$ (and a minimum at $\varphi = 180°$).

The studies of the 3-dim. 1-body distribution performed for baryons can be done for π^+ and π^-. This is a specific advantage of Diogene which can measure the charged pions with large statistics. We studied the pion azimuthal distributions and in-plane transverse momentum $< p_x > (y)$, using the baryons for estimating the reaction plane azimuth[24,82,88,89]. Hence, such a study is based on the measurement of pion-baryon correlations. The most striking result is that $< p_x >(y)$ is *always positive* even for backward rapidities, for the asymmetric (Ne or Ar) + (Nb or Pb) systems. This is mainly due to the fact that, at any rapidity, more pions are emitted towards positive p_x than towards negative p_x as can be seen of figure 4 where the distribution of the pions azimuth, φ', is never minimal at $\varphi' = 0°$ whatever the rapidity is, contrary to $\tilde{p}$'s (see refs. 39,40).

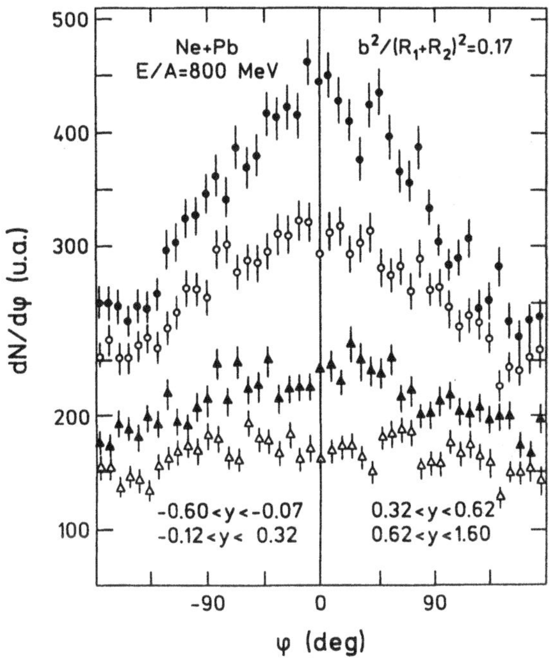

Figure 4 Azimuthal distribution of positive pions emitted in four rapidity bins for Ne+Pb collisions at E/A=800 MeV, and for a selection on $\tilde{b}$ (see §*1.2*). The estimated participant center of mass rapidity is $y_{c.m.} = 0.24$.

Those results can be interpreted in terms of "target shadowing": the incoming projectile makes a hot zone *on one side* of the target, from which the pions are emitted, and the pions that go towards the other side are partly absorbed in the target. Hence, the pion "flow" would be a way to solve the ambiguity of the baryon flow *sign*[90-92]. The target shadowing hypothesis is confirmed by the the fact that the measured $<< p_x >>$ (the double averaging means that $< p_x > (y)$ is averaged over y) increases with A_P/A_T, and is larger for π^+ than for π^-. The latter result can be explained by the target isospin asymmetry. Note that charge exchange in the target (i.e. $\pi^{\pm} \rightarrow \pi^0$) can simulate absorption and even "anti-absorption" (i.e. $\pi^0 \rightarrow \pi^{\pm}$). We made a very simple geometrical model with a point-like source in the target, shifted by the estimated impact parameter from the target center, radiating pions, with an absorption mean free path λ as the only parameter. This model together with our data yields $\lambda = 4-6\,fm$. Similar values were obtained from the analysis of p+A→π+X data[93] for the (rather) low average energies of the pions we measure (around 100 MeV). Intranuclear cascade[5] predicts

$<< p_x >> \simeq 0$, hence pion absorption would be underestimated in this model, which could explain why the cascade overpredicts the pion multiplicities[19,23-25]. On the contrary, QMD predicts some pion flow[94].

References

1. H.H. Gutbrod et al., *Phys. Rev. Lett.* **37** (1976) 667 ; G.D. Westfall et al., *Phys. Rev. Lett.* **37** (1976) 1202 ; J. Gosset et al., *Phys. Rev.* **C16** (1977) 629 ; S. Nagamiya et al., *Phys. Lett.* **81B** (1979) 147 ; A. Sandoval et al., *Phys. Rev.* **C21** (1980) 1321 ; S. Nagamiya et al., *Phys. Rev.* **C24** (1981) 971 ; K.L. Wolf et al., *Phys. Rev.* **C26** (1982) 2572.
2. S. Nagamiya and M. Gyulassy, High-energy nuclear collisions, in: *Advances in Nuclear Physics*, Vol. **13**, eds. J.W. Negele and E. Vogt (Plenum Press, New York-London, 1982) pp. 201-320.
3. S. Das Gupta and A.Z. Mekjian, *Phys. Rep.* **72** (1981) 131.
4. J. Knoll, *Phys. Rev.* **C20** (1979) 773.
5. J. Cugnon, *Phys. Rev.* **C22** (1980) 1885 ; J. Cugnon, T. Mizutani and J. Vandermeulen, *Nucl. Phys.* **A352** (1981) 505 ; J. Cugnon and D. L'Hôte, *Nucl. Phys.* **A452** (1986) 738 ; J. Cugnon and M.-C. Lemaire, *Nucl. Phys.* **A489** (1988) 781.
6. Y. Yariv and Z. Fraenkel, *Phys. Rev.* **20** (1979) 2227 ; Y. Yariv and Z. Fraenkel, *Phys. Rev.* **24** (1981) 488.
7. K.K. Gudima and V.D. Toneev, *Dubna Report* **E2**-12644 (1979).
8. H. Stöcker and W. Greiner, *Phys. Rep.* **137** (1986) 278.
9. R.B. Clare and D. Strottman, *Phys. Rep.* **141** (1986) 177.
10. H.H. Gutbrod, A.M. Poskanzer and H.G. Ritter, *Rep. Prog. Phys.* **52** (1989) 1165 ; and K.H. Kampert, *J. Phys. G* **15** (1989) 691.
11. A. Sandoval et al., *Phys. Rev. Lett.* **45** (1980) 874.
12. J.P. Alard et al., *Nucl. Instr. Meth.* **A261** (1987) 379.
13. GSI Nachrichten 11-89, Darmstadt.
14. C. Cavata et al., submitted to *Phys. Rev. C.*
15. T.C. Awes and C.K. Gelbke, *Phys. Rev.* **C27** (1983) 137; J. Cugnon and D. L'Hôte, *Nucl. Phys.* **A397** (1983) 519.
16. LNS experiment N° **192**.
17. M. Trzaska et al., Report GSI 90-1 (1990), p. 51.
18. J.J. Lu et al., *Phys. Rev. Lett.* **46** (1981) 898.
19. J.W. Harris et al., *Phys. Lett.* **153B** (1985) 377 ; J.W. Harris et al., *Phys. Rev. Lett.* **58** (1987) 463.
20. S. Hayashi et al., *Phys. Rev.* **C38** (1988) 1229.
21. R. Szwed, Proc. 5th. conf. Rel. Heavy Ion, Berkeley, Report LBL-12652 (1981) 371 ; SKM-200 collab., *Nucl. Phys.* **A362** (1981) 376 ; M. Kh. Anikina et al., *Sov. J. Nucl. Phys.* **41** (1985) 452.
22. G.N. Agakishiev et al., *Sov. J. Nucl. Phys.* **34** (1981) 842 ; G.N. Agakishiev et al., *Sov. J. Nucl. Phys.* **40** (1984) 767 ; G.R. Gulkanyan,

Sov. J. Nucl. Phys. **40** (1984) 479.
23. D. L'Hôte et al., *Phys. Lett.* **198B** (1987) 139 ; D. L'Hôte, Thèse à l'Université Paris-Sud (1987), unpublished.
24. C. Cavata, Thèse à l'Université Paris-Sud (1989), note CEA-N-2629 (1990).
25. R. Stock, *Phys. Rep.* **135** (1986) 259.
26. D. L'Hôte, in: Nuclear Matter and Heavy Ion Collisions, eds. M. Soyeur, H. Flocard, B. Tamain, M. Porneuf, NATO ASI Series B Physics Vol. 205 (Plenum Press, New York-London, 1989), pp. 323-342.
27. R. Brockmann et al., *Phys. Rev. Lett.* **53** (1984) 2012.
28. H. Stroebele et al., in: Proc. 7th. High Energy Heavy Ion Study, Darmstadt, Oct. 1984, Report GSI-85-10 (1985).
29. S. Nagamiya et al., *Phys. Rev. Lett.* **45** (1980) 602.
30. K.L. Wolf et al., *Phys. Rev.* **C26** (1982) 2572.
31. J. Miller et al., *Phys. Rev. Lett.* **58** (1987) 2408.
32. N. Bastid et al., *Nucl. Phys.* **A506** (1990) 637.
33. J.P. Alard et al., "Proton-proton correlations in relativistic nuclear collisions at Saturne", this volume.
34. P. Dupieux et al., *Phys. Lett.* **200B** (1988) 17.
35. L. Fraysse, Thèse à l'Université Clermont II (1989), unpublished.
36. O. Valette et al., Contributed papers of the third international conference on Nucleus-Nucleus collisions (St Malo, June 1988), eds C. Estève, C. Grégoire, D. Guerreau and B. Tamain (Caen, 1988) p. 141.
37. J. Gosset et al., Proc. Int. Workshop on gross properties of nuclei and nuclear excitations XVIII (Hirschegg, Austria, 1990), ed. H. Feldmeier, p. 143.
38. M.C. Lemaire et al., Proc. 6th. Winter workshop on nuclear dynamics (Jackson Hole, USA, Feb. 1990), ed. J. Randrup, p. 40.
39. M. Demoulins et al., *Phys. Lett.* **241B** (1990) 476.
40. M. Demoulins, Thèse a l'Université Paris-Sud, note CEA-N-2628 (1990).
41. B. Friedman, V.R. Pandharipande, and Q.N. Usmani, *Nucl. Phys.* **A372** (1981) 483.
42. B.D. Serot and J.D. Walecka, *Adv. Nucl. Phys.* **16** (1986).
43. T.L. Ainsworth et al., *Nucl. Phys.* **A464** (1987) 740.
44. B. ter Haar and R. Malfliet, *Phys. Rev.* **C36** (1987) 1611.
45. B. ter Haar and R. Malfliet, *Phys. Rep.* **149** (1987) 207.
46. G.E. Brown, *Nucl. Phys.* **A488** (1988) 689c.
47. G.E. Brown, *Phys. Rep.* **163** (1988) 167.
48. M. Schönhofen et al., *Nucl. Phys.* **A504** (1989) 875.
49. L. Ferreira et al., Nuclear matter with delta degrees of freedom, in: *Proc. Int. Workshop on Nucl. Dynamics* (2-7 April 1990, Marciana Marina, Italy), Ed.: I. Iori.
50. G.E. Brown et al., *Nucl. Phys.* **A505** (1989) 823.
51. F. De Jong, B. ter Haar and R. Malfliet, *Phys. Lett.* **220B** (1989) 485.
52. D. Contardo et al., *Phys. Lett.* **168B** (1986) 331 ; D. Bachelier et al.,

Phys. Lett. **172B** (1986) 23 ; C. Ellegaard et al., *Phys. Rev. Lett.* **59** (1987) 974 ; M. Roy-Stephan, *Nucl. Phys.* **A482** (1988) 373c.
53. J. Arends et al., *Phys Lett.* **98B** (1981) 423 ; J. Ahrens et al., *Phys. Lett.* **146B** (1984) 303 ; P. Carlos et al., *Nucl. Phys.* **A431** (1984) 573.
54. P. Barreau et al., *Nucl. Phys.* **A402** (1983) 515 ; Z. E. Meziani et al., *Phys. Rev. Lett.* **54** (1985) 1233 ; J. S. O'Connell et al., *Phys. Rev. Lett.* **53** (1984) 1627 ; R.M. Sealock et al., *Phys. Rev. Lett.* **62** (1989) 1350 ; D.T. Baran et al., *Phys. Rev. Lett.* **61** (1988) 400.
55. T. Ericson and W. Weise, Pions and Nuclei, Clarendon Press, Oxford, 1988.
56. K. Klingenbeck, M. Dillig and M.G. Huber, *Phys. Rev. Lett.* **41** (1978) 387.
57. R.F. Sawyer, *Nucl. Phys.* **A335** (1980) 315.
58. Y. Horikawa, M. Thies and F. Lenz, *Nucl. Phys.* **A345** (1980) 386.
59. K. Dreissigacker et al., *Nucl. Phys.* **A375** (1982) 334.
60. E. Oset, H. Toki and W. Weise, *Phys. Rep.* **C83** (1982) 281.
61. G. Chanfray and M. Ericson, *Phys. Lett.* **141B** (1984) 163.
62. V.F. Dimitiev and T. Suzuki, *Nucl. Phys.* **A438** (1985) 697.
63. J. Koch and N. Otsuka, *Nucl. Phys.* **A435** (1985) 765.
64. P.A.M. Guichon and J. Delorme, Proc. 5^e Journees d'etudes Saturne (Piriac, France, 1989), Ed. P. Radvanyi.
65. G. Bertsch et al., *Nucl. Phys.* **A490** (1988) 745.
66. Y. Kitazoe et al., *Phys. Rev. Lett.* **58** (1987) 1508.
67. H. Kruse, B.V. Jacak and H. Stöcker, *Phys. Rev. Lett.* **54** (1985) 289.
68. G.F. Bertsch, H. Kruse and S. Das Gupta, *Phys. Rev.* **C29** (1984) 673.
69. C. Gale, *Phys. Rev.* **C36** (1987) 2152.
70. J. Aichelin et al., *Phys. Rev.* **C37** (1988) 2451.
71. P.J. Siemens et al., *Phys. Rev.* **C40** (1989) 2641.
72. W. Botermans and R. Malfliet, *Phys. Lett.* **171B** (1986) 22.
73. W. Cassing and U. Mosel, *Prog. Part. Nucl. Phys.* **25** (1990).
74. H. Feldmeier, M. Schönhofen and M. Cubero, *Nucl. Phys.* **A495** (1989) 321c.
75. C.M. Ko, *Nucl. Phys.* **A495** (1989) 321c.
76. R. Fauser, A. Bonasera and H.H. Wolter, Proc. Int. Workshop on gross properties of nuclei and nuclear excitations XVIII (Hirschegg, Austria, 1990), ed. H. Feldmeier, p. 148.
77. L.P. Csernai and J.I. Kapusta, *Phys. Rep* **131** (1986) 223.
78. H.W. Barz et al., *Z. Phys.* **A302** (1981) 73.
79. G. Odyniec et al., Proc. 8th. High energy heavy ion study (LBL, Berkeley, USA, 1987), report LBL-24580 (1987) ; S.I. Chase et al., Proc. Workshop on nuclear dynamics VI (Jackson hole, USA, 1990), rep. LBL-28709.
80. D. Hahn and N. K. Glendenning, *Phys. Rev.* **C37** (1988) 1053.
81. H. Stöcker, A.A. Oglobin, and W. Greiner, *Z. Phys.* **A303** (1981) 259.
82. C. Cavata et al., Proc. Int. Workshop on gross properties of nuclei and nuclear excitations XVIII (Hirschegg, Austria, 1990), ed. H. Feldmeier,

p. 148.
83. M.M. Sternheim and R.R. Silbar, *Phys. Rev.* **D6** (1972) 3117, and references therein.
84. P. Danielewicz and G. Odyniec, *Phys. Lett.* **B157** (1985) 146.
85. H.A. Gustafsson et al., *Phys. Rev. Lett.* **52** (1984) 1590.
86. H.H. Gutbrod et al., *Phys. Lett.* **216B** (1989) 267.
87. R.E. Renfordt et al., *Phys. Rev. Lett.* **53** (1984) 763.
88. J. Gosset et al., *Phys. Rev. Lett.* **62** (1989) 1251.
89. J. Poitou et al., Proc. XXVIIIth. Winter meeting on nuclear physics (Bormio, Italy, 1990), ed. I. Iori, p. 54.
90. G.F. Bertsch et al., *Phys. Lett.* **189B** (1987) 384.
91. M.B. Tsang et al., *Phys. Rev. Lett.* **57** (1986) 559.
92. J.P. Sullivan et al., LPC Caen preprint, LPCC 90-04.
93. D.A. Sparrow et al., *Phys. Rev.* **C10** (1974) 2215.
94. C. Hartnack et al., Proc. Int. Workshop on gross properties of nuclei and nuclear excitations XVI (Hirschegg, Austria, 1988).

Dilepton Production in Heavy–Ion Collisions‡

Gy. Wolf*, G. Batko, W. Cassing, U. Mosel, K. Niita, and M. Schäfer
Institut für Theoretische Physik, Universität Giessen
D–6300 Giessen, West Germany

ABSTRACT

We study the dynamics of heavy–ion collisions by use of transport equations. Our model evolves phase–space distribution functions for nucleons, Δ's, N*(1440)–resonances and pions with their isospin degrees of freedom. The equations are solved by the test–particle simulation method. We apply this model to proton–nucleus and nucleus–nucleus collisions from 400 MeV/A to 2.1 GeV/A. Our approach is found to correctly describe the spectra of baryons and pions. Furthermore, we calculate the production of dileptons considering proton–neutron and pion bremsstrahlung, Δ and N* decay and $\pi^+\pi^-$ annihilation. We compare the theoretical predictions for p+^{9}Be and for ^{40}Ca+^{40}Ca at the bombarding energies of 1 GeV/A and 2 GeV/A with experimental data. Finally, we discuss possible in–medium effects on the dilepton spectra and propose to measure the proton electromagnetic formfactor by dilepton spectra.

1. Introduction

Heavy–ion collisions (HIC) offer a unique possibility to study nuclear matter far from its ground state. The main goals of this research are to extract a nuclear equation of state (EOS) and to find phase transitions. Furthermore, HIC's may also provide information about baryon and meson selfenergies at high density.

Electromagnetic observables (γ's and dileptons) are promising probes to provide such information[1], since after their production they interact only weakly with the baryonic environment. Furthermore, the photon production rate is an increasing function of the density and the available energy such that

‡Work supported by BMFT and GSI Darmstadt.
*On leave from Central Research Institute for Physics, Budapest.

it may serve as a probe for the early high–density stage of the collisions.

There is another strong motivation for doing dilepton–spectroscopy: according to several theoretical calculations[2-4] the pion dispersion relation $\omega_{\pi}(k_{\pi})$ strongly deviates from the free relation in a dense nuclear medium. The first calculations[5-7] of the dilepton yield from $\pi^+\pi^-$ annihilation and proton–neutron bremsstrahlung in thermodynamical models have indicated a measurable effect of the pion dispersion relation. (However, a recent calculation[8] shows that the effect is strongly reduced if the pion dispersion relation is taken into account also in the calculation of the annihilation matrix elements.) The main conclusion of these works, namely that in the dilepton spectra for ^{40}Ca+^{40}Ca a clear pion annihilation signal could be seen, was confirmed in recent calculations[9-12] including nonequilibrium dynamics.

Above several hundred MeV/A bombarding energy the pions play a very important role in HIC. By subsequent absorption and reemission, they help to thermalize the system. Moreover, the pions also serve as important secondary observables and influence other particle production yields as well (e.g. dileptons). Therefore, besides the baryons in our model also the pions appear as explicit degrees of freedom. In the present study the pions propagate as free particles; this allows us to investigate whether the present experimental data require any deviation from the free dispersion relation.

In Section 2 we describe the BUU model, while in Section 3 we study dilepton production in proton–nucleus and nucleus–nucleus collisions. Section 4, finally, is devoted to a brief summary of our work.

2. Description of the BUU approach

2.1 The transport equations

We study the dynamics of HIC by use of transport equations. One of the most powerful methods for the description of heavy–ion reactions is the BUU model. The details of our model are described in ref. 12; here we give only a brief summary.

At the energies of our interest the colliding system contains no longer only nucleons, but also pions, deltas and higher resonances as well. Therefore, distribution functions and evolution equations have to be introduced for each type of particles. The Δ's and N*'s are propagated in the same (Skyrme–type) potential[12,13] as the nucleons, whereas pions are treated as free particles. This mean–field potential used in our calculations generates an EOS with $m^* = 938$ MeV and compressibility $K = 240$ MeV.

We solve the equation system by the test–particle method, where the continuous distribution functions are substituted by a finite number of test–particles. The test–particles move on classical trajectories in the mean–field potential. The collisions are handled as in a cascade code; the only modification appears for fermions in the final state, where a Pauli–blocking factor $(1-f(r,p,t))$ is included.

We use the parallel ensemble algorithm[14] in which collisions are only allowed between particles in the same ensemble.

2.2 The collision term

As noted before we include in our transport approach nucleons (N), deltas (Δ(1232)), N*(1440)'s and π's, using different distribution functions for different isospin states. In the collision term we include the following reactions (B denotes a baryon and R denotes either a Δ or a N*):

$$
\begin{array}{lllll}
1. & B_i + B_j & \rightarrow & B_i + B_j & \\
2. & N + N & \rightarrow & N + R & \\
3. & N + R & \rightarrow & N + N & \\
4. & R & \rightarrow & N + \pi & \\
5. & N + \pi & \rightarrow & R & \qquad (2.1)
\end{array}
$$

For all baryon–baryon elastic cross–sections (reaction 1 in (2.1)) we use the Cugnon–parametrization[14,15]. For the total cross–section of reaction 2 in (2.1) we adopt the parametrization of VerWest and Arndt[16].

The angular distribution of reaction 2. in (2.1) is usually assumed to be isotropic. However, the Δ is produced in a p–wave, so its creation cross–section is forward peaked. We use an angular distribution fitted to experimental data[17]. The masses of the resonances are randomly distributed according to the Breit–Wigner distribution function using an energy dependent width[12,18].

The cross–section for reaction 3 in (2.1) are determined by the law of detailed balance[14,15] from the cross–section of reaction 2.

In the case of π absorption on nucleons, (reaction 5 in (2.1)) we assume resonance absorption and the maximum cross–sections are taken from the particle data tables[12].

The decay probability of the resonances is determined by an exponential decay law using their width and their proper time; the decay is assumed to be isotropic in their rest frame.

2.3 Results on πNΔ dynamics

We now compare the above model with experimental data. In fig. 1a the calculated transverse kinetic energy spectra of protons are compared with the experimental data[19] at 800 MeV/A bombarding energy. The agreement is good.

Since we are going to study $\pi^+\pi^-$ annihilation it is very important to reproduce the pion spectra and yields. In fig. 1b and fig. 2 we compare the

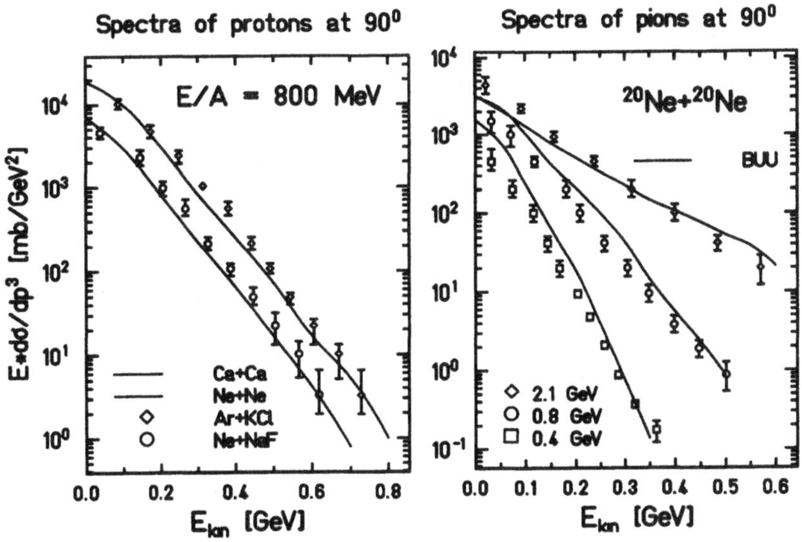

Fig. 1 a) Transverse energy spectra of protons at 800 MeV/A.
b) Transverse energy spectra of pions for Ne+Ne.
Solid lines represent the calculations. Experimental data from ref. 19.

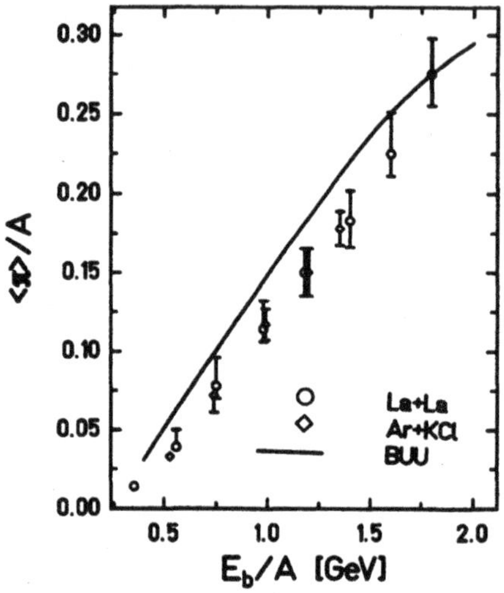

Fig. 2 Pion yield per participating nucleons as a function bombarding energy. Solid lines represent the calculations. Experimental data from ref. 20 and 21.

experimental spectra[19] and yields[20,21] with our calculations. In the whole energy region (from 400 MeV/A to 2.1 GeV/A) we reproduce the shape of the spectra quite well. The cross–sections are somewhat overestimated at low pion momenta. Since the differences are not large, we may conclude that this model is appropriate to study HIC in the energy region from 400 MeV/A to 2.1 GeV/A.

3. Dilepton production

In the following we study the $\pi^+\pi^-$ annihilation using free pions, the neutron–proton and pion–nucleon bremsstrahlung contribution and the Dalitz–decay of Δ and N^* as elementary processes for e^+e^- production in the course of a nuclear reaction. To get comparable results with the DLS data[22-24], we use the experimental acceptance filter[25] on the calculated results.

3.1 $\pi^+\pi^-$ annihilation

Since the cross–section for pion annihilation is very small for the reactions considered here we use a perturbative method to calculate the dilepton yield[12]. In this method one calculates the probability for producing dileptons, without, however, taking their energy out of the nucleus–nucleus simulation. This is in contrast to the treatment of pions which are produced at a much larger rate and, therefore, treated explicitly.

The pion annihilation proceeds through the ρ meson which decays into a virtual massive photon by vector dominance. The cross–section is known to be

$$\sigma^{\pi^+\pi^-\to e^+e^-}(M) = \frac{4\pi}{3}\left[\frac{\alpha}{M}\right]^2 \sqrt{1-\frac{4m_\pi^2}{M^2}}\;|F_\pi(M)|^2$$

$$|F_\pi(M)|^2 = \frac{m_r^4}{(M^2-m_r'^2)^2 + m_r^2\Gamma_r^2} \tag{3.1}$$

where M is the dilepton invariant mass and α is the fine structure constant,

$$m_r = 775 \text{ MeV},\quad m_r' = 761 \text{ MeV} \quad \text{and} \quad \Gamma_r = 118 \text{ MeV}. \tag{3.2}$$

3.2 e^+e^- bremsstrahlung

When charged particles collide, they can radiate a photon or –via a virtual photon– produce a dilepton pair. In our case, the relevant bremsstrahlung contribution comes from proton–neutron and charged pion–nucleon collisions.

Since these microscopic cross–sections have not been measured, theoretical calculations are needed. One method is based on a One–Boson–Exchange (OBE) model[26,27]. In an effective OBE approximation the bremsstrahlung cross–section was calculated taking into account both the internal and the external radiation, the coupling constants and formfactors were fitted to the proton–neutron elastic cross–section. The same method has been applied for real photons and has given good fits to the experimental data[1]. This method has the practical disadvantage that the coupling constants and formfactors are energy dependent.

A less ambitious, but cheaper way to get these cross–sections, is the soft–photon approximation, where the radiation from internal lines is neglected and the strong interaction vertex is assumed to be on–shell. This cross–section has to be corrected by reducing the remaining phase–space for the colliding particles in their final state[26]. Since the applicability of these approximations is not quite clear, it was compared with the results from the Born–approximation in ref. 12 and found to be surprisingly good above 100 MeV dilepton invariant mass. The phase–space corrected soft–photon cross–section reads

$$\frac{d\sigma}{dy\, d^2q_t\, dM} = \frac{\alpha^2}{6\,\pi^3}\,\frac{\overline{\sigma(s)}}{M\, q_o^2}\,\frac{R_2(s_2)}{R_2(s)}$$

$$R_2(s) = \sqrt{1 - (m_1+m_2)^2/s}$$

$$s_2 = s + M^2 - 2\, q_o \sqrt{s}$$

$$\overline{\sigma(s)} = \frac{s - (m_1+m_2)^2}{2\, m_1^2}\,\sigma(s) \tag{3.3}$$

where m_1 is the mass of the accelerating charged particle, M is the invariant mass, q_o is the energy, q_t is the transverse momentum, and y is the rapidity of the dilepton pair. In our actual calculations we have therefore used the cross–section (3.3). The same formula was used also for pionic bremsstrahlung.

3.3 Dalitz–decays

Any particle that has a decay to photons can also radiate dileptons (Dalitz–decay). The width of the decay to dileptons of mass M is determined by QED[28]. Since the invariant mass of a dilepton pair from π^o decay cannot be larger than the π^o mass, the π^o Dalitz–decay does not contribute at large M of interest here.

An important contribution, furthermore, comes from Δ decay. The $\Delta N\gamma$ vertex used here is given in ref. 12. The coupling constant was fitted to the photonic decay width $\Gamma_0(0) = 0.72$ MeV. Note that only Δ^+ and Δ^o can decay into dilepton pair. (The Dalitz–decay of N^* is handled in the same fashion.)

3.4 Results

We calculate the dilepton yields for p+^{9}Be at 1.0 and 2.1 GeV and for ^{40}Ca+^{40}Ca at 1.0 and 2.0 GeV/A bombarding energies and compare our results in fig. 3 and 4 with the experimental data of the DLS collaboration. Below about 200 Mev invariant mass, where the experimental filter has a significant effect, we obtain much more cross-section and a different shape

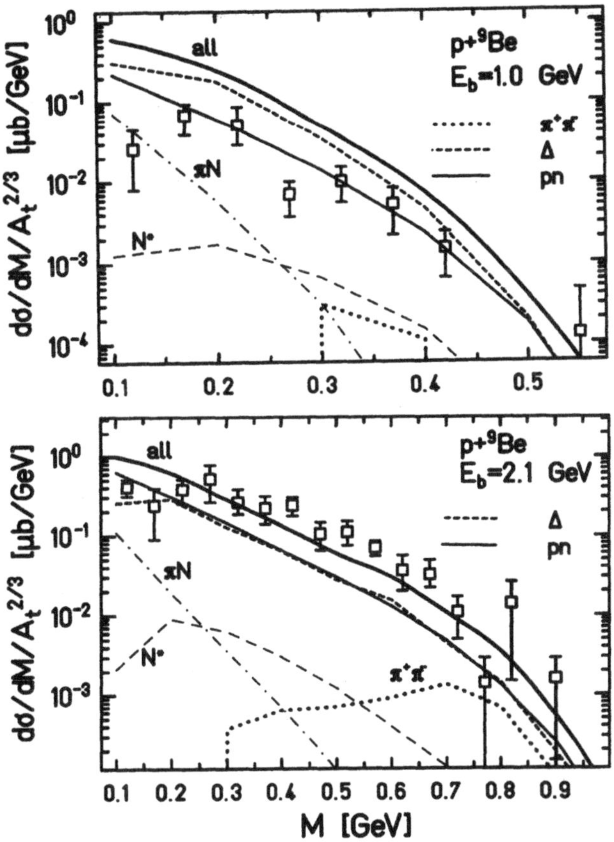

Fig. 3 Comparison of the calculated dilepton yield with the data[22] for p+^{9}Be at 1 GeV and 2.1 GeV bombarding energy. Solid line: pn bremsstrahlung; dotted–dashed: π–nucleon bremsstrahlung; dotted: $\pi^+\pi^-$ annihilation; dashed: Δ–decay; thick solid line: sum of all contributions.

than measured. In the following we discuss only the higher invariant mass range.

In p+^{9}Be at 1 GeV our calculations overestimate the data by a factor of two to three. At 2.1 GeV, on the other hand, the agreement is quite good.

For the Ca+Ca case (fig. 4) the agreement is similarly good and there is clear evidence for a pion annihilation component in the spectrum; for M > 400 MeV the dominant contribution to the total yield is that from $\pi^-\pi^+ \to e^-e^+$.

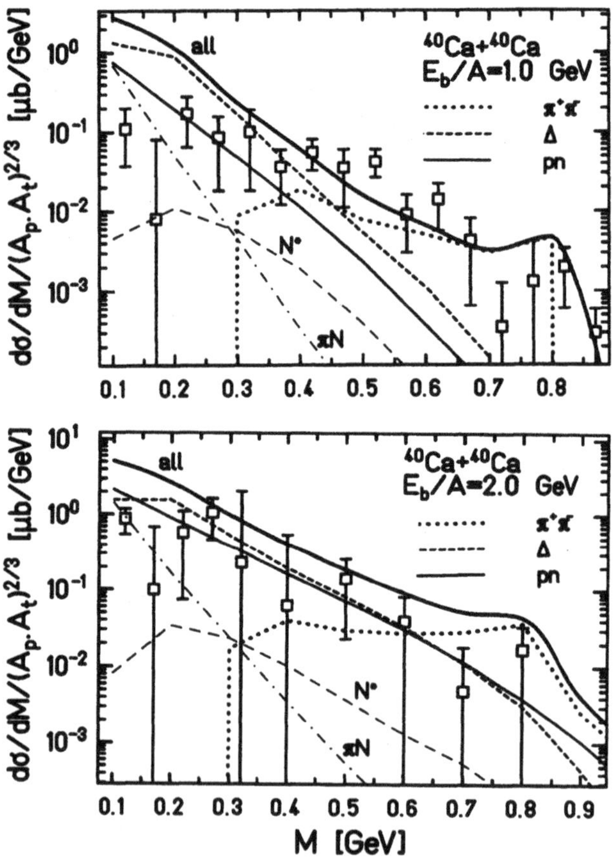

Fig. 4 Comparison of the calculated dilepton yield with the data[23] for ^{40}Ca+^{40}Ca at 1 and 2 GeV bombarding energy. Solid line: pn bremsstrahlung; dotted–dashed: π–nucleon bremsstrahlung; dotted: $\pi^+\pi^-$ annihilation; dashed: Δ–decay; thick solid line: sum of all contributions.

Comparing the relative importance of the different contributions the π–nucleon bremsstrahlung and the Dalitz–decay of the N* can be neglected for each system. The annihilation contributions are negligible for p+Be, as one would expect. At 1 GeV bombarding energy and masses below about 0.4 GeV the Dalitz–decay contribution is again too high as in the p+^{9}Be case.

Our results for Ca+Ca at 1 GeV/A bombarding energy more or less agree with a recent calculation of Xiong et al[11]. The most prominent difference is found to be in the Dalitz–decay contribution of the Δ which turns out to be larger in our calculation. In this context it is worth mentioning that the pion yield in ref. 11 is about twice as high as the experimentally measured one[20], whereas our pion yield agrees within about 30% with the data (cf. fig. 2).

All the calculations discussed up to now have been performed without using electromagnetic formfactors for the nucleons because these are experimentally not known for time–like photons in the mass range of interest here. In order to see the possible effect of such formfactors we have also performed the calculation for p+^{9}Be with an electromagnetic formfactor for the baryons as obtained from the vector–meson–dominance model, i. e. with a formfactor of the form (3.1). Whereas this has virtually no effect at 1 GeV, it raises the

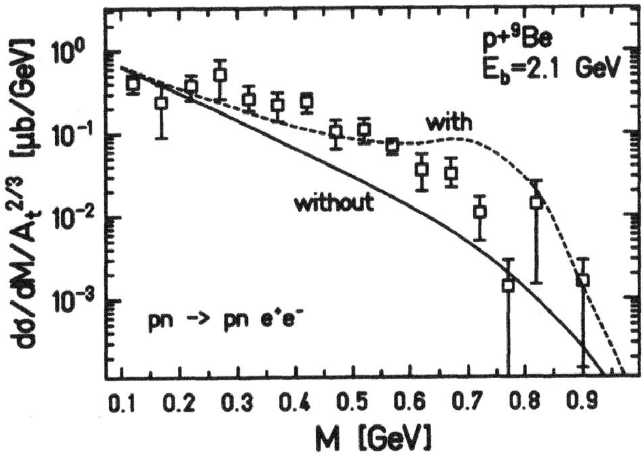

Fig. 5 Comparison of the calculated dilepton yield with the data[22] for p+^{9}Be at 2.1 GeV. Shown are the results of the calculation from pn bremsstrahlung without (solid line) and with an electromagnetic formfactor (dashed curve) for the baryons obtained from the vector–meson–dominance model.

cross–section for 2.1 GeV above the data up to a factor of three at M ≃ 700 MeV (see fig. 5). For the $^{40}Ca+^{40}Ca$ reaction, for which our calculations give already quite high cross–sections, the inclusion of the formfactor raises the calculations above the data up to a factor of 20 at M ≃ 800 MeV.

Our calculations have shown that a large part of the observed dilepton yield is due to nucleon–nucleon bremsstrahlung and to the Dalitz–decay of the Δ. It will, therefore, not be easy to detect evidence for a non–trivial pion dispersion relation in the $\pi^-\pi^+$ events. One of the signals of a non–trivial dispersion relation would be a shift of the annihilation threshold to lower invariant masses. However, at 1 GeV bombarding energy the annihilation contribution at about 300 MeV invariant mass (the threshold range) is much smaller than the bremsstrahlung component so that this shift will probably not be observable. Although our calculations with a free dispersion relation seem to describe the data, it is impossible to draw any definite conclusions by now since the experimental error bars are still very large, and in the higher invariant mass range other mesons and resonances can also contribute.

With increasing bombarding energy the situation become worse (see fig. 4) the annihilation contribution becomes important only at higher invariant mass

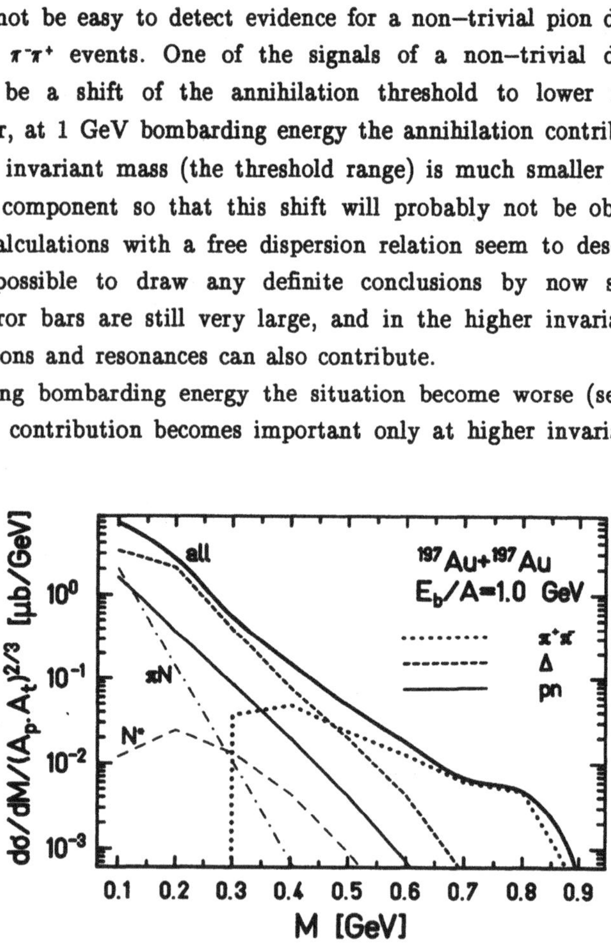

Fig. 6 Comparison of the calculated dilepton yield for $^{197}Au+^{197}Au$ at 1 GeV bombarding energy. Solid line: pn bremsstrahlung; dotted–dashed: π–nucleon bremsstrahlung; dotted: $\pi^+\pi^-$ annihilation; dashed: Δ–decay; thick solid line: sum of all contributions.

range. A model calculation for Au+Au at 1 GeV/A bombarding energy shows (fig. 6) that an increase of the size of the system does not change the ratio of the components.

One possibility to enhance the $\pi^+\pi^-$ dilepton signal would be the selection of central collisions. The impact parameter weighted proton–neutron bremsstrahlung and the annihilation contribution can be seen in fig. 7 for $^{40}Ca+^{40}Ca$ at 1 GeV/A as a function of the impact parameter b. A high

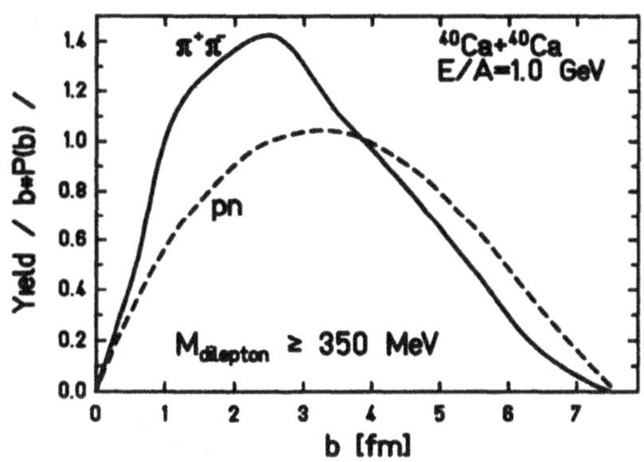

Fig. 7 Impact parameter dependence of the pn bremsstrahlung contribution (dashed line) and the $\pi^+\pi^-$ annihilation (solid line) for $^{40}Ca+^{40}Ca$ at 1 Gev/A.

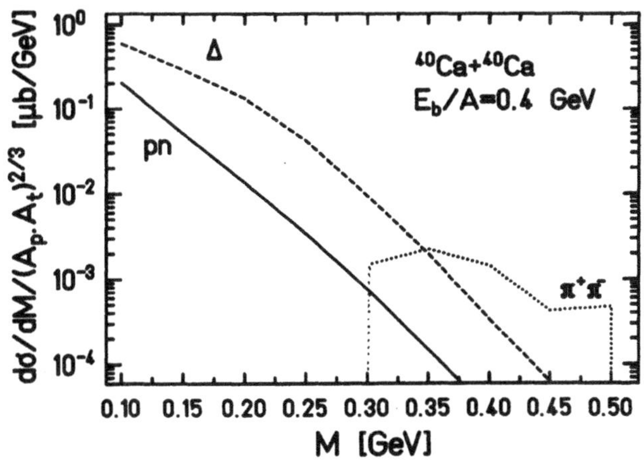

Fig. 8 Invariant mass spectra of dileptons for $^{40}Ca+^{40}Ca$ at 0.4 GeV/A bombarding energy. Solid line: pn bremsstrahlung; dotted: $\pi^+\pi^-$ annihilation; dashed: Δ–decay.

multiplicity cut could suppress the bremsstrahlung component by almost 50%.

Dileptons may also provide information about the properties of Δ in hot dense matter. For $^{40}Ca+^{40}Ca$ at 1 GeV/A bombarding energy the Δ contribution is quite high compared to the other processes, at 400 MeV/A bombarding energy it is even more prominent (see fig. 8).

4. Summary

We have studied heavy–ion dynamics on the basis of a microscopical model of the BUU type. In this model besides the nucleons also two baryonic resonances, the Δ(1232) and the N*(1440) as well as the pions are explicitly propagated. The effect of the medium on the resonance energies are not taken into account. For the Δ and N* production we have used a non–isotropic cross–section, the angular dependence of which is fitted to experimental data. This angular dependence gives about 20% less stopping at 2 GeV/A than the previously used isotropic cross–section. The isospin degrees of freedom are explicitly included and the isospin dependence of the cross–sections are taken into account.

The model correctly reproduces the nucleon and the pion transverse energy spectra and thus seems to describe the hadronic processes quite reliably. We have applied this model to dilepton production by taking contributions from proton–neutron bremsstrahlung, pion–nucleon bremsstrahlung, the Dalitz–decay of the Δ and N* and $\pi^-\pi^+$ annihilation into account. It is found that at all bombarding energies considered the pn–bremsstrahlung and the Δ Dalitz–decay account for most of the observed cross–section. This opens the possibility to study properties of the Δ in dense and hot nuclear matter. It will be particularly interesting to study both the self–energy and the width of the Δ in this environment; such studies thus can yield complementary information to that obtained from photo–absorption experiments on nuclei and from charge–transfer reactions both of which test the Δ–properties only at or below the equilibrium density.

One could speculate that also the properties of the ρ–meson in dense matter could possibly be explored in dilepton experiments because they effect the baryonic electromagnetic formfactors. In this context we would like to point out that a precise measurement of the pn → pn e^+e^- cross–section could for the first time determine the nucleonic electromagnetic formfactors for time–like momentum transfers up to a few GeV/c which are otherwise experimentally not accessible.

We have also shown that – in agreement with earlier findings – the data exhibit a clear $\pi^+\pi^-$ annihilation signal at high invariant dilepton mass so that these experiments can be used to study also the properties of the pions in dense matter.

It will furthermore be interesting to investigate if particular contributions to the total yield can be enriched by proper experimental set–ups. As an example, we have pointed out, that a multiplicity trigger might help to enhance the pion annihilation contributions over the bremsstrahlung components.

References

1) W. Cassing, V. Metag, U. Mosel and K. Niita, *Phys. Rep.* **188** (1990) 363
2) A.B. Migdal, *Rev. Mod. Phys.* **50** (1978) 107
3) G.F. Bertsch et al., *Nucl. Phys.* **A490** (1988) 745
4) T. Ericson and W. Weise, *Pions and Nuclei* (Clarendon Press, Oxford, 1988)
5) C. Gale and J. Kapusta, *Phys. Rev.* **C35** (1987) 2107
6) L.H. Xia, C.M. Ko, L. Xiong and J.Q. Wu, *Nucl. Phys.* **A485** (1988) 721
7) C.M. Ko, L.H. Xia and P.J. Siemens, *Phys. Lett.* **B231** (1989) 16
8) C.L. Korpa and S. Pratt, *Phys. Rev. Lett.* **64** (1990) 1502
9) G. Batko et al., in: *Proc. NATO ASI The Nuclear Equation of State* (Peniscola, Spain, 1989) ed. by W. Greiner, Pergamon Press, in press
10) L. Xiong et al., *Phys. Rev.* **C41** (1990) R1355
11) L. Xiong, Z.G. Wu, C.M. Ko and J.Q. Wu, *Nucl. Phys.* A (1990) in press
12) Gy. Wolf et al., *Nucl. Phys.* A (1990) in press
13) D. Vautherin and D.M. Brink, *Phys. Rev.* **C5** (1972) 626

14) G.F. Bertsch and S. Das Gupta, *Phys. Rep.* **160** (1988) 189
15) J. Cugnon, T. Mizutani and J. Vandermeulen, *Nucl. Phys.* **A352** (1981) 505
16) B.J. VerWest and R.A. Arndt, *Phys. Rev.* **C25** (1982) 1979
17) F. Shimizu et al., *Nucl. Phys.* **A389** (1982) 445
18) J.H. Koch, E.J. Moniz and N. Ohtsuka, *Ann. Phys.* **154** (1984) 99
19) S. Nagamiya and M. Gyulassy, *Advances in Nuclear Physics* **13** (Plenum, 1984) p. 201
20) J.W. Harris et al., *Phys. Lett.* **153B** (1985) 377
21) J.W. Harris et al., *Phys. Rev. Lett.* **58** (1987) 463
22) G. Roche et al., *Phys. Rev. Lett.* **61** (1988) 1069
23) C. Naudet et al., *Phys. Rev. Lett.* **62** (1989) 2652
24) G. Roche et al., *Phys. Lett.* **B226** (1989) 228
25) J. Carroll and G. Roche, private communication
26) M. Schäfer et al., *Phys. Lett.* **B221** (1989) 1
27) K. Haglin, J. Kapusta and C. Gale, *Phys. Lett.* **B224** (1989) 433
28) B.E. Lautrup and J. Smith, *Phys. Rev.* **D3** (1971) 1122

IV. Contributed Papers Not Selected for Oral Presentation

Detecting equilibrium prior to multifragmentation using mass or charge partition probabilities.

P. Désesquelles, A.J. Cole, A. Giorni, D. Heuer and A. Lleres.

ISN Grenoble (IN2P3-CNRS)/Université J. Fourier, 38026 Grenoble Cedex, France.

ABSTRACT

A pilot study using 3D bond percolation in a simple cubic lattice indicates that equilibrium of a nucleus prior to disassembly may be detectable via the study of partition probabilities. The absence of correlations (interpreted as a signature of equilibrium) is concretised by the success of multinomial forms in reproducing the observed (experiment or model) partition probabilities.

1 - Introduction

Theoretical studies[1-4] of small impact parameter heavy ion collisions at intermediate energies (approximately in the range 30 to 100 MeV/nucleon) suggest that the initial compression and heating of the projectile target (parent) complex is followed by an expansion phase which results in multifragmentation of the parent. For energies above about 40 MeV/nucleon mean field models predict that the expansion continues unchecked leading directely to (dynamical) multifragmentation[5] whereas at lower energies the expansion is followed by a contraction leading to a monopole oscillation.

Several works[5-10] have attempted to treat the latter scenario by supposing that, at some point, the dynamically evolving system can be replaced by an equilibrated parent complex at less than normal nuclear matter density (typically $\rho/\rho_0 = 0.5$) whose decay can be studied using standard techniques of statistical mechanics.

It should be emphasized that experimental investigation of multifragmentation is well underway partly through photographic emulsion experiments[11-12] and partly due to the use of large solid angle hodoscopes[13] or better still 4 Π multidetector set-ups such as AMPHORA[14]. However, the principle experimental difficulties remain unchanged. They are :

1) detection of complete events,
2) the isolation of small impact parameter (central) collisions,

3) the signature of multifragmentation events as opposed to the sequential decay ("boiling off") mode of highly excited parents.

Thus it seems fair to say that at the present time, despite some good agreement between calculations and observations (particularly for inclusive yields[15]) that evidence for a true multifragmentation (dynamical or statistical) decay mode comes mainly from theory.

It was recently shown[16] that several model predictions for statistical multifragmentation of atomic nuclei as well as experiment data sets for light nuclei[17-18] could be accurately simulated using simple multinomial expressions for the partition probabilities.

Thus the probability to observe a partition of a given fixed initial mass M_t or charge Z_t (we will use the mass variable herein) can be written :

$$P^{M_t}(n_1,n_2...n_{M_t}) = G(N)\ \Pi_k\ (X_k^{n_k})/n_k!)\ \delta(\Sigma\ k.n_k - M_t) \qquad (1)$$

where the partition $\{n_1, n_2 ... n_{M_t}\}$ is defined as the set of n_1 fragments of mass 1, n_2 fragments of mass 2 etc... In equation (1) the multiplicity $N = \Sigma\ n_k$ and the total mass $M_t = \Sigma\ k.n_k$.

An intriguing possibility arises from the fact that the model systems studied in ref. 16 imply equilibrium prior to multifragmentation. The success of the multinomial approximation may then be due to the presence of a special kind of initial equilibrated state i.e. to the absence of correlations (other than those implied by explicit conservation of the total mass M_t).

Among existing models the percolation approach to multifragmentation is well adapted for the investigation of this hypothesis[19]. Similarities, as well as some differences, between the predictions of such models and experimental data have been discussed by Campi[20] and by Gross and collaborators[21].

In the bond percolation models, disassembly is provoked by breaking bonds in a finite lattice (in our case simple cubic) and extracting the connected fragments. Typically 10^6 such "events" are generated. Equilibrium is simulated by using a constant bond-breaking probability throughout the lattice. In Fig. 1a. we show partition probabilities for disassembly of mass 27 together with the multinomial approximation. The partition channels (abscissae) are defined simply by ordering the partitions so that the multinomial predictions form a monotonically decreasing set. The parameters X_k of equation (1) were obtained by requiring the M_t mean values $<n_k>$ to be accurately reproduced (to within one part in 10^3) by equation (1). Thus :

$$\frac{\Sigma\, n_k . P^{M_t}(n_1,n_2 \ldots n_{M_t})}{\Sigma\, P^{M_t}(n_1,n_2 \ldots n_{M_t})} = <n_k>_{exp} \qquad (2)$$

where the sums are carried out over all partitions (3010 for mass 27) and the suffix "exp" refers to the computer simulation...

In this sense the multinomial formula is to be thought of not as a model calculation but rather as a parameterisation of the "data" in which quadratic and higher order correlations of physical origin are suppressed.

We have attempted to simulate a non-equilibrium situation by using a variable bond breaking probability in the percolation calculation. In fact we have simulated a rather extreme case in which, for each Monte-Carlo event, bonds in one half of the 3x3x3 cube are broken with probability P_1 and in other half with a different probability P_2. the parameters X_k were fixed as before by requiring the detailed reproduction of the mean values $<n_k>_{exp}$.

As can be seen in Fig. 1b this procedure fails to reproduce an accurate description of the partition probabilities generated in the computer experiment and this finding remains true despite several attempts to improve the description by modifying the multiplicity dependent factor G(N). Other studies made, for example, by adding preequilibrium (free) nucleons to each partition generated lead to the same observation.

In conclusion we have described a pilot study using 3D bond percolation in a simple cubic lattice wich indicates that equilibrium of the parent complex prior to multifragmentation may be detectable via the absence of correlations in partition probabilities. The absence of correlations (quadratic or higher order) is concretised by the success of multinomial forms in reproducing the observed (experiment or model) partition probabilities (see ref. 16 for details).

In parallel studies[22], wich are not developed herein for lack of space, we have also shown that multidetector efficiencies and sequential evaporation from excited primary fragments do not significantly alter this conclusion.

For the multifragmentation of heavy nuclei care is needed in the extrapolation of this result since even for an initial state in statistical equilibrium quadratic and higher order correlations may be sufficiently strong to invalidate the multinomial approximation.

Despite the obvious limitations of our study (notably the relavence of the percolation approach to multifragmentation) we believe that investigations along these lines should contribute significantly to the understanding of the concept of equilibrium in small short lived nuclear systems.

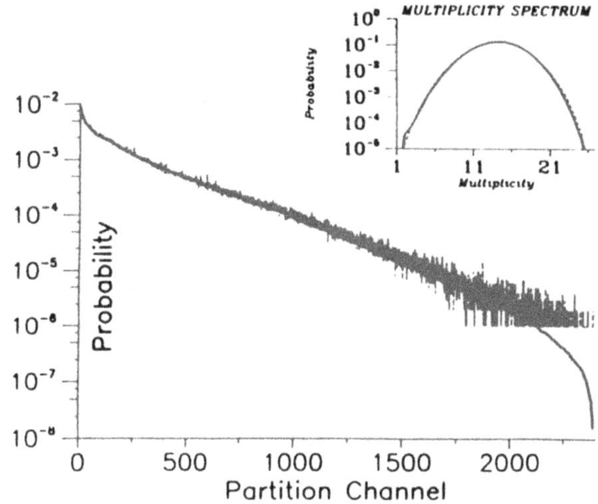

Fig. 1a

Equilibrium simulation. Multinomial partition probabilities (solid line) compared with results (dotted line) from the 3x3x3 cubic bond breaking Monte Carlo calculation with constante bond breaking probability p=0.24. The partition channel correspond simply to an ordering of the partitions so that the multinomial predictions form a monotonically decreasing set.. (χ^2=2.8)

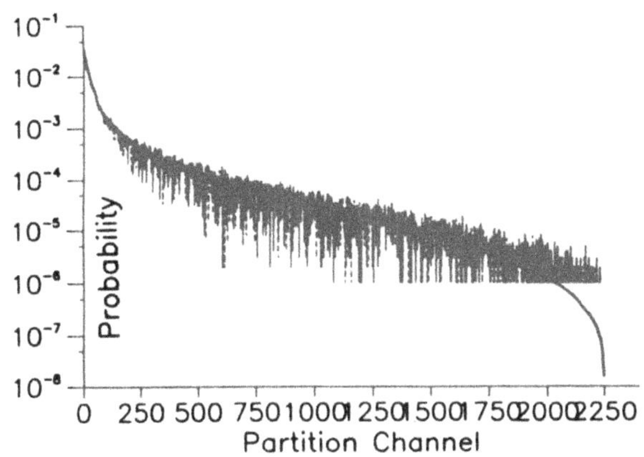

Fig. 1b

Out of equilibrium simulation. As in Fig. 1a but with two bond breaking probabilities p1=0.2, p2=0.6 used in each half of the cubic lattice for each Monte Carlo event. (χ^2=273)

LIST OF REFERENCES

1) A. Dhar and S. Das Gupta, *Phys. Rev.* **C30**, 1545 (1984).
2) C. Ngo, H. Ngo, S. Leray and M.E. Spina, *Nucl. Phys.* **A499**, 148 (1989).
3) C.J. Pethick and D. Ravenhall, *Nucl. Phys.* **A471**, 19c (1987).
4) C. Gregoire, B. Remaud, F. Sebille, L. Vinet and Y. Raffray, *Nucl. Phys.* **A456**, 317 (1987).
5) K. Sneppen and L. Vinet, *Nucl. Phys.* **A480**, 342 (1988).
6) J.P. Bondorf, R. Donangelo, I.N. Mitsushin and H. Schulz, *Nucl. Phys.* **A444**, 410 (1985).
7) Zhang Xiao-Ze, D.H.E. Gross, Xu Shu-Yan and Zheng Yu-Ming, *Nucl. Phys.* **A461**, 668 (1987).
8) J. Randrup and S. Koonin, *Nucl. Phys.* **A356**, 223 (1981).
9) G. Fài and J. Randrup, *Nucl. Phys.* **A404**, 551 (1983).
10) A.Z. Mekjian, *Phys. Rev. Lett.* **64**, 2125 (1990).
11) C.J. Waddington and P.S. Freier, *Phys. Rev.* **C31**, 1888 (1985).
12) B. Jakobsson, G. Jonsson, L. Karlsson, V. Kopljar, N. Norén, K. SOderstrom, E.Monnand, F Schussler, H. Nifenecker, G. Fài, J.P. Bondorf and K. Sneppen, *Nucl. Phys.* **A509**, 195 (1990).
13) R. Trockel, K.D. Hildenbrand, U. Lynen, W.F.J. Muller and H.J. Rabe *Phys.Rev.* **C39**, 729(1989)
14) D. Drain et al, *Nucl. Instr. and Methods* **A281**, 528 (1989).
15) Zhang Xiao-Ze, D.H.E. Gross, Xu Shu-Yan and Zheng Yu-Ming, *Nucl. Phys.* **A461**, 668 (1987).
16) A.J. Cole and P. Désesquelles, Zeits. Phys. **A**, (to be published).
17) J. Pouliot, Y. Chan, A. Dacal, D.E. Gregorio, B.A. Harman, R. Knop, C. Moison, L. Potvin, C. Rioux and R. Roy, *Phys. Lett.* **223**, 16 (1989).
18) A.J. Cole, H.J. Crawford, P.J. Lindstrom and B.G. Harvey, *Phys. Rev.* **C39**, 891 (1989).
19) see for example C. Ngo in Nuclear Matter and Heavy Ion Collisions "Les Houches", Plenum Press 1989
20) X. Campi, *J. Phys.* **A19**, L917 (1986).
21) H.R. Jaquaman, G. Papp and D.H.E. Gross, *Nucl. Phys.* (to be published).
22) A.J. Cole Internal report ISN90-12 (unpublished).

MEASUREMENTS OF COLOURED LIFETIMES

MARTIN ERDMANN

Fakultät für Physik,
Albert-Ludwigs-Universität Freiburg i.Br.,
Hermann Herderstr.3, 7800 Freiburg, FRG

ABSTRACT

A recently proposed method to measure coloured lifetimes is briefly reviewed and results are compared with lifetimes found in Bose-Einstein interferometry.

Size and lifetime of coloured sources can be measured by particle interferometry. An alternative method to measure coloured lifetimes has recently been suggested [1,2]. The observed object is a nucleon before and after an inelastic interaction. The picture of the analysis is the following: The nucleon looses colour neutrality at the beginning of the interaction. Due to a constant confinement force as seen f.e. in charmonium spectroscopy the coloured nucleon gets accelerated. After some specific time the coloured nucleon is neutralized and can escape freely from the interaction zone. It has been shown[2] that the proper time τ which the nucleon stays coloured is proportional to the rapidity change $y_{initial} - y_{final}$ of the nucleon which can be measured by experiments:

$$\tau = \frac{m\,c}{\kappa} \; | \, y_{initial} - y_{final} \, | \tag{1}$$

κ is the strength of the confinement force ($\kappa =$[3] $0.93\ GeV/fm$), c is the velocity of light and m is the rest mass of the coloured nucleon assumed to be approximately the mass of the colour neutral nucleon. The distribution of the coloured lifetime should follow an exponential decay with mean time τ_0. In pp→pX experiments the rapidity change of protons has in fact exponential form[2]. An exponential fit to these data $exp(- \mid y_{initial} - y_{final} \mid \frac{mc}{\kappa} \, / \, \tau_0)$ using equation (1) and assuming above values for κ and m gives the mean lifetime of the coloured proton:

$$\tau_0 = (3.9 \pm 0.3) \times 10^{-24}\, sec = (1.2 \pm 0.1)\, fm/c. \tag{2}$$

This result can be compared with lifetime measurements from pp and e^+e^- interferometry experiments. Taking the diameter of the coloured source gives a lifetime of $\tau_0 = 2R/c \approx$[4] $1.3\ fm/c$ in these experiments, which looks like an encouraging agreement between the two methods.

REFERENCES

1. M.Erdmann, thesis, University Freiburg i.Br. (Mar-1990)
2. W.Busza, T.Dreyer and M.Erdmann, "Direct study of the lifetime and acceleration of a coloured nucleon", MIT-LNS-1904, THEP90/5 (Apr-1990), subm. to *Zeit.f.Phys.C*
3. E.Eichten et. al.: *Phys. Rev.* **D21**, 203 (1980)
4. B.Lörstad, see this conference.

PROTON-PROTON AND DEUTERON-DEUTERON CORRELATIONS IN INTERACTIONS OF RELATIVISTIC HELIUM NUCLEI WITH PROTONS

DUBNA-KOSICE-MOSCOW-TBILISI-WARSAW COLLABORATION
J.Galazka-Friedman, T.Sobczak, J.Stepaniak, I.P.Zieliński
Institute for Nuclear Studies and Institute of Physics, Warsaw Technical University, WARSAW, POLAND
V.V.Glagolev, R.M.Lebedev, G.D.Pestova
Joint Institute for Nuclear Research, DUBNA
M.Báno,J.Hlaváčova,G.Martinská,J.Patočka,M.Seman,L.Šandor,J.Urban
Institute of Experimental Physics Slovac Academy of Sciences and University of P.J.Šafarik, KOŠICE, CSFR
K.U.Khairetdinov
P.N.Lebedev Physical Institute of the Academy of Sciences, MOSCOW, USSR
A.K.Kacharava, M.S.Nioradze, Z.S.Salukvadze
Tbilisi State University, TBILISI

1 Abstract

The reactions ^{4}He p→p+p+X, ^{3}He p→p+p+X and ^{4}He p→d+d+X have been investigated. The correlations were measured between protons and between deuterons with small relative momenta. Strong correlation have been observed and compared with theoretical predictions.

2 Introduction

The proton-proton correlations at small relative momenta can be due to the symmetry properties of the wave function, strong nucleon-nucleon final state interactions and Coulomb pp interaction [1,2,3]. There is also a possibility of preexistence of the proton clusters in nuclei [11]. A two-proton cluster may also preexists in helium nucleus.

The Coulomb repulsion prevents the two protons from having the same momenta. It results in anti-correlation for small relative momenta $(\vec{p_1} - \vec{p_2}) \rightarrow 0$. Due to the spatial anti-symmetry of the pp wave function in the spin triplet state the anti -correlation arises. The large positive correlation is due to the attractive singlet S-wave nuclear interaction. The resultant correlation function depends on the space-time dimensions of the generation region .

The proton-proton correlations at small relative momenta have been observed and used as a probe of the space-time picture of the collision in hadron-nucleus and nucleus-nucleus interactions in wide energy range and for different nuclei [4,5,6,7,

8,9,10]. Deuteron - deuteron correlations were also investigated e.g. in ^{16}O -^{197}Au and ^{40}Ca -^{40}Ca reactions [20,21].

Here we present the results on the correlations between nucleons emitted from the interactions of light nuclei - helium isotopes with protons. The situation is to some extent simpler than for heavy nuclei because there are no evaporation nucleons and the combinatorial background is smaller. One can also assume that the particles are emitted almost simultaneously, whereas in the interactions with heavy nuclei the time structure of the interaction is complicated e.g. the possibility of the compound nucleus formation should be taken into account.

Both proton-proton and deuteron-deuteron correlations can give information on the structure of the helium nuclei wave function (e.g. deuteron-deuteron component of the D-state).

3 Experimental procedure

The 1 meter JINR hydrogen bubble chamber was irradiated in the beams of the ^{3}He and ^{4}He nuclei. The momentum of the ^{3}He was 13.5 GeV/c. The momenta of ^{4}He were 8.6 and 13.6 GeV/c. Scanning for all topology events was performed. The events were measured and kinematically reconstructed. The reaction hypotheses consistent with visual estimation of the ionisation on tracks have been retained. Due to the inverse geometry used (fast incoming nucleus) the reaction channel is well determined for majority of events. In this paper two prong events have been excluded from analysis. Details concerning the experiment can be found in previous publications [12,13,14].

In the table below the statistics used for proton-proton correlation have been summarised. In the first column the number of events with at least two protons is shown, in the second one - the number of the proton - proton combinations.

	$p(GeV/c)$	N_{ev}	N_{comb}
^{4}He	8.6	17710	29687
^{4}He	13.6	13491	24716
^{3}He	13.5	16980	34713

In both experiments with ^{4}He beam 677 events with two deuterons in the final state have been found.

4 Results

4.1 Proton-proton correlation

The correlations between two protons with small relative momenta have been measured for ^{4}He - proton interactions at 8.6 and 13.6 GeV/c incident momenta of the alpha particle and for ^{3}He - proton at 13.5 GeV/c. The two nucleon correlation function $R(p_1, p_2)$ is defined as

$$R(p_1,p_2) = \frac{\sigma(p_1,p_2)}{\sigma(p_1)\sigma(p_2)} \tag{1}$$

Here we plot R as a function of four momenta difference k^*, where

$$k^* = \sqrt{(\vec{p_1} - \vec{p_2})^2 - (E_1 - E_2)^2}\ /2 \tag{2}$$

The measured values of the correlation function R is presented in figs 1 and 2 for these three sets of the data. The "background" distribution has been obtained by mixing nucleons from different events of the same class (e.g. with the same number of protons). The distribution was then normalised to the experimental one in the interval of the $k^*(100, 300) MeV/c$. Number of background combinations was about hundred thousands each time.

The clear positive correlation is observed for small relative momenta. There is no significant dependence of R on the incident momentum of ^{4}He (figs 1,2). The correlation is weaker for ^{3}He interactions (fig.3).

The correlation function calculated according to the formulae proposed by Lednicky and Lyuboshits [3] is also shown in the figures. The r_0 parameter characterises the dimensions of the emission region. The distribution of the points of protons generation has been taken in the form:

$$W(r,t) = (2\pi)^{-4} r_0^{-3} \tau_0^{-1} \exp[-\frac{r^2}{2r_0^2} - \frac{t^2}{2\tau_0^2}] \tag{3}$$

The root mean square radius of the generation region is $\sqrt{3}$ times larger than r_0 parameter. The pp scattering length in $^1S_0$ state being the most important at small relative momenta has been taken to be equal $7.8 fm$, the effective radius $2.78 fm$.

The parameter that characterises the time dimension of the emitting region τ_0 was taken to be equal $1 fm$ in our calculations. It was natural to assume the value of this order, as it corresponds to the travelling time of relativistic particle through the nucleus. The results are not sensitive to the exact value of τ_0. The effect of the variation of τ_0 from 1 to $2 fm$ is the same as the variation of r_0 by about 5-8% in the region of interest (see fig.4). In this figure the R value at the maximum (for $k^* = 20$ MeV/c) is plotted as a function of r_0 for different τ_0 .

The theoretical correlation function only weakly depends on the velocity of the proton pair $\bar{v} = (\vec{p_1} + \vec{p_2})/(E_1 + E_2)$. The maximum of the correlation function decreases by about 5% when v changes from $0.1c$ to $0.4c$. The curves in figs 1-3 have been calculated for $v = 0.2c$ and $\tau_0 = 1 fm$.

The best agreement between the theory and the experiment has been obtained for $r_0 = 1.0 + 0.2 fm$ in the case of ^{4}He and $r_0 = 1.5 + 0.2 fm$ for ^3He. It corresponds to the root mean square radius of the emitting region $1.7 fm$ and $2.1 fm$, when the radii of the helium isotopes are known to be equal 1.67 fm and 1.88 fm respectively [19]. The discrepancies between theory and the experiment in the first interval of k^* are due to the finite resolution of the momentum measurement.

In order to investigate the dependence of the correlation function on the proton momenta we have selected two subsamples - one with at least two slow protons (momentum of each $p < 200$ MeV/c) and the second with two fast ($p > 200$ MeV/c) protons in the helium nucleus rest frame.The average momentum for the first class is about 0.08 GeV/c for ^{4}He and 0.065 GeV/c for ^{3}He. For the second class it is equal to about 0.5 GeV/c for both incident nuclei. These average values were calculated for events in the relative four momentum k^* interval $0 - 0.3$ GeV/c.

The proton-proton correlation functions for events with both slow and both fast nucleons are presented in fig.6 for ^{4}He→p interactions and in fig.7 for ^{3}He→p interactions. The correlation between faster nucleons are stronger for both projectiles.

Similar observation of stronger correlations for faster protons has already been reported in the interactions with heavier nuclei [7][8][15][16][17]. Two possible reasons were discussed to explain such effect [15][18]: (i) the correlation between the lifetime of the source and the protons momenta (ii) the decrease of the perpendicular size of the emitting region with the proton momentum and consequently to the non-spherical volume of the proton emission volume. The first explanation is improbable in our case.

4.2 Deuteron-deuteron correlations

The correlation function for deuterons was calculated using the same formula as for the protons. The data for both incident energies were combined since no difference between correlation curve has been observed within errors (fig.6).The background was constructed from all possible combinations of deuterons pairs from various interactions.The theoretical curves for $r_0 = 4fm$ and $r_0 = 3fm$ based on calculations performed by D.H.Boal [20][21]. Our experimental points are below these two curves, indicating that the size of the source of deuterons coming from the reaction ^{4}Hep is smaller than $3fm$ and can be of the same order as the value obtained from pp correlation.

5 Conclusions

Strong correlations are observed between protons emitted at small relative momenta from the ^{3}He and ^{4}He interactions with protons. The strength of the correlations was found to be independent on the incident energy of incoming ^{4}He nuclei between 8.6 and 13.6 GeV/c. The dimensions of the emission region have been found close to the helium nucleus size. The best value of the r_0 parameter, that characterises the gaussian source is about $1.0 + 0.2fm$ and $1.5 + 0.2fm$ for ^{4}He and ^{3}He respectively.

The correlation is stronger for faster protons. It can not be explained by the time structure of the source. For helium nuclei the time of interaction can not differ substantially for different proton momenta.

6 Acknowledgments

We would like to thank Dr R.Lednicky for discussion, helpful suggestions and for making available the computer program that calculates the theoretical pp correlation function. The authors also thank Dr D.H.Boal for providing the results of the two-deuteron correlation function calculations.

References

[1] S.Koonin , Phys.Lett., 70B (1977) 43

[2] G.I.Kopylov and M.I.Podgoretsky, Yadernaya Fizika 15 (1972) 392

[3] R.Lednicky, V.L.Lyuboshits, Yadernaya Fizika 35 (1982) 1316

[4] T.Siemiarczuk and P.Zielinski, Phys.Lett. 24B (1967) 675.

[5] S.A.Azimov et al., Yadernaya Fizika 19 (1974) 317.

[6] B.S.Aladashvili et al., Acta Physica Polonica B7 (1976) 707.

[7] N.Angelov et al., Yadernaya Fizika 32 (1980) 1357.

[8] S.A.Azimov et al., Phys. Rev. D29 (1984) 1304.

[9] Yu.D.Bayukov et al., Phys.Lett. B189 (1987) 291.

[10] Z.Chen et al., Phys Rev. C36 (1987) 2297

[11] M.Biyajima., Phys.Lett. 132B (1983) 299

[12] V.V.Glagolev et al., Nucl.Phys. A445 (1985) 572

[13] N.A.Buzdavina et al., JINR Report 1-81-530

[14] P.Zielinski et al., Sov. Journal Nucl.Phys. 40 (1984) 482

[15] J.Pluta , JINR Report E1-88-754, Dubna 1988 and references therein

[16] Yu.D.Bayukov et al., ITEP 88-2, Moskow 1988

[17] W.G.Lynch et al., Phys.Rev.Lett. 51 (1983) 1850

[18] M.I.Allaberdin et al, Yadernaya Fizika 46 (1987) 1785

[19] Y.C.Yang, R.C.Herndon, Phys.Lett. 25B (1967) 307

[20] C.B. Chetwood et al., Phys. Rev. Lett. 54 (1985) 302

[21] D.H.Boal et al.,Phys. Rev. C,33 (1986) 549

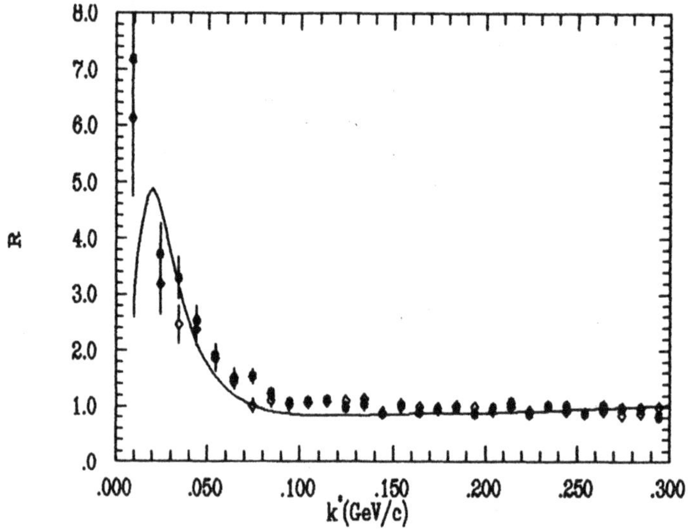

Fig. 1. Proton-proton correlation function for ^{4}Hep interactions at 8.6 GeV/c (black points) and at 13.5 GeV/c (open circles) The curve calculated as described in the text (r_o=1 fm)

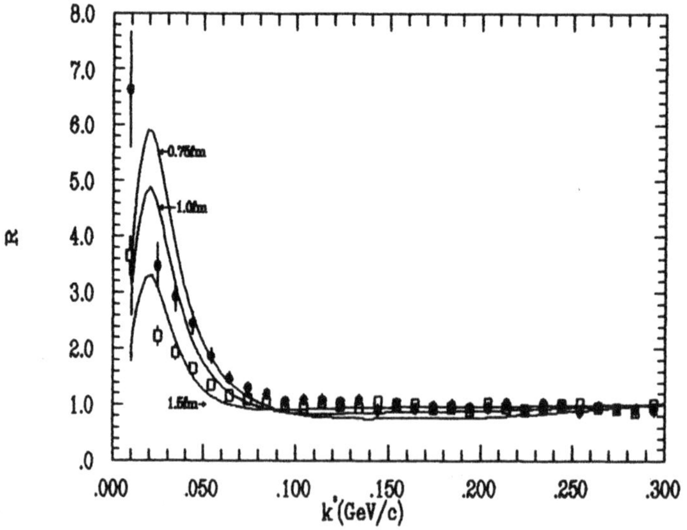

Fig. 2. Proton-proton correlation function; black points - ^{4}Hep open squares - ^{3}Hep interactions. Curves for r_o equal to 0.75, 1.0 and 1.5 fm.

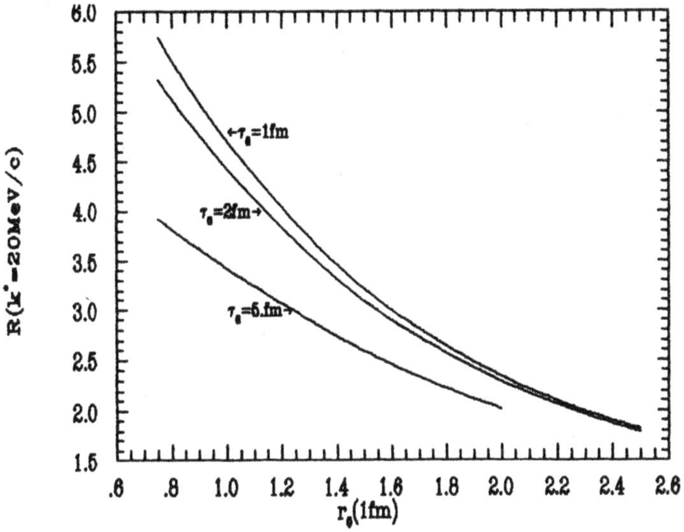

Fig. 3. The pp correlation function value at k^{*} = 20 MeV as a function of r_{o}, for different time dimensions of the emitting source.

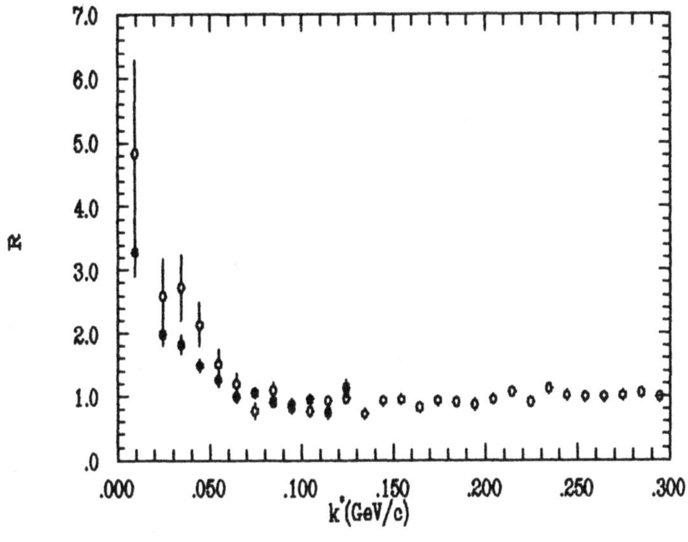

Fig. 4. The proton-proton correlation function for slow and fast protons in the ^{4}Hep interactions. Full points - events with both proton momenta smaller than 200 MeV/c, open points - momenta larger than 200 MeV/c.

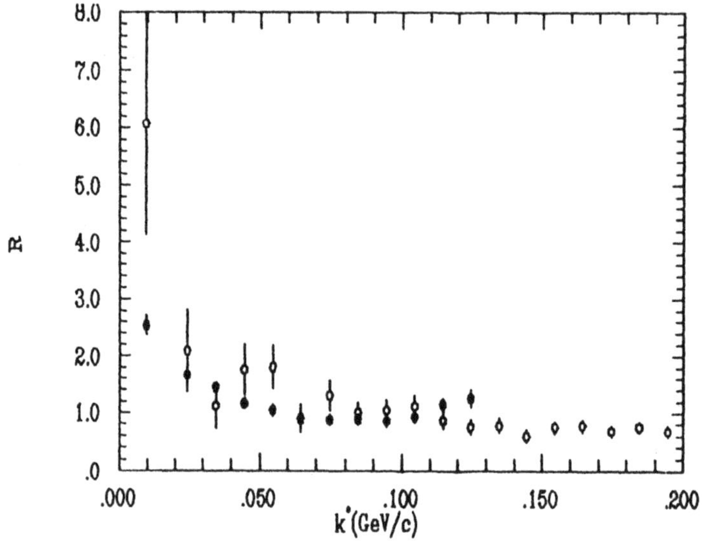

Fig. 5. The pp correlation function for slow and fast protons in the ^{3}Hep interactions. Full points - events with both proton momenta smaller than 200 MeV/c, open points - momenta larger than 200 MeV/c.

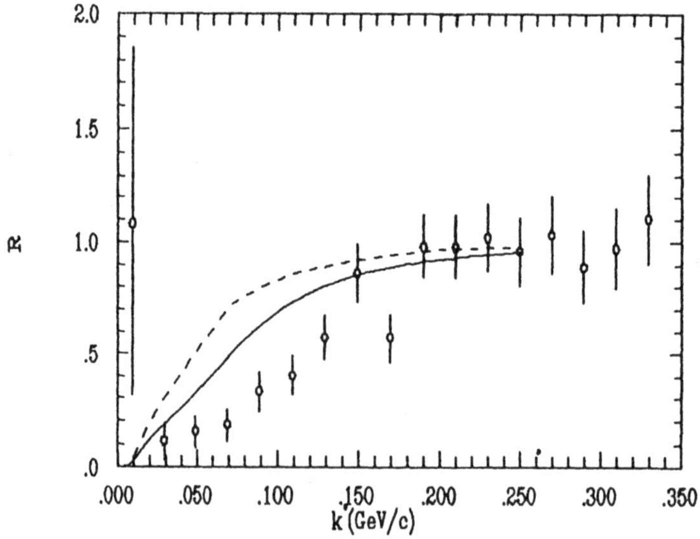

Fig. 6. The dd correlation function in the ^{4}Hep interactions. Theoretical curves: dotted line - r_o = 4 fm , solid line - r_o = 3 fm.

AZIMUTHAL ANGLE CORRELATIONS FOR ALPHA PARTICLE EMISSION IN THE REACTION $^{40}Ar + {}^{nat}Cu$ AT 25 MeV/NUCLEON

A. Giorni, A. Lleres, P. Désesquelles, A.J. Cole, H. El Hage, D. Heuer, C. Morand, H. Nifenecker, F. Schussler, P. Stassi, J.B. Viano

ISN Grenoble (IN2P3-CNRS)/Université J. Fourier, 38026 Grenoble cedex, France

and

F. Benrachi, B. Chambon, B. Cheynis, D. Drain, C. Pastor

IPN Lyon (IN2P3-CNRS)/Université C. Bernard, 69622 Villeurbanne cedex, France

and

J.P. Wessels

G.S.I. D-6100 Darmstadt 11, West Germany

ABSTRACT

Angular distributions and correlations of alpha particles detected in coincidence with a projectile or target-like residue have been measured using the AMPHORA multidetector for the reaction $^{40}Ar + Cu$ at 25 MeV/nucleon. A simulation of peripheral collisions in which α particles are emitted sequentially from excited projectile and target-like fragments gives generally a good account of the observations provided that angular momentum alignment effects are included. However, the simulation does not explain the amplification of alpha correlations observed perpendicular to the reaction plane.

1 - Introduction

Investigation of correlations involving light particles and heavier fragments has, in recent years, provided a new method for studying the reaction mechanism in heavy ion collisions at intermediate energies[1]. Detection of a projectile or target-like fragment in coincidence with such particles[2-4] provides an indication of the reaction plane of the collision but limits the study insofar as it selects mainly peripheral rather than central collisions. For heavy targets, the detection of fission fragments rather than PLF or TLF provides a method for studying more central collisions[5]. The present work describes the results obtained from a study of peripheral collisions in the reaction $^{40}Ar + {}^{nat}Cu$ at 25 MeV/nucleon. This is the first step of a more complete investigation carried out using the 4π AMPHORA detector at SARA.

2 - Experimental results

Detection of projectile and target-like fragments was achieved using an ionisation chamber[6] coupled with a phoswich Ne102/Ne115, set at $\theta = 7°$ with respect to the beam axis (referred to hereafter as the master). Coincident light particles were detected by the AMPHORA CsI detectors[7]. The alpha energy threshold was about 15 MeV. Offline analysis selected events in which the charge detected in the master was greater than 5.

The polar angular distribution (with respect to the beam axis) of alpha particles is shown in Fig. 1. For each polar angle, the azimuthal angular distribution, with respect to the reaction plane defined by the master, was extracted (Fig. 2). It was found that these distributions could be well described by the function :

$$N(\theta , \Phi) = \sigma(\theta) [1 + a(\theta).\cos(\Phi) + b(\theta).\cos(2\Phi)] \quad (1)$$

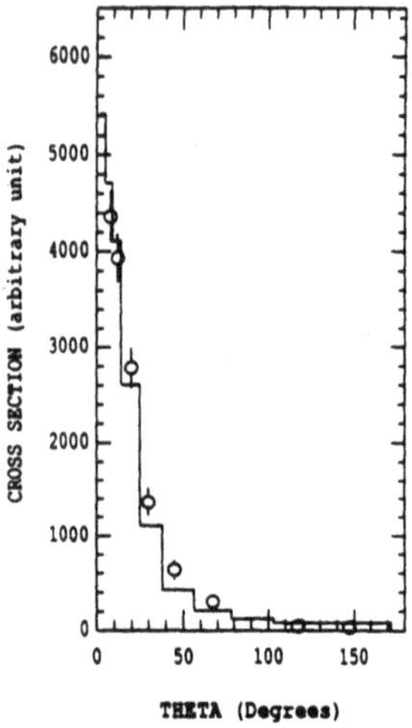

Fig. 1 : Polar angular distribution of alpha particles emitted in coincidence with a residue of charge Z≥5 at θ=7°. The histogram is the result of the simulation (see text).

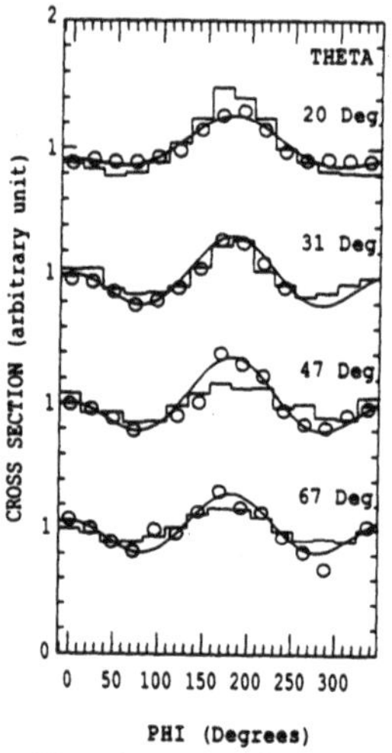

Fig. 2a : Azimuthal distributions of coincident alpha particles for various polar angles. They are normalized so that their mean value is equal to 1. The lines represent the fit of the data by equation (1). The histograms are the results of the simulation.

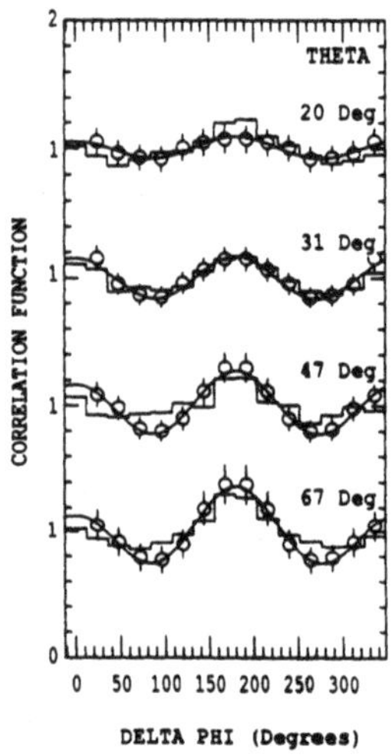

Fig. 2b : same as 2a but for correlation function. Delta-phi is the angle between the two alpha particules

For events in which at least two α particles were detected at the same polar angle θ, triple correlations (master, α_1, α_2) were extracted ; we have chosen to represent the data as a correlation matrix defined by :

$$M(\theta, \Phi_1 ; \theta, \Phi_2) = \sigma(\theta, \Phi_1 ; \theta, \Phi_2) / \sigma(\theta, \Phi_1).\sigma(\theta, \Phi_2) \quad (2)$$

where $\sigma(\theta,\Phi_1;\theta,\Phi_2)$ is the cross section for the correlated emission of two alpha particles and $\sigma(\theta,\Phi)$ the one particle emission cross section. A typical correlation matrix is displayed in Fig. 3 (similar studies involving α particles detected at different polar angles were also made but are not presented herein).

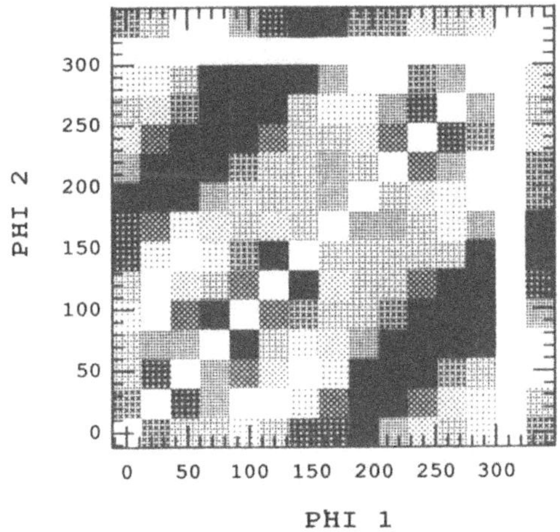

Fig. 3 : Correlation matrix of alpha particles emitted at θ=67° in coincidence with the master.

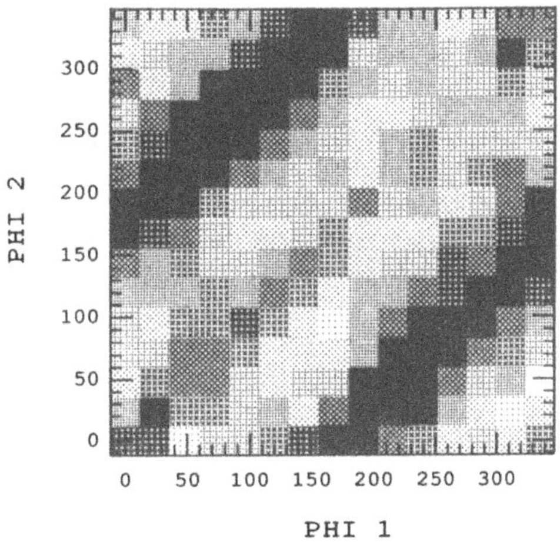

Fig. 4 : Correlation matrix of alpha particles at θ=67° obtained from the simulation.

An interesting result is obtained when two α particles are emitted in the same plane (with $\Phi_1 - \Phi_2 = 180°$). The corresponding spectrum (Fig. 5) exhibits an amplification of the correlation when the emission plane is perpendicular to the reaction plane. Furthermore, this feature does not seem to be explained by the simulation (see below).

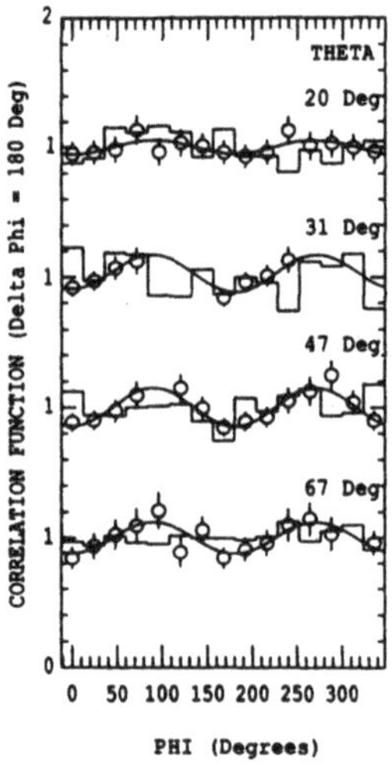

Fig. 5 : Correlation functions of alpha particles emitted in the same plane ($\phi_1 - \phi_2 = 180°$) for different polar angles. They are normalized so that their mean value is equal to 1. The lines represent the fit by equation (1). The histograms are the results of the simulation (see text).

3 - Simulation

A Monte Carlo simulation of the reaction was made based on the PERCUTE code[8]. The essential features of the code are that the mass of the PLF and TLF primary reaction products as well as the dissipated kinetic energy depend on the number of nucleon-nucleon collisions taking place between projectile and target nucleons along the trajectory of relative motion. The (classical) deflection of the projectile and target fragments is calculated using a folded potential added to a point Coulomb repulsion.

In the present work, the evaporation stage of the reaction (normally treated using multistep Hauser-Feschbach or sequential binary decay modules) was simplified by treating the primary PLF and TLF essentially as sources of α particles emitted with a Maxwellian spectrum above the Coulomb barrier. It was found not to be possible to reproduce the observations using an isotropic emission for the alphas evaporated by both sources. Ericson and Strutinsky[9] proposed a simple way to include an angular momentum alignment effect via a parameter $a = J L \hbar^2 / I T$ in which J is the angular momentum of the emitting nucleus, L the orbital angular momentum of the emitted particle, T the temperature and I the moment of inertia of the residual nucleus (the angular distribution of particles emitted with the angle Ψ with respect to the direction of I is described by the Bessel function $I_0[a.\sin(\Psi)]$). With this modification and using a constant (average) value of a = 3 , the main features of the data were quite well reproduced (see Fig. 2, 3 and 4). The code also reproduces quite accurately the polar angular distribution of emitted α particles (Fig. 1).

On the other hand the simulation seems to be unable to reproduce the out of plane correlation mentionned above (see Fig. 5). We tentatively interpret this result as evidence for correlated α particles produced in the primary step of the reaction. Indeed the observation is consistent with a quasi-free α-α scattering process shadowed by the presence of the interacting ions. More work is however necessary in order to substantiate this result particularly as regards the evaporation stage of the simulation and the neglect of central collisions (the simulation covers about 60% of the total reaction cross section).

References :

1) D. Ardouin et al., *Nucl. Phys. A* (in press).
2) H.J. Rabe et al., *Phys. Lett.* **B196** (1987) 439.
3) C.B. Chitwood et al., *Phys. Rev.* **C34** (1986) 858.
4) M.B. Tsang et al., *Phys. Lett.* **B148** (1984) 265.
5) M.B. Tsang et al., *Internal Report* **MSU-716**.
6) J.P. Wessels, *Internal Report* **GSI 86-16**.
7) D. Drain et al., *Nucl. Instr. and Meth.* **A281** (1989) 528.
8) A.J. Cole, R. Cherkaoui-Tadili and J. Alarja, *Phys. Rev.* **C40** (1989) 1265.
9) T. Ericson and V. Strutinsky, *Nucl. Phys.* **8** (1958) 284.

LIFETIME EFFECTS OF AN EVAPORATIVE SOURCE VIA TWO-PROTON CORRELATIONS

D. Goujdami, F. Guilbault, C. Lebrun, H. Dabrowski,
D. Ardouin,
Laboratoire de Physique Nucléaire, IN2P3-CNRS et Université de Nantes,
2 rue de la Houssinière - F-44072 Nantes Cedex 03 - France.

S. Pratt
Department of Physics, University of Wisconsin, Madison WI 53706, U.S.A.

P. Lautridou, R. Boisgard, J. Quebert
Centre d'Etudes Nucléaires de Bordeaux Gradignan, IN2P3-CNRS et Université de Bordeaux I, F-33170 Gradignan, France.

A. Péghaire
Laboratoire GANIL, IN2P3-CNRS, BP 5027, F-14021 Caen Cedex, France.

Abstract : The directional dependence of the correlation-function of two particles emitted from an evaporative source is shown to offer a weak sensitivity to the source lifetime. No evidence is found for an expanding emission source ; rather a sequential type of particle emission could be postulated from our observation.

Measurements of two-particle correlations between light-particles emitted in heavy-ion collisions can be used to provide information on the spatial-time extent of the emitting source [1-4]. In principle, a two-dimensional study of the correlation-function vs. relative momentum and energy variations should be used for that purpose. However, due to statistical requirements and difficulties to correctly describe the time dependence when final state interactions develop, superimposed to quantum interference effects, most of the analyses have been carried out assuming a zero (or fixed) lifetime of the source, leading then to an overestimated size. Several interesting systematics of the data nevertheless seem to contain the fingerprint of the lifetime of the emission. The first one is provided by the multiplicity dependence : larger effective radii are measured [5,6] for higher multiplicities. If one can make a link with impact parameter, Landau-Vlasov calculations do not indicate [7] such variations of the interacting zone with the impact parameter for intermediate heavy-ion collisions. Rather, a variable collision-time is predicted with the impact parameter. The second possible fingerprint is the observation that lower energy particles always correspond to larger apparent sizes (see ref. 3,4 as an example). Indeed, slower particles are predicted [7,8] to be associated with the last steps of the interaction. Alternatively, a spatial expansion of the source can produce [9,10] the same effect. The same expansion effect corresponding to smaller freeze-out densities for complex fragments emission has been also invoked to account for the larger effective sizes measured for light fragments [4,11]. Similarly, finite time developments for complex fragment emission are certainly difficult to disentangle from pure spatial evolutions. It thus appears necessary to find other fingerprints for the time effects. Recent measurements of proton-proton correlations at backward angles have been proposed to contain such a sensitivity [12,13]. On the other hand, the direction of the relative momentum between the two particles as compared to the mean direction of emission has been also predicted by Koonin[1] to present such a dependence. Following this idea, Pratt et al[9,10] have postulated that one could observe an elongated appearance in the direction of emission for a long-lived emission. This appearance is caused by the quantum statistical effects between the two correlated fermions producing a weaker negative interference along a direction where the emission time difference will result in a larger separation of the pair. A recent investigation [15] failed to observe this dependence from which the existence of an extended emission source was concluded rather than a cooling evaporative compound system. As explained below, we however think that these conclusions were biased by several experimental and physical conditions. So ,the goal of our work was to search for this effect under appropriate conditions.

Our experiment was performed at the Laboratoire GANIL at Caen. A Ag target was irradiated by an Ar beam of 44 MeV/nucleon incident energy. Protons were identified and detected by a multi-detector ensemble of 24 closed CsI scintillators [12] 10cm-thick. The angular separation between adjacent elements was 5-6°. These detectors were set at a backward average polar angle of 130°. The energy calibrations were made by detecting velocity calibrated protons resulting from reaction products of an Ar beam on a thick C target and are accurate within 2%. An energy threshold was set at 5 MeV. The study of heavy residues and coincident light-particles for the same system between E/A = 39 and 60 MeV has shown clearly[14] that, in this energy domain, the incomplete fusion followed by residue evaporation dominates the cross-section. The single proton spectra measured at backward angles in the present experiment reflect these conclusions : they present a characteristic maxwellian shape which can be well accounted for by a unique moving source fit with a velocity v_s = 0.0034 c and a slope parameter T = 5.6 MeV. The two-proton correlations built with these evaporative proton distributions then appear as a favorable condition to test the above postulated time effects.

The correlation-function R(K,q) is extracted from the ratio of the coincidence yield $Y_{12}(p_1,p_2)$ to the product of the single yields $Y_1(p_1)xY_2(p_2)$:

$$1 + R(K,q) = C \times [\Sigma\, Y_{12}(p_1\ p_2)/\Sigma\, Y_1(p_1)\ Y_2(p_2)] \qquad (1)$$

where p_1 and p_2 are the two proton momenta, K and q are the total and relative momenta of the pair respectively. C is a normalization constant which is chosen in order to give an average value R = 0 in the range q = 40 - 90 MeV/c when summing over all K values.

Fig. 1 shows the directional dependance of the correlation for momentum gates of K = 300 - 400 and 500 - 700 MeV/c. These selections represent the lower and upper halves of the observed total momentum spectrum. For each momentum selection, the correlation is shown for a relative momenta q either parallel (ψrel = 0° ± 30° , 180° ± 30°) or perpendicular (ψrel = 90° ± 30°) to the total momenta K. Within statistics, we see that the correlation at low-relative momenta (q < 30 MeV/c) is slightly enhanced in the case of the parallel selection : the effect is of the order of 20%, but does not present a significant dependence upon the observed momentum range. These results have to be compared with recent similar measurements[15] made by Awes et al. using the ^{32}S+Ag reaction at 22 MeV/nucleon. Following there conclusions, no difference was found between the observed directions. However, we think that there is no inconsistency between both data for several reasons : (i) the lowest proton energies considered in the S + Ag work were considerably above the Coulomb barrier where the evaporation and time scale should manifest themselves, (ii) our choice of looking at very backward angles eliminates most of the emission origins other than an evaporation process, like for instance the pre-equilibrium emission mostly associated with the lighter partner motion in the collision[16], (iii) the estimated weakness[10] of the postulated effect requires high data statistics in order to be observed.

In order to compare our data with theoretical predictions, we have used the evaporative model of Friedman and

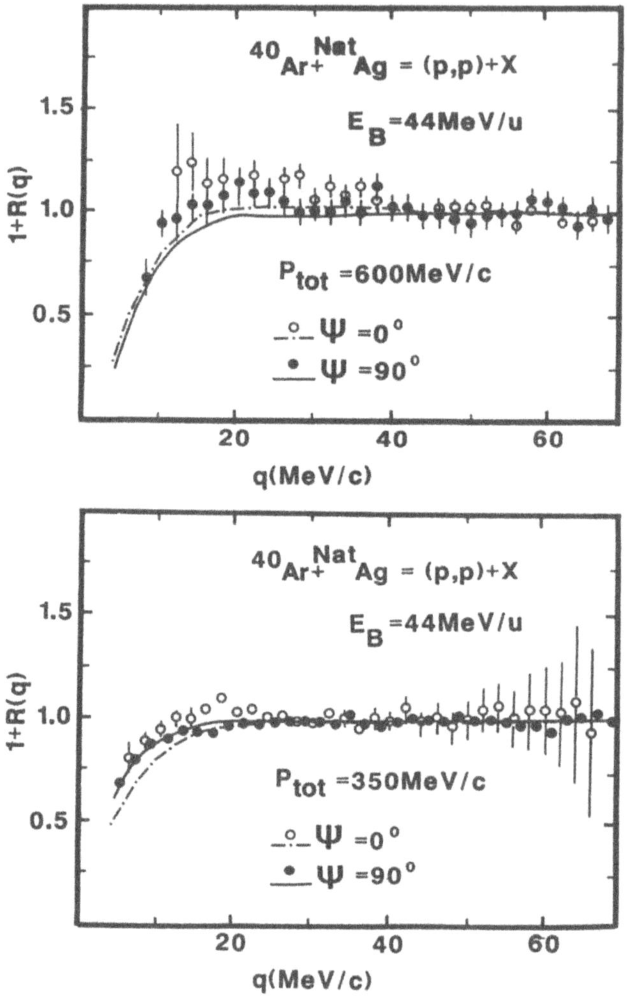

Figure 1 : Two-proton correlations for the Ar + Ag reaction at E/A = 44 MeV. The average angle of the measurement is $\theta = 130^\circ$. Data are presented for two P_{tot} regions : the lower part corresponds to 300 MeV/c < P_{tot} < 400 MeV/c and the upper (part corresponds to 500 MeV/c < P_{tot} < 700 MeV/c. In each case the data are shown for a relative momenta q parallel (ψ = 0) or transverse ($\psi = 90^\circ$) direction with respect to the total momentum K.

Lynch [17]. It describes the emission of nucleons from a hot liquid drop that cools as it evaporates particles. The proton emission probability dN/dtdE comes from time-reversal argument and depends on the temperature, baryon number and charge of the nucleus :

$$dN_p(E_p,t)/dE_p dt = \sigma.\theta(E_p - V_{coul.}).(E_p - V_{coul})\exp[S(E^* - E_p)S(E^*)] \quad (2)$$

where V_{coul} is the coulomb barrier, S the entropy of the compound nucleus, E_p is the energy of a proton with momentum p and σ is the geometrical cross-section for formation of the compound nucleus. The time evolution of the nucleon number and the excitation energy per nucleon are shown in Fig. 2. The emission points can be taken uniformly from the surface of the emitting nucleus, thus yielding the emission probability per unit volume and per unit time for protons of any momentum.

From these probabilities, single-particle Wigner functions are obtained :

$$f(p,r,t) = \int_{-\infty}^{t} dt' \times dN_p(p, r - V_p(t-t'), t')/dt' \; d^3r \; d^3p \quad (3)$$

which can be folded together :

$$F_K(r) = \lim_{t\to\infty} \int d^3R \; f((K/2, R+r/2, t) \; f(K/2, R-r/2, t) \quad (4)$$

in order to get the two-particle correlation-function :

$$1 + R(K,q) = \int d^3r \times F_K(r) \; \phi^*(q,r) \; \phi(q,r) / \int d^3r \; (F_K(r)) \quad (5)$$

where ϕ is the antisymmetrized wave-function of the relative motion.

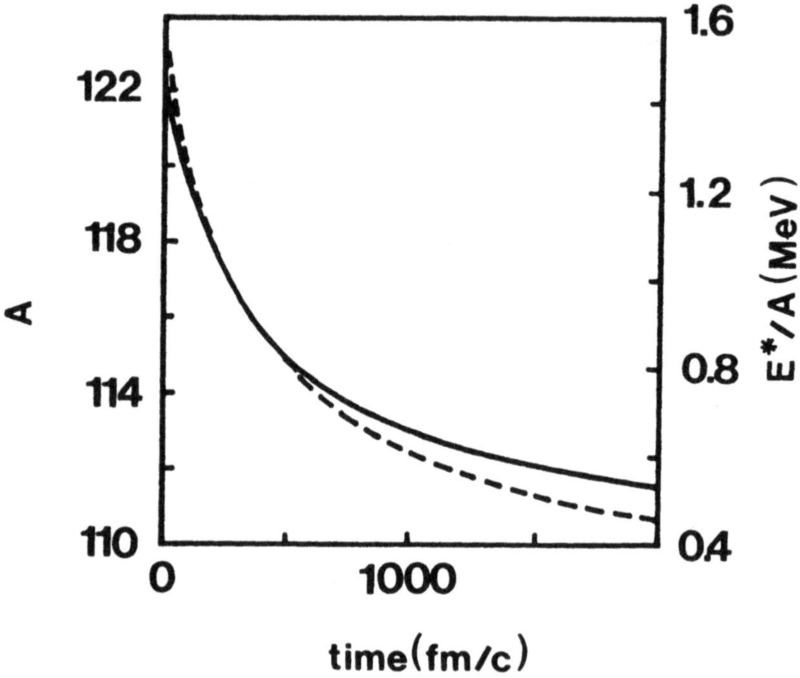

Fig. 2 : Evaporative calculation showing the time evolution of the nucleon number A (full line) and the excitation energy per nucleon E*/A (dotted line). The initial conditions are A = 122, Z = 52, E*/A = 1.5 MeV. Half of the protons are emitted in the first 250 fm/c and 90 % of the emission has occured by 2000 fm/c.

The best fit has been obtained for an initial thermal energy $E^*/A = 1.5$ MeV which represents only a fraction of the available excitation energy. This type of difference has been already discussed[18] in this energy domain and is reflected as well in temperature measurements[19] performed at backward angles : it can be attributed to the contributions of preequilibrium emission and collective motion of the source.

For this thermal energy, the calculated results are shown if Fig.1 for the corresponding average momentum and direction of emission. The calculations have been made for an evaporative nucleus of initial size A = 122, Z = 52 which corresponds to exclusive measurements[14] based on recoil velocity determinations. The theoretical angular dependence of the correlation function is weak. At 600 MeV/c the prediction presents the same trends in angular variation than the experimental one. For the P_{tot} = 350 MeV/c case the tendency is reversed in the low relative momentum domain.

In conclusion, this work seems to demonstrate the ability of the two-particle correlation functions and their directional dependence to offer a weak but observable sensitivity to the time development of the emission sequence, as it was theoretically predicted [9-10]. This predicted effect appears visible with high data statistics and cannot be observed in forward-angle measurements with a light and fast projectile[15] due to a contamination with non-equilibrium emission. Indeed, backward angle measurements allow to observe mostly an evaporative pattern where the time sequence of emission allows quantum interference effects to give rise to the observed effect. In terms of heavy-ion collision mechanisms, contrary to earlier conclusions[15], one cannot rule out a sequential type of emission caused by the formation of a cooling source.

The authors thank M. Rio and J. Ropert for a generous assistance at various steps of this work.

REFERENCES

1 S. Koonin, Phys. Lett. B 70, 43 (1977)
2 G. Kopylov and M. Podgoretskii, Sov. J. Nucl. Physics 18, 336 (1984)
3 W. Lynch et al, Phys. Rev. Lett. 51, 1850 (1983)
4 J. Pochodzalla et al, Phys. Rev. C 35, 1695 (1987)
5 A. Kyanowski et al, Phys. Lett. B 181, 43 (1986)
6 H. Gustafson et al, Phys. Rev. Lett. 53, 544 (1984)
7 C. Grégoire et al, Nucl. Phys. A 465, 317 (1987)
8 D. Boal, H. DeGuise, Phys. Rev. Lett. 57, 2901 (1986)
9 S. Pratt and M.B Tsang, Phys. Rev. C 36, 2390 (1987)
10 S. Pratt, Phys. Rev. Lett. 53, 1219 (1984)
11 D.A. Cebra et al, Phys. Lett. B 227, 601 (1989)
12 J. Québert et al, in Proceedings of the XXVII Winter Meeting on Nuclear Physics, (Bormio - Italy January 1989) ed. by I. Iori (Universita Degli Studi, Milano, 1989) Supp. n. 69, p.220.
13 D. Ardouin et al, Nucl. Phys. A 495, 57c (1989)
14 G. Bizard et al, in Proceedings of the Texas A & M. Symposium on Hot Nuclei, (College Station - Texas, December 1987) ed. by S. Shlomo, R.P. Schmidt, J.B. Natowitz (World Scientific, Singapore, 1988), p. 262.
B. Tamain, in Proceedings of the CXI Enrico Fermi School, Varenna, Italy - July 1989 (to be published)
15 T. Awes et al, Phys. Rev. Lett. 61, 2665 (1988)
16 H. Morgenstern et al, Phys. Rev. Lett. 52, 1104 (1984)
17 W.A. Friedman and W.G. Lynch, Phys. Rev. C. 28, 16 (1983)
18 E. Suraud, C. Grégoire and B. Tamain, GANIL-Preprint 89-04 an to be published in Progress in Part. and Nucl. Phys. (1989)
19 H. Dabrowski et al, Phys. Letters B, (in press).

ROUND TABLE DISCUSSION

Discussion Leaders :

B. REMAUD, K. GELBKE

Discussion Panel :

J. BARTKE, B. LORSTAD, S. PRATT, P. SCHUCK, R. REINER

TENTATIVE SUMMARY OF THE POINTS WHICH WERE DISCUSSED IS HERE PRESENTED

INTRODUCTION

The various speakers, that we have heard for two and half days during the CORINNE meeting, have shown the scientific wealth of the correlation studies in order to investigate the properties of nucleon or particle systems. As underlined in Labeyrie's introductory talk, the bases of such studies are quite different from those of interferometry methods in astronomy, where the information on far-lying objects can be found in the interferences in the detection apparatus itself ; in astronomy, there are good reasons to suppose that protons are emitted incoherently from the stars or more precisely, the correlation length of the process of light emission is negligibly small when compared to the size of the stars.

Furthermore, the lifetime of the emitting source is so long, that one can easily assume that it keeps constant its properties during the emission process.

In nuclear and particle physics, the situation is certainly quite different, since the size of the excited emitter of particles is very small and one cannot a-priori exclude coherence effects in the emission of particle pairs. As to the lifetime of the emitter, two extreme limiting cases can exist, from the very short flash of preequilibrium particles ($< 10^{-22}$s) to the statistical emission from a relatively well defined compound nucleus, with time scales longer by 3 or 4 orders of magnitude. In real experiments, all intermediate situations between these extremes can occur ; and in all cases, one cannot barely assume that the source characteristics are kept constant during the emission. From these statements, emerges the necessity of a knowledge of the source formation before making any comparison (see for instance two neutron or two-proton data presented at this conference).

Nuclei (and particle systems) are very correlated objects ; nuclear matter studies show that pairs of nucleons are strongly correlated by quantum statistics (they are fermions feeling the Pauli exclusion principle) and nuclear interaction. It is certainly a main output of this meeting to have put in evidence this fundamental question : in correlation functions experimentally measured, are there signatures of the genuine pair correlation function in nuclei, or are the correlation built during the transport from the emitting nuclear system to the experimental apparatus. One thinks that this question cannot be completely answered yet ; at least, one suspects that this answer may differ depending on the time duration of the emission process itself. If one considers the considerable advance of theories concerning the treatment of final correlations (quantum statistics, nuclear interaction and Coulomb interaction) and the coupling of such theories to dynamical models of nuclear collisions, the challenging opened question remains the development of theories including the initial correlation of the source.

The experimental corollary of this theoretical challenge is the need for very precise measurements of correlations at very small relative momentum ; in particular, very specific predictions of the correlation function at vanishing small Δp's, when quantum statistics or pure Coulomb interaction are considered, yield possibilities to know if all initial correlations are washed out during the transport of particles to the detectors.

THE DISCUSSION TOPICS

The various questions which emerge at the end of this meeting can be gathered together in several chapters :

a) Boson/Fermion Interferometry

The discussion precises the points already mentioned in the preceding introduction : effects of antisymmetrisation of particle wave packets on the correlation function (Feldmeier) ; limits of correlation function at $\Delta p \rightarrow 0$ (Pratt) and what we can learn of it , interest of interferometry including neutral particles to remove the presumably overwhelming effects of the Coulomb field.

b) Problems in boson interferometry

The discussion concerns the important question of the balance between coherence and chaoticity in the emission processes. The problem has been particularly investigated in pion interferometry (Weiner). The absence of coherent effects for fermion studies is expected to bring an interesting comparison with boson interferometry. New and valuable insight is expected to come from photon-photon interferometry experiments now being analysed in heavy-ion collisions.

c) Validity of variable determinations

The minimal information that one can learn from correlation measurements concerns the spatial extension of the source (eventually coupled to its lifetime). If an isolated measurement of a source size may be difficult to interpret, much can be learned from systematics which compare the deduced sizes with projectile/target geometrical scales ; this is seen in particular, in systematics of pion interferometry for heavy-ion collisions at high energies. Such systematics begin to be available for proton-proton correlations showing clearly the strong influence of lifetime effects for low-energy emission domains.

d) Experimental problems

First measurements at small Δp's show very instructive different behaviours between heavy-ion reactions and proton collision on deuterium (Kox). However, the energy thresholds and geometrical granularity of the detectors have to be carefully considered and analyzed, in particular with Monte Carlo simulations.

Another well discussed problem concerns the normalization of the correlation functions, which is conventionally taken to unity at large Δp's. This normalization can be based on the assumption that correlations vanishes for particles with large relative momentum ; this assumption is far from being well funded (two-body collisions and particle exchange mechanisms produce correlations at large momenta). The effects of normalization procedures on the resulting observables (Lorstad) can be very important ; further studies of this problem are mandatory to get unbiased information.

e) Future of correlation studies

The convergence of important advances in detection techniques and in theories provides an unique opportunity to fully exploit the interferometry methods in nuclear and particle physics (Pratt).

(Summary written by B. REMAUD and D. ARDOUIN)

Remarks on Bose-Einstein Correlations (BEC)

R.M. Weiner
University of Marburg, F.R.G.

1. Why should we study BEC?

   (i) To learn something about the existence of a coherent state in the hadronic sources (fields). This is important because it would reflect some collective properties of the system.

   (ii) To learn something about the multiparticle dynamics, fluctuations of the system and in particular about the multiplicity distributions P(n) or their moments. This is so because the moments are the integrals of the BEC, e.g.

$$\langle n(n-1)\rangle = \int G_2(x_1,x_2)\, dx_1\, dx_2 \qquad (1)$$

   where G_2 is the second order correlation function = $\langle I(1)I(2)\rangle$ here I is the intensity.

   (iii) To get source information about radii and lifetimes. In my view, the order in which these topics were enumerated corresponds to the order of their importance.

2. The application of coherent states in particle physics and in BEC in particular meets sometimes resistance because of some misunderstandings, two of which will be quoted below:

   (a) Conservation laws like e.g. charge conservation.
   These would contradict the existence of coherent states for which the number of particles is undefined. The error in this argumentation lies in the fact that conservation laws apply for the global system, but not for parts of it, for which coherent states can certainly be used.

   (b) A large system like a nucleus can hardly be coherent.
   Nature has provided us with many counter examples of macroscopic systems which are in coherent states like superconductors, ferromagnets etc.

3. Treatment of BEC by using the symmetrized wave function of free particles, misses an important aspect of the problem, connected with the multiparticle dynamics. The proper treatment of BEC is through quantum statistics (QS) which starts from the correlation of (chaotic) fields π

$$\langle \pi(x_1)\, \pi(x_2) \rangle = \langle n_{ch} \rangle \exp - |x_1 - x_2|^2/L^2$$

where $\langle n_{ch} \rangle$ is the average number of particles in the chaotic state and L the correlation length in x space. The fluctuations involved in this contain the Bose-Einstein correlations as a particular case. They introduce a new length scale L in the problem, on top of the space-time (geometrical) length scale R.

In BEC we thus measure a combination of L and R. In the particular case of spherical space-time distribution this combination reads [1]

$$\tilde{R}^2 = R^2 - R^2L^2/(R^2+L^2).$$

Another QS element is $\langle n_c \rangle$, the average number of coherent particles leading to the chaoticity

$$p = \frac{\langle n_{ch} \rangle}{\langle n_c \rangle + \langle n_{ch} \rangle} .$$

The same two quantities $\tilde{R}$ and p govern also the multiplicity distributions. Finally the (partially) coherent state changes the form of BEC, replacing the one exponential formula

$$1 + \lambda \exp(-|k_1 - k_2| \cdot R^2) \qquad (2)$$

by a two exponential formula

$$1 + \lambda_1 \exp\left(-|k_1 - k_2|^2 \cdot \tilde{R}^2\right) + \lambda_2 \exp\left(-2|k_1 - k_2|^2 \cdot \tilde{R}^2\right) \qquad (3)$$

with λ_1 and λ_2 given functions of p, k_1, k_2.

This implies that measurements of R based on eq. (2) have to be qualified.

[1] I.V. Andreev and R.M. Weiner, Proc. of the Int. Workshop on Correlations and Multiparticle Production (CAMP), 1990, World Scientific, to appear.

SYSTEMATICS OF SOURCE SIZES FROM PARTICLE INTERFEROMETRY

Jerzy BARTKE

High Energy Physics Laboratory
Institute of Nuclear Physics
ul. Kawiory 26A
30-055 Cracow, Poland

1. PION SOURCE RADII

In ref.[1] a compilation of pion source radii has been given for relativistic nuclear collisions at projectile energies in the range of 1-4 GeV/nucleon (LBL and Dubna). In order to make a valid comparison of various experiments, size parameters published by different authors were converted to r.m.s. radii R_{rms}, as suggested earlier in ref. [2]. It has been found that the r.m.s. radius increases linearly with $A_P^{1/3}$, A_P denoting the mass number of the projectile nucleus ($A_P \leqslant A_T$). Moreover, the values of the r.m.s. radius come close to the "effective nuclear radius" of the projectile: $R_{rms} \cong R_I = 1.21\ A_P^{1/3}$. This extends the $A^{1/3}$ scaling law found by Zajc et al.[3] for collisions of nuclei of the same mass to the case of non-equal-mass nuclei. Some data points for heavy nuclear systems have been added recently to the plot given in ref.[1]: Ar+KCl, Fe+Fe, Nb+Nb by the JANUS spectrometer group [4] and Ar+KCl, Ar+La, Xe+La by the HISS spectrometer group[5].These new data do not change the overall tendency, however the JANUS results lie lower than those of other authors. A similar plot, but with all size parameters rescaled to "Gaussian radii" R_G, can be found in ref.[6]. In view of the relation $R_{rms} = \sqrt{3}\ R_G$, the result quoted is $R_G = 0.7\ A_P^{1/3}$ instead of $1.21\ A_P^{1/3}$. Thus, at energies of a few GeV per nucleon the picture is rather simple: the pion source radius corresponds to the size of the smaller nucleus, as one would expect from the geometrical model with the projectile cutting through the target a cylindrical tunnel, the rest of the target nucleus remaining unexcited ("spectator matter").

At the SPS energies the situation looks different. Very large radii obtained at midrapidity in O+Au and S+Ag central collisions at 200 A GeV by the NA35 Collaboration [7] do not fit the above systematics - they are about twice too large. It seems relevant to relate them to the freeze-out volume for pion production. The simplest approach would be to consider freeze-out as the situation in which the source (fireball) is big enough as to contain all produced particles. This leads to the dependence $R \sim (dN/dy)^{1/3}$ where dN/dy is the rapidity density (multiplicity) in the rapidity interval considered. This approach has been employed in the compilation of ref.[6]. Another approach is to assume that the pion distribution should freeze out when the size of the system is of the order of the scattering mean free path $\lambda_{\pi\pi}$ (then the interactions cease)[8]. Equating R and $\lambda_{\pi\pi}$ one obtains the dependence $R \sim (dN/dy)^{1/2}$. The square-root dependence also follows from hydrodynamical models including transversal expansion [9]. This approach has been employed for data presentation in refs [9-11]. Unfortunately, the existing data are not yet precise enough as to distinguish between the square-root and cube-root dependences. More statistics from NA35 and the new dedicated experiments NA44 at CERN and E859 at BNL promise results of higher accuracy.

2. PROTON SOURCE RADII

Proton-proton correlations are dominated by final-state interactions and the source size parameters are, in fact, determined from the study of these interactions, and not from "interferometric" effects.

At 4.2 A GeV/c it has been found that in collisions of p, d, He and C projectiles with a carbon target the p-p correlations measure the size of the target nucleus [12], and not that of the projectile as in the case of pion-pion correlations. The above conclusion has been confirmed in other experiments. A compilation of proton source radii in hadron-nucleus and nucleus-nucleus collisions at projectile energies between about 0.1 and 10 GeV per nucleon has been presented if ref. [13]. For protons with momenta exceeding 0.3 GeV/c the values of the r.m.s. radii obtained from p-p correlations are compatible with the r.m.s. radii of the target nuclei. The new data from the DIOGENE detector at SATURNE [14] for Ne and Ar projectiles at energies between 0.2 and 1.0 A GeV, and for various target nuclei, show the proportionality of the proton source radius to that of the geometrical overlap region of the colliding nuclei. For a given system, the radius is independent of the incident energy.

So far there is no data on p-p correlations at ultrarelativistic energies.

REFERENCES

1. J.Bartke, Phys.Lett. B174, 32 (1986), see also: J.Bartke, Int.Journ.Mod.Phys. A4, 1319 (1989).
2. J.Bartke and M.Kowalski, Phys.Rev. C30, 1341 (1984).
3. W.Zajc et al., Phys.Rev. C29, 2173 (1984).
4. A.Chacon, Ph.D.Thesis, LBL-28149, Berkeley 1989.
5. W.Christie et al., Paper presented as poster at the QUARK MATTER 90 conference, Menton 1990.
6. B.Lörstad, Int.Journ.Mod.Phys. A4, 2861 (1989).
7. For references see: J.Bartke, these proceedings.
8. A.Bamberger et al., Phys.Lett. B203, 320 (1988).
9. R.Stock, Preprint IKF 90-3, Frankfurt 1990, to be published in Annalen der Physik.
10. W.Zajc, QUARK MATTER 90, Menton 1990, to be published in Nucl.Physics A.
11. B.Lörstad, these proceedings.
12. G.N.Agakishiev et al., Z.Phys. A327, 443 (1987).
13. J.Pluta, JINR E1-88-745, Dubna 1988, see also: J.Pluta, these proceedings.
14. P.Dupieux et al., Phys.Lett. B200, 17 (1988); J.P.Alard et al., these proceedings.

CORINNE FOR POETRY

Corine, je te prie, approche :
Couchons-nous sur ce tapis vert,
Et pour être mieux à couvert,
Entrons au creux de cette roche.

(...)

Ma Corine, que je t'embrasse !
Personne ne nous voit qu'Amour;
Vois que même les yeux du jour
Ne trouvent point ici de place.

Les vents, qui ne se peuvent taire,
Ne peuvent écouter aussi,
Et ce que nous ferons ici
Leur est un inconnu mystère.

La Solitude from Theophile DE VIAU (1590-1626)
after a compilation by Jean Peter, Chairman
of the last session of the workshop.

LIST OF PARTICIPANTS

ABE Yasuhisa	Kyoto (Japan)
ALARD Jean-Pierre	LPC Clermont-Ferrand (France)
ALEXANDER	Stony Brook (USA)
ALM Thomas	Rostock (RDA)
ARDOUIN Daniel	LPN Nantes (France)
BARTKE Jerzy	INP Krakow (Poland)
BASRAK Zoran	RBI Zagreb (Yougoslavia)
BASTID Nicole	LPC Clermont-Ferrand (France)
BECK Christian	CRN Strasbourg (France)
BONDORF Jakob	NBI Copenhagen (Denmark)
BOYD Graham	Cape Town (South Africa)
CARJAN Nicolas	CENBG Bordeaux (France)
CHARLET Martine	Nijmegen (Netherlands)
COLE Antony	ISN Grenoble (France)
CUGNON Joseph	Liège (Belgium)
DABROWSKI Henryk	LPN Nantes (France) and Krakow (Poland)
DE LA MOTA Virginia	LPN Nantes (France) and Buenos Aires (Argentina)
DE YOUNG Paul	Hope College (USA)
DESESQUELLES Pierre	ISN Grenoble (France)
DIAZ MEDINA José	Valencia (Spain)
DOUBRE Hubert	GANIL (France)
DUNNWEBER Wolfgang	München (RFA)
DUPIEUX Pascal	LPC Clermont-Ferrand (France)
DURAND Dominique	LPC Caen (France)
DURAND Mireille	ISN Grenoble (France)
ERAZMUS Barbara	LPN Nantes (France)

ERDMANN Martin	Freiburg (RFA)
EUDES Philippe	LPN Nantes (France)
FARINE Michel	LPN Nantes (France)
FELDMEIER Hans	GSI Darmstadt (RFA)
FERRAGUT Alain	GANIL (France)
FLEURIEL Olivier	LPN Nantes (France)
FUCHS Hermann	HMI - Berlin (RFA)
GALIN Joël	GANIL (France)
GARCIAS Francesca	Palma de Mallorca (Espagne)
GELBKE Konrad	Michigan St. U. (USA)
GHISALBERTI Claude	LPN Nantes (France)
GIORNI Alain	ISN Grenoble (France)
GIOVANNINI Alberto	Torino (Italy)
GONTHIER Peter	Hope College (USA)
GRISHIN Valentin	JINR Dubna (URSS)
GUILBAULT François	LPN Nantes (France)
HANAPPE Francis	Louvain (Belgium)
HOURDEQUIN Emmanuel	LPN Nantes (France)
IDIER Dominique	LPN Nantes (France)
JAENICKE Judith	Heidelberg (RFA)
KOROLIJA Milorad	RBI Zagreb (Yougoslavia)
KOX Serge	ISN Grenoble (France)
KRYGER Robert	Notre-Dame (USA)
KUNDE Gerd	GSI Darmstadt (RFA)
LABEYRIE A.	Nice (France)
LAUTRIDOU Pascal	CENBG Bordeaux (France)
LAVILLE Jean-Louis	LPC Caen (France)
LEBRUN Claude	LPN Nantes (France)

LEDNICKY Richard	JINR Dubna (URSS)
LEMAIRE Marie-Claude	Saturne, Saclay (France)
LERAY Sylvie	Saturne, Saclay (France)
LERUSTE Philippe	LPC Collège de France (France)
L'HOTE Denis	Saclay (France)
LORSTAD Bengt	Lund (Sweden)
LONGEQUEUE J.P.	IN2P3 (France)
MERCHEZ Fernand	ISN Grenoble (France)
MORRISSEY David	Michigan St. U. (USA)
NARJOUX Jean-Louis	LPC Collège de France (France)
NOREN Birgitta	Lund (Suède)
OBERSTEDT Andreas	GSI Darmstadt (RFA)
OSTENDORF R.	GANIL (France)
PEITZMANN Thomas	Münster (RFA)
PETER Jean	LPC Caen (France)
PLUTA Jan	WARSAW (Poland)
POCHODZALLA Josef	GSI Darmstadt (RFA)
PODGORESTKII Mikhail	JINR Dubna (URSS)
POLLS Arturo	Barcelona (Spain)
PRATT Scott	Wisconsin U. (USA)
QUEBERT Jean	CENBG Bordeaux (France)
RAFFRAY Yves	LPN Nantes (France)
RAMSTEIN Bruno	LPN Nantes (France)
REBREYEND Dominique	ISN Grenoble (France)
REMAUD Bernard	LPN Nantes (France)
ROPKE Gerd	Rostock (RDA)
ROSSNER Hermann	Hahn Meitner Institut (RFA)
ROY René	GANIL (France)

ROYER Guy	LPN Nantes (France)
SAID Pierre	LPN Nantes (France)
SAINT-LAURENT François	GANIL (France)
SCHICKER Reiner	GSI Darmstadt (RFA)
SCHUCK Peter	ISN Grenoble (France)
SCHUTZ Yves	GANIL (France)
SEBILLE François	LPN Nantes (France)
SELTZ Raymond	CRN Strasbourg (France)
SEZAC Laurent	LPN Nantes (France)
TAMAIN Bernard	LPC Caen (France)
TAMISIER Roland	LPN Nantes (France)
VOUTIER Eric	IPN Lyon (France)
WADA Takahiro	Riken (Japan)
WEINER Richard	Marburg (RFA)
WOLF Gyuri	Giessen (RFA)
ZLOMANCZUK Jozef	Warsaw (Poland)

www.ingramcontent.com/pod-product-compliance
Ingram Content Group UK Ltd.
Pitfield, Milton Keynes, MK11 3LW, UK
UKHW061932100926
8556IPUK00007B/881

9 789810 203177